D0571642

Gene Function

E. coli and its heritable elements

Gene Function

E. coli and its heritable elements

Robert E. Glass

University of California Press
Berkeley and Los Angeles 1982

UNIVERSITY OF CALIFORNIA PRESS

ISBN 0-520-04619-6
　　　0-520-04654-4 Pbk

Library of Congress Catalog Card Number 81-69893

Printed in Great Britain

Reprinted 1983

Contents

List of Tables
Preface

PART ONE: INTRODUCTION   1

1   The Bacterial Cell   3
1.1   The Nature of the Genetic Material   5
        1.1.1   Structural Considerations   5
        (a)   The Composition of Nucleic Acid
        (b)   The Double Helix
        (c)   Chromosome Structure
        1.1.2   The Biological Role of Nucleic Acid: A Preview   15
        (a)   Gene Expression
        (b)   Replication
        (c)   Genetic Recombination
        (d)   DNA Repair
        (e)   Symmetrical Recognition Sequences
        1.1.3   Phenotype and Genotype   26

1.2   Cell Composition   28
        1.2.1   The Cell Envelope   29
        1.2.2   Cellular Appendages   34

1.3   Bacterial Growth   36
        1.3.1   An Introduction to Metabolism   36
        (a)   Production of the Chemical Intermediate, ATP
        (b)   Biosynthetic Reactions
        (c)   Pathway Regulation
        1.3.2   Growth of Populations   39
        (a)   Growth Requirements
        (b)   Culture Characteristics
                (i) Lag Phase; (ii) Exponential Phase; (iii) Stationary Phase
Bibliography   46

PART TWO: GENE EXPRESSION   49

2   **RNA and Protein Production**   51
2.1   DNA Transcription   55
       2.1.1 The Elements of Transcription   56
       (a)   DNA-Dependent RNA Polymerase
       (b)   Transcription Factors
       2.1.2 The Mechanism of Transcription   58
       (a)   The Transcription Cycle
              (i) Initiation; (ii) Elongation; (iii) Termination
       (b)   Control: RNA Synthesis and Decay
       (c)   Antibiotic Inhibitors of Transcription

2.2   RNA Translation   71
       2.2.1 The Elements of Translation   71
       (a)   The Genetic Code
       (b)   Transfer RNA
       (c)   The Ribosome
       2.2.2 The Mechanism of Translation   90
       (a)   The Ribosome Cycle
              (i) Initiation; (ii) Elongation; (iii) Termination
       (b)   Post-Translational Modification
       (c)   Control
       (d)   Antibiotic Inhibitors of Translation

Bibliography   107

3   **Mutation**   110

3.1   Mutation Classification   111
       3.1.1 Types of Point Mutation   113
       (a)   Samesense Mutation
       (b)   Missense Mutation
       (c)   Nonsense Mutation
       (d)   Frameshift Mutation
       3.1.2 Conditional Mutants   114

3.2   Mutagenesis   116
       3.2.1 The Molecular Basis of Mutagenesis   117
       (a)   Spontaneous Mutation
       (b)   Induced Mutation
              (i) Radiation; (ii) Chemical Mutagens
       (c)   Mutational Hotspots
       3.2.2 The Application of Mutagenesis to Bacterial Systems   132

      (a)   Isolation of Independent Mutants that Carry Single Lesions
      (b)   Mutant Selection
           (i) Replica-plating; (ii) Penicillin Enrichment

3.3   Suppression   135

      3.3.1   Intragenic Suppression   138
      3.3.2   Intergenic Suppression   140
      3.3.3   Phenotypic Suppression: The Role of the Ribosome
             in Misreading   146

3.4   Summary: Mutation Identification   150

Bibliography   153

PART THREE: GENE TRANSFER   157

4 Plasmids   159

4.1   The F Plasmid   160

      4.1.1   Vegetative Replication   160
      4.1.2   F Transfer   161
      (a)   The Mechanism of Transfer
      (b)   Barriers to Transfer
      4.1.3   F-Mediated Transfer   166
      (a)   F integration : Hfr Formation
           (i) Insertion; (ii) Hfr Transfer; (iii) The One Hundred
           Minute Map
      (b)   F Excision : F-Prime Formation

4.2   A Guide to Naturally-Occurring Plasmids   184

      4.2.1   R Plasmids   187
      4.2.2   Col Plasmids   187

Bibliography   191

5 Bacteriophages   193

5.1   Bacteriophages and their Life-Cycles   193

      5.1.1   Anatomical Considerations   194
      (a)   Phage Morphology
      (b)   The Viral Genome
      5.1.2   Phage Infection   197
      (a)   The Lytic Response
      (b)   Morphogenesis

(c) The Lysogenic Response
5.1.3 Phage Methodology   205
  (a) Titration: The Plaque Assay
  (b) Production of Phage Lines

5.2 Bacteriophage-Mediated Gene Transfer   209
  5.2.1 Generalised Transduction   211
  5.2.2 Specialised Transduction   215

Bibliography   227

6 Reactions of DNA   230

6.1 DNA Replication   231
  6.1.1 The Elements of Replication   234
    (a) DNA-Dependent DNA Polymerases
    (b) Protein Components of the Replication Fork
  6.1.2 The Mechanism of Bacterial Replication   243
    (a) Bidirectional Replication of the *E. coli* Chromosome
         (i) Initiation; (ii) Elongation; (iii) Termination
    (b) The Role of the Cell Envelope in Replication
    (c) Control of DNA Replication
    (d) Antibiotic Inhibitors of Replication
  6.1.3 Replication of Small Genetic Elements   260
    (a) Plasmid Production
    (b) Viral DNA Synthesis

6.2 Genetic Recombination   268
  6.2.1 General Recombination   271
    (a) The Elements of General Recombination
         (i) The *recA* Gene Product; (ii) The Involvement of Other
         Gene Products in Recombination
    (b) Recombination Pathways
         (i) *recA*-mediated Pathways; (ii) The *rpo* Pathway
    (c) The Mechanism of General Recombination
         (i) Strand Breakage; (ii) Strand Pairing; (iii) Strand Invasion;
         (iv) Chiasma Formation; (v) Breakage and Reunion;
         (vi) Mismatch Repair
  6.2.2 Non-Homologous Recombination   292
    (a) Site-Specific Recombination
         (i) Bacteriophages; (ii) Insertion Sequences and Transposons
    (b) Illegitimate Recombination
  6.2.3 Phylogenetic Implications of Recombination   307

6.3　DNA Repair　312

　6.3.1　The Elements of DNA Repair　313
　　(a)　Specific Repair Enzymes
　6.3.2　The Mechanism of DNA Repair　318
　　(a)　Pre-Replication Repair
　　　(i) Excision Repair; (ii) Photoreactivation;
　　　(iii) Base Replacement
　　(b)　Post-Replication Repair
　　　(i) Daughter-strand Gap Repair; (ii) Error-prone Induced Repair

Bibliography　325

7　Investigation of Gene Structure and Function　329

7.1　Genetic Analysis of Bacteria and their Viruses　329

　7.1.1　The Elements of Genetic Analysis　330
　　(a)　The *cis-trans* Complementation Test: The Gene as a
　　　Unit of Function
　　(b)　Recombination: An Indication of Genetic Distance
　7.1.2　Mapping Bacterial Genes　336
　　(a)　Conjugation
　　　(i) Hfr Mating; (ii) F-Prime Plasmids
　　(b)　Transduction
　　　(i) P1 Generalised Transduction; (ii) Specialised Transduction
　　(c)　Mutagenesis by DNA Insertion Elements
　　　(i)　Bacteriophages as Transposable Genetic Elements;
　　　(ii) Insertion Sequences and Transposons
　　(d)　Pathway Analysis
　7.1.3　Mapping Phage Genes　350

7.2　Monitoring Gene Function　353

　7.2.1　Protein Biosynthesis　353
　7.2.2　RNA Production　356

7.3　Physical Analysis of Gene Structure　358

　7.3.1　DNA Isolation　358
　7.3.2　Heteroduplex Analysis　359
　7.3.3　Restriction Technology　360
　　(a)　Restriction Cleavage Maps
　　(b)　DNA Cloning

Bibliography　369

PART FOUR: GENE REGULATION   371

8   Operon Control   373
8.1   Transcriptional Control   373
       8.1.1   Control at RNA Chain Initiation   374
       (a)   Promoter Control
             (i) The Elements of Promoter Control; (ii) Promoter
             Control Circuits
       (b)   *Cis-* and *Trans*-Acting Regulatory Mutations
       8.1.2   Control at RNA Chain Termination   381
       (a)   Polarity and its Suppression
       (b)   Attenuator Control

8.2   Post-Transcriptional Control   388
       8.2.1   Control at the Level of the Transcript   388
       8.2.2   Translational Control   388

Bibliography   389

9   Control of Bacterial Gene Expression   390

9.1   Catabolite-Controlled Operons: The Lactose System   391
       9.1.1   The *lac* Operon   392
       9.1.2   *lac* Promoter Control   395
       (a)   Induction and Repression
       (b)   Catabolite Control

9.2   Attenuator-Controlled Operons: The Tryptophan System   401
       9.2.1   The *trp* Operon   401
       9.2.2   *trp* Control   405
       (a)   *trp* Promoter Control
       (b)   *trp* Attenuator Control

9.3   Multivalent Control of Transcriptional-Translational Operons   411
       9.3.1   Organisation of the Genes of the Transcription-Translation
               Apparatus   411
       (a)   RNA Polymerase Operons
       (b)   RNA Operons
             (i) Ribosomal RNA Operons; (ii) Transfer RNA Operons
       (c)   Ribosomal Protein Operons
       9.3.2   Control of Transcriptional-Translational Operons   419
       (a)   Metabolic Control
       (b)   Stringent Control
       (c)   Co-Transcriptional and Post-Transcriptional Control

Bibliography   423

10   Control of Extrachromosomal Genetic Elements   425

10.1   Regulation of F Plasmid Transactions   427

    10.1.1 Organisation of the F Plasmid   427
      (a)   The Transfer Region
      (b)   The Replication Region
      (c)   The Insertion Region
    10.1.2 Interaction between F and F-like R Plasmids   431
    10.1.3 The 'Life-Cycle' of the F Plasmid   434

10.2   Regulation of Phage *Lambda* Development   434

    10.2.1 Organisation of the Intracellular Viral Chromosome   435
      (a)   The Regulatory Region
      (b)   The Site-Specific Recombination Region
      (c)   Replication Elements
      (d)   The Morphogenetic Region
    10.2.2 Lysogen Formation   449
      (a)   Establishment of Lysogeny: P_E-promoted Transcription
      (b)   Maintenance of Lysogeny: P_M-promoted Transcription
    10.2.3 Prophage Induction   452
      (a)   Repressor Inactivation
      (b)   Immediate-Early Transcription: P_L- and P_R-promoted Events
      (c)   Delayed-Early Transcription: P_L- and P_R-promoted
         Readthrough Past the Early Terminators t_L, t_{R1} and t_{R2}
      (d)   Late Transcription: P_R-promoted Readthrough
    10.2.4 The Lysogenic-Lytic 'Decision'   455

Bibliography   457

Index   459

Tables

1.1 Nucleic Acid Nomenclature   7
1.2 The Size of Some Genetic Elements   11
1.3 Some Non-Essential and Essential *E. coli* Genes   28
1.4 A Typical Synthetic Growth Medium for a Wild-Type Strain of *E. coli*   41
2.1 Subunit Composition of *E. coli* RNA Polymerase Holoenzyme   57
2.2 Transcription Factors   57
2.3 The Genetic Code   74
2.4 Degeneracy of the Genetic Code   77
2.5 Some Common Modified Nucleosides of tRNA   80
2.6 Wobble Pairing: Non-Standard (and Standard) Base Pairing between the 5′-Proximal Anticodon Base and 3′-Proximal Codon Base   83
2.7 Genetic Nomenclature for Components of the Translation Apparatus   92
3.1 Mutagen Action   118
3.2 Some Common Carcinogenic Sources   122
3.3 Suppressor Classification   138
3.4 Nonsense Suppressors   148
3.5 Interaction of the Ribosomal Proteins S12 *(rpsL)* and S4 *(rpsD)*   151
3.6 Intrinsic Ambiguity in Codon-Anticodon Pairing as Shown by the Range of Nonsense Suppression   151
3.7 Characterisation of Structural Gene Lesions   153
4.1 A Guide to Naturally-Occurring Plasmids   185
4.2 A Comparison of Chromosomal and Plasmid-Mediated Antibiotic Resistance   189
5.1 Comparative Phage Anatomy   194
5.2 Some Phage Receptors   198
6.1 Properties of the Three *E. coli* DNA Polymerases   235
6.2 Chromosomal Genes Involved in Replication of the Bacterial Genome   245
6.3 Alleles of *recA* and *lexA*   276
6.4 Chromosomal Genes Involved in Genetic Recombination   280
6.5 A Comparison of Four Models for General Recombination   295
6.6 Properties of Some Transposable Genetic Elements   300
6.7 Chromosomal Genes Implicated in DNA Repair   314

7.1 Properties of Some Type II Site-Specific Restriction
    Endonucleases   367
8.1 Characteristics of Negative- and Positive-Control Systems   378
9.1 Mutations Affecting the Lac Phenotype   393
9.2 Mutations that Alter the Trp Phenotype   403
9.3 Components of the rRNA Polycistronic Operons   416
10.1 Some *Lambda* Genes and their Role in Phage Development   438

Preface

My aim in writing *Gene Function* has been to present an up-to-date picture of the molecular biology of *Escherichia coli*. I have not attempted a chronological description, believing that a mechanistic account is more useful for such a highly developed field.

I have divided the book into four parts. Part I is a general introduction to bacterial systems, their genetic material, structure, composition and growth. It has seemed desirable to include herein a brief preview of the remaining text, to introduce the nomenclature and to help place subsequent chapters in perspective. The expression of genetic material and its perturbation through mutation is considered in Part II. Part III discusses how the transfer of prokaryotic genetic material can be mediated by plasmids and bacteriophages. It describes the DNA transactions involved (replication, recombination and repair) and ends with a description of the genetic and biochemical techniques employed in the study of gene organisation. Finally, Part IV considers the control of expression of bacterial, plasmid and phage genes. Key reviews are listed at the end of each chapter.

I should like to express my gratitude to the following: Richard Hayward, Michael Hunter, Robert Lloyd, David McConnell, Vishvanath Nene and Terek Schwartz for invaluable advice and criticism of the text; Michael Billett, Ray McKee and Philip Strange for comments on certain areas; Tony Brown for good counsel; Bryan Clarke for a 'sensible' title; Colin Wilde for constructing the index; Gill Burgess and Rosalyn Chapman for patiently typing the manuscript; Sally Smith for her superb graphic work; Ray McKee for supplying material for photographing; Kate Kirwan and the audio-visual department of the Queen's Medical Centre for photographic work.

Robert E. Glass
Nottingham

To my father,
in memory.

Part I
Introduction

1 The Bacterial Cell

What is a bacterium? Of the multitude of different cellular organisms present on our earth, all can be conveniently divided into two main groups: *eukaryotes*, those that carry their genetic material physically retained within a membrane, separate from the cytosol; and *prokaryotes*, those that lack a distinct nuclear membrane. Bacteria are microscopic, predominantly unicellular species, ubiquitous in nature, belonging to this latter class. They come in many shapes and sizes: spherical, rod-shaped (straight or curved); some form a mycelium. They have in common the ability to divide asexually by fission (certain species, such as *Bacillus subtilis*, also survive by the generation of spores that are aerially dispersed).

Why *Escherichia coli?* Prominent in any list of the advantages of research on bacteria must be the rapid growth and limited nutritional requirements of these organisms. This allows the ready production of large quantities of cells — a homogeneous population, moreover. Also important is that results are obtained in a matter of hours (certainly no more than one or two days). Bacteria are easily handled — they require minimal sterile techniques, being fast growers. Since the strains of *E. coli* generally used are non-pathogenic, little (if any) containment is necessary. The fact that *E. coli* grows on a defined medium is crucial for genetic studies and for the investigation of biochemical pathways. It seems fortuitous that this harmless gut bacterium was initially chosen for study since *E. coli* (unlike *Salmonella typhimurium*) carries a functional system for lactose utilisation, study of which led to formulation of the operon model for regulation of expression of prokaryotic genetic material. Moreover, comparison of the conjugative plasmids — extrachromosomal genetic elements able to transfer between bacteria — present in different species suggests that the high transfer efficiency of the F plasmid of *E. coli* was instrumental in its discovery.

The present discussion, indeed the remaining text, is restricted to *E. coli* and its heritable elements. This is a reflection of the mass of research that has gone into a single organism. Already, we know the complete structure of the genetic material of certain of its simple viruses. The processes responsible for their replication and propagation are likely to be elucidated completely in the near future. The next level of challenge is, surely, a whole bacterium. Perhaps half of the *E. coli* genetic material has been characterised in terms of coding potential, and a number of regions have been sequenced. A unified, frontal attack on a single species, albeit at many different levels, therefore has its advantages. Critics of such prokaryotic studies have cited the clear disparity with higher organisms and the apparent lack of relevance to clinical problems. Yet bacterial research (and not just that on *E. coli*) has led to models for regulation of gene expression, to an understanding of plasmid-mediated drug resistance and of mechanisms for DNA repair, and to *in vitro* dissection of both prokaryotic and eukaryotic genetic material.

It should not be thought that all attention has been placed on the bacterium *E. coli*. Considerable diversity — both at the biochemical level and in terms of chromosome organisation — exists among bacterial species. *S. typhimurium*, for example (also a member of the family Enterobacteriaceae,[1] though not closely related to *E. coli* biochemically), offers a useful parallel, perhaps even giving an indication of divergence in chromosome evolution since its genetic material shows striking similarities to that of *E. coli*. Moreover, it was studies on this species which led to the discovery of prokaryotic genetic exchange mediated by bacterial viruses. *B. subtilis*, a nonenteric organism, forms spores under adverse conditions such as nutritional deficiency and is, thus, a potential model for the study of differentiation, albeit at the prokaryotic level. Actinomycetes of the genus *Streptomyces* are of great importance clinically as major producers of small, antibacterial agents such as streptomycin, chloramphenicol and tetracycline. These *antibiotics* (and their semi-synthetic derivatives) are active against a diverse range of microbial genera. The genetic system(s) involved in antibiotic production is a promising area of investigation. Studies on *Pseudomonas*, another soil micro-organism, aimed at developing antibiotics active against the highly pathogenic *P. aeruginosa* — responsible for fatal infection in patients with severe burns — have identified a source of extrachromosomal genetic elements responsible for resistance to various antibiotics. The pseudomonad *plasmids* are highly promiscuous, transferring across wide taxonomic boundaries. The ability of other plasmids to bring about the degradation of unusual organic substances such as camphor or octane makes them of particular commercial value — as bacterial 'scavengers' for pollution control or as a cheap source of protein. Finally, the hope of dinitrogen (N_2) fixation by natural means (rather than the use of nitrate-containing fertilisers), clearly of great agronomic importance, has led to two main branches of research: the isolation of the genetic region responsible *(nif)* from *Klebsiella* with the aim of its introduction into plants; and the study of the symbiotic association between strains of *Rhizobium* and legumes such as peas, with the aim of extending the range of host plants.

1.   In *Bergey's Manual of Determinative Bacteriology*, bacteria (other than cyanobacteria) are separated into nineteen parts. Part 8, the Gram-negative, facultatively anaerobic rods consists of two main families: Enterobacteriaceae and Vibrionaceae (facultative anaerobes are capable of growing in the presence or absence of oxygen). The former is further subdivided into 12 genera which include *Escherichia, Salmonella, Shigella, Klebsiella* and *Proteus. Escherichia, Salmonella* and *Shigella* are commonly part of the intestinal flora of man and other vertebrates. Whereas *E. coli* is only exceptionally pathenogenic, the genera *Salmonella* and *Shigella* give rise to a wide variety of enteric diseases in both man and animals. Various laboratory strains of *E. coli* are now available. *E. coli* K12 and B, for example, are closely related strains of the same species. *E. coli* B/r is a radiation-resistant derivative of strain B.

1.1 The Nature of the Genetic Material

Nucleic acid as genetic material can be viewed either as the 'programme' that dictates cellular growth or as the sole cellular constituent that is transferred intact, and *in toto*, to succeeding generations. Thus, in the first case, it is an entity of function; in the second, one of heredity. These operational definitions might, at first sight, appear distinct. Consider, however, the ability of a particular species to carry out a cellular process. The region of genetic nucleic acid responsible for that property, and hence the property itself, will be passed to later generations whereas those bacteria that lack the function in the first instance are unable to impart it to daughter cells.

Since one or more macromolecules of genetic nucleic acid can coexist within a single cell and since each of these *chromosomes* may be distinct, each responsible for particular genetic traits, the term *genome* is used to denote the total hereditary material — the total chromosome content. *E. coli* contains a single chromosome, normally present in one to two copies per cell (depending upon the growth conditions). Certain strains carry, in addition, small, supernumerary, *extrachromosomal genetic elements:* these mini-chromosomes are termed *plasmids*.

1.1.1 Structural Considerations

(a) The Composition of Nucleic Acid. Nucleic acid is made up of three major components: pentose sugars, phosphoric acid and nitrogen-containing aromatic bases (Figure 1.1). The first two constituents are invariant (apart from the presence — *ribose* — or absence — *deoxyribose* — of an hydroxyl group on the sugar residue at the 2'-position): only the latter may differ. There are five common nitrogenous bases:[2] the purines, adenine and guanine, and the smaller pyrimidines, uracil, thymine and cytosine (abbreviated to A, G, U, T and C, respectively; Table 1.1). They are joined to the sugar moiety through an N-glycosyl bond (previously termed N-glycoside), to form a *nucleoside* (Figure 1.1). Each *nucleotide* carries, in addition, a phosphate attached to the sugar 5'-OH group (a phosphoester bond). Each has, therefore, a 'blocked' 5'-OH but free 3'-OH; these two hydroxyls (as well as the 2'-OH) are important in dictating the properties of each nucleotide (see below). There can be one, two or three 5'-phosphate groups — the additions made through pyrophosphate linkage to the 5'-terminal phosphate. Thus, the term nucleotide is a generic one and includes, in addition to *nucleoside monophosphates*, both *nucleoside diphosphates* and *triphosphates*. Two ribonucleotides, adenosine triphosphate (ATP) and guanosine triphosphate (GTP), play an important role *per se* in expression of the

2.   There are also variations on the theme. 5-Hydroxymethyl-cytosine (HMC), for example, replaces cytosine in the genome of bacteriophage T4 (see Section 5.1.1 (b)). Methylation, a common modification, protects the genetic material against host-specific degradative systems (see Section 7.3.3). Modified bases found in one form of nucleic acid, ribonucleic acid, are discussed in Section 2.2.1 (b).

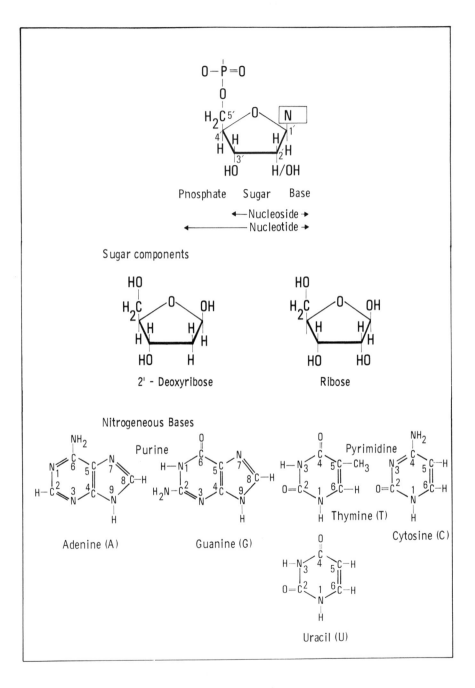

Figure 1.1: The Components of Nucleic Acid

Table 1.1 Nucleic Acid Nomenclature

Base	Nucleoside		Nucleotide (nucleoside 5' - monophosphate ^{a})	
	Ribose	Deoxyribose	Ribose	Deoxyribose
adenine (A)	adenosine	deoxyadenosine	aderylate (adenosine 5'-monophosphate, AMP)	deoxyadenylate (deoxyadenosine 5'-monophosphate, dAMP)
guanine (G)	guanosine	deoxyguanosine	guarylate (guanosine 5'-monophosphate, GMP)	deoxyguanylate (deoxyguanosine 5'-monophosphate, dGMP)
thymine (T)	-b	deoxythymidine	-b	deoxythymidylate (deoxythymidine 5'-monophosphate, dTMP)
uracil (U)	uridine	-b	uridylate (uridine 5'-monophosphate, UMP)	-b
cytosine(C) c	cytidine	deoxycytidine	cytidylate(cytosine 5'-monophosphate, CMP)	deoxycytidylate (deoxycytosine 5'-monophosphate, dCMP)

a   Other common nucleoside 5'-phosphates include diphosphates and triphosphates (for example, ADP and ATP).

b   Uracil normally replaces thymine in ribonucleic acid; thymine is also present as a minor base in certain RNA species (see Transfer RNA, Section 2.2.1 (b)).

c   5-Hydroxymethyl cytosine (HMC) is found in the place of cytosine in some viral genomes (Section 5.1.1 (b)).

genetic material (see Chapter 2) and in DNA replication (Section 6.1), as well as in the metabolic processes of the cell (discussed in Section 1.3.1 (a)). However, it is as a macromolecule, as a polyribonucleotide or polydeoxyribonucleotide, that nucleic acid makes up the genetic material.

In the polymerised form, each nucleoside is linked, via a 3', 5'-phosphodiester bond: a single phosphate group connects adjacent sugar residues at their 3'- and 5'-hydroxyls (Figure 1.2). This generates a regular sugar-phosphate 'backbone' — alternating ribose sugar rings and phosphate groups — from which extends a purine or pyrimidine nitrogenous base (the chain itself is unbranched). The structure has clear polarity because the backbone carries one 5'-terminal phosphate and one 3'-hydroxyl free of a sugar linkage. When the sugar present is ribose, the macromolecule is termed *ribonucleic acid (RNA):* its counterpart, carrying a deoxyribose sugar, is called *deoxyribonucleic acid (DNA)*. In addition to the above distinction — the presence or absence of a 2'-hydroxyl[3] — the pyrimidine base uracil is present in the place of thymine (5-methyluracil) in RNA.

In conclusion, each major type of nucleic acid consists of only four common bases: two purines (A, G) and two pyrimidines (T/U,C). It is noteworthy that a small oligonucleotide consisting of just 12 nucleotides — at least 1200 nucleotides are required to carry enough information for an average-sized protein of molecular weight 40 000 — may be arranged into 1.68×10^7 (4^{12}) different permutations. There is, therefore, ample room for variation, based upon a meagre language of four words, for the programming function of genetic nucleic acid. This is exemplified by the enormous size of these macromolecules. In prokaryotes, RNA molecules are normally between 100 and 6000 nucleotides in length (about 2.5×10^4 to 200×10^4 daltons) whereas DNA species may consist of approximately 5000 to 5 million bases (see Table 1.2).

(b) The Double Helix. Whereas RNA is usually present in a single-strand form (Figure 1.2), DNA is more commonly found as a duplex (in either case, nucleic acids are long, thin threads). The two strands of deoxyribonucleic acid are associated through weak, non-ionic *hydrogen bonds* (Figure 1.3). Specific interactions, termed *base pairing*, occur between the large purine bases and the smaller pyrimidines: the 6-amino group of adenine with the thymine 4-keto group (as well as hydrogen bonding involving ring nitrogens), and the amino and keto groups of guanine with their counterparts on cytosine (a further bond is formed through ring nitrogens, giving the G·C pair of three hydrogen bonds more stability than an A·T pair). Thus, each strand of DNA is associated with a complementary copy: the sequence of one being automatically dictated by the other. The two chains of this duplex run in an antiparallel orientation since the 5'-terminus of one chain is associated with the 3'-end of the complementary

3.   It is the presence of the free 2'-OH group in RNA that makes this nucleic acid particularly susceptible to extreme alkaline conditions. Alkaline hydrolysis is, thus, useful for distinguishing RNA from DNA.

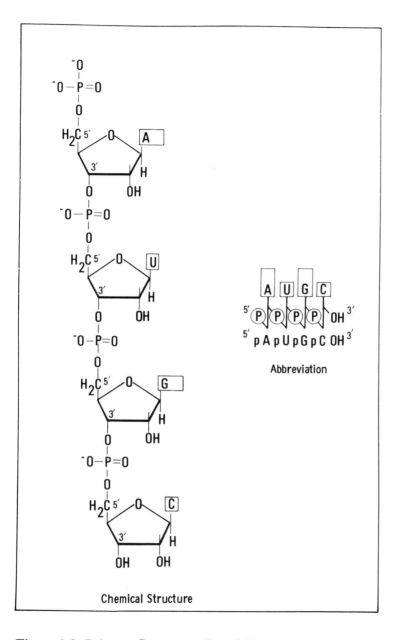

Chemical Structure

Figure 1.2: Primary Structure of an Oligoribonucleotide
Note the terminal 5'-phosphoryl and 3'-hydroxyl groups

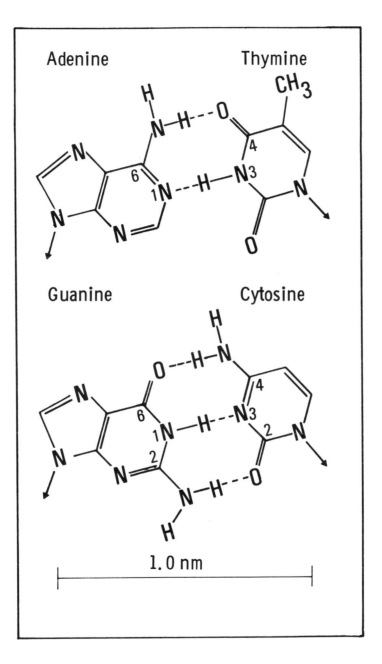

Figure 1.3: Hydrogen Bonding between Nitrogenous Bases
Hydrogen bonds are indicated by dotted lines; arrows represent the N-glycosyl bonds between base
and sugar residues. In each base pair, a large purine hydrogen bonds with a small pyrimidine

Table 1.2 The Size of Some Genetic Elements[a]

Genetic Element		Prototype		Size	
				Mdal	kb
Viral chromosomes					
ss linear	RNA	MS2		1.3	3.6
ss circular	DNA	ØX174		1.8	5.4
ds linear	DNA	T7		26.5	40.4
ds linear/circular [b]	DNA	lambda		31.9	48.6
Plasmids					
ds circular	DNA, 'small'	ColEl		4.2	6.4
ds circular	DNA, 'large'	F		62	94.5
Bacterial chromosomes					
ds circular	DNA, E.coli	Gram -ve[c]		2.5×10^3	3.8×10^3
ds circular	DNA, B.subtilis	Gram +ve[c]		1.3×10^3	2.0×10^3

a   Abbreviations: ss, single-stranded; ds, double-stranded; Mdal, 10^6 daltons; kb, 10^3 nucleotide bases or base pairs.

b   The genome of phage *lambda*, though linear, circularises upon injection (Section 5.2.2).

c   Bacteria can be conveniently grouped according to whether they take up a particular stain or not (Gram-positive or negative, Section 1.1.2).

molecule (and *vice versa*, see Figure 1.4). Similarly, the internucleotide linkage is 3′ → 5′ in one strand but 5′ → 3′ in the other. The polarity in the backbone generates four different types of base pairs rather than two. A 3′ G 5′·5′ C 3′ pair is, for example, clearly different from a 3′ C 5′ · 5′ G 3′ pair.

It follows from the base-pairing rules that in any one duplex, adenine and thymine must be present in equimolar amounts: the same is true for guanine and cytosine. That this equality does not dictate the proportion of total adenine plus thymine (or guanine and cytosine)[4] is indicated by the wide variation in the values of per cent of (G + C) for different DNAs — some organisms are (G + C)-rich while others have a higher proportion of (A + T) ((G + C) per cent values for *B. cereus*, *E. coli* and *P. aeruginosa* are 37, 51 and 68, respectively).

Base pairing — a large purine base with a smaller pyrimidine — is essential for the regular secondary structure of DNA (RNA conformation is discussed in Section 2.2). The two strands of this macromolecule are intertwined, twisted around the same axis, to form a *double helix*, in which both components are 'right-handed' (Figure 1.5). Each turn is separated by 10 nucleotide pairs, a

4.   Consider A=T=x and G=C=y (where x and y are the amounts of the respective bases). Whereas A/T=G/C=1, (G+C)/(A+T)=y/x. Clearly, the (G+C) content of a DNA molecule is totally unrelated to the proportion of (A+T).

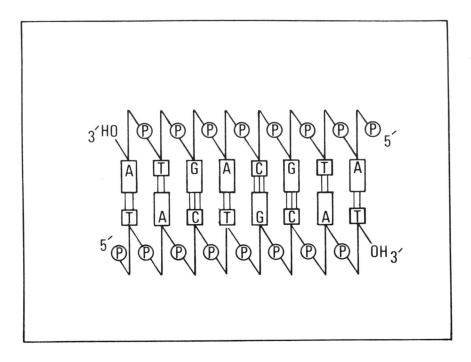

Figure 1.4: Schematic Representation of a Complementary Sequence
An eight base-pair stretch of double-stranded DNA is shown (see Figure 1.2 for abbreviations). Note the 5′ → 3′/3′ → 5′ antiparallel chains. Each base pair consists of a large purine with small pyrimidine (A · T, G · C) though there are different numbers of hydrogen bonds (two and three, respectively). See also Figure 1.3

distance, on the molecular level, of 3.4 nm (in the unstrained case but see Supercoils, below). The diameter of the helix is about 2 nm. Phosphate groups are arranged around the outside, with the bases turned inwards towards the axis of the molecule (stacking of the flat base pairs stabilises the structure).

Does direct interaction of exogenous factors with the bases involve melting (unwinding) of the DNA duplex? Although only the uniform sugar-phosphate backbone (essentially hydrophilic in nature) is most readily accessible in the DNA double helix, base-specific groups protrude into two grooves — the *major* (or wide) *groove* and the *minor* (or small) *groove* — and are available for interaction. Sufficient informational variation is available in the grooves to allow distinction between the four different types of base pairs. The major groove appears to be most involved in base-specific interactions of DNA with proteins (see Regulatory Sequences, Section 2.1.2 (ai) and Part IV).

DNA as the genetic material may be enormous in size (Table 1.2). The fact that such apparently complex material (in terms of function) can be both simple in composition and regular in structure is of paramount importance for our present-day understanding of genetic systems at the molecular level. The

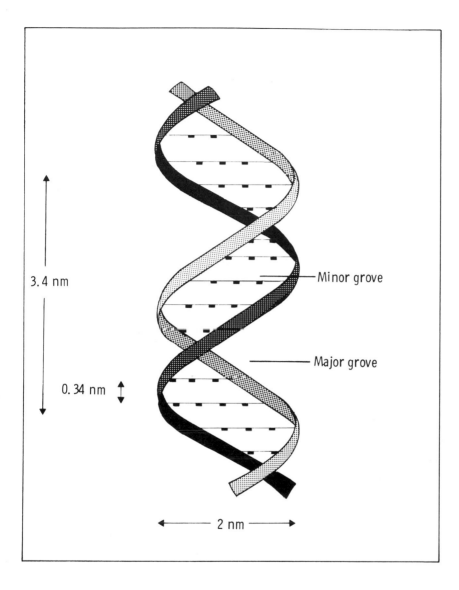

3.4 nm

0.34 nm

Minor grove

Major grove

2 nm

Figure 1.5: The Secondary Structure of DNA: The Double Helix.
The high-humidity, or regular B, form of DNA has a pitch of 3.4 nm and an axial repeat distance of
0.34 nm, corresponding to 10 base pairs per turn of the helix. DNA in a less solvated state or in the
presence of different cations has different helix characteristics. The A-conformation, for example,
found under conditions of low water content and ionic strength, has 11 rather than 10 base pairs per
turn and tilted bases, a structure, in fact, taken up by RNA and RNA·DNA heteroduplexes. A new
helical form, Z DNA, has recently been found that is left-handed instead of right-handed and has 12
base pairs per turn. It is unlikely to be present under normal physiological conditions because of the
high salt concentrations required for the B to Z transition

elucidation of the structure of DNA — the double helix — by Watson and Crick has been hailed as the dawn of molecular biology.

(c) Chromosome Structure. Both the size and form of prokaryotic genetic material may vary (Table 1.2). Within the kingdom Prokaryota and including the bacterial viruses, chromosome size ranges over approximately three orders of magnitude, about 5×10^3 to 5×10^6 base pairs. (Note the two abbreviations used in conjunction with nucleic acid molecules: kb for 10^3 nucleotide bases and Mdal for a megadalton, 10^6 daltons.) The maxim that the complexity of an organism increases with chromosome content is exemplified by the bacterial viruses (discussed in Chapter 5). These bacteriophages, the most simple of species, are at the lower end of the scale in terms of chromosome molecular weight. Since they consist of just a protein coat encapsidating the viral genome, bacteriophages require the metabolic processes of the host cell for propagation. They are generally lethal to their host.

Not all genetic material consists of double-stranded DNA. Some bacteriophages contain single-stranded molecules — of either DNA or RNA (Table 1.2). Moreover, the form that this nucleic acid takes is also species dependent. The viral chromosome of bacteriophage *lambda*, for example, is a linear double-stranded DNA molecule. Phage ϕX174, on the other hand, contains a circular single strand of DNA — the two 'ends' are covalently joined. Even within a single species, chromosome form is by no means invariant. It may change, for example, during replication (witness the double-stranded replicative intermediate of phage ϕX174 present during its propagation). Also present in some bacterial cells are extrachromosomal genetic elements — elements that exist free in the cytoplasm (plasmids are discussed in Chapter 4). These small supernumerary chromosomes consist of covalently-closed circular molecules of duplex DNA (see Figure 1.6). They are apparently non-viral in origin and are innocuous to the cell. Though dispensable, some plasmids impart beneficial properties, for instance drug resistance. Finally, the chromosome of the bacterium *E. coli* (like those of other simple bacteria) is also a covalently-closed circular molecule of double-stranded DNA. In fact, the prevalence of circularity in the genomes of bacteria, these viruses and other extrachromosomal genetic elements, has been termed the 'rule of the ring' (Thomas 1967).

Some helical DNA molecules have a definite tertiary structure. Consider, for example, the superhelical form of covalently-closed circular chromosomes (and mini-chromosomes such as extrachromosomal genetic elements). In these molecules, the circular duplex has become twisted to form a *supercoil*, in which the axis of the helix is itself helical (Figure 1.6). The overall base-pair repeat number may also be altered (from 10 nucleotide pairs per turn in the normal B form, see Figure 1.5). DNA molecules that are 'overwound', twisted in the direction of the double helix, are described as *positive* supercoils: naturally occurring, circular DNA is *negatively* supercoiled (though unwinding during replication removes these twists to generate positively coiled molecules). It has

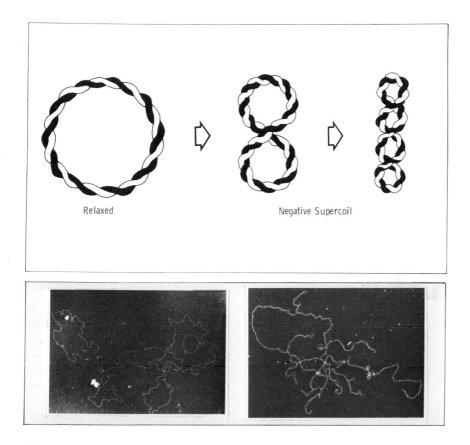

Figure 1.6: Supercoiling of DNA
Although circular DNA is negatively supercoiled, unwinding during replication produces positively coiled molecules (see Figure 6.5). Supercoiling may also occur in linear duplexes if the two ends are 'restrained'. The two electron micrographs show a 100 Mdalton plasmid (relaxed and supercoiled, respectively) from *Halobacterium halobium*. From Weidinger *et al.* (Weidinger, G., Klotz, G. and Goebel, W. (1979) *Plasmid, 2,* 377-386) with permission of the publishers and authors; photographs by courtesy of Professor G. Klotz

been estimated that the bacterial chromosome has about one hundred negative supercoiled loops. Supercoils may be separated from their *relaxed* circular counterparts — where at least one phosphodiester bond is broken — by making use of the property of ethidium bromide to intercalate between adjacent bases: the amount of dye bound is much reduced since intercalation is inhibited by the covalently closed state.

1.1.2 The Biological Role of Nucleic Acid: A Preview

The role of the hereditary material is not restricted to a passive passage from one generation to the next. Chromosomal nucleic acid is the genetic programme. For it is this substance that acts as a template both for self-replication and for the production of the myriad components necessary for cellular growth. Nucleic acid is used in three major processes: the expression of information contained within the genetic material; chromosomal replication; and genetic exchange. These processes are introduced below and are treated in detail in later Chapters.

(a) Gene Expression. Whereas nucleic acid is made up of a regular sugar-phosphate backbone and four different bases, all protein molecules — whether enzymes or not — consist of amino acids linked by peptide bonds (there are 20 different naturally occurring amino acids, Figure 1.7). Clearly, then, nucleic acid cannot fulfil the role of polypeptide species (it participates directly, in fact, in few biochemical reactions). Information for the biosynthesis of each different protein (or RNA) is contained *(encoded)* in the genetic material. The chromosomal region responsible for a particular product (be it protein or ribonucleic acid) is termed a *gene* (or *cistron*). Thus, the macromolecule(s) of nucleic acid that make up the entire genetic content of an organism may be considered as consisting of a number of different genes (perhaps two to three thousand in the case of the circular chromosome of *E. coli*), separated by controlling regions (see below).

The realisation of the information encoded in genetic material — *gene expression* — involves two main steps, the production of an RNA intermediate *(DNA transcription)* and its use as a template for protein biosyntheis (*RNA translation*, Figure 1.8). Complementarity between bases — a single-stranded region of DNA and substrate ribonucleotides in the case of RNA production — is important in the transcription process (as well as in DNA replication, below). Transcription is catalysed by the enzyme *DNA-dependent RNA polymerase* (DNA acts as the template) using ribonucleoside triphosphates as substrate. Some RNA species have a direct role in gene expression and are not translated: regions of DNA that encode protein products are referred to as *structural genes* in the text (Figure 1.8 (a)). Whether translated or not, each molecule of RNA (and each protein) is encoded by a particular gene — each is a *gene product*. The existence of enzyme molecules that consist of more than one different polypeptide chain has led to the 'one gene-one polypeptide' hypothesis (an extension of the earlier proposal that one gene is responsible for one enzyme). Each different subunit (protomer) of a multimeric enzyme is encoded by a separate gene. Though every protein molecule requires a specific RNA template, one particular RNA may, in fact, give rise to a number of different protein products upon translation. Such an RNA template is termed *polycistronic* (or polygenic) since it is the result of transcription of several genes rather than just one (Figure 1.8 (b)). Nevertheless, it is, in general, true to say that within each long transcript, protein-encoding sequences are contiguous rather than overlapping. Transcription and translation will be covered in Chapter 2.

In addition to coding regions, genetic elements also carry specific DNA sequences necessary for gene expression. These are *regulatory sequences* since they contain information for the initiation and modulation of transcription.

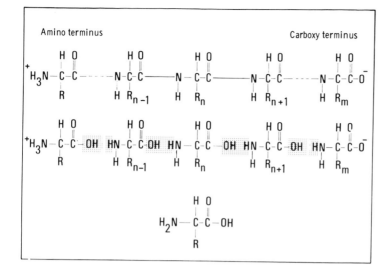

Amino terminus

Carboxy terminus

Amino acid	Symbol	R	Amino acid	Symbol	R

Aliphatic side chains

Glycine — Gly – H

Alanine — Ala – CH_3

Valine — Val – $CH(CH_3)_2$

Leucine — Leu – $CH_2CH(CH_3)_2$

Isoleucine — Ile – $CH_2CH(CH_3)CH_2CH_3$

Acidic side chains

Aspartate — Asp – $CH_2CO_2^-$

Glutamate — Glu – $(CH_2)_2CO_2^-$

Amide side chains

Asparagine — Asn – CH_2CONH_2

Glutamine — Gln – $(CH_2)_2CONH_2$

Basic side chains

Lysine — Lys – $(CH_2)_4NH_2$

Arginine — Arg – $(CH_2)_3NHCNH(NH_2)$

Histidine — His –

Hydroxy side chains

Serine — Ser – CH_2OH

Threonine — Thr – $CH_2(OH)CH_3$

Sulphur-containing side chains

Cysteine — Cys – CH_2SH

Methionine — Met – $(CH_2)_2SCH_3$

Aromatic side chains

Phenylalanine — Phe – CH_2

Tyrosine — Tyr – CH_2 —OH

Tryptophan — Trp – CH_2

Imino acids

Proline — Pro — HO CO_2^-

Figure 1.7: The Composition of Protein

Only naturally occurring, unmodified amino acids are shown. Note that the structure presented for proline is the complete molecule and not just the side group

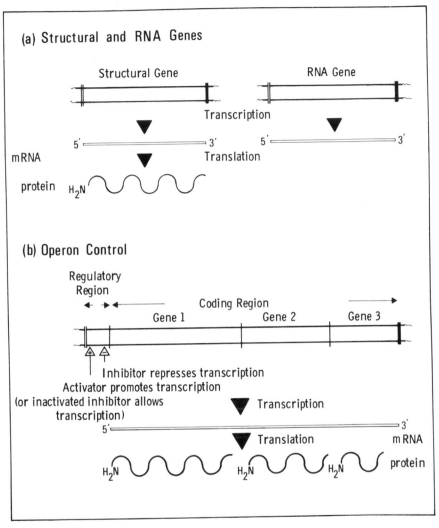

Figure 1.8: The Process of Gene Expression

Whereas a structural gene encodes a protein product and, therefore, requires both transcription and translation for expression, DNA transcription is suffient for realisation of the information contained within a RNA coding gene (a). Expression of genetic material is controlled primarily at the transcriptional level in prokaryotes, either by negative-acting repressor proteins that inhibit transcription or positive-functioning polypeptide species which actually promote transcription (b). A coding region and its control sequence is termed an operon; a polycistronic operon contains two or more genes that are subject to a common control site

Modulation is effected both by activator molecules that promote transcriptional initiation and repressors which act to inhibit this process: in each case, the active component is a polypeptide species encoded by a *regulatory gene*. Termination can also be controlled. Moreover, the siting of transcriptional regulatory sequences in prokaryotes adjacent to the gene or genes that they control allows direct coordinate regulation of expression of a series of gene products, be they RNA or protein in nature (Figure 1.8 (b)). A coding region and its associated regulatory sequences is termed an *operon* (Jacob and Monod 1961). Operon control, both for bacterial and bacteriophage genomes, is discussed in Part IV.

Correct gene expression relies, of course, upon an unaltered template. Changes in the genetic material may affect both the RNA transcript and the protein product (in the case of a structural gene). Moreover, such changes, termed *mutations* (or *lesions*), are heritable: they will be passed onto future generations. Lesions at a single site may affect a single function (consider, for example, a base change in a structural gene leading to a mutant protein product). Alternatively, they may be *pleiotropic*, altering the expression of many gene products. Regulatory mutations — alterations in control sequences or in regulatory genes — exemplify this latter class (although impairment in functions that are essential for cellular growth may also elicit such a response). Chapter 3 deals with the incidence of mutational changes and their effect on gene expression.

(b) Replication. The specificity of base pairing implies that DNA *per se* contains information for its replication. Irrespective of the enzyme systems used (DNA synthesis is one of the most complex known examples of nucleic acid-protein interactions), a key aspect of DNA replication is the local unwinding of the DNA to offer single-stranded templates for synthesis of each complementary chain (Figure 1.9). Such a mode of replication is termed *semi-conservative* since each daughter duplex consists of one parental strand and a newly synthesised molecule (fully conservative replication would require the production of a duplex carrying two new chains, yet preserving the parental duplex, Figure 1.9 (a)). Use of a single parental strand as a template for DNA replication ensures that the daughter molecules are faithful replicas. Specificity in base pairing is crucial for accurate copying.

The enzyme directly responsible for the polymerisation of deoxyribonucleotides is termed *DNA-dependent DNA polymerase*. Not only are there a number of different DNA polymerases in *E. coli* itself, but also bacteriophages may encode their own system for DNA replication. These enzymes have in common the limitation that each new nucleotide is added at the free 3'-end. That is, DNA polymerase(s) can synthesise DNA in only the 5' → 3' direction (Figure 1.9 (b)). The fact that DNA duplexes are anti-parallel in nature — a 5' → 3' chain paired with a 3' → 5' strand — imposes a further constraint upon enzyme function. Consider, then, an enzyme molecule that starts replication from one end of a linear double-stranded DNA molecule: only one of the two strands — that with a 3'-terminus — is immediately available for template function. As replication progresses, it exposes the other strand of the parental duplex, allowing this 5' → 3' specific polymerase to synthesise a short stretch of DNA, termed an *Okazaki fragment*, on the exposed template (in the reverse direction). Migration of the *replication fork* is necessary before the 5'-terminal strand can

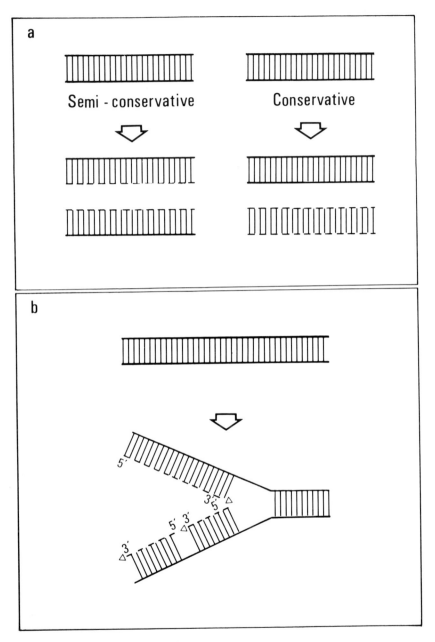

Figure 1.9: DNA Replication

Only one strand is conserved in semi-conservative replication (the dotted line in (a) represents the newly synthesised daughter strand). Since all DNA polymerases function in the 5′→3′ direction, synthesis on only one strand (the 3′ → 5′ strand) can be continuous, while on the other strand replication is discontinuous, requiring unwinding of the DNA duplex to expose a single-strand template (the open arrowheads in (b) indicate the leading 3′-OH of each nascent DNA chain) be further copied. Thus, whereas DNA polymerase may proceed continuously down one chain, there is *discontinuous replication* of the other strand. Moreover, replication of the two polynucleotide chains proceeds in opposite directions.

Circular molecules of DNA that are covalently-closed offer yet another order of topological complexity. Two main mechanisms, θ *(theta)* or *Cairns*-type replication and the *rolling-circle* model, have been proposed for their replication (Figure 1.10). In the first, replication is initiated at a fixed point and proceeds bidirectionally to a defined terminus (variations of this model include unidirectional movement of the replication fork). On the rolling-circle model, one strand is nicked and peels off the intact molecule (in a manner analogous to paper passing over a printing drum), displaced by chain extension. It then acts as a template for synthesis of the complementary strand. The rolling-circle mode can potentially give rise to a *concatemer*: a number of replicons covalently attached in tandem (*concatenane*, on the other hand, refers to interlocked circles of DNA). DNA replication is discussed in Section 6.1.

Clearly, then, different types of DNA molecule may act as templates for replicative DNA synthesis. The term *replicon* was initially coined by Jacob, Brenner and Cuzin (1963) for a genetic unit of replication, a segment of DNA that is maintained whole during cell growth and division. A replicon carries its own system for independently controlling the initiation of replication (extra-chromosomal replicons normally rely on the host for some or all replication enzymes but multiply autonomously).

(c) Genetic Recombination. The seemingly indivisible genetic material can be considered functionally as a number of different genes. As a string of nucleotides, this macromolecule is, in fact, susceptible to chemical degradation *in vitro*. Moreover, certain enzymes termed *nucleases* attack nucleic acid *in situ*, cleaving phosphodiester bonds (those specific for DNA are termed *deoxyribonucleases* or *DNase*'s, while enzymes acting on RNA are called *ribonucleases* or *RNase*'s). *Exo*nucleases require a free end, whereas *endo*nucleases produce a nick in the intact DNA duplex. These activities are important both in replication of chromosomal material (see Section 6.1) and in exchange of DNA between different molecules. The fact that *genetic recombination* can occur both between *(intergenic)* and within *(intragenic)* coding regions indicates that chromosomal material is further divisible, beyond the unit of inheritance (or function), the gene.

Recombination may take many different forms. It requires, nevertheless, at least one crossover point — one site at which strand exchange takes place. Synapsis is thought to occur between DNA regions of homology. It is here, presumably, that strand exchange is initiated. Such *general* (or *homologous*) recombination requires a functional bacterial recombination system, in particular the protein encoded by the *recA* gene (the nomenclature for *E. coli* genes is given in Section 1.1.3). The product of this biological reaction, the *recombinant(s)*, consists of one — in a single crossover event — 'hybrid' molecule that

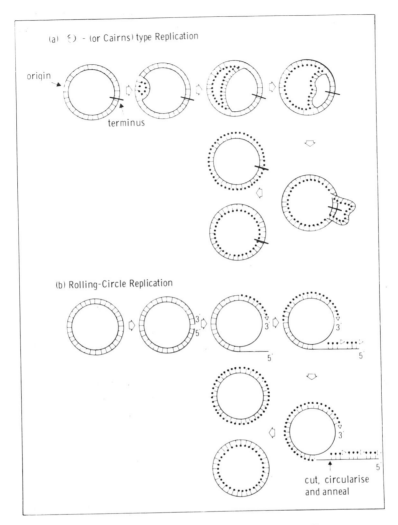

Figure 1.10: Modes for Replication of Circular Chromosomes

In Cairns-type replication (a), DNA synthesis initiated from a fixed point, the origin, proceeds unidirectionally or bidirectionally around the circular element to stop at the terminus; the latter may be a defined sequence or simply the site at which the replication point(s) meet. The scheme shows bidirectional replication from the origin, the two replication forks terminating opposite the start point. Rolling-circle replication (b) is initiated with a nick in one of the DNA strands, with the resultant 3'-OH serving as a primer for DNA extension. Replication on the displaced chain is discontinuous since all DNA polymerases can only polymerise dNTPs in the 5' → 3' direction (the *open arrowheads*, representing the 3'-OH of each short chain synthesised discontinuously, have been left in to emphasise the discontinuous nature of replication on this strand though, in fact, these chains are subsequently ligated together). Replication on the broken strand generates a concatemer, which has to be severed to produce unit chain lengths (circularisation of the newly synthesised DNA completes the process). Similarly, synthesis around the covalently-closed circular single-stranded molecule is completed by nicking at the junction of the 'old' and 'new' DNA and ligation of the juxtaposed 5'- and 3'-termini carries part of the DNA from one chromosome and part from the other. In a *reciprocal* event between linear chromosomes both strands are exchanged, producing two hybrid molecules (note that certain recombination systems, for example, that of bacteriophage *lambda*, catalyse only a non-reciprocal process). Clearly, the types of products formed depend upon the nature of the 'substrate' — whether recombination takes place between linear or covalently circular DNA molecules — as well as the number of crossover events (see Figure 1.11).

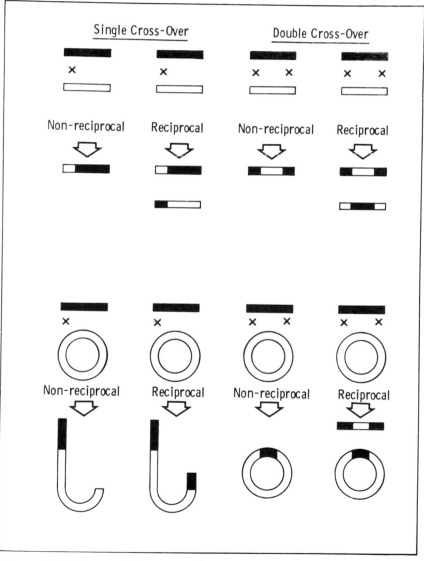

Figure 1.11: Types of Crossover Events in Genetic Recombination

It is this 'jumbling' of genetic material through recombination, in conjunction with the occurrence of mutations in DNA, that is important in increasing genetic variation. Recombination, both homologous and non-homologous, is considered in Section 6.2.

(d) DNA Repair. The fidelity of DNA synthesis is not always perfect: occasionally an error occurs, resulting in the introduction of an incorrect base into the genetic material. Such events are termed *spontaneous mutations*. Not all *genetic lesions* are of spontaneous origin — irradiation and certain chemical mutagens are powerful inducers of mutations — nor do they necessarily affect only one nucleotide pair. Different mutations in the same gene generate a family of related genes or *alleles*.

It should not be thought that the bacterial chromosome and its heritable elements, prime targets owing to their large size, exist unprotected against DNA damage. The 'proof-reading' function associated with all prokaryotic DNA polymerases, the $3' \rightarrow 5'$ exonuclease activity, serves, in fact, to keep the spontaneous mutation rate down to a very low level. Moreover, *E. coli* encodes a number of different repair mechanisms that protect against the action of mutagens. DNA repair is covered in Section 6.3.

(e) Symmetrical Recognition Sequences. It is pertinent to conclude with a brief description of a major type of DNA sequence thought to be involved in nucleic acid-protein interactions. Certain proteins, in particular those that participate in, or modulate, gene expression, appear to act at specific nucleotide sequences which have definite symmetry (a notable exception to this rule is the RNA-polymerase binding site). These sequences are referred to as palindromes or inverted repeats.

There may be some confusion over the use of the term *palindrome* with reference to nucleic-acid sequences. Although the original meaning of this term is a word that reads the same whether backwards or forwards (for instance, 'rotator'), it implies, in the biological sciences, a DNA sequence which is the same (or very similar) in opposite directions of complementary strands, with the property of rotational (dyad) symmetry. Thus, for example, the sequence

$$\overrightarrow{\text{A B C D D'C'B'A'}}$$

$$\underleftarrow{\text{A'B'C'D'D C B A}}$$

(where A', B', C' and D' are complementary bases to A, B, C, D, respectively) is said to be a palindrome. It is, in fact, a perfect *inverted repeat*. The latter term is generally used for DNA segments duplicated in reverse nucleotide sequence orientation but separated by a region of mismatch, i.e. a region with hyphenated dyad symmetry. The complementary flanking sequences have the potential to

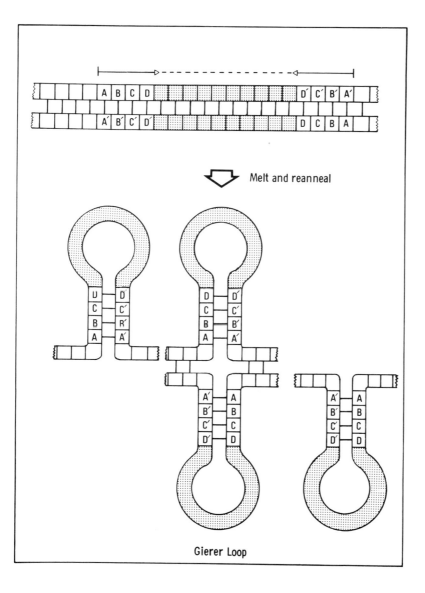

Figure 1.12: Hairpin and Gierer Loops

Though initially proposed as a mechanism for recognition of specific DNA sequences by regulatory proteins, looped structures in DNA are not now thought to be important. However, the structures formed *in vitro* — as viewed in the electron microscope — when two sequences are present in an inverted repeat are useful for the identification of specific genetic elements (see Section 7.3.2) anneal in an intrastrand event to form *hairpin loops* (termed *Gierer loops* in the case of a double-stranded structure), with the complementary sequence present in the stalk of the structure (Figure 1.12).

1.1.3 Phenotype and Genotype

When a bacterial gene is expressed, it imparts some property to the cell, for instance the ability to grow in the absence of the aromatic amino acid tryptophan, or to utilise the sugar lactose. The observable effect of gene expression on an organism is termed the *phenotype*. A three-letter symbol in Roman type, the first of which is capitalised, is used to describe the phenotype of a bacterial strain. In this system of Demerec *et al.* (1966), the triplet symbol is normally a mnemonic for function (or an acronym), thus Trp for tryptophan biosynthesis and Lac for lactose utilisation. A superscript, normally ' + ' or ' – ', is affixed to the three-letter symbol to indicate expression of a particular phenotype. A normal wild-type bacterium carries, by definition, no apparent lesion.[5] It will be, phenotypically (among other things), Trp^+ Lac^+, as indicated by growth in the absence of exogenous tryptophan and utilisation of the six-carbon sugar disaccharide lactose. In general, only the mutant phenotype(s) is indicated since normal functions are, clearly, in great excess.

According to the Demerec convention, each *genetic locus* is similarly represented by a three-letter, lower-case italicised symbol, thus *trp, lac.* In some cases, a number of gene products, linked or otherwise, are required for a particular function. This may arise from the presence of an oligomeric enzyme or a pathway that requires several components. The different loci, mutations in which may give rise to the same mutant phenotype, are distinguished by the addition of an italicised capital letter immediately following the three-letter, lower case italicised symbol. Thus, tryptophan synthetase, which consists of two different subunits, α and β, is encoded by two genes, *trpA* and *trpB*. Similarly, *lacZ* and *lacY* specify the enzyme β-galactosidase and the lactose permease, respectively, required for lactose utilisation.

The term *genotype* refers to the condition of an organism at the level of the gene. A prototrophic wild-type strain carries no apparent mutation(s) (but may carry cryptic ones); its genotype with respect to tryptophan synthetase and lactose-utilising enzymes, for instance, may be written $trpA^+$, B^+ $lacZ^+$, Y^+, the plus ' + ' superscript indicating the wild-type allele. The presence of a functional gene(s) is, in fact, not usually indicated (since the *E. coli* chromosome is likely to code for more than 2000 different gene products, a large space would be required just to indicate all the functional genes!). Rather, only the absence of a functional gene is noted, using a minus ' – ' superscript.[6] The particular mutation and hence

5.   Note that a strain may be apparently wild-type but carry, in fact, a *cryptic* mutation(s). These are 'silent' genetic defects since they have no (observable) effect on phenotype.
6.   Mutant alleles are normally not given a minus superscript in the scientific literature. Thus, a strain with the genotype *trpA lacZ* has a mutation(s) at each of these loci. A minus (and where necessary, plus) superscript has been retained in this text for clarity.

allele is designated by placing an italicised number after the locus symbol, for instance *trpA1⁻*, *lacZ53⁻* (the number has no significance, normally, aside from designating the allele).[7] Note that when two (or more) bacteria have identical makeup, when they are genotypically the same, they are said to be *isogenic*. Some genetic and phenotypic symbols used in the text are shown in Figure 1.13 and Table 1.3, and are listed in later chapters.

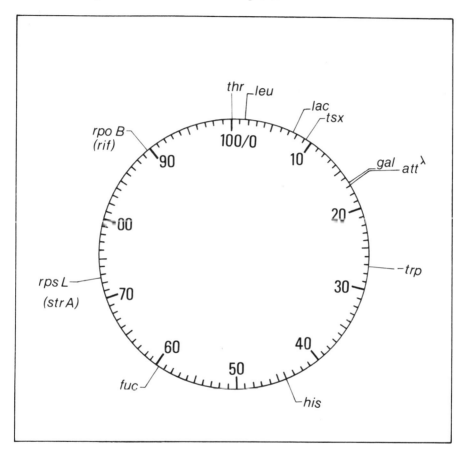

Figure 1.13: Location of Some Genes on the Bacterial Chromosome
These loci are listed in Table 1.3 (for nomenclature, see text). The 0 to 100 minute markings are explained in Chapter 4

7.   When the particular gene carrying the genetic lesion has not been identified, the capital letter is replaced by a hyphen; *trp-49⁻*, for example, would indicate a mutation residing in one of the five structural genes responsible for tryptophan biosynthesis.

Which allele is expressed if both mutant and wild-type are present? Strains may be constructed that are *partial diploid* (or *merodiploid*) for a region of the bacterial chromosome (Section 4.1.3 (b)). Genetic manipulation allows the isolation of merodiploid organisms that carry both mutant and wild-type alleles (Section 7.1). When the phenotype of one allele appears to be preferentially expressed in such a partial diploid strain, this allele is said to be *dominant:* the other allele is, accordingly, *recessive*. Thus, since *lacZ*$^+$ is dominant over *lacZ53*$^-$, the *lacZ*$^+$/*lacZ53*$^-$ merodiploid will be phenotypically Lac$^+$. However, this dominance-recessive character refers only to a particular pair of alleles, as exemplified by the dominance of other *lacZ*$^-$ mutations over *lacZ*$^+$. The phenotype of the recessive allele which is 'silent' in the presence of a dominant allele, will be expressed in a haploid or homozygous merodiploid organism (*lacZ53*$^-$ or *lacZ53*$^-$/*lacZ53*$^-$ strains, for instance, are both phenotypically Lac$^-$).

Table 1.3 Some Non-Essential and Essential *E. coli* Genes

Gene Symbol	Mnemonic For	Map Location (min) [a]	Mutant Phenotype
Non-essential			
his	histidine	44	unable to grow in the absence of the amino acid
leu	leucine	2	"
thr	threonine	0	"
trp	tryptophan	27	"
fuc	fucose	60	unable to utilise the sugar
gal	galactose	17	"
lac	lactose	8	"
tsx	T-six	9	resistant to phage T6
att$^{\lambda}$	attachment	17	lacks integration site for phage lambda
Essential			
rpo B	RNA polymerase	90	lethal[b]
rps L	ribosomal protein, small	72	lethal[b]

a The position of chromosomal markers is defined in units of time (between 0 and 100 min) making use of the F conjugative plasmid to transfer the bacterial chromosome; the entire genetic element may be transferred in 100 minutes (Section 4.1.3 (a)). See Figure 1.13.

b. Although lesions that inactivate essential genes are lethal, certain mutations may alter rather than impair function. Thus, resistance to the antibiotics rifampicin and streptomycin (Rif-r and Str-r phenotype) is due to changes in *rpoB* and *rpsL*, respectively.

1.2 Cell Composition

It will be recalled that two processes, transcription and translation, are necessary for production of cellular proteins. The protein molecules produced (about one half of the dry cell weight) serve as structural elements as well as in a multitude of different biochemical reactions. In addition to their role in the biochemical pathways of the cell, biological catalysts — enzymes — are involved in expression, and in replication and repair, of the bacterial chromosome. It is, therefore, not surprising that some 1100 different polypeptide species have been identified in *E. coli*. A large amount of nucleic acid is also present: both template RNA, and RNA that though not translated has a key role in the process (total RNA represents about 6 per cent of the dry cell weight) as well as DNA, in the form of the bacterial chromosome (about 1 per cent of the cell weight).

The majority of the bacterial cell is composed of water (about 70 per cent) containing various metal cations (for instance, Na^+, K^+, Ca^{2+}, Mg^{2+}, Mn^{2+} and $Fe^{2+/3+}$) and inorganic anions (PO_4^{3-}, SO_4^{2-}, NO_3^{2-} and Cl^-). The latter also serve as a source of phosphorus, sulphur and nitrogen in biosynthetic reactions (see Section 1.3). The metal ions, in conjunction with certain trace elements such as zinc and copper, play an important role in enzyme function — acting as cofactors or as part of non-protein prosthetic groups. Thus, zinc (as Zn^{2+}) is found in DNA-dependent RNA polymerase, the enzyme responsible for transcription of DNA templates and, indeed, in other nucleic-acid enzymes. Iron, on the other hand, either in the reduced or oxidised state (Fe^{2+} and Fe^{3+}, respectively), is present in haem, a nonprotein group attached to the cytochromes of the electron transport chain. Other nonprotein groups that are only loosely associated may be also required for enzyme activity. Consider, for example, coenzyme A (CoA), which is used in acyl group transfer such as in the pyruvate dehydrogenase reaction (pyruvate → acetyl CoA). Apart from protein, the structural components of the cell are made up of lipopolysaccharide, phospholipid and peptidoglycan. Certain small organic molecules, nucleotides and amino acids, are required for nucleic acid and protein biosynthesis. Others are the product of energy-producing and utilising systems.

This vast array of components, essential constituents of a living cell, is retained by the *cell envelope*. This rigid structure maintains the rod-like shape of the bacterium *E. coli* (about $2\mu m$ in length and $1\mu m$ in diameter, Figure 1.14).

1.2.1 The Cell Envelope

An early method for differentiation between bacteria has led to their division into *Gram-positive* and *Gram-negative* types (Figure 1.15; see also footnote 1 on Bergey's classification, page 4). It is based upon the extent to which stain binds (or penetrates), an effect that depends upon the nature of the bacterial cell envelope. Thus, Gram-negative bacteria — those that do not retain the stain — such as *E. coli* and *S. typhimurium* are characterised by their impervious membrane. They are intrinsically more resistant to antibiotics. (In the case of the semi-synthetic drug rifampicin, for example, the minimum inhibitory concentration is about $1-10\mu g/ml$ for *E. coli* but $0.01\mu g/ml$ for *Staphylococcus aureus*, a Gram-positive bacterium).

The *E. coli* cell envelope represents at least 20 per cent of the bacterial dry weight, of which about one half is protein (perhaps about 100 different polypeptide species are present). It consists of an *outer* and *inner* membrane, separated by the *periplasmic* region and rigid peptidoglycan or murein layer, which is the target of the antibiotic penicillin (Figures 1.16 and 1.17). Gram-

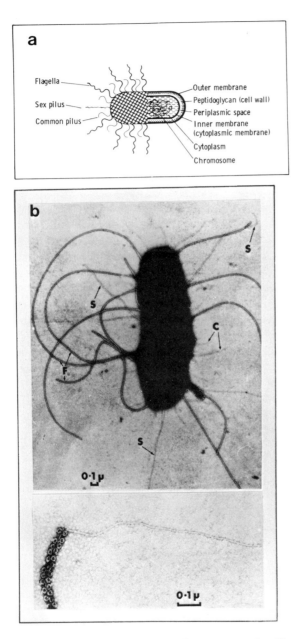

Figure 1.14: The Principle Structures of a Gram-Negative Bacterium

Schematic representation (a) and electron micrographs (b) of *E. coli*. The latter shows the whole cell with F sex pili (S), common pili (C) and flagella (F) as well as an F pilus heavily 'stained' with phage MS2 (along the side) and a single filamentous phage M13 adsorbed to the tip. From Meynell *et al.* (Meynell, E., Meynell, G.G. and Datta, N. (1968) *Bacteriol. Rev.*, *32*, 55-83), reproduced with permission of the publishers and authors

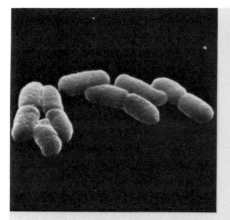

Escherichia coli

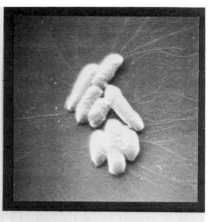

Salmonella typhimurium

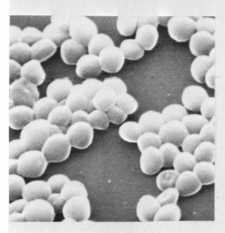

Staphylococcus aureus

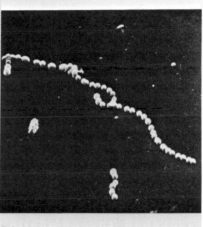

Streptococcus pyogenes

Figure 1.15: Scanning Electron Micrographs of Some Gram-Negative and Gram-Positive Bacteria

Top row, Gram-negative; bottom, Gram-positive (the bar represents 1 μm). Electron micrographs from Greenwood and O'Grady (Greenwood, D. and O'Grady, F. (1972, 1975) *J. Gen. Microbiol., 70*, 263-270 and *91*, 110-118), reproduced with permission of the publishers and authors, photographs by courtesy of Dr D. Greenwood

positive bacteria have only a single, teichoic acid-containing membrane but an enlarged external peptidoglycan layer; they lack lipopolysaccharide. It is the rigid peptidoglycan, the *cell wall*, that contains the cytosolic osmotic pressure.

The two membranes may be isolated by electrophoresis or density gradient centrifugation. Both have a lipid bilayer structure — a bimolecular sheet of

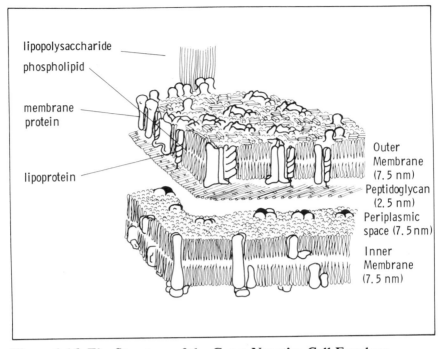

Figure 1.16: The Structure of the Gram-Negative Cell Envelope
After DiRienzo *et al.* (DiRienzo, J.M., Nakamura, K., and Inouye, M. (1978) *Ann. Rev. Biochem.*, *47*, 481-532)

phospholipids (mainly phosphatidylethanolamine, phosphatidylglycerol and cardiolipin), their polar groups forming the external surfaces — interdispersed with protein molecules (Figure 1.16). The fluid mosaic model postulates a dynamic structure in which the globular proteins are 'embedded' in, some even extending through, the lipid bilayer. The outer membrane carries, in addition, lipopolysaccharide; a complex molecule consisting of a lipid moiety covalently linked to a heteropolysaccharide (variation in the sugar component in different strains is responsible for the serological classes of O antigens). It is largely devoid of enzymes. The *cytoplasmic* (or inner) *membrane* is the site of oxidative phosphorylation, and harbours the systems for active transport and for synthesis of the major membrane structural components, phospholipid, lipopolysaccharide and peptidoglycan. Both the lipid and protein composition vary with growth conditions (witness the inducible nature of most active transport systems). Patches of contact between the outer and inner membranes are apparently present *in vivo* (200-400 sites per cell). There is some speculation that these *adhesion points* may function in the entrance of certain viral DNAs or allow transport of *de novo* synthesised proteins to the outer membrane.

　　In terms of its protein content, the outer membrane has a simpler composition

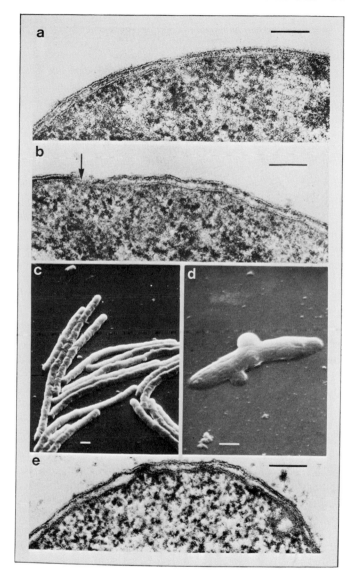

Figure 1.17: Electron Micrographs of the Cell Envelope

Sections showing the *E. coli* outer and inner membrane (a, b and e) and scanning electron micrographs (c and d) after treatment with a β-lactam antibiotic ($> 60\,\mu g$/ml, b and d; 8-$16\,\mu g$/ml, c) and EDTA/lysozyme (e) (the bar represents $0.1\,\mu m$ in a, b and e and $1\,\mu m$ in c and d). At low concentrations of antibiotic, the cell filaments due to lack of septation (c) while at 'high' amounts, spheroplasts are formed (a spheroplast can be seen emerging in d); the arrow in b indicates an apparent division between a normal and altered cell wall. Note that after EDTA/lysozyme treatment (e) the periplasmic space seems vacant (of peptidoglycan). From Murray *et al.* (Murray, R.G.E., Steed, P., and Elson, H.E. (1965) *Can J. Microbiol., 11,* 547-560) and Greenwood and O'Grady (Greenwood, D. and O'Grady, F. (1973) *J. Infect. Dis., 128,* 211-222 and 791-794) reproduced with permission of the publishers and authors; photographs by courtesy of Dr D. Greenwood than that of the cytoplasmic membrane: there are about five major outer membrane proteins and about 10-20 minor ones (some, if not all, proteins destined for the cell wall are synthesised as a precursor, carrying an N-terminal 'signal sequence' responsible for their translocation). The *lpp* gene product, a lipoprotein (molecular weight 7200) present in about 7×10^5 molecules per cell, is covalently linked to the peptidoglycan and may serve to hold the two layers together. (This major outer membrane protein does not seem to be essential for cellular survival.) A number of cell wall proteins, in addition to lipopolysacccharide, have been implicated as receptor sites for phage adsorption (see below). The fluidity of the structure is demonstrated by the incorporation of certain foreign proteins (for example, phage-coded species) and its ability to accommodate increased levels of the minor constituents.

The external membrane carries pores for passive diffusion of small hydrophilic molecules. It is an effective barrier against substances greater than about 500-600 daltons in size, thus imparting low-level resistance to many antibiotics. The entrance of hydrophobic antibiotics of larger size, for example, actinomycin D (1255 daltons), is attributed to their partial solubilization in the non-polar phase of the phospholipid bilayer (the exclusion limit for Gram-positive bacteria is estimated to be about 100 000 daltons). Uptake of specific growth agents (metal ions, sugars and vitamins) requires participation of the outer membrane. This membrane also carries *receptors* for certain noxious substances such as bacteriophages and colicins — antibacterial proteins — as well as for cell-cell contact for DNA transfer (Sections 5.1.2, 4.2.2 and 4.1.3). Several of the minor outer membrane proteins are involved in the receptor-dependent translocation of hydrophilic molecules above the critical size (about 600 daltons). Uptake of vitamin B_{12} (molecular weight 1327), for example, appears to require the participation of at least two different proteins, the *btuB* and *tonB* gene products, though it binds specifically only to the former (present in about 200 copies per cell). Interestingly, membrane receptors frequently serve dual functions, allowing the passage of nutrients as well as the adsorption of noxious substances, in the form of bacteriophages and colicins (phage BF23 and the E colicins in the case of the *btuB* receptor). These toxic substances have probably evolved by making use of bacterial receptor sites (rather than the reverse!). Since these lethal agents are known to act in a disparate manner at the molecular level — the injection of viral nucleic acid on the one hand and a toxic protein on the other — there is no indication that their uptake is mediated through a common mechanism. The multivalent property of such receptors is a fascinating area of membrane research; so is the part played by the *tonB* gene product (36 000 daltons) in the function of a number of different receptors.

1.2.2 Cellular Appendages

The outer membrane has by no means a featureless surface. There are, in addition to lipopolysaccharide, certain proteinaceous appendages (Figure 1.14). These fall into two classes: structures responsible for motility (and chemotaxis), *flagella*, and thin filaments termed *pili*.

Flagella, composed of the protein *flagellin*, the *flaF* gene product (molecular weight 54 000), consist of a helical filament about 20nm in diameter and many times the length of the cell (16-22 μm). There are between 1 and 100 per cell. The filament is attached to the cell through a complex structure — a hook and four rings (at least 28 different genes are involved in the synthesis of the flagellum apparatus, Figure 1.18). Flagellin, synthesised in the cytoplasm, appears to be transported through the central cavity of the filament to the tip, where self-assembly takes place (a polar event since growth is only at this distal end). The hook serves as a joint between the flagellum and basal structure (the four rings), allowing bacterial movement through rotation of the appendage. A three-

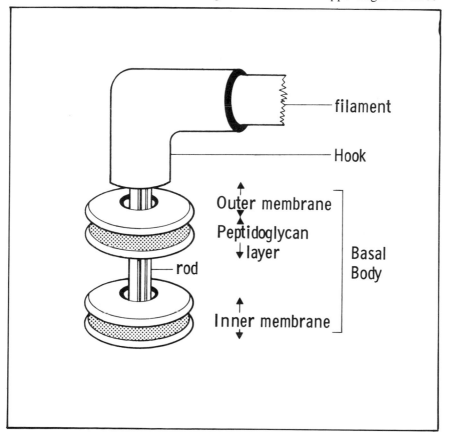

Figure 1.18: The Structure of the Terminal Portion of the Gram-Negative Flagellum
After DePamphilis and Adler (DePamphilis, M.L. and Adler, J. (1971) *J. Bacteriol., 105,* 384-395)

dimensional random walk is generated by a combination of straight runs (smooth swimming) and tumblings: suppression of the latter allows movement towards a source of attractant (for instance, sugars or certain amino acids). It is the direction of rotation of the propeller-like appendage that dictates the mode of swimming movements. Bacterial chemotaxis is mediated by protein modification, the addition and removal of methyl groups.

E. coli carries a number of *common pili* (or *fimbriae*). They are generally both shorter in length ($<1\,\mu m$) and thinner in diameter ($<10nm$) than flagella (Figure 1.14). Such proteinaceous appendages may be present in a large number. Bacterial pili appear to be responsible for tissue 'stickiness' and, thus, for pathogenicity of virulent species.

Certain strains of *E. coli* also carry a few long, thin filaments in addition to the common pili. These *sex pili* (Figure 1.14) participate in the transfer of genetic material between bacteria (their role in conjugation is discussed in Section 4.1.2 (a)). Their synthesis is genetically determined by extrachromosomal DNA rather than by the bacterial genetic apparatus itself. Sex pili have the doubtful distinction of acting as receptors for certain bacteriophages (see Sections 4.1.2 (a) and 5.1.2).

1.3 Bacterial Growth

Apart from certain metal ions, the bacterial cell is composed entirely of carbon-based molecules (bonded to hydrogen, nitrogen, oxygen, sulphur and phosphorus atoms, the latter through an oxygen link). Though some of these constituents are enormous substances (witness the genetic material itself, a macromolecule of nucleic acid; the various proteins; and the complex sugar-based molecules), they are built up from relatively simple units: nucleic acids are synthesised from four (or five) different nucleotides, proteins from 20 non-identical amino acids and polysaccharides from a small number of sugars. Some micro-organisms, the *autotrophs*, are able to synthesise the basic 'building-blocks' — the starting materials for these macromolecules — from carbon dioxide (consider, for example, the photosynthetic bacteria that fix CO_2 in a light-dependent reaction). Others, such as *E. coli*, are *heterotrophs* since they obtain carbon for biosynthesis from preformed organic molecules (and derive energy from breakdown of these organic substrates). In essence, heterotrophs make use of the ability of autotrophic organisms to trap solar energy by conversion of CO_2 into more complex (and more reduced) derivatives.

1.3.1 An Introduction to Metabolism

Metabolism deals with the chemical reactions of the cell — the myriad biochemical pathways involved in the creation and destruction of organic matter. Biosynthesis, the breaking and making of chemical bonds and the assembly of complex organic molecules, is a form of work, albeit at the cellular level; other energy-dependent processes include mechanical and osmotic work (as exemplified by flagella movement or active transport, respectively). Work, however, has to be paid for.

(a) Production of the Chemical Intermediate, ATP. There is a dynamic coupling between biosynthetic reactions *(anabolism)* and degradative processes *(catabolism):* catabolism 'drives' anabolism. In essence, certain foodstuffs, notably fatty acids, sugars and amino acids are degraded to produce ATP (only a portion of available energy is used in this way). It is this crucial intermediate that allows the energetically unfavourable biosynthetic reactions to occur (ATP is also used in other work processes).[8] Thus, ATP is the currency of cellular energy, with catabolism as its source and anabolism as one of its end users.

Energy requirements for biosynthetic work are obtained by oxidising (degrading) a few common organic compounds. Other catabolic reactions include the turnover of cellular constituents, notably polymeric substances such as nucleic acids and proteins, that supply a substrate pool for biosynthetic processes. Macromolecules are first degraded to their constituents (nucleotides, amino acids, sugars and fatty acids). Each then undergoes a series of reactions, feeding into the *tricarboxylic acid (TCA) cycle* at various points. A number of pathways, including the breakdown of sugars to pyruvate in *glycolysis*, enter the cycle through the common endpoint acetyl CoA.

ATP may be produced during this breakdown by *substrate level phosphorylation*, direct coupling of ATP formation with pathway reactions (and uncoupled from electron flow, see Oxidative Phosphorylation, below). *E. coli*, a facultative anaerobe, is able to sustain the supply of ATP through anaerobic reduction of pyruvate, a process termed *fermentation*. However, the majority of ATP is not made directly by substrate-level phosphorylation in the cytosol. The main supplier of this essential intermediate, albeit indirectly, is the oxidative decarboxylation of acetyl CoA in the TCA cycle.

Reduced electron donors, NADH and $FADH_2$ (reduced nicotinamide adenine dinucleotide and flavin adenine dinucleotide), produced in the oxidation of acetyl CoA to CO_2 (and in the preceding reactions) are used to 'fuel' the *electron transport chain* (associated with the cytoplasmic membrane). This respiratory chain consists of a series of *electron carriers*[9] that are alternatively oxidised and reduced allowing the donation of electrons from NADH and $FADH_2$ to oxygen, to return these important cellular constituents to their oxidised form (NAD^+ and FAD). Most importantly, ATP production occurs during this electron flow; coupling of electron transport to phosphorylation is

8. Reference is often made to ATP as a 'high-energy compound'. High energy is ascribed to the two terminal acid anhydride (or pyrophosphate) bonds. There is no doubt that ATP hydrolysis is, itself, a strongly *exergonic* reaction (that is, it liberates, rather than requires, free energy). However, it would be wrong to consider that direct hydrolysis of the 'high-energy compound' drives energetically unfavourable reactions. Rather, such processes proceed through the creation of a phosphorylated intermediate of high free energy.
9. It is the last component of the electron transport chain, cytochrome oxidase (or cytochrome O), that is responsible for susceptibility of the bacterium to the metabolic poisons cyanide and azide.

termed *oxidative phosphorylation*. The reduction of the terminal electron acceptor, molecular oxygen, to water provides for the reoxidation of the respiratory chain components.

(b) Biosynthetic Reactions. Biosynthesis of many small organic molecules is dependent upon intermediates of glycolysis and the TCA cycle. These pathways have, therefore, a central role in both anabolic and catabolic processes. Biosynthetic pathways are usually similar to, but distinct from, degradative ones (consider, for example, glucose production from pyruvate, *gluconeogenesis:* although the majority of reactions are common to those of glycolysis, the three irreversible steps are bypassed). It is these anabolic processes that require ATP and a specialised form of reducing power; the latter is supplied by NADPH (nicotinamide adenine dinucleotide phosphate, a derivative of NADH).

Low molecular weight substances of related structure — organic molecules that are frequently repeat units in larger macromolecules — are often synthesised from common precursors. Thus, condensation of phosphoenolpyruvate (itself a metabolite of glucose) with erythrose-4-phosphate is the starting point for the biosynthesis of the aromatic amino acids tryptophan, tyrosine and phenylalanine. The nucleotides also have similar origins: purines from glutamine and ribulose-5-phosphate (the latter produced in the pentose-phosphate shunt); pyrimidines from condensation of aspartate with carbamoyl phosphate. Steps in pathways common to different end products, therefore, require only one enzyme. It is at the branch-point, where the pathways diverge, that different biological catalysts become necessary. This use of similar enzymes in both anabolic and catabolic reactions, and for the production of related molecules, reduces the total number of enzyme species necessary (and, thus, the genetic load). Moreover, a common catalytic centre normally functions in the polymerisation of similar monomeric units into macromolecules of varying chain length and composition (the synthesis of the two major classes of biological polymers, nucleic acids and proteins, is a case in point; Chapters 2 and 6).

(c) Pathway Regulation. It is worthwhile, at this point, considering briefly the mechanisms involved in pathway regulation. Enzymes involved in metabolism are prime targets for control, either with respect to their function or at the level of their production.

In the former, the effector molecule responsible for the change in enzyme activity is, in general, structurally unrelated to the normal substrate. It binds to the enzyme at a region — termed the *allosteric site* — distinct from the active site. All such *allosteric proteins* are multimeric in nature. Anabolic processes are characterised by feedback inhibition. Tryptophan synthesis is an example of end-product inhibition in that the conversion of chorismate to anthranilate — the first step specific to tryptophan — is sensitive to the amino acid product. Degradative reactions are also subject to feedback control. In this case, the energy status of the cell, in terms of the ratio of ATP to ADP and the reducing potential available, is also very important.

Gene expression is controlled primarily at the transcriptional level in prokaryotes (see Figure 1.8 (b)). It is mediated by proteins and effectors, regulatory molecules that are substrates, end products or other metabolites showing some direct biochemical relationship to the pathway itself. Whereas allosteric control may act rapidly (but wastefully since enzyme synthesis continues), genetic regulation is a slower means of control. It results, however, in a long-lasting effect — inhibition of enzyme synthesis — with the concomitant dilution of residual molecules upon cell division. Moreover, control may be coordinate, affecting more than one gene product. The enzymes responsible for tryptophan biosynthesis, from the chorismate conversion step onwards, are all subject to this type of control. Expression of these enzymes is inhibited by a repressor protein under conditions of high tryptophan concentration. Enzymes involved in the degradation of more novel carbon sources such as the disaccharide lactose, or galactose or arabinose (six- and five-carbon sugars, respectively) are also controlled at the level of synthesis. In this case, the respective enzymes are expressed only during growth of the bacterium on the sugar.

These various control systems serve to reduce wasteful flow of substrates through a pathway and to inhibit non-essential synthesis of enzymes, thereby avoiding unnecessary energy consumption under conditions of high end-product concentration or in the absence of a particular substrate. Not all reaction steps, however, are controlled. Many essential enzymes, for instance those concerned with aerobic energy metabolism, are synthesised independently of the physiological state of the cell — they are *constitutive*.

Control at the level of gene expression is discussed in Part IV.

1.3.2 Growth of Populations

Bacterial growth is asexual. New cell envelope is laid down as the cell elongates, outgrowth apparently proceeding from one end of the parent cell. It is at this growth point that the *septum* — a rigid cross-wall dividing the cell into two — is formed. Cell division is thought to occur at a fixed time after completion of a round of chromosome replication (the period required for each round of division is termed the *generation time*, Figure 1.19). The two 'new' daughter cells are assured of an exact copy of the parental genome through correct segregation of the accurately replicated genetic material prior to septum formation. A single bacterium, therefore, gives rise to two faithful copies upon division; these progeny, in turn, each produce two further cells. Since every round of growth involves all the daughter cells from the previous division cycle, increase in bacterial cell number is exponential rather than arithmetic.

(a) Growth Requirements. The biochemical pathways of a 'normal', wild-type bacterial cell direct the synthesis of all the essentials of life from a few simple molecules (Table 1.4): glucose as a carbon (and energy) source; ammonia as a nitrogen source; phosphate, sulphate and chloride ions present as salts of sodium, potassium and magnesium (buffering at about pH7); and water (also,

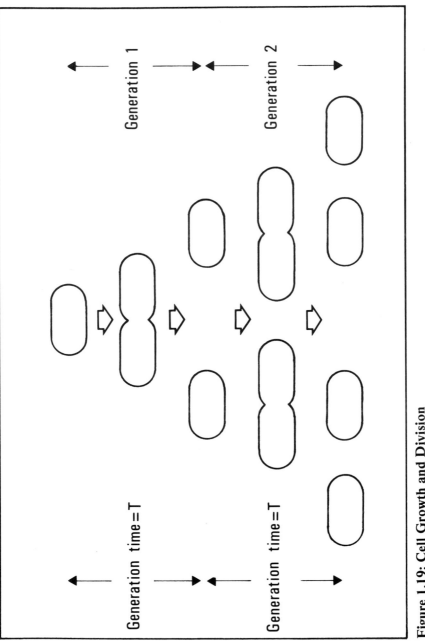

Figure 1.19: Cell Growth and Division
The diagram shows the course of cell elongation, septation and division for two cell cycles. The time taken for one bacterium to produce two daughter cells, the generation time T (see also Figure 1.21), can be used to define the number of cells after time t, N_t, for a starting number, N_0; $N_t = N_0 2^{t/T} = N_0 2^n$ (where n is the number of generations)

Table 1.4 A Typical Synthetic Growth Medium for a Wild-Type Strain of
E. coli

Component	Concentration
KH_2PO_4	0.022 M
Na_2HPO_4	0.049 M
NH_4Cl	0.019 M
$MgSO_4$	0.002 M
Glucose	0.011 M

A convenient medium for bacterial growth is, in fact, not chemically defined. *Nutrient* (or *rich*) medium is made from meat infusions, yeast extract or peptide mixtures obtained from protein hydrolysates. It promotes rapid growth since it contains a large number of substances that are normally synthesised (many biosynthetic pathways are accordingly shut off, Section 1.3.1. (c)). Nutrient media cannot, of course, be used for testing growth requirements

fortuitously a source of trace metals). A medium containing just these ingredients is called *minimal* since it has the minimum substances necessary for growth of wild-type *E. coli*. The fact that the chemical constituents are defined allows the ready determination of additional growth requirements that may arise from a mutational change.

Bacteria that multiply on minimal medium contain all the essential enzymes: they are *prototrophs*. Such strains have no further nutritional requirements. Moreover, wild-type bacteria have the capacity to make use of a range of different substances as carbon sources in the absence of glucose (for example, lactose, galactose, even the amino acid proline). Damage to the genetic material may lead to a mutant species that, lacking a particular enzyme function, has changed growth requirements. Mutations affecting biosynthetic pathways lead to the requirement for supplements in the growth medium for survival (for instance, amino acids, vitamins, nucleic acid precursors): these nutritional mutants are termed *auxotrophs*. It is lesions in genes responsible for catabolic reactions that result in the cell's inability to utilise certain carbon sources. These derivatives and auxotrophic mutants have proved of great importance for the analysis of biochemical pathways concerned with degradative and biosynthetic processes (see Pathway Analysis, Section 7.1.2 (d)) and for studies on the regulation of a gene expression (discussed in Part IV).

In addition to a preferred pH of near neutrality, and essential growth requirements, *E. coli* has an optimum growth temperature of 37°C (above or below this, growth rate drops). It may, however, be cultured at sub-optimal conditions and even at temperatures as high as 45°C without apparent permanent damage. This flexibility in culture conditions allows the isolation (and analysis) of strains that exhibit a mutant phenotype at one temperature but not at another. A *temperature-sensitive* mutant with a lesion, for example, in one of the *trp* genes will be unable to survive in the absence of exogenous tryptophan at 43°: it will, however, grow on unsupplemented minimal media at 30°. *Cold-sensitive* mutants that exhibit the opposite response have also been isolated. These *conditional mutants* have proved of great value for the investigation of essential gene products, such as those involved in the expressive machinery or DNA replication. However, since these 'extremes' in temperature are of restricted employment, culture conditions in all future discussion will be optimal, that is at 37°C, unless otherwise stated. (Similarly, it should be assumed that all glassware and media are sterilised as a matter of course before use.)

(b) Culture Characteristics. Bacteria are grown in liquid or on semi-solid media. The constituents are the same (Table 1.4); the gelling agent, agar,[10] is simply added in the latter case (to give a final concentration of about 1-1.5 per cent). While molten, the sterile agar solution (either minimal medium alone, or containing additional nutrients, see Table 1.4) is poured into petri-dishes. Bacteria are spread (or 'streaked') out on the solid surface with a metal wire or loop, separating out the mass of the initial *inoculum* to give single cells. Alternatively, the bacterial suspension is serially diluted, prior to its separation with a glass spreader, to reduce the final cell number to manageable proportions (no more than, say, 500 per plate). Though individual bacterial cells are invisible to the naked eye (microscopic examination at 400 × magnification is necessary for direct visualisation), their position on the agar plate may be visualised making use of their rapid growth characteristics.

If the minimal (or nutrient) plate is incubated overnight (a period of about 20 hours is usually sufficient), each single cell will have multiplied at its fixed site on the agar plate some twenty times (for a generation time of 60 minutes). In itself, this number of generations may seem unremarkable: it must be remembered that at the completion of each successive round of division, the whole population doubles in size (Figure 1.19, but see page 44). A single cell gives rise, in fact, to a little over one million progeny in a period of time sufficient for 20 doublings. Since the cells remain at their point of origin, the *colony* formed is a tight, opaque mass (Figure 1.20). It is readily visible and easy to handle (it may, for instance, be picked up with a metal loop for transfer to liquid or solid media). Moreover, all the cells in one colony are identical (in the absence of a mutational event); they are *clones*. This property is immensely useful since a single colony can be tested on other media with the knowledge that each cell in that colony has exactly the same genetic makeup.

10.   Agar is obtained from seaweed. In the pure state (agarose), it is not a substrate for any of the bacterial enzymes. Though the microbial-grade reagent contains a small amount of impurities, it too is still largely inert.

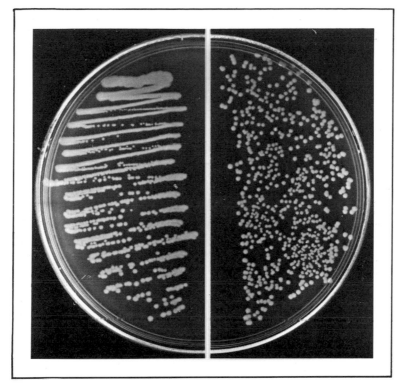

Figure 1.20: Single Colonies
Well-separated colonies may be obtained either by 'streaking out' with a metal loop (left) or by spreading after serial dilution (right). The latter is commonly used for cell viability estimation whereas genetic manipulation involves streaking out

Bacterial number can be monitored either by the determination of actual cell number or by measurement of the resultant turbidity. The direct method involves the transfer of a small volume of cell suspension to a special minute counting vessel, a Petroff-Hauser chamber. Cell density is obtained from microscopic counting of individual cells present in a known volume (alternatively, an electronic particle counter is used). The indirect method makes use of the light-scattering property of particles and, thus, bacterial cells in suspension. Ideally, this is measured with a nephelometer: in practice, a simple spectrophotometer or colourimeter that measures transmitted light is often used (at a fixed wavelength of, say, 550nm). There is a reasonable correlation between optical density and cell number at low densities (though cell lysis may confuse the issue). Note that neither the direct or indirect method distinguishes between a cell capable of dividing — a *viable* cell — and one that is not.

The proportion of viable cells is enumerated as follows. The total number of cells is first obtained directly using a counting chamber. Samples are then diluted and spread onto agar plates to give viable counts. If, say, 100 cells — as estimated by microscopic counting — were spread, yet only forty colonies are observed after incubation, this would indicate that only 40 per cent of the cell population were viable. The accuracy of such calculations hinges mainly upon the error in direct microscopic counting.

When growing liquid cultures, a single colony is frequently used to inoculate the culture flask (but see below). Although perhaps 10^6 bacteria are added, the medium appears clear since a minimum density of 10^7 cells/ml is necessary for the turbidity to be visible. The increase in turbidity after overnight incubation (shaking ensures adequate aeration) indicates bacterial multiplication: the final concentration can be as high as 10^9 cells/ml. But what are the stages in growth of a bacterial population? When inoculating a liquid medium for growth studies, the addition of one colony is normally insufficient to create a change in optical density. Moreover, such a culture tends to be very slow in starting growth. For these reasons, a liquid overnight culture — diluted about 1 part in 20 to give a measurable turbidity — is used as inoculum. Growth is monitored every 30 or 60 minutes by optical density measurement. Only a small volume of culture liquid is present in each flask (say, 25 ml per 250 ml flask) to maintain efficient aeration for 'healthy' growth (see below). There are three main stages of bacterial growth in liquid culture (Figure 1.21):

(i) *Lag Phase*. For some hours, growth appears absent or minimal. The extent of the *lag phase* reflects the inoculum — the conditions of previous culture, the nature of the strain — and the present culture conditions (a single colony or a stationery culture, see below, gives an extended lag).

(ii) *Exponential Phase*. As growth takes off, it is soon apparent that cell mass does not increase arithmetically. Rather, a straight plot is only obtained with a log function (of either optical density or actual cell number, Figure 1.21). This *exponential phase* is characteristic of dividing cells that exhibit a doubling in total cell number every generation time (if the culture was to be diluted at this stage, there would be only a brief lag, if any, before exponential growth resumed). The *generation time* of the bacterial population — the period required for a doubling in cell mass — varies, depending on both the genetic makeup of the strain and the nature of the culture medium. Thus, bacteria grow fast in rich solutions but poorly in minimal media, in which all cellular components other than glucose have to be synthesised by the bacterium itself. [Growth of bacterial populations is normally asynchronous. The overall increase in cell number with time reflects cells at different stages in the division cycle. In cultures that have been artificially synchronised, bacterial growth is 'stepped' — a period of constant cell number, followed by one of growth — until synchrony is lost.]

(iii) *Stationary Phase*. The rapid growth period, the exponential phase, continues for some time (again controlled by the strain, the constituents of the culture medium and by growth conditions). The generation time then appears to

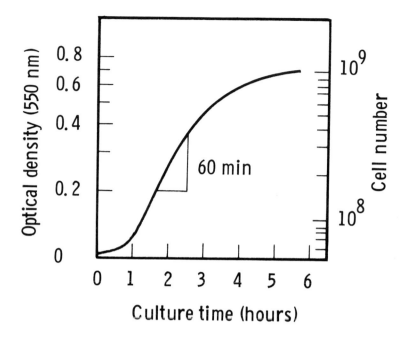

Figure 1.21: Growth Profile in Liquid Medium

A fresh overnight culture was diluted 20-fold into glucose minimal medium (see Table 1.4) and shaken at 37°C; values for optical density and cell number are plotted on a semilogarithmic scale. Note the lag, exponential, and stationary phases (about 0-1h, 1-3h, and 3h onwards). The generation time for the total cell population is obtained from the slope of the log plot — it is the time taken for the cell mass and, thus, the optical density to double. Typical values for wild-type strains in minimal medium containing acetate, succinate, glucose, casamino acids and broth as carbon source are 120, 67, 40, 28 and 22 minutes, respectively (equivalent to 0.5, 0.9, 1.5, 2.1 and 2.7 doublings per hour). The $recA^-$ culture shown has a doubling time of 60 (rather than 40) minutes since $recA^-$ mutants show reduced viability (see Section 6.2). The spontaneous mutation rate (about 10^{-4} per genome replication, Section 3.2.1 (a)) becomes significant in a liquid overnight culture because of the large number of cells involved (about 10^9 cells/ml). There is, therefore, a strong likelihood that a small proportion of such a bacterial population carries deleterious mutations (Chapter 3). In this case, the culture will no longer be completely homogeneous

increase. This occurs as the growth medium becomes depleted of essential nutrients as well as, perhaps, through accumulation of toxic metabolites. The tailing-off in growth rate culminates in a final cessation as the *stationary phase* is reached. At this point, there is little, if any, bacterial division: the major metabolic pathways appear to be shut down (or 'idling') and the cell envelope is morphologically altered. Although a large proportion of cells are viable at this stage (spontaneous lysis may reduce the population somewhat), stationary cultures exhibit long lag phases when diluted into fresh medium — presumably, because of depletion of essential cellular constituents.

Bibliography

Brill, W.J. (1979) 'Nitrogen Fixation' in J.R. Sokatch and L.N. Ornston (eds.), *The Bacteria, VII: Mechanisms of Adaptation* (Academic Press, New York), pp. 85-109.

Buchanan, R.E. and Gibbons, N.E., (eds.) (1974) *Bergey's Manual of Determinative Bacteriology*, 8th edn (The Williams and Wilkins Co., Baltimore).

Chakrabarty, A.M. (1976) 'Plasmids in *Pseudomonas*', *Ann. Rev. Genet., 10*, 7-30.

Doi, R.H. (1977) 'Genetic control of sporulation', *Ann. Rev. Genet., 11*, 29-48.

Holloway, B.W., Krishnapillai, V. and Morgan, A.F. (1979) 'Chromosomal genetics of *Pseudomonas*', *Microbiol. Rev., 43*, 73-102.

Hopwood, D.A., Chater, K.F., Dowding, J.E. and Vivian, A. (1973) 'Advances in *Streptomyces coelicolor* genetics', *Bacteriol. Rev., 37*, 371-405.

Hopwood, D.A. and Merrick, M.J. (1977) 'Genetics of antibiotic production', *Bacteriol. Rev., 41*, 595-635.

Levinthal, M. (1974) 'Bacterial genetics excluding *E. coli*', *Ann. Rev. Microbiol., 28*, 219-230.

Lowbury, E.J.L. (1975) 'Ecological importance of *Pseudomonas aeruginosa*: medical aspects' in P.H. Clarke and M.H. Richmond (eds.), *Genetics and Biochemistry of Pseudomonas* (John Wiley, New York), pp. 37-65.

Sanderson, K.E. (1976) 'Genetic relatedness in the family Enterobacteriacea', *Ann. Rev. Microbiol., 30*, 327-349.

Sanderson, K.E. and Hartman, P.E. (1978) 'Linkage map of *Salmonella typhimurium*, edition V', *Microbiol. Rev., 42*, 471-519.

Schmidt, E.L. (1979) 'Initiation of plant root-microbe interactions', *Ann. Rev. Microbiol., 33*, 355-376.

1.1 The Nature of the Genetic Material

1.1.1 Structural Considerations

Olby, R. (1974) *The Path to the Double Helix* (Macmillan, London).

Sundaralingam, M. (1975) 'Principles governing nucleic acid and polynucleotide conformations' in M. Sundaralingam and S.T. Rao (eds.), *Structure and Conformation of Nucleic Acids and Protein-Nucleic Acid Interactions* (University Park Press, Baltimore), pp. 487-524.

Thomas, C.A. (1967) 'The rule of the ring', *J. Cell. Physiol., 70*, suppl. 1, 13-34.

Watson, J.D. (1968) *The Double Helix* (Weidenfeld and Nicolson, London).

Watson, J.D. and Crick, F.H.C. (1953) 'Molecular structure of nucleic acids. A structure for deoxyribose nucleic acid', *Nature, 171*, 737-738.

1.1.2 The Biological Role of Nucleic Acids: A Preview

see relevant section

1.1.3 Phenotype and Genotype

Demerec, M., Adelberg, E.A., Clark, A.J. and Hartman, P.E. (1966) 'A proposal for a uniform nomenclature in bacterial genetics', *Genetics, 54*, 61-76.

1.2 Cell Composition

1.2.1 The Cell Envelope

Braun, V. (1975) 'Covalent lipoprotein from the outer membrane of *Escherichia coli*', *Biochim. Biophys. Acta, 415*, 335-377.

———— (1978) 'Structure function relationship of the Gram-negative bacterial cell envelope', *Symp. Soc. Gen. Microbiol., 28*, 111-138.

Costerton, J.W. (1979) 'The role of electron microscopy in the elucidation of bacterial structure and function', *Ann. Rev. Microbiol., 33*, 459-479.

Croonan, J.E. (1978) 'Molecular biology of bacterial membrane lipids' *Ann. Rev. Biochem., 47*, 163-189.

Davis, B.D. and Tai, P.-C. (1980) 'The mechanism of protein secretion across membranes', *Nature, 283*, 433-438.

DiRienzo, J.M., Nakamura, K. and Inouye, M. (1978) 'The outer membrane proteins of Gram-negative bacteria: biosynthesis, assembly, and functions', *Ann. Rev. Biochem., 47*, 481-532.

Inouye, M. (ed.) (1979) *Bacterial Outer Membranes: Biogenesis and Functions* (John Wiley, London).

Neilands, J.B. (1978) 'Transport functions of the outer membrane of enteric bacteria', *Horizons Biochem. Biophys., 5*, 65-98.

Orskov, I., Orskov, F., Jann, B. and Jann, K. (1977) 'Serology, chemistry, and genetics of O and K antigens of *Escherichia coli*', *Microbiol. Rev., 41*, 667-710.

Raetz, C.R.H. (1978) 'Enzymology, genetics and regulation of membrane phospholipid synthesis in *Escherichia coli*', *Microbiol. Rev., 42*, 614-659.

Saier, M.H. (1979) 'The role of the cell surface in regulating the internal membrane', in J.R. Sokatch and L.N. Ornston (eds.), *The Bacteria, VII: Mechanisms of Adaptation* (Academic Press, New York), pp. 167-227.

Tipper, D.J. and Wright, A. (1979) 'The structure and biosynthesis of bacterial cell walls', in J.R. Sokatch and L.N. Ornston (eds.), *The Bacteria, VII: Mechanisms of Adaptation* (Academic Press, New York), pp. 291-426.

Tonn, S.J. and Gander, J.E. (1979) 'Biosynthesis of polysaccharides by prokaryotes', *Ann. Rev. Microbiol., 33*, 169-199.

Wright, A. and Tipper, D.J. (1979) 'The outer membrane of Gram-negative bacteria', in J.R. Sokatch and L.N. Ornston (eds.), *The Bacteria, VII: Mechanisms of Adaptation* (Academic Press, New York), pp. 427-485.

1.2.2 Cellular Appendages

Hazelbauer, G.L. and Parkinson, J.S. (1977) 'Bacterial chemotaxis', in J.L. Reissig (ed.), *Receptors and Recognition. Microbial Interactions* (Chapman and Hall, London), pp. 59-98.

Iinio, T. (1977) 'Genetics of structure, function of bacterial flagella', *Ann Rev Genet, 11*, 161-182.

Koshland, D.E. (1979) 'Bacterial chemotaxis' in J.R. Sokatch and L.N. Ornston (eds.), *The Bacteria VII: Mechanisms of Adaptation* (Academic Press, New York), pp. 111-166.

Macnab, R.M. (1978) 'Bacterial motility and chemotaxis: the molecular biology of a behavioural system', *CRC Crit. Rev. Biochem., 5*, 291-341.

Sokatch, J.R. (1979) 'Roles of appendages and surface layers in adaptation of bacteria to their environment' in J.R. Sokatch and L.N. Ornston (eds.), *The Bacteria, VII: Mechanisms of Adaptation* (Academic Press, New York), pp. 229-289.

1.3 Bacterial Growth

1.3.1 An Introduction to Metabolism

Gottschalk, G. (1979) *Bacterial Metabolism* (Springer-Verlag, New York).

Haddock, B.A. and Jones, C.W. (1977) 'Bacterial respiration', *Microbiol. Rev., 41*, 47-99.

Harold, F.M. (1977) 'Membranes and energy transduction in bacteria', *Curr. Topics Bioenerg., 6*, 83-149.

Umbarger, H.E. (1978) 'Amino acid biosynthesis and its regulation', *Ann. Rev. Biochem., 47*, 533-606.

1.3.2 Growth of Populations

Meynell, G.G. and Meynell, E. (1970) *Theory and Practice in Experimental Bacteriology* (Cambridge University Press, London).

Miller, J.H. (1972) *Experiments in Molecular Genetics* (Cold Spring Harbor Laboratory, New York).

Part II
Gene Expression

2 RNA and Protein Production

Cellular processes are not carried out directly by the hereditary material. The main agents required for growth are the biological catalysts, enzymes, protein molecules that carry, in some instances, organic prosthetic groups. In addition, non-enzymatic proteins have a major structural role. Many hundreds, perhaps as many as one to two thousand different proteins in the case of *E. coli*, one of the simplest of unicellular organisms, are necessary for growth. Thus, the genetic material of an organism contains the information necessary for survival — it is the 'programme' for cellular growth — without being directly involved. How is this information realised? Gene expression consists of two major stages, transcription and translation.

The first, *transcription*, is the production of a complementary single strand of nucleic acid, ribonucleic acid; a D-ribose, rather than 2-deoxy-D-ribose, sugar is present in the backbone and thymine is replaced by the pyrimidine base, uracil. The enzyme *DNA-dependent RNA polymerase* catalyses this reaction; other factors are required for transcription of specific genes. Each gene encodes an RNA transcript (where several genes are transcribed as a single message, such transcripts are said to be *polycistronic*). The same genetic information is, therefore, expressed in a similar but slightly altered language. Whereas a single gene is a fraction the size of, yet covalently joined to, the massive bacterial chromosome (*E. coli* carries two to three thousand different genes), the complementary RNA transcript is a discrete entity with potential mobility. Moreover, it is a single-stranded polymer.

Transcription of genetic material is, in general, a highly asymmetric process. RNA polymerase binds to, and initiates at, a specific site on the DNA template, termed the *promoter*, and synthesises a complementary RNA strand in the 5′ → 3′ direction (the 5′-end, therefore, carries a triphosphate group and is chemically distinguishable from the 3′-OH terminus). Only one strand, the *sense* (or *codogenic*) strand, is normally transcribed.[1] RNA polymerase terminates at specific sites, releasing the nascent transcripts into the cytoplasm. Genetic

---

1. The complementary DNA strand that is not expressed is referred to as the anti-sense strand. Its transcription leads to the production of an *anti-messenger* RNA *(anti-mRNA)* molecule. In some cases, both strands are transcribed but from opposite directions, either starting from a single point (*divergent* transcription) or from separate sites (*convergent* or bidirectional transcription); see, for example, P_M/P_R-promoted events and convergent transcription from P_R and P_E (Section 10.2, Regulation of Phage *Lambda* Development).

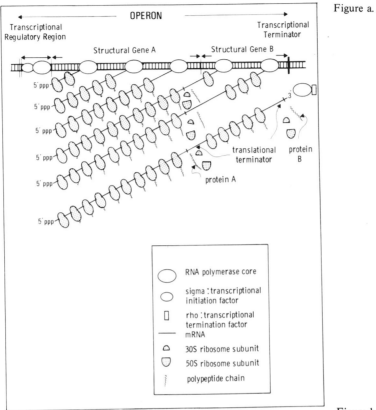

Figure a.

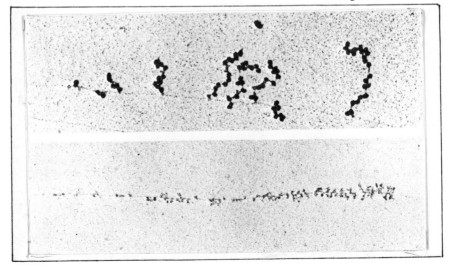

Figure b.

Figure 2.1: The Major Processes Involved in Gene Expression.

Schematic representation (a) and electron micrograph (b, *top*) of *E. coli* structural gene(s) being expressed ((a) represents a polycistronic operon); an electron micrograph showing transcription of a putative ribosomal operon — the transcript is devoid of ribosomal particles — is also included ((b) *bottom*). Note the polarity of movement, transcription (and translation) clearly moving from 'left' to 'right'. Ribosomes traverse the nascent RNA strand at a comparable speed to DNA transcription (55 to 60 nucleotides per second) resulting in a polypeptide elongation rate of about 16-20 amino acids per second. An average-sized gene of 1200 bp, coding for a protein of 40 000 daltons (about 400 residues), is, therefore, transcribed and translated in a maximum of 40 seconds (the minimum time may be as low as 20 seconds since transcription and translation occur concurrently in prokaryotes). Micrographs from Miller *et al.* (Miller, O.L., Hamkalo, B.A. and Thomas, C.A. (1970) *Science, 169,* 392-395) reproduced with permission of the publishers and authors (copyright 1970 by the American Association for the Advancement of Science); photographs by courtesy of Professor O.L. Miller

material that is expressed (some may be functionally inactive) is, thus, bounded by DNA sequences, *promoters* and *terminators*, responsible for the control of transcription (Figure 2.1).

There are two main types of RNA. Some ribonucleic acid, tRNA and rRNA (see below), has a functionally direct role in gene expression; the minority (in molar terms) is biologically inactive *per se* since translation is required for expression of the encoded genetic information. This latter class has been termed *messenger RNA* (or *mRNA*) by Jacob and Monod (1961).[2] A region of DNA that encodes a protein product and is, hence, expressed initially as a mRNA species is referred to as a *structural gene*.

Translation, the second stage in gene expression, involves a new language — amino acids rather than nucleotides. The genetic information in a mRNA molecule is encoded in groups of three nucleotides, *codons*. Each triplet codes for one, and only one, particular amino acid. These RNA codons are not, however, translated directly. Rather, an adaptor molecule, which is itself a small RNA species, is covalently attached to each amino acid prior to translation; these adaptors are termed *transfer RNAs* (or *tRNAs*). The translational matrix, the *ribosome*, is a large cytoplasmic organelle consisting of 52 different low molecular weight proteins and three non-identical RNA molecules *(rRNAs)*.

Translation, like transcription, proceeds in the $5' \rightarrow 3'$ direction, i.e. the mRNA template is both produced and utilised in this direction (Figure 2.1). These two processes occur simultaneously in prokaryotes, organisms which have, by definition, no nuclear membrane (the genetic material is not physically compartmentalised). Ribosomes initiate translation upon a RNA template prior to its completion. In addition, a number of ribosomes may translate one mRNA molecule, forming a *polyribosome* (or *polysome*). Both these latter processes

2.   Jacob and Monod (1961) proposed the existence, in addition to stable rRNA, of a hyperlabile population of RNA, *messenger* RNA, as an intermediate in gene expression: information contained within a gene — the DNA 'message' — is transferred in this manner from the nuclear apparatus to the ribosome. The concept, though derived from studies of bacterial systems, allowed for the site of protein synthesis to be physically distinct from the nucleus.

mRNA is the minority in terms of RNA accumulation (about 3%) but can represent about half the total ribonucleic acid synthesised at any one time.

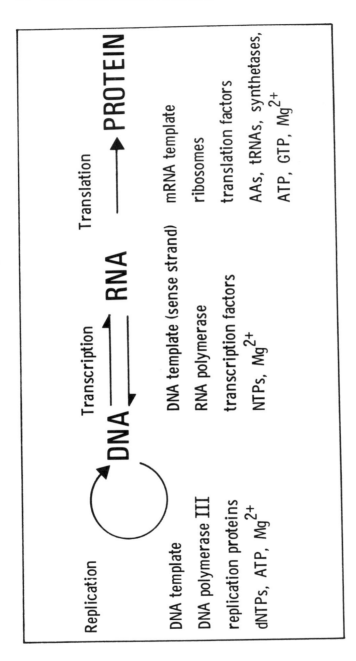

Figure 2.2: The Central Dogma
The direction of flow of genetic information is unidirectional — from nucleic acid to protein (abbreviations: AAs, amino acids; synthetases, aminoacyl-tRNA synthetases). Note that 'reverse transcription', the production of DNA strands on RNA templates, has not been found in prokaryotes.

allow efficient translation of prokaryotic mRNA molecules which are, themselves, notably unstable *in vivo*. Gene expression is, clearly, a dynamic process; within the cytoplasm its elements are in a constant state of flux as genes are transcribed and translated, and their products are turned over.

It is worthwhile, at this stage, considering the milieu in which gene expression takes place. The cytosol is not, in fact, a dilute aqueous solution. Rather, it is crowded with subcellular organelles and macromolecules. The 25 000-30 000 ribosomes occupy about 25 per cent of the cellular interior. In addition, approximately 2000 molecules of RNA polymerase produce 300 000-400 000 tRNAs (about 6000 copies of 60 different species) and perhaps 2000 non-identical mRNAs of average size, 1200 nucleotides. The bacterial genome is present in the cytoplasm as a highly condensed mass, the nucleoid, attached to the inner membrane: in the uncoiled state it would be some 800 times the length of the bacterium. Thus, the components of the expressive machinery occupy the majority of space within the cytosol. This topological complexity clearly exerts a constraint on cytoplasmic mobility, yet within the dense 'soup' a particular gene is rapidly transcribed and translated in response to a defined physiological signal (see legend to Figure 2.1).

Gene expression can be seen to proceed from DNA to protein via an RNA intermediate (Figure 2.2). There is, therefore, a *unidirectional* transfer of genetic information from nucleic acid (be it DNA or RNA) to the protein product. This flow of information terminates at translation since no protein molecule encodes the ability to produce itself.

Transcription and translation are described in Sections 2.1 and 2.2. Gene disfunction — perturbations that affect the genetic structure or one of the stages of gene expression — is discussed in Chapter 3. Gene regulation is covered in Part IV.

2.1 DNA Transcription

The enzyme DNA-dependent RNA polymerase catalyses the formation of 3'-5' phosphodiester bonds on DNA templates using the four common ribonucleoside triphosphates as substrates.[3] The bacterial enzymes from a number of different Gram-negative and Gram-positive organisms are remarkably similar in structure (Section 2.1.1 (a)). Obligate parasites such as bacterial viruses may carry the genetic information for an entirely new and much simpler enzyme (bacteriophage T7 RNA polymerase, for example, is a single polypeptide of about 110 000 daltons). Alternatively, these parasites may control their host metabolism, in part, by the synthesis of phage-coded factors that alter the host RNA polymerase specificity (see bacteriophage *Lambda*, Chapter 10). Studies with inhibitors of

3.   *In vitro, E. coli* RNA polymerase also catalyses both DNA-independent polymerisation of ribonucleotides (unprimed homopolymer ribonucleotide synthesis) and to some extent polymerisation on RNA templates. The biological significance, if any, of these reactions is unknown.

the bacterial polymerase (Section 2.1.2 (c)) indicate that a single enzyme is responsible for all RNA synthesis in *E. coli* (except for some priming of DNA synthesis).[4]

2.1.1 The Elements of Transcription

(a) DNA-Dependent RNA Polymerase. Bacterial RNA polymerases are complex enzymes consisting of at least four different proteins, $\alpha \beta$, β', and σ (Table 2.1.). The β and β' subunits are the largest known monomeric polypeptides present in *E. coli* (apart from a DNA-dependent ATPase, see Section 6.2.1 (a)). Affinity labelling studies have implicated β in substrate binding. The fact that rifampicin, a β-specific antibiotic (Section 2.1.2 (c)), inhibits ribonucleoside triphosphate extension of the initiation complex and blocks RNA chain extension at high concentration indicates a role at both the initiation and elongation stages. The β subunit is also likely to be involved in the termination event since certain genetic lesions which result in an altered β polypeptide affect this process. The β' subunit may be involved in binding the RNA polymerase to template DNA although its basic nature would impart the property of non-specific binding to polyanions; genetic studies also support a role in promoter selection. Modification of the α subunit upon bacteriophage T4 infection and the concomitant decrease in expression of bacterial genes indicates that this subunit is important for the specificity of transcription. Finally, the σ subunit is a transcription specificity factor (below and Section 2.1.2 (a)).

Two major forms of this enzyme exist: *core* RNA polymerase ($\alpha_2\beta\beta'$) and *holoenzyme* ($\alpha_2\beta\beta'\sigma$). Core (molecular weight 382 000) transcribes genetic material but does not initiate efficiently at promoter sites since it binds non-specifically to DNA templates. The presence of sigma in holoenzyme (465 000 daltons) imparts selectivity in RNA chain initiation. This dissociable subunit is, therefore, a specificity factor.

(b) Transcription Factors. RNA polymerase is responsive to a variety of genetic signals that regulate site selection and RNA chain termination (these DNA sequences are discussed in detail in Part IV). In addition, certain physiological factors — regulatory proteins as well as organic molecules (notably nucleotide derivatives) — mediate in the transcription of specific genes. Whereas core enzyme can transcribe any region of DNA, albeit at low efficiency, these specificity factors enhance or reduce expression of certain genes — they impart selectivity of transcription.

The sigma subunit has no catalytic activity *per se*. In combination with the core complex, it allows both site and strand selection. Sigma is important in reducing the general affinity of core enzyme for DNA to allow, presumably,

---

4.   The *dnaG* product, DNA primase, is responsible for some RNA synthesis — that required for the initiation of replication of certain replicons and for priming Okazaki fragment synthesis (Section 6.1). It is resistant to the antibiotic rifampicin to which the bacterial RNA polymerase is sensitive (by virtue of the affinity of the β subunit for this drug, Section 2.1.2 (c)).

rapid selection of promoter sites: in its absence, free DNA would 'mop up' cytoplasmic core. Only a proportion of RNA polymerase molecules isolated from *E. coli* carries a sigma subunit. This lack of overall stoichiometry is in keeping with the synthesis of polymerase subunits — sigma is produced at about 30 per cent the rate of the core polypeptides (Table 2.1). The fact that the specificity factor plays only a brief part in the transcription process (*sigma cycle*, Section 2.1.1 (a)) rationalises these observations.

Table 2.1 Subunit Composition of *E. coli* RNA Polymerase Holoenzyme

Subunit	Size	Structural Gene	Map Location (min)	Stoichiometry (relative to β)	Molecules/cell [a]	Function in Transcription Initiation	Elongation	Termination
α	36,000	rpo A	72	2	4,000	+		
β	150,000	rpo B	90	1	2,000	+	+	+
β'	160,000	rpo C	90	1	2,000	+		
σ	83,000	rpo D	67	1	600	+		

a. RNA polymerase concentration varies with growth rate and, hence, culture conditions (Section 9.3.2 (a)). Figures given are for glucose minimal medium (Table 1.4).

Does expression of each gene (or group of genes) require its own transcriptional specificity factor? The initial publication on sigma (Burgess *et al.*, 1969) suggested just this. However, the supposition has not been substantiated: there are a number of putative transcription factors but the majority are, at best, poorly characterised. Two proteins, *CAP* (catabolite activator protein) and *rho*, have a definite and demonstrable role in transcription (Table 2.2). The endoribonuclease, RNase III (originally termed 'sizing factor') acts at the post-transcriptional level and is important for RNA processing (discussed in Section 9.3.1 (b)).

Table 2.2 Transcription Factors

Factor	Size	Structural Gene	Map Location (min)	Site of Action	Function in Transcription Initiation	Termination	Post-Transcription
sigma (σ)	83,000	rpo D	67	RNA polymerase	+		
CAP [a]	22,500 (x2)	crp	73	promoter	+		
rho (ρ)	50,000 (x6)	rho	84	transcription complex		+	
RNase III [b]	50,000	rnc	55	transcript			+

a. CAP has also been referred to as cyclic adenosine monophosphate receptor protein (CRP) or catabolite gene activator (CGA).

b   Strictly speaking, RNase III is not a transcription factor since it acts after the transcription event. That is, it affects the product rather than the process.

CAP factor (a homodimer of 22 500 dalton subunits) potentiates expression of certain operons, those mainly affecting the use of 'unusual' sugars as carbon and energy sources (see The *lac* Operon, Section 9.1). These operons have in common the property that their expression is reduced by the presence of glucose in the growth medium. They have been termed *catabolite repressible* (*catabolite controlled* is a more accurate description; Section 9.1). CAP, in conjunction with the ubiquitous cyclic nucleotide cAMP, functions by 'assisting' transcriptional initiation, in particular the initial binding of RNA polymerase to catabolite repressible promoters, possibly by facilitating local melting of promoter DNA. Rho factor, on the other hand, a hexamer of identical subunits (molecular weight 50 000), acts at the termination stage. It prevents further elongation of RNA polymerase at certain sites and augments release of RNA at other terminators (Section 2.1.2 (a) and Part IV).

The main *E. coli* genes involved in transcription are shown in Figure 2.3.

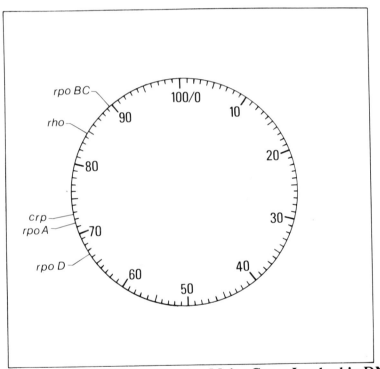

Figure 2.3: *E. coli* **Map Showing Major Genes Involved in DNA Transcription**
Note how these loci are clustered around the origin of replication at 83 min. Properties of the encoded gene products are listed in Tables 2.1 and 2.2. Numerous other components are necessary for gene control (considered in Part IV). Map locations from Bachman and Low (Bachman, B.J. and Low, K.B. (1980) *Microbiol. Rev., 44,* 1-56)

2.1.2 The Mechanism of Transcription

Asymmetric transcription by RNA polymerase implies a polarity of movement — the enzyme travels from the promoter to the adjacent gene, moving from the region encoding the polypeptide's N-terminus to the carboxy-terminal portion of the gene. The terms *upstream* and *downstream* have been used to describe regions within such transcribed areas in relation to the direction of 'flow' of RNA polymerase (the promoter, for instance, is *upstream* from the gene it controls, while the 5'-end of the codogenic DNA strand — the region encoding the polypeptide's C-terminus — is *downstream* from the 3'-end).

Considering the fidelity of transcription, about 1 error in 10^3-10^4 ribonucleoside triphosphates polymerised, the basic reaction requires remarkably few constituents: DNA template, substrate ribonucleoside triphosphates, RNA polymerase and divalent metal cations (compare, for example, with the myriad components of the translational apparatus, Section 2.2.1). This accuracy depends, in part, on base-pairing specificity.

The bacterial RNA polymerase is readily amenable to isolation thanks to its high cellular concentration and large size (Table 2.1). The basic polymerase reaction, the conversion of acid-soluble ribonucleoside triphosphates (one of which may be radiolabelled) into an insoluble form, is a convenient assay during its purification. Moreover, this assay requires both few and, most importantly, defined components. Crude, heterologous templates such as calf thymus or salmon sperm DNA are routinely employed at this stage (for economic reasons, and see below). Further refinement is necessary for the study of specificity proteins. The specificity factor σ, for example, can be conveniently investigated using phage T7 DNA as template since transcription from the 'early' viral promoters (other promoters are only accessible to the phage-coded RNA polymerase) is entirely dependent upon the holoenzyme form. The RNA transcripts produced from these sites are well characterised (methods used for the investigation of RNA production are discussed in Section 7.2.2). It was, in fact, the low activity of RNA polymerase preparations obtained from phosphocellulose chromatography on phage T4 DNA as compared with the calf thymus template that led to the identification of the σ subunit. In conclusion, pure DNA from bacterial viruses in combination with biochemical methods such as electrophoresis, DNA-RNA hybridization and restriction technology (Sections 7.2.2 and 7.3.3) offer powerful tools for the investigation of the transcription process.

(a) The Transcription Cycle. Like most, if not all, nucleic-acid polymerase-mediated reactions and, indeed, most reactions dependent upon *processive enzymes* (proteins that do not dissociate after each successive catalytic event), RNA polymerase-promoted transcription on DNA templates can be described in terms of three main steps: initiation, elongation and termination, although such a division does not necessarily imply a series of discrete steps. (The similarity with other nucleic-acid enzymes is extended by the dependence of this process upon divalent metal ions and the fact that RNA polymerase is a zinc metallo-enzyme; see Section 6.1.1 (a).) *RNA chain initiation* includes binding of the enzyme to the promoter, formation of a stable binary RNA polymerase-DNA complex and catalysis of the first 3'-5' internucleotide bond (giving a

ternary complex consisting of dinucleoside tetraphosphate, enzyme and DNA).
Further phosphodiester bond formation and, thus, enlargement of the nascent
RNA component of the ternary complex, *RNA chain elongation*, requires
translocation of RNA polymerase along the DNA template. Transcriptional
termination involves dissociation of the complex. The template is left unaltered
upon release of the product of this reaction, a polyribonucleotide complementary
and antiparallel in sequence to the DNA sense strand. The overall process has
been referred to as the *transcription cycle* (Figure 2.4) since once released, the
enzyme is capable of transcribing other genes.

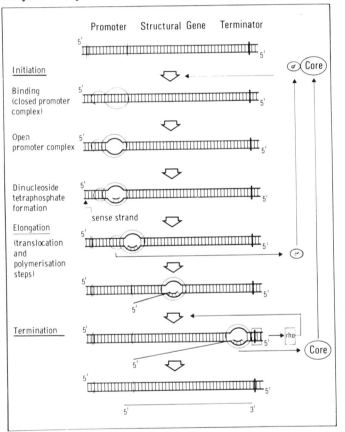

Figure 2.4: The Transcription Cycle

DNA transcription is a processive reaction — the enzyme responsible for this process, the DNA-
dependent RNA polymerase, catalyses the polymerisation of ribonucleoside monophosphates
through translocation along the DNA template. It can, therefore, be described in terms of three main
stages, RNA chain initiation, elongation and termination. The few protein components required for
RNA production — core enzyme, the transcriptional specificity factor, σ and ρ termination factor —
are recycled. Note that whereas σ is necessary for specific initiation at promoter sites, RNA chain
termination may be rho-independent or dependent, depending upon the nature of the transcriptional
termination sequence. The various elements that control transcriptional initiation and termination
— the DNA sequences themselves, regulatory proteins and organic effector molecules — are
discussed in Part IV

(i) Initiation. Two main DNA sequences, centred at about –35 and –10 base pairs from the transcriptional start point (that is, 35 and 10 bp upstream), have been implicated in promoter function:[5]

5′ TTGACA   and   5′ TATAATR
   AACTGT 5′   ATATTAY 5′

(where R and Y in the –10 sequence represent a purine and pyrimidine base, respectively). The AT-rich regions have been referred to as the *–35* or *recognition site* and the *–10 site* or *Pribnow 'box'*. The –35 site allows the initial, essential interaction between promoter and DNA-dependent RNA polymerase. The term *recognition site* stems from the original observation that though this region is not apparently protected against DNase digestion when complexed with RNA polymerase (but see below), it is crucial for enzyme binding in the first instance. The Pribnow box is so called for the manner in which the author indicated a sequence common to several promoter sequences centred at –10. That the recognition site is required *in vivo* is indicated by promoter-down and up mutations, lesions altering transcription levels, that result in base changes in this region. The formation of a stable complex (see below) could, therefore, involve lateral movement of holoenzyme from the recognition site to the Pribnow box. Alternatively, the enzyme covers both sites but not sufficiently to shield the recognition sequence *in toto* from DNase digestion. Certainly, the *E. coli* RNA polymerase is physically large enough to stretch over such distances. This 'static' model finds support from recent protection studies that show an RNA polymerase interaction extending some 54 base pairs upstream from the RNA initiation point. Promoter sequences are discussed in relation to the regulation of gene function in Part IV.

What is responsible for these specific protein-nucleic acid interactions? Although the conserved sequences found at transcriptional start sites, the –35 and –10 sequences, are shielded from direct interaction with RNA polymerase, association is possible through base-specific groups in the major and minor grooves. RNA polymerase appears to interact solely with one side of the DNA helix upstream of the Pribnow box. The secondary or tertiary DNA structure, in addition to the primary sequences, may also be important in promoter function. Binding of RNA polymerase is, however, unlikely to be the result of Gierer loop formation (Section 1.1.2 (e)) since there is a general absence of palindromic sequences at promoter sites. The superhelical nature of circular genetic elements, rather than a unique structure, appears to play a role in transcriptional initiation. Thus, negatively supercoiled DNA is a better template for RNA polymerase *in vitro*. Moreover, DNA gyrase, the *E. coli* enzyme responsible for introducing most, if not all, negative supertwists *in vivo*, has been directly

---

5.   These are *consensus sequences*. The –35 and –10 regions differ slightly from one promoter to another: only T at –8 is invariant.

implicated in RNA initiation from certain promoters (see Section 6.1.1 (b)). It would seem likely that the strained supercoiled configuration promotes transcription by driving RNA polymerase-DNA interactions (a reaction, perhaps, initially mediated through groove contacts). On this hypothesis, those promoters particularly dependent upon DNA gyrase may be most resistant to helix unwinding.

A *closed promoter complex*, in which bound DNA is unmelted, is initially formed. Subsequently, upon suitable conditions (the presence of divalent metal ions, and a temperature above the critical transition point specific for a particular promoter), helix destabilisation is triggered to create an *open promoter complex*. About 11 base pairs — one turn of the helix — is opened from the middle of the Pribnow box to just past the initiation site. RNA polymerase is thus, in this instance, a sequence-specific melting protein.

It is this open complex that allows tight binding of RNA polymerase and, in the presence of ribonucleoside triphosphates, initiation of RNA synthesis (the very stability of open promoter complexes, in fact, facilitates the isolation and sequencing of regions responsible for tight binding). Transcription is initiated some 10 base pairs downstream from the binding site (note that unlike DNA replication, Section 6.1, initiation does not require a primer). Both the -10 and *initiation* sites are protected by a single enzyme molecule against DNase digestion *in vitro* suggesting that RNA polymerase movement is not required once the tight-binding open complex has been formed (the recognition site also appears to be covered under these conditions, Figure 2.5).

At least two sites on the RNA polymerase holoenzyme complex have been postulated to explain catalysis of RNA synthesis on DNA templates, an *initiation* (or *product terminus*) site and an *elongation* (or *substrate*) site. At initiation, the product terminus site normally carries a purine nucleoside triphosphate since the 5′-terminal ribonucleotide is, almost always, a purine (pppA more often than pppG), while the substrate site is responsible for accurate location of the second residue. Interestingly, the two ribonucleoside triphosphates generally used in the transcriptional initiation, ATP and GTP, are those that are maintained at a high concentration for energy transduction and protein biosynthesis. During elongation, substrate molecules also enter the ternary complex by this latter site (see below). It is customary to define transcriptional initiation as the formation of the first internucleotide bond, a process that generates a dinucleoside tetraphosphate of structure pppRpN (where R = purine, Figure 2.6) and results, presumably, in translocation of the substrate binding site by one nucleotide. Subsequent events — successive additions of nucleoside monophosphates to the 3′-OH terminus with concomitant enzyme translocation — are, therefore, considered RNA chain elongation.

(ii) Elongation. At some stage, apparently after the formation of only a small number of phosphodiester bonds (perhaps less than ten), sigma is released. Further elongation is, thus, catalysed by core enzyme alone. The released

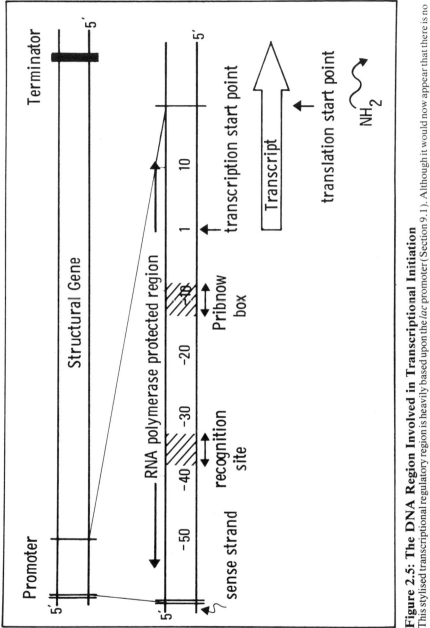

Figure 2.5: The DNA Region Involved in Transcriptional Initiation

This stylised transcriptional regulatory region is heavily based upon the *lac* promoter (Section 9.1). Although it would now appear that there is no such thing as a 'normal' promoter, these control regions have two main elements in common: similar sequences centred at −35 (recognition site) and −10 (Pribnow box) nucleotides from the transcriptional start point (the −35 sequence has been implicated in the protein-mediated potentiation of promoter function, Sections 9.1.2 (b) and 10.2.1). Note that in the present example (a situation true, in fact, for many operons), a portion of the transcript, the leader sequence, is not translated. That is, translational initiation is an internal event

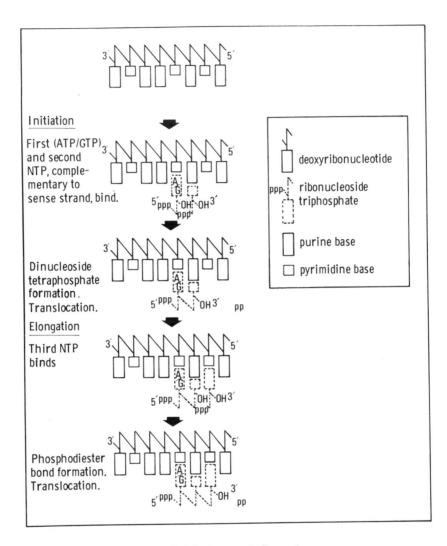

Figure 2.6: RNA Chain Initiation and Growth

Transcription almost always initiates with a purine base (rare instances of CTP starts have been reported). On the basis that there are two sites on the oligomeric enzyme, one for the product (carrying in this instance the purine nucleoside triphosphate, ATP or GTP) and one for the substrate, pairing of the second base fills both sites. Translocation is, therefore, necessary to allow entrance of the third NTP (after phosphodiester bond formation). Further elongation is accomplished by translocation and bond formation steps. Note that the transcript is extended in the 5′ → 3′ direction (RNA polymerase moving 3′ → 5′ along the DNA sense strand)

cytoplasmic protein factor can then be reutilised by a free core molecule. Cyclic use of sigma (Figure 2.7) explains the less than stoichiometric amounts of specificity factor in relation to the core components (Table 2.1).

Once initiated, transcription proceeds at a uniform rate of about 55-60

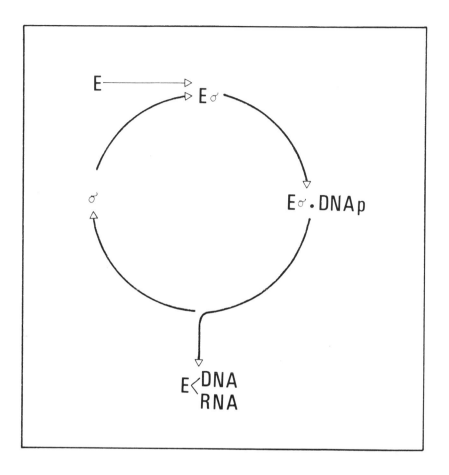

Figure 2.7: The Sigma Cycle
On combining with core RNA polymerase, σ allows specific initiation at promoter sites *(DNAp)*. Transcription, started within this regulatory region, generates a ternary complex consisting of core enzyme, DNA and transcript $(E <^{DNA}_{RNA})$, the specificity factor apparently dissociating shortly after initiation. The fact that σ is rapidly recycled — rather than stably associated with each core molecule throughout its transcription 'life-cycle' — suggests that equimolar amounts of σ and core are not necessary (see Table 2.1)

nucleotides per second at 37° in *E. coli*. Irrespective of the nature (and destiny) of the transcript, RNA chain growth is always in the 5′ → 3′ direction. RNA polymerase is thought to promote local melting of the DNA duplex close to the site of the bound core molecule: the helix reforms in the wake (upstream) of the enzyme. The net effect is, thus, lateral displacement of the DNA 'bubble'. There appears to be some transient interaction between a small region of the nascent transcript and the DNA region being transcribed to form a DNA · RNA hybrid (about 10-12 ribonucleotide residues are protected against ribonuclease degradation *in vitro*).

At first sight, RNA synthesis on DNA templates may appear energetically wasteful as the transcript is normally larger than the structural gene itself. The promoter-proximal region of mRNA, the *leader sequence*, carries, in fact, crucial information necessary for translation of the coding region — it signals initiation of polypeptide synthesis (see Section 2.2.2 (a)). The length of this sequence varies from one species of mRNA to another. The reason for this variation is not known but, certainly, absence of the leader sequence is an extreme form of translational control as ribosomes translate such templates with apparent inefficiency (P_M-promoted expression of bacteriophage *lambda* cI gene is a case in point, Section 10.2.2 (b)).

(iii) Termination. Transcriptional termination occurs at specific sites, notably runs of adenylate residues, such that the 3'-end of the completed transcript frequently carries the sequence ——— (UUUU)UUUU(A) OH (where bases in parenthesis are variable). Moreover, deceleration sequences — a string of contiguous G·C base pairs — may precede the natural terminator by about one turn of the helix; RNA polymerase appears to transcribe G · C pairs more slowly than A · T pairs, presumably because of the difficulty in unwinding the DNA helix in such a region (the former has three rather than two hydrogen bonds) or because of increased stability of the nascent RNA · DNA hybrid in GC-rich sequences. The almost invariable presence of inverted repeat sequences preceding the termination point — RNA chain termination occurs about 20 bp from the centre of the region of hyphenated dyad symmetry — suggests a role for hairpin loops in transcriptional termination. Certainly, RNA secondary structure appears to be important in the case of certain transcripts. Termination is discussed with respect to gene regulation in Part IV.

An *E. coli* protein, *rho factor* (Table 2.2), is required for specific RNA chain termination at certain terminators.[6] It has no nucleolytic function. An NTPase activity — RNA-dependent phosphohydrolysis of nucleoside triphosphates — associated with this transcription factor is thought to be concerned with 'driving' the termination rection (see Section 8.1.2 (a)). The transcription factor is likely to interact directly with RNA polymerase since rho-mediated termination is sensitive to certain genetic lesions that result in alterations in the β subunit (Section 2.1.1 (a)). It would seem that rho, like sigma, is recycled.

Whether transcriptional termination is rho-factor dependent or independent, the result is the same: termination of RNA chain elongation, dissociation of a discrete transcript of defined length, and core enzyme release. Detachment of the nascent RNA chain renders the complete sequence accessible to ribosomal

---

6.   Rho may also be involved in transcriptional termination in leader sequences of certain operons (attenuation, see Sections 8.1.2 (b) and 9.2.2 (b), The *trp* Operon). The bacterial translation apparatus regulates this process. There are, in addition, bacteriophage-coded positive regulators (phage *lambda N* and *O* protein, Section 10.2.3) that reduce rho-mediated termination of transcripts initiated from specific phage promoters.

influence (though RNA and protein biogenesis normally occur concurrently, the mRNA region interacting with RNA polymerase is, clearly, not available for translation). The free core complex can then recombine with sigma factor in the cytoplasm to produce holoenzyme and, thus, allow another round of transcriptional initiation.

(b) Control: RNA Synthesis and Decay. The RNA population within the bacterium can be divided, conveniently, into two major groups with respect to their decay rate *in vivo:* stable and unstable RNA. The former is fairly permanent once synthesised (Figure 2.8). This classification is useful for a functional approach to cellular ribonucleic acids as *stable RNA* consists mainly

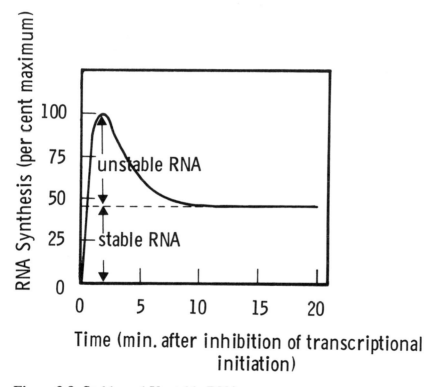

Time (min. after inhibition of transcriptional initiation)

Figure 2.8: Stable and Unstable RNA
Time course for RNA synthesis after inhibition of transcription with the antibiotic rifampicin (see Section 2.1.2 (c)). Since further rounds of transcriptional initiation are inhibited, residual RNA synthesis (the increase from t = 0 to t = 2 min) represents RNA production by RNA polymerase molecules involved in chain elongation at the time of the antibiotic block. Whereas messenger RNA decays (an average half-life of 1.5 min is obtained from the exponential slope between t = 2 and t = 10), ribosomal RNA is stable under these conditions. The proportion of stable to unstable RNA can, therefore, be determined by extrapolation (about 50 per cent). Adapted from Pato and von Meyenburg (Pato, M.L., and von Meyenburg, K. (1970) *Cold Spring Harbor Symp. Quant. Biol.,* *35,* 497-504)

of rRNA and tRNA species (and is, thus, part of the translational machinery), whereas *unstable RNA* is messenger RNA. The difference in lability may lie in the fact that stable RNA is either present as a ribonucleoprotein complex or has profound secondary structure, properties that shield the 5'-terminus (and, perhaps, susceptible internal targets) from attack by endogenous ribonucleases (see below). [mRNAs for ribosomal proteins and certain outer membrane proteins are more stable than the bulk of mRNAs. RNA that is not translated is also susceptible to post-transcriptional modification. In fact, this processing is crucial for function since only the mature molecules are fully active. Processing includes cleavage at specific sites to produce shortened RNA species as well as modification of RNA bases *in situ* (Sections 9.3.1 (b) and 2.2.1 (a)). Processing of certain bacterial mRNAs as well as phage species has also been established. Under certain conditions, such as amino acid starvation, stable RNA fails to accumulate in the cell. This *stringent response* is apparently not due to degradation of existing (r + t) RNA but a switch from transcription of stable RNA to mRNA (Section 9.3.2 (b)).]

Although the degradation rate for each mRNA is unique, and independent of its size or function, messages decay exponentially in *E. coli*. Turnover of unstable RNA is rapid. The *half-life* — the time taken for the RNA population to be reduced by half — of an average-sized mRNA molecule of 1200 nucleotides (coding for a protein of 40 000 daltons) is about 40 seconds at 37° in *E. coli* (depending upon the growth conditions). This rapid degradation has the property of 'flushing' RNA molecules out of the cytoplasm. Moreover, when transcription of a particular gene is halted, there is rapid, functional inactivation of remaining mRNA molecules.

The overall direction of degradation occurs in the 5' → 3' direction — the direction of transcription and translation — yet there is clear evidence for internal inactivation events (distal messages, for example, from a polycistronic operon may decay more rapidly than proximal ones and *vice versa*). Both endoribonucleolytic and exoribonucleolytic activities would, therefore, appear to be involved. However, only 3' → 5' exoribonucleases have been characterised to date and the 3'-terminus (the nascent end) is likely to be protected by association with the DNA template and RNA polymerase. Perhaps an initial, random endonucleolytic cut in a mRNA molecule near to the 5'-end or in intercistronic sequences (between regions coding for protein products) activates exonucleolytic degradation. This model explains the overall 5' → 3' direction of mRNA degradation (see Figure 2.9). The presence of ribosomes would afford some protection against the initial cut (each ribosome covers about 35 ribonucleotides, see Section 2.2.2 (a)).

The components involved in this degradation process have not, as yet, been identified. RNase III, an endoribonuclease involved in both ribosomal and messenger RNA processing (Section 9.3.1 (b)) has been implicated in bacterial messenger stability. It is also now believed that exposure of mRNA strands to cytoplasmic factors as a result of premature or natural termination of translation

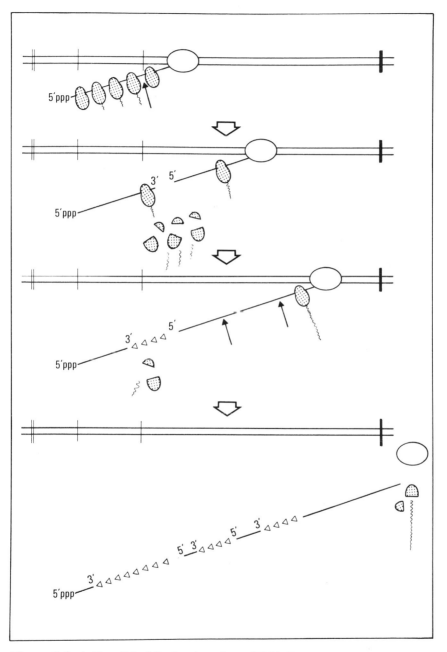

Figure 2.9: A Possible Mechanism for mRNA Degradation
The *large black arrows* indicate the position of endonucleolytic attack at an intercistronic region; the *open arrowheads* show the direction of 3′ → 5′ exonucleolytic attack. Overall, there is 5′ → 3′ degradation of unstable RNA

can affect mRNA chain elongation due to activation of transcriptional termination (see Section 8.1.2 (a)).

The rapid turnover of mRNA and the comparative stability of protein clearly point to transcriptional control as being the major form of regulation of gene expression in prokaryotes. However, the RNA chain elongation rate is independent of cell doubling time. The rate-limiting step in messenger production is all the more likely to be chain initiation, in particular the formation of the initiation complex, since RNA starts are slow (0.5-25 initiations per minute) in relation to the elongation rate. Prokaryotic transcription and, in fact, gene expression is regulated mainly by control of transcriptional initiation. The promoter for a particular operon thus plays a major role in determining the degree of expression of each gene product encoded (other elements important in control are discussed in Part IV).

(c) Antibiotic Inhibitors of Transcription. Three well-studied groups of antibiotics affect DNA transcription directly: actinomycin D; the ansamycins, rifampicin (a semi-synthetic derivative of the naturally occurring rifamycin) and streptovaricin; and streptolydigin, a compound with certain similarities to the previous group. The mode of action of each group is different.[7]

Actinomycin D consists of co-planar heterocyclic rings (similar to those of the acridines, Section 3.2.1 (b)) carrying two cyclic oligopeptide side groups (Figure 2.10). Many natural and synthetic actinomycins are now available. Actinomycin D blocks transcription on DNA templates by binding to the template, to guanine residues. Inhibition would, thus, appear to be due to steric interference by the chromopeptide antibiotic. Numerous models for the antibiotic-DNA complex have been proposed. They have in common positioning of the cyclic oligopeptide in the minor groove of helical DNA. More recent crystal studies suggest intercalation betwen G · C base pairs in the duplex by the phenoxazone ring system; the pentapeptide chains are likely to be important in the sequence specificity of actinomycin binding.

The target of the ansamycins — a group of compounds with a flat aromatic nucleus bridged by a long aliphatic molecule, the ansa chain (Figure 2.10) — and the related antibiotic, streptolydigin, is the transcribing enzyme, itself, DNA-dependent RNA polymerase. These antibiotics bind specifically to the β subunit (*E. coli* strains resistant to these substances have a drug-resistant enzyme with an altered β polypeptide: the genetic lesion responsible for this change in property, therefore, defines *rpoB*). The two ansamycins, rifampicin and streptovaricin, have been referred to as inhibitors of RNA chain initiation, whereas streptolydigin appears to inhibit the elongation step. In fact, contrary to earlier beliefs, rifampicin, the most studied ansamycin, does not block formation

7.   The Gram-negative bacterial membrane is largely impermeable to actinomycin D and streptolydigin. *E. coli*, therefore, exhibits partial resistance to these antibiotics. Treatment with the chelator EDTA reduces this natural barrier.

of the first phosphodiester bond: the dinucleoside tetraphosphate is definitely synthesised in the presence of inhibitory concentrations. Rather, rifampicin now appears to prevent the first translocation step, possibly by steric blockage, and is, therefore, an inhibitor of productive initiation (at high concentrations of the drug, RNA chain elongation is also inhibited).

Binding is a condition *sine qua non* for inhibition. Rifampicin binds tightly to RNA polymerase (molar ratio of 1 : 1) at any step prior to formation of a trinucleotide[8] (attachment to DNA affords some protection, possibly by shielding the drug binding sites). Thus, transcription is susceptible to inhibition up to this stage but refractory thereafter; the ternary elongation complex is, clearly, unaffected by the antibiotic (except at high concentrations). However, those enzymes molecules 'frozen' at the initiation stage by rifampicin binding may act as a temporary barrier to potentially functional, unbound holoenzyme.

In addition to the above antibiotics, the naturally occurring nucleoside, cordycepin (3'-deoxyadenosine, 3'-dA), isolated from culture filtrates of *Cordyceps militaris*, acts at the level of polyribonucleotide chain elongation. Incorporation of the analogue into a growing RNA chain results in premature transcriptional termination since it lacks a 3'-OH group and cannot be extended (Figure 2.10). The triphosphate derivative of cordycepin (3'-dATP) is, therefore, a chain terminator.

2.2 RNA Translation

Genetic material can be expressed either as RNA or protein. Structural genes encode protein products and, therefore, require both transcription and translation. Decoding of mRNA, the conversion of genetic information contained within the ribonucleotide sequence into protein by the translational machinery, involves a change in the language base. It forms the second major stage of gene expression. The absence of a distinct nuclear membrane in prokaryotes allows concomitant transcription and translation. In fact, coupling between RNA and protein biogenesis — the absence of ribosomes from messengers exposes these RNA species to the action of endogenous ribonucleases and sometimes even affects their production — makes translation of messenger molecules almost obligatory for their survival.

2.2.1 The Elements of Translation
The synthesis of protein requires three main components, the mRNA template, ribosomes and transfer RNA species. In addition, extraribosomal protein factors are required. These constituents are common throughout the bacterial

---

8.   Structure-function studies of semi-synthetic rifamycin derivatives indicate that large parts of the ansa chain and the chromophore — the aromatic nucleus — are involved in tight binding between RNA polymerase and antibiotic. This is not by covalent bonds. Rather, some non-polar association seems likely. The drug does not seem to bind to the free β subunit but only to core or holoenzyme.

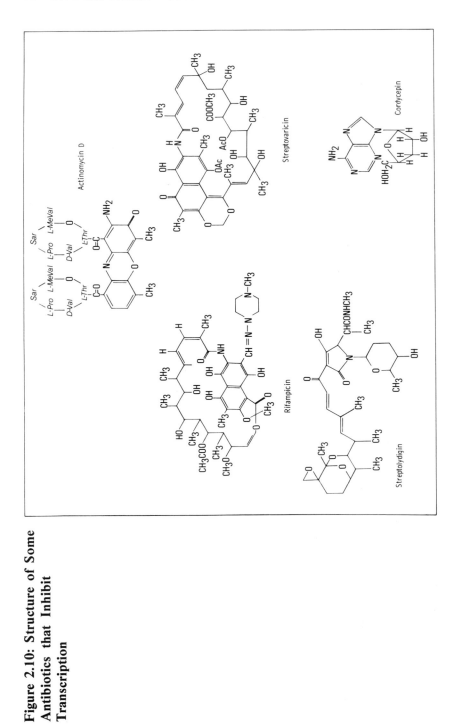

Figure 2.10: Structure of Some Antibiotics that Inhibit Transcription

kingdom and have close parallels with the eukaryotic translational apparatus. Moreover, the genetic code — the means by which groups of ribonucleotides encode specific amino acids — is very largely the same irrespective of the type of organism (but see legend to Table 2.6).

(a) The Genetic Code. Sufficient information is contained within a single strand of messenger RNA for the polymerisation of particular amino acids in a specific sequence. The same, unique protein product is synthesised each time by the translational matrix. The basic coding unit, or *codon*, is a group of three adjacent ribonucleotides. The total number of permutations of these triplets, where each constituent of a codon can be one of four bases, is sixty-four (4^3). This would suggest an excess coding capacity since there are only twenty different, common amino acids.

In fact, the code is *degenerate:* some amino acids are encoded by as many as six different codons (consider, for example, the codons for leucine, UUA, UUG, CUU, CUC, CUA and CUG; Tables 2.3 and 2.4). Sixty-one of the 64 triplets are *sense codons*, coding for specific residues. The remaining three, UAG (amber), UAA (ochre) and UGA (opal), have no sense. These *nonsense codons* are not translated but rather code for translational termination. In addition, three sense codons, AUG (Met), GUG (Val) and UUG (Leu), have dual roles since, depending on the adjacent nucleotide sequence, they either specify the insertion of the indicated amino acid or act as *initiation codons* for methionine at the start of polypeptide chain formation (AUG is most commonly used). Though translation is punctuated by start and stop signals, the code itself is 'comma-free'. Thus, once the reading-frame has been defined (see below), no additional signals are required to signify the end of one codon and the beginning of another.

Each mRNA can, in principal, be translated in three different *reading-frames* (Figure 2.11) and, thus, code potentially for three different protein products (though translation in two of these frames would usually 'uncover' termination codons, resulting in the synthesis of foreshortened polypeptides). It is the initiation codon that dictates the translational reading-frame: after a ribosome has initiated, the mRNA is read in non-overlapping[9] groups of three nucleotides from this start point. Whichever reading-frame is translated, the code is *colinear* since the linear order of triplets and corresponding amino acids is the same.

The structure of the genetic code is, itself, interesting. There is a high and varied degree of degeneracy (even the initiators and terminators show redundancy). Three amino acids (Arg, Leu and Ser) have six codons and five (Ala, Gly, Pro, Thr and Val) have four different codons (Table 2.4). The largest group (Asn, Asp, Cys, Gln, Glu, His, Lys, Phe and Tyr) is composed of those coded by two triplets. The two residues with the chemically most complex side-chains,

---

9.   It has recently been shown that structural sequences may rarely be translated in all three different phases. This economy of coding allows the small single-stranded bacteriophage $\phi \cdot 174$ (5375 bases) to promote the synthesis of 11 different polypeptides.

Table 2.3 The Genetic Code

First Base (5')	Second Base: U	Second Base: C	Second Base: A	Second Base: G	Third Base (3')
U	UUU ⎤ Phe UUC ⎦ UUA ⎤ Leu UUG ⎦ b	UCU ⎤ UCC ⎥ Ser UCA ⎥ UCG ⎦	UAU ⎤ Tyr UAC ⎦ UAA Stop a UAG Stop a	UGU ⎤ Cys UGC ⎦ UGA Stop a UGG Trp	U C A G
C	CUU ⎤ CUC ⎥ Leu CUA ⎥ CUG ⎦	CCU ⎤ CCC ⎥ Pro CCA ⎥ CCG ⎦	CAU ⎤ His CAC ⎦ CAA ⎤ Gln CAG ⎦	CGU ⎤ CGC ⎥ Arg CGA ⎥ CGG ⎦	U C A G
A	AUU ⎤ Ile AUC ⎥ AUA ⎦ AUG Met b	ACU ⎤ ACC ⎥ Thr ACA ⎥ ACG ⎦	AAU ⎤ Asn AAC ⎦ AAA ⎤ Lys AAG ⎦	AGU ⎤ Ser AGC ⎦ AGA ⎤ Arg AGG ⎦	U C A G
G	GUU ⎤ GUC ⎥ Val GUA ⎥ GUG ⎦ b	GCU ⎤ GCC ⎥ Ala GCA ⎥ GCG ⎦	GAU ⎤ Asp GAC ⎦ GAA ⎤ Glu GAG ⎦	GGU ⎤ GGC ⎥ Gly GGA ⎥ GGG ⎦	U C A G

The code is written in terms of ribonucleotides (the chemical structures of the 20 standard amino acids and their abbreviations are given in Figure 1.7). Clearly, the equivalent sequence in DNA will contain deoxyribonucleotides, and uracil will be replaced by thymine. Note that the DNA sense strand — the polynucleotide chain used as a template for transcription — is complementary to the corresponding RNA sequence. Thus, the Met codon 5' AUG 3' will be encoded by the DNA triplet 3' TAC 5'.

There is some correlation between codon position (in the above table) and the properties or metabolism of the amino acid encoded. Hydrophobic residues, for example, are mainly encoded by codons in the left half of the table while the remainder is responsible for a number of charged amino acids. Similarly, clustering of triplets according to the amino acid biosynthetic family occurs to a certain extent (amino acid families are discussed by Umbarger (1978)).

a. UAA (ochre), UAG (amber) and UGA (opal) code for termination of translation.
b. AUG (Met), and UUG (Leu) and GUG (Val) are responsible for polypeptide chain initiation.

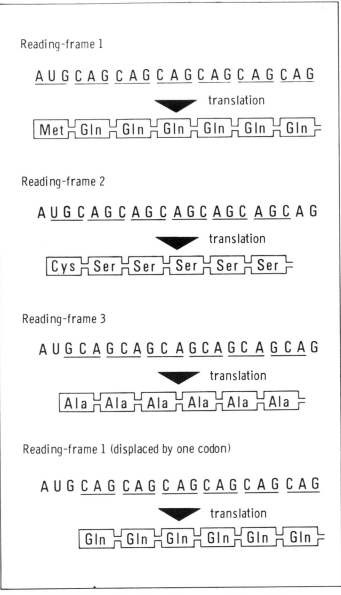

Figure 2.11: Three Reading-Frames of Translation
Depending upon which three ribonucleotides are employed as the first codon, the translation product differs markedly (note that under normal physiological magnesium concentrations, translational initiation and, thus, the reading-frame is, in fact, dictated by the siting of the initiation codon, AUG). Displacement by one or two bases is equivalent to a −1 or −2 frameshift (see Section 3.1.1 (d)): displacement by three-bases is equivalent to initiation at the second triplet — that is, the reading-frame is unchanged — except that the N-terminal amino acid is lost. Clearly, the correct reading-frame is crucial for accurate translation

Table 2.4ᵃ Degeneracy of the Genetic Code

Degeneracy	Amino Acid	Codons	Degeneracy	Amino Acid	Codons
6	Arg	CGX,AGR	2	Asn	AAY
	Leu	CUX,UUR		Asp	GAY
	Ser	UCX,AGY		Cys	UGY
4	Ala	GCX		Gln	CAR
	Gly	GGX		Glu	GAR
	Pro	CCX		His	CAY
	Thr	ACX		Lys	AAR
	Val	GUX		Phe	UUY
3	Ile ᵇ	AUA,AUY		Tyr	UAY
	Start	UUG,RUG	1	Met	AUG
	Stop	UGA,UAR		Trp	UGG

a   Abbreviations: X, A, G, U or C; R, A or G (purine); Y, U or C (pyrimidine).
b   Start codons also code for Met (AUG), Leu (UUG) or Val (GUG) and are, therefore, represented twice in the table.

methionine and tryptophan (amino acids that are absent from certain simple proteins) are specified by single codons. There is some relationship between the frequency of occurrence of amino acids and the degeneracy of their codons (leucine, for example, is the most common residue in *E. coli* proteins).

Degenerate codons retain, generally, the first two ribonucleotides: the four serine codons, for example, can be written UCX, where X = U, C, A or G. The first two bases of such tetradegenerate codons, therefore, solely dictate the amino acid inserted. The third base is important for those residues encoded by two triplets, though there is no discrimination among purines (A/G) or pyrimidines (U/C). This very degeneracy allows organisms with genomes of vastly different base ratios (a range of 25-80% G + C in bacteria) to employ the same code.[10] The code is, in fact, universal (but see legend to Table 2.6). Degeneracy in the code buffers against mutational change (Chapter 3) since an alteration in the third base is unlikely to result in the insertion of a changed amino acid. Such mutations are neutral events to the cell. Moreover, the clustering of triplets — codons are, to some extent, grouped according to the properties or metabolism of the amino acids for which they code (see Legend to Table 2.3) —

10. Consider, for example, an organism where A+T >> G+C. Despite the probable low frequency of AUC (Ile) codons, isoleucine may still be encoded by two other triplets, AU_U^A.

further minimises the chance of a deleterious mutation.

(b) Transfer RNA. Translation of messenger RNA does not proceed by direct interaction between RNA codons and their cognate amino acids (ribonucleotides are too similar in structure to each other, and distinct from amino acids, to allow specific pairing). Rather, the code is read by ribosomes making use of adaptor molecules, to which an amino acid is covalently attached. These adaptors are small RNA species, termed *transfer RNAs* (or tRNAs);[11] their charging enzymes are the aminoacyl-tRNA synthetases. Transfer RNAs, therefore, interface between the two 'languages'.

The structure of these adaptors is remarkably conserved in nature. Each tRNA consists of a sequence, between 70-100 ribonucleotides,[12] which can be arranged in a characteristic *cloverleaf* form through hydrogen-bonding of both normal and modified bases (as much as 16 per cent of tRNA bases may be modified, Table 2.5 and Figure 2.12); various non-standard base pairs such as $G \cdot U$, $G \cdot \psi$ and $\psi \cdot A$ have also been implicated.

In this secondary structure (Figure 2.12), the two ends of the RNA molecule are together. The 3'-terminal sequence CCA-OH is found universally among tRNA species. It is responsible for the covalent attachment of the cognate amino acid through an ester bond formed between the 3'- (or 2'-; see below) hydroxy group and the residue's carboxy-terminus. The double-stranded region of this *amino acid acceptor stem* commonly consists of 7 base pairs. The *anticodon loop*— an unpaired region of 7 nucleotides — has a stem of 5 base pairs (termed collectively an *arm*) and is at the other end of the proposed clover-leaf structure. The anticodon is involved in precise anti-parallel base pairing between the two RNA chains, tRNA and mRNA; it consists of three bases (either normal or modified) complementary to the triplet coding for the cognate amino acid. The anticodon of tRNA^{Met}, for example, is 3' UAC 5' (5' AUG 3' codes for methionine; see Table 2.3 and Figure 2.15). Two other major features of the secondary structure are the thymine-pseudouracil-cytosine, TψC (or pentanucleotide) *arm* (7 nucleotides in the loop, 5 base pairs in the stem) and the *dihydrouridine* or *D arm* (7-10 nucleotides in the loop, 3-4 base pairs in the stem). In addition, transfer RNAs can also have a *variable* (or extra) arm, some 4-21 nucleotides in length (the differences in tRNA size reside in this region and in the D loop). RNA sequences of different tRNAs indicate that certain bases, presumably crucial for tRNA function, are conserved (Figure 2.12).

X-ray crystallographic analysis of yeast tRNA^{Phe} suggests a more compact

11.   A note on tRNA nomenclature: the amino acid by which a tRNA is charged is indicated in the superscript position, thus tRNA^{Leu} for a leucine accepting species. Different tRNAs that accept the same residue — isoaccepting species — are distinguished using a lower case number. tRNA$^{Tyr}_1$ and tRNA$^{Tyr}_2$, for example, are both charged with tyrosine (tRNA^{Met} and the initiator species, tRNA$^{Met}_f$, are an exception to this nomenclature).

12.   Transfer RNAs (and indeed rRNA) are synthesised as longer molecules, precursors, and are subsequently processed; maturation also involves base modification. Transcription and conversion of primary tRNA and rRNA transcripts to their mature species is discussed in Section 9.3.

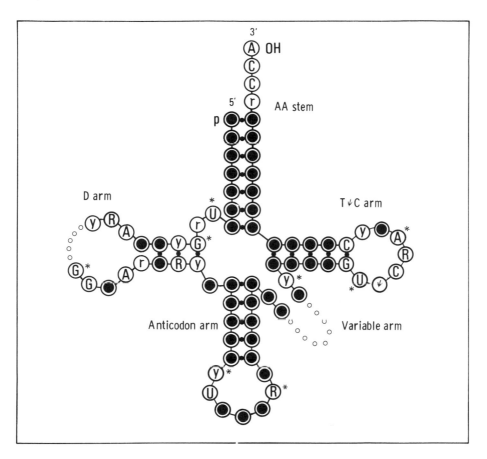

Figure 2.12: The Generalised Cloverleaf Structure of tRNA
Positions of conserved bases (abbreviations: R, purine; Y, pyrimidine; a small letter indicates less frequent occurrence) and common sites of modified bases (*) are shown. Note that all *E. coli* tRNAs carry the sequence TψC in the TψC loop

arrangement — a flat 'L-shaped' tertiary structure — than one might presume from the clover-leaf model (Figures 2.13 and 2.14). The dihydrouridine and TψC helical arrays interdigitate and hydrogen bond with each other to create two, approximately perpendicular, domains of base stacking. Hydrogen bonding between invariant (or semi-invariant) bases is responsible for this specific three-dimensional structure; 2'-hydroxyl groups of ribose sugars and phosphate oxygens are also involved. The two obvious functional domains — the acceptor stem and the anticodon loop — are about 8 nm apart, well separated from the rest of the molecule (they even, perhaps, have some flexibility). The conserved sequence TψC, situated at the 'corner', is approximately 6 nm from both these regions.

What is the number of different tRNAs in *E. coli*? The presence of

Table 2.5: Some Common Modified Nucleosides of tRNA

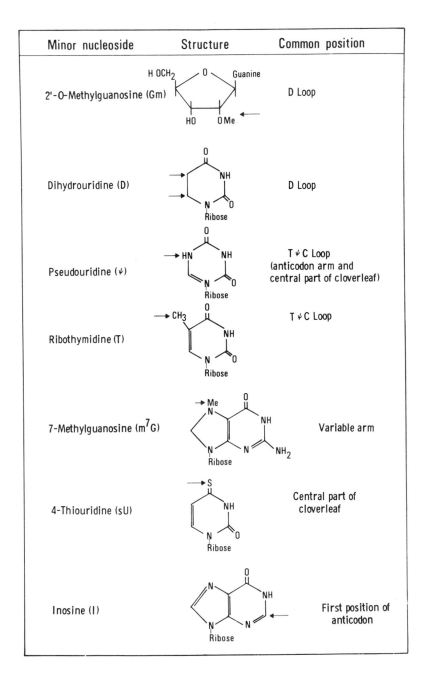

Minor nucleoside	Structure	Common position
2'-O-Methylguanosine (Gm)		D Loop
Dihydrouridine (D)		D Loop
Pseudouridine (ψ)		T ψ C Loop (anticodon arm and central part of cloverleaf)
Ribothymidine (T)		T ψ C Loop
7-Methylguanosine (m^{7}G)		Variable arm
4-Thiouridine (sU)		Central part of cloverleaf
Inosine (I)		First position of anticodon

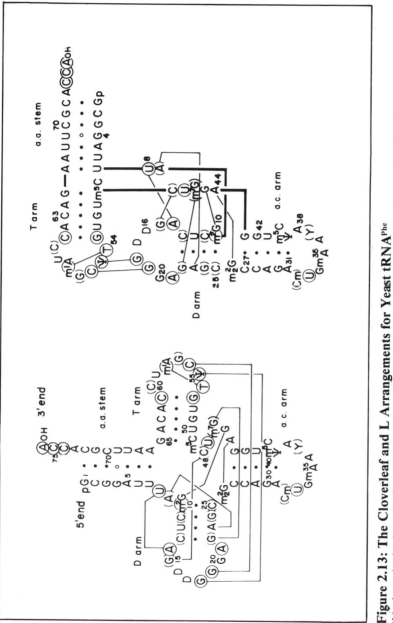

Figure 2.13: The Cloverleaf and L Arrangements for Yeast tRNA[Phe]

Y in the anticodon loop signifies wybutosine (a complex derivative of guanosine). Note that the majority of invariant and semi-invariant bases (*circled* and in *parentheses*, respectively; see also Figure 2.12) are sited in the middle of L structure as well as being involved in tertiary hydrogen bonds (*thin lines*). Each extension of the L is about 6 nm long, with a diameter of approximately 2 nm; their two ends are separated by a distance of some 8 nm (the diagonal in the right-hand structure). See the space-filling model determined by X-ray crystallographic methods (Figure 2.14). From Kim (Kim, S.-H. (1978) in A. Meister (ed.) *Advances in Enzymology* (John Wiley and Sons, New York) pp. 279–315) with permission of the publishers and author

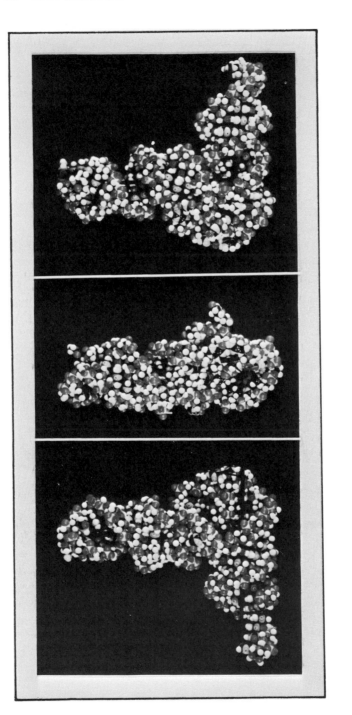

Figure 2.14: Space-Filling Model of Yeast tRNA[Phe]

This model is viewed by successive 90° rotations about the vertical axis (compared with the L arrangement in Figure 2.13). The tertiary structure of yeast tRNA[Phe] is based upon X-ray crystallography. This is the only tRNA species that has been examined at high resolution (0.25 nm). The ability to arrange all tRNAs (over 100 different molecules) sequenced to date in the cloverleaf form containing conserved bases and structures (Figure 2.12) suggests that all transfer RNAs have the same tertiary conformation. There is now strong evidence that the solution conformation is the same as the crystal form. From Holbrook *et al.* (Holbrook, S.R., Sussman, J.L., Warrant, R.W. and Kim, S.-H. (1978)*J. Mol. Biol., 123,* 631-660) with permission of the publishers and authors (copyright by Academic Press Inc. (London) Ltd.); photograph by courtesy of Professor S.-H. Kim *isoaccepting tRNA species* — different tRNAs that are charged by the same amino acid — allows for degeneracy in the code. There are, for example, at least five distinct transfer RNAs that accept leucine. However, the size of the tRNA population is not dictated entirely by the number of sense codons since interaction between anticodon and codon lacks the restriction of standard base pairing. In fact, as initially proposed by Crick (1966), the third base in the anticodon shows a reduced specificity of interaction (Table 2.6). This *wobble* in the third position allows pairing of 5' G in the anticodon with both C or U in the

Table 2.6: Wobble Pairing: Non-Standard (and Standard) Base Pairing between the 5'-Proximal Anticodon Base and 3'-Proximal Codon Base[a]

Pairing of the Third Base

Anticodon (5'-base)	Codon (3'-base)
A	U
G	C,U
U	A,G
C [b]	G
I	U,C,A

More recent evidence suggests that there is greater flexibility in pairing between codon and anticodon than permitted by the original wobble hypothesis. Translational initiation in *E. coli* at AUG, GUG and UUG codons (Table 2.3) suggests a reduction in specificity of pairing of the first base of the codon with the initiator tRNA, tRNA$_f^{Met}$. That misreading occurs in the second position is shown by the range of nonsense suppression (Table 3.6). The apparent readthrough of UGA terminators by the mitochondrial machinery (tryptophan seems to be inserted at these sites) and the translation of AUA (Ile) for methionine may be ascribed to the more conventional wobble in the third position; *E. coli* tRNA^{Trp} (anticodon 3'ACC) translates both UGG (Trp) and UGA (with very low efficiency) so that UGA mutants are leaky. A more extraordinary finding is that in yeast mitochondria, threonine is inserted in response to the leucine codon, CUA.

On the basis of the wobble hypothesis (and taking into account the UGA/Trp and AUA/Met equalities), a minimum of 31 different tRNAs are required to translate all 61 sense codons. The 'two out of three' hypothesis, with U in the wobble position recognising all possible tetradegenerate codons (that is, where 4 codons encode a single amino acid), reduces this number to 24. The finding that mammalian mitochondria encode this number of tRNAs suggests an 'exception' to the universality of the genetic code. Clearly, these subcellular organelles have evolved to make the most of their limited coding potential with the result that the tRNAs encoded are more flexible in their anticodon-codon pairing characteristics.

a   After Crick (1966).
b   I, inosine, the nucleoside of hypoxanthine.

codon, U with A or G, and inosine (the nucleoside form of hypoxanthine) with U, C or A. Recognition of the two phenylalanine codons 5′ UU_C^U 3′ by *E. coli* tRNA^{Phe} (anticodon 3′ AAG 5′) demonstrates wobble pairing of 5′ G in the anticodon. The reduction in anticodon-codon specificity is necessarily limited as too large a drop in fidelity would result in misreading of codons by non-cognate tRNAs. Wobble acts to increase the number of codons read by a particular tRNA molecule, and, therefore, reduces the theoretical number of necessary isoaccepting species. *E. coli* tRNA$_3^{Leu}$ (anticodon 3′ GAG) is a case in point: it recognises two of the six leucine codons, 5′ CU_C^U. The actual number of individual tRNAs is less than 61, probably between 40 and 60 in *E. coli* (the average 6000 copies of each tRNA, the majority of which are in the charged form, represent about 1 per cent of the dry cell weight).

The terms 'major' and 'minor' with respect to tRNAs refer to cellular abundance (gene number appears to be responsible for determining the contents of tRNAs in *E. coli*). Analysis of coding sequences for various prokaryotic genes has provided evidence for a non-random usage of synonymous codons. Generally speaking, codons recognised by minor tRNAs are employed infrequently.

It is rare that different tRNAs have the same anticodon. The two *E. coli* species charged with tyrosine, tRNA$_1^{Tyr}$ and tRNA$_2^{Tyr}$, only differ in two bases (Figure 2.15) and most likely arose from a gene duplication event and subsequent mutation (see Section 6.2.3). However, there are, clearly, two distinct species of tRNA^{Met} in *E. coli* that share the same anticodon but have quite different primary structures (Figure 2.15); tRNA$_f^{Met}$ (the 'f' refers to the fact that the methionine residue is formylated once charged, Section 2.2.2 (a)) acts only at the initiation stage of translation, whereas the function of tRNA^{Met} is limited to peptide chain elongation.

Amino acids are charged to their cognate tRNAs in a reaction requiring ATP and a specific aminoacyl-tRNA synthetase (one for each residue in *E. coli*). An ester bond links their C-terminus to the vicinal hydroxyls, the 3′- (or 2′-)[13] OH of the end adenosine in the tRNA acceptor stem (Figure 2.16). A non-concerted, possibly two-step mechanism, involving an aminoacyladenylate intermediate is favoured at present (Figure 2.17) though a concerted reaction that occurs in the absence of such an intermediate cannot be completely ruled out. It should be stressed that aminoacylation is crucial for the overall fidelity of translation; an error in charging would result in the insertion of an incorrect amino acid into the growing polypeptide chain. Two processes appear to mediate in charging

13.   The 3′- or 2′-hydroxyl group is aminoacylated; only L-amino acids are activated. Once charged, there can be rapid equilibration, suggesting that the initial position may be unimportant. As the specificity of aminoacylation resides in the enzyme itself, and since there is a correlation between codon structure and aminoacylation position, overall charging specificity may be maintained by groups of aminoacyl synthetases and tRNAs for amino acids of like properties (say, with certain hydrophilic side groups). On this 'secondary cognition' model, deleterious errors and, thus, the degree of translational infidelity are minimised by the insertion of competent though incorrect amino acids.

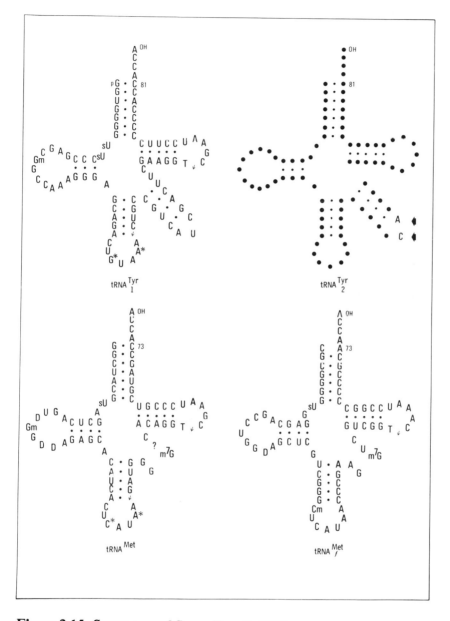

Figure 2.15: Sequences of Some *E. coli* tRNAs

The tRNA$^{Tyr}_2$ diagram shows only the nucleotide differences from tRNA$^{Tyr}_1$. Modified bases in tRNA^{Tyr}: G*, queuosine, a derivative of guanosine; A*, ms^{2}i^{6}A, 2-methylthio-N^6-isopentyl adenosine. Modified bases in tRNA^{Met}: C*, ac^{4}C, N^4-acetyl cytidine; A*, t^{6}A, N^6-(N-threo-nylcarbonyl) adenosine. For the more common modified nucleosides, see Table 2.5. Sequences from Sprinzl *et al.* (Sprinzl, M., Grueter, F., Spelzhaus, A. and Gauss, D.H. (1980) *Nucl. Acid Res., 8,* r1-r22)

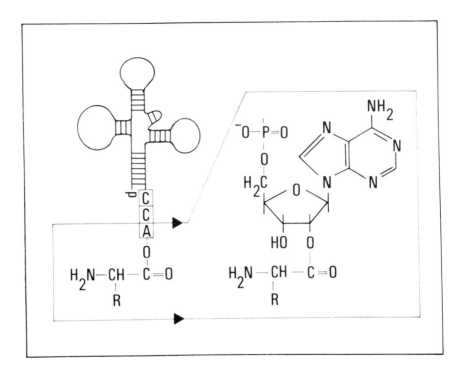

Figure 2.16: Structure of Aminoacyl-tRNA
Although the amino acid is shown attached via the 2'-position, both the 2'- and 3'-ribose hydroxyls of the terminal adenosine residue may be used (see text)

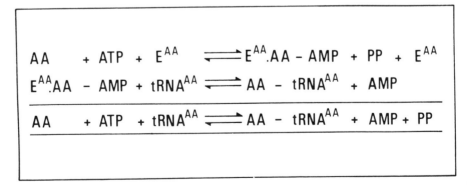

Figure 2.17: A Possible Two-Step Non-Concerted Mechanism for Amino-acylation of tRNA
In the proposed scheme, each amino acid (AA) is attached to its cognate tRNA $(tRNA^{AA})$ by an aminoacyl-tRNA synthetase specific for that residue (E^{AA}) in a reaction involving an aminoacyl-adenylate intermediate $(E^{AA} AA\text{-}AMP)$

specificity, synthetase-tRNA binding and aminoacylation. The deacylation activity associated with synthetases (a reaction that proceeds in the absence of AMP and pyrophosphate) has also been implicated in the removal of misactivated tRNAs. This proof-reading function is dependent upon the vicinal hydroxyls of the terminal adenosine *in vitro*.

This charging process offers a useful system for the study of nucleic acid-protein interactions (transfer RNAs can be readily sequenced because of their small size). However, attempts to define a single tRNA domain responsible for the specificity of aminoacylation have proved unsuccessful — the tRNA regions so far attributable to recognition tend to fall on the open side of the L-shaped molecule (the studies are complicated by the variation in size and oligomeric nature of the enzymes responsible). One attractive hypothesis suggests specificity is dependent upon the pseudo two-fold symmetry of the tRNA and synthetase. This 'symmetry recognition' model allows for the wide disparity in synthetases.

(c) The Ribosome. This large multicomponent assembly, the 'power house' of translation, consists of two major, non-identical components, a small 30S and large, 50S subunit.[14] The sedimentation coefficient of the intact ribosome, a stoichiometric complex of one of each of these subunits, is 70S. Although at one time there appeared to be a heterogenous population of ribosomes, it now seems likely that only one kind is present in *E. coli*.

The complete, 70S ribosome is a complex structure made up of 52[15] different ribosomal proteins *(r-proteins)* and three non-identical RNAs (rRNA): the small subunit contains 21 r-proteins and a 16S rRNA species (1541 nucleotides in length) while the large subunit has 33 r-proteins and two different RNAs, 23S rRNA and the small 5S rRNA (2904 nucleotides and 120 nucleotides, respectively, see legend to Figure 2.18). These components are present in only one copy per ribosome (apart from L7/L12 and L26/S20). The r-proteins are small entities (molecular weight in the range 5000-26 000, but see S1 below), which are basic in character and are mainly elongated in shape. A number carry modified amino acids (acetyl or methyl group).

These components are apparently condensed into a compact structure (Figure 2.18). The ribosomal three-dimensional architecture is one of an 'embryo-like' 30S subunit lying across a 50S subunit that resembles an

14. The sedimentation coefficient, s, of a substance is determined from its velocity of sedimentation in the applied gravitational field of an ultracentrifuge; it is expressed in Svedbergs, S. s values are dependent upon both molecular size and shape and are, therefore, only approximately additive (exemplified by the sedimentation coefficient of the intact ribosome as compared with its subunits).
15. 'S' and 'L' designate protein components of the small and large subunits, respectively. Thus, the ribosome consists of proteins S1-S21 and L1-L34 (the numbering system is based upon the position of these r-proteins on two-dimensional electrophoretograms). Of these 54 proteins (there is no L8 r-protein), only 52 are different since L26 is identical to S20 and L7 is the acetylated form of L12, beginning N-acetyl serine; this latter protein is present in several copies (4 or more per ribosome). No extensive sequence similarities have been found among r-proteins apart from these two pairs.

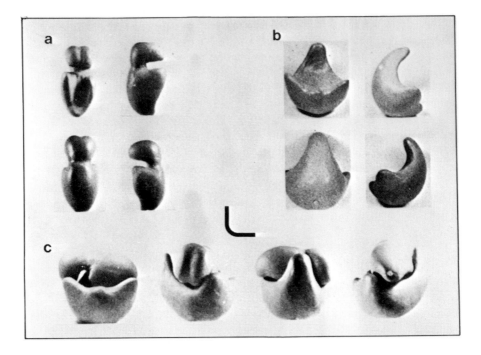

Figure 2.18: A Three-Dimensional Model of the *E. coli* 70S Ribosome

Models of the 30S subunit (a), 50S subunit (b) and 70S ribosome (c) are shown with successive 90° rotations about the centre axis (the 'L'-shaped insert in (c) represents a single tRNA molecule, each arm being only one-third the diameter of the intact ribosome, see Figure 2.14 for tRNA dimensions). The 70S ribosome, of molecular weight 2.7 Mdal (50S, 1.8 Mdal; 30S, 0.9 Mdal), has an estimated particle volume of about 5×10^3 nm³. About two-thirds of the mass of the ribosome consists of RNA: one molecule each of 16S rRNA (1541 nucleotides, 0.50 Mdal), 23S rRNA (2904 nucleotides, 1.00 Mdal) and 5S rRNA (120 nucleotides, 0.04 Mdal). The 16S and 23S, but not the 5S, RNAs contain modified bases and are tightly packed. rRNA both provides a scaffold for assembly of ribosomal proteins as well as participating directly in the translation process (Section 2.2.2 (a)). There are a total 54 r-proteins, of which 52 are non-identical (see text). To date, the primary structures of 19 ribosomal proteins have been determined. The tunnel, and the cleft formed between the central protruberance and the head, protect about 30 nucleotides of mRNA from endoribonuclease attack *in vitro*. This allows the isolation of RNA sequences responsible for polypeptide chain initiation (Section 2.1.2 (ai)). From Stöffler *et al.* (Stöffler, G., Bald, R., Kastner, B., Lührmann, R., Stöffler-Meilicke, M., Tischendorf, G. and Tesche, B. (1980) *Ribosomes: Structure, Function and Genetics*, Chambliss, G., Craven, G.R., Davies, J., Davis, K., Kahan, L. and Nomura, M. (eds.), pp. 171-205, University Park Press, Baltimore) with permission of the publishers and authors

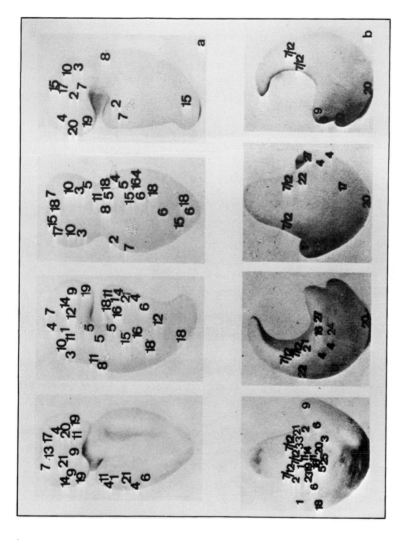

Figure 2.19: Positioning of Ribosomal Proteins on the 30S and 50S Ribosome

In the main, these proteins have been located on the 30S subunit (a) and 50S subunit (b) by immune electron microscopy and by protein cross-linking. The former technique involves the examination of ribosome dimers formed by binding a purified, bivalent IgG-antibody, specific to a single r-protein. From Stöffler *et al.* (Stöffler, G., Bald, R., Kastner, B., Lührmann, R., Stöffler-Meilicke, M., Tischendorf, G. and Tesche, B. (1980) *Ribosomes: Structure, Function and Genetics*, Chambliss, G., Craven, G.R., Davies, J., Davis, K., Kahan, L. and Nomura, M. (eds.), pp. 171-205, University Park Press, Baltimore) with permission of the publishers and authors 'armchair'. A channel is thus created (about 10 nm) through which the mRNA strand is presumed to proceed. There is overall asymmetry.

The majority of ribosomal proteins have been located on this three-dimensional structure (Figure 2.19). L7/L12, for example, seems to be associated with the central protuberance. Localisation of r-proteins situated close to rRNA is complicated by the dependence of RNA-protein interaction on the method used to isolate the nucleic acid, and the protein conformation: numerous proteins appear to be associated with rRNA. Reassociation of ribosome subunits from their constituents is a powerful tool for the analysis of functional domains in protein biosynthesis. Despite extensive studies, the role of the individual components is, as yet, unclear (some pertinent results are discussed below and in Section 2.2.2 (a)). The enzyme peptidyl transferase, which is responsible for both peptide bond formation and release of the completed polypeptide at termination (Section 2.2.2 (a)), is located on the large subunit. About 11 r-proteins of this subunit have been implicated in peptidyl transferase action (though some are likely to be neighbours rather than components of the transferase centre, this enzyme is, clearly, an integral part of the 50S subunit). These proteins are situated in a band close to, or at, the 50S-30S subunit interface.

The main genes involved in translation are shown in Figure 2.20/Table 2.7.

2.2.2 The Mechanism of Translation

Polypeptide chain elongation occurs in the same direction as RNA chain extension: the ribosome binds to the initiator region at the 5'-end of mRNA (or 5'-proximal region in the case of an internal initiation event on, say, a polycistronic message) and moves in the 5'-3' direction. Since the 5'-terminal region of the message encodes the protein's N-terminus, amino acid polymerisation entails the sequential addition of residues to the carboxy-terminal end. This asymmetric process is completed upon translation of the carboxy-terminal residue and release of the product from the polysome. Note that whereas transcription must occur upon the genetic material, RNA translation need not suffer such restriction; it may occur anywhere within the bacterium (where the participants have access) as well as concomitantly with DNA transcription.

mRNA-dependent protein biosynthesis is a highly complex reaction. It involves, in addition to the multicomponent ribosome, a 'language' of 20 different 'words' — expressed in amino acids rather than ribonucleotides. There is a specific aminoacyl-tRNA synthetase for each residue, and between 40-60 different tRNAs. In addition, cytosolic protein factors are required at different stages of translation (see below). Not surprisingly, therefore, a defined cell-free protein synthesising system has not as yet been developed.

Antibiotics, and antibodies directed against ribosomal proteins, have played a key role in the analysis of protein biosynthesis. Ribosomes depleted of particular constituents (through dissociation-reassociation procedures) are also employed. The overall elongation reaction can be studied by following the conversion of

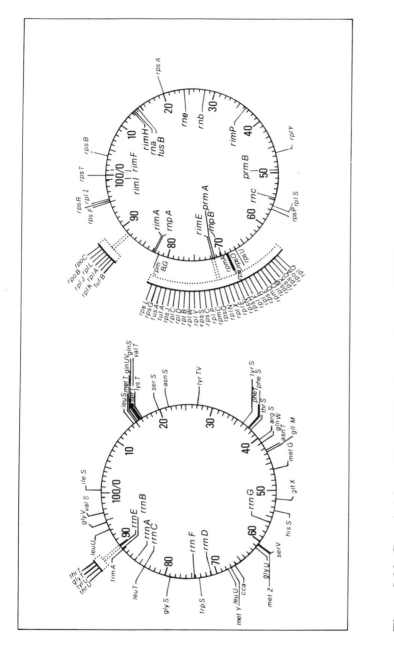

Figure 2.20: Genetic Map of *E. coli* Showing the Genes of the Translational Apparatus

Left, tRNA and synthetase genes (*external*) and ribosomal RNA operons (*internal*); *right*, ribosomal protein genes (*external*) and genes for accessory factors (*internal*). The nomenclature for these genes is given in Table 2.7 (see also Section 9.3). Note how the loci for rRNA and r-proteins are clustered around the origin of replication at 83 min. Map locations from Bachman and Low (Bachman, B.J. and Low, K.B. (1980) *Microbiol. Rev., 44*, 1-56)

Table 2.7 Genetic Nomenclature for Components of the Translation Apparatus[a]

Gene Symbol	Gene Product
Ribosome Components	
rplA to rplY	50S ribosomal subunit protein L1 to L25
rpmA to rpmG	50S ribosomal subunit protein L27 to L33
rpsA to rpsU [b]	30S ribosomal subunit protein S1 to S21
rrnA to rrnG	rRNA polycistronic operon
rrf	5S rRNA (encoded by rrn)
rrl	23S rRNA (encoded by rrn)
rrs	16S rRNA (encoded by rrn)
Accessory Factors	
cca	tRNA nucleotidyl transferase
fusA	translation factor EF-G
prmA, B	methylation of ribosomal proteins
rimA to rimJ	maturation of ribosome
rna to rnp	ribonuclease (I to P)
trmA to trmD	tRNA methyltransferase
tufA, B	translation factor EF-Tu
Transfer RNAs and their charging [c]	
alaT, U	alanine tRNA$_{1B}$
valT	valine tRNA$_1$
etc.	
alaS	alanyl-tRNA synthetase
valS	valyl-tRNA synthetase
etc.	

a   From Bachmann and Low (1980)
b   The following components of the rRNA polycistronic operons have been identified to date (see also Section 9.3.1 (b)).

rrnA :	*rrsA,*	*alaT,*	*ileT,*			
rrnB :	*rrsB,*	*gltT,*		*rrlB,*	*rrfB,*	
rrnC :	*rrsC,*	*gltU,*		*rrlC,*	*rrfC,*	*aspT,* *trpT*
rrnD :	*rrsD,*	*alaU,*	*ileU,*	*rrlD,*		
rrnE :	*rrsE,*	*gltV*				
rrnF :	*rrsF,*					
rrnG :	*rrsG,*			*rrlG,*		

c   Generally, capitals of T and further in the alphabet indicate a tRNA gene, whereas S and earlier letters are used for synthetases.

acid soluble, radiolabelled amino acids into the insoluble, polymeric form; it requires partially-purified cell extracts, and ribosomal and tRNA fractions. Such *in vitro* systems are routinely primed with defined synthetic RNA templates or the viral genome of small RNA phages such as MS2 or Qβ (Section 5.1). The latter are particularly useful since these viral 'messages' are readily available in high purity and encode only three proteins.

Interactions of the translational components, divorced from the elongation reaction, are amenable to analysis making use of the triplet-binding assay (first devised for elucidation of the genetic code). In this test system, trapping of charged tRNAs on nitrocellulose filters is dependent upon a functional 70S ribosome and a cognate triplet, a synthetic oligoribonucleotide consisting of just three bases; in the absence of the appropriate trinucleotide, ribosomes, but not tRNA, bind. Similarly, association of the translational initiation factor, IF-2, to the 30S or 70S ribosome requires fMet-tRNA$_f^{Met}$ and the AUG codon. A modified version of the binding assay, the formylmethionine release assay, monitors translational termination by release of fMet from an fMet-tRNA$_f^{Met}$· AUG · ribosome intermediate in the presence of terminator trinucleotides. Retention of termination-specific proteins, release factors, in the triplet binding system is also dependent upon terminator trinucleotides.

Approximately one amino acid in 10^4 polymerised is inserted incorrectly. Two main processes — specific aminoacylation and ribosomal discrimination of charged tRNAs — are responsible for this translational fidelity. Accuracy in the aminoacyl-tRNA synthetase reaction is necessary for overall fidelity since mischarging would result in the insertion of an incorrect amino acid; several different proof-reading schemes have been proposed to explain this. The ribosomal complex itself, in particular the decoding region, appears to screen charged tRNAs (see Section 3.3.3). Codon-anticodon interaction through base pairing, clearly, plays a major role in the accuracy of reading. The invariant tRNA sequence, TψCG,[16] of the TψC loop and the complementary tetranucle otide, AAGC (residues 46-43), on 5S rRNA have been implicated in binding of charged tRNAs (at the A site, see below). Moreover, the ternary tRNA structure is also likely to be important in tRNA discrimination. The ribosomal proteins S1, S4, S5, S12 and S17 (among others) and magnesium ions are involved in controlling translational fidelity.

(a) The Ribosome Cycle. Synthesis of protein upon mRNA templates is, like RNA production, a processive reaction. Translation can, thus, be described in terms of three main steps, polypeptide chain initiation, elongation and termination. The site on mRNA at which the ribosome binds to form a stable complex defines the initiator region; it encodes the amino-terminus and carries, in addition to the initiation codon, specific sequences involved in initiation. Translocation of the ribosome along the template, concomitant with codon reading and amino acid insertion, ensures elongation of the nascent peptide chain. Translational termination is required for release of the completed polypeptide from the ribonucleoprotein complex; the message is left unchanged though it is highly susceptible to digestion by endogenous ribonucleases, particularly in the ribosome-free state. Ribosome release and/or dissociation into its component subunits is necessary for translation of subsequent mRNA molecules. The *ribosome cycle* allows efficient use of the translational apparatus

16. All three bases, T, ψ, C, are hydrogen bonded, either internally or to the D-loop in the tertiary structure (see Figure 2.13), suggesting that a conformational change is required for interaction with a tRNA sequence. Recent evidence suggests that the 5S rRNA, itself, may not be accessible without rearrangement.

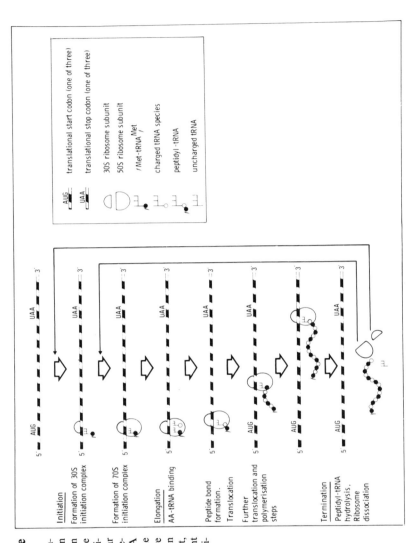

Figure 2.21: The Ribosome Cycle

Polypeptide synthesis on RNA templates is a processive reaction (Section 2.1.2 (a)) and, thus, can be described in terms of three main steps: polypeptide chain initiation, elongation and termination. Such a process is cyclic in so far as the translational machinery is 're-usable', both with the original mRNA template and other RNA species. In the present scheme, AUG and UAA are shown as the initiation and termination codons, respectively. There are, in fact, three different triplets for each event (initiation, AUG, GUG, UUG; termination, UAG, UAA, UGA)

(Figure 2.21). The cytosol factors involved in chain initiation (IF-1, IF-2 and IF-3), elongation (EF-Tu, EF-Ts and EF-G) and termination (RF-1, RF-2 and RF-3) are also re-utilised. Elucidation of the mechanism of protein biosynthesis has been greatly aided by the discovery of antibiotics that block specific stages in translation (Section 2.2.2 (d)).

(i) Initiation. Polypeptide chain initiation appears to occur in discrete steps: the 30S subunit, mRNA and N-formylmethionyl-tRNA$^{Met}_f$ (fMet-tRNA$^{Met}_f$, see

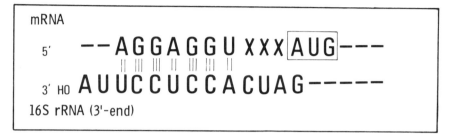

Figure 2.22: The Translational Initiation Sequence: Ribosome-Protected Regions and their Complementarity with the 3'-Terminus of 16S rRNA
Approximately 30 ribonucleotides are protected from ribonuclease digestion in a 70S ribosome initiation complex. Complementarity between this ribosome binding sequence and the 3'-terminus of the 16S rRNA varies between 3 and 9 bases, centred about 10 nucleotides from the initiation triplet. The sequence presented is, therefore, a consensus one: (A)GGAG(G) (or components thereof) is most commonly found (to date, 74 different mRNA initiator regions have been sequenced). That such an mRNA-rRNA interaction occurs has been verified by isolating actual complexes between the Shine-Dalgarno sequence and the 3'-end of 16S rRNA

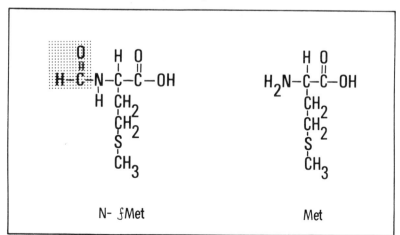

Figure 2.23: Structure of N-Formylmethionine
Methionine charged to the initiator tRNA, tRNA^{Met}, is subsequently formylated by the cytosolic enzyme, formyl-THF-met-tRNA transformylase (making use of 10-formyl-tetrahydrofolate). This effectively blocks the amino group of the N-terminus of all nascent polypeptide chains in prokaryotes. The blocking group, or even the whole methionyl residue, may be removed at a later stage (see Post-Translational Modification, Section 2.2.2 (b))

Figures 2.15 and 2.23) combine in the *30S initiation complex*, which then allows binding of the 50S subunit to form the *70S initiation complex*. Thus, the initiation process itself is reponsible for bringing the small and large subunits together (assembly is, apparently, impeded by IF-1 and IF-3). Moreover, since efficient initiation seems to require the free subunits, dissociation of 70S ribosomes at termination would appear to be an absolute requirement for the supply of a pool of subunits to be used in translational initiation. The specificity of translational initiation is important both for ribosome binding at an initiator AUG, GUG or UUG triplet and for alignment of its decoding region in the correct reading-frame.

Ribosome binding involves a specific interaction between two RNA species, the 16S rRNA of the small subunit and the initiation region on the message. Sequence analysis of this region from different prokaryotic messages (about 30-40 ribonucleotides are protected from endoribonuclease attack by the 70S (or 30S) initiation complex) suggests a common polypurine stretch, A G G A G G U (termed the *Shine-Dalgarno* sequence after its discoverers), centred approximately 10 nucleotides from the initiation codon. [Once again (see footnote to page 61), this is a consensus sequence; variation is observed between Shine-Dalgarno sequences.] It is complementary to the 3′-terminal sequence of 16S rRNA (Figure 2.22). The apparently redundant RNA of messenger leader sequences is, consequently, required to effect ribosomal positioning for accurate and efficient translational initiation (the involvement of RNA secondary structure in translational control is discussed below).[17] In general, translation is initiated within the leader sequence rather than at the 5′-end. The Shine-Dalgarno sequence presumably allows unambiguous distinction between initiation and internal AUG, GUG or UUG codons. Formation of the 30S initiation complex has an absolute requirement for fMet-tRNA$^{Met}_f$ (Figure 2.24). A second RNA-RNA interaction, namely codon-anticodon pairing, would, thus, appear to be involved in stabilisation of the 30S complex.

Specific cytosol proteins, the initiation factors IF-1, IF-2 and IF-3 (molecular weights 9000, 115 000 and 22 000, respectively) are also required; these proteins are not normal residents of active ribosomes. One molecule of each acts co-operatively to allow formation of the 30S initiation complex. The exact order of events — whether the 30S subunit first binds to mRNA or fMet-tRNA$^{Met}_f$ — and the precise role of each initiation factor is still unclear (a possible sequence is outlined in Figure 2.24). The initiation factors are re-used upon dissociation. The large S1 protein (68 000 daltons in size) is also involved in binding of the 30S subunit to mRNA, possibly by stabilising base pairing between the 3′-end of the 16S rRNA and the complementary region on mRNA. The finding that S1,

17.   The initiation codon alone is, in fact, sufficient for polypeptide chain initiation, albeit at a low level, under certain conditions. Consider, for example, P_M-promoted expression of the phage *lambda* repressor producing a message whose 5′-terminus is the start codon itself (See Section 10.2.2 (b)). The presence or absence of the Shine-Dalgarno sequence as well as variation in its length may thus exert a fine control on translational initiation.

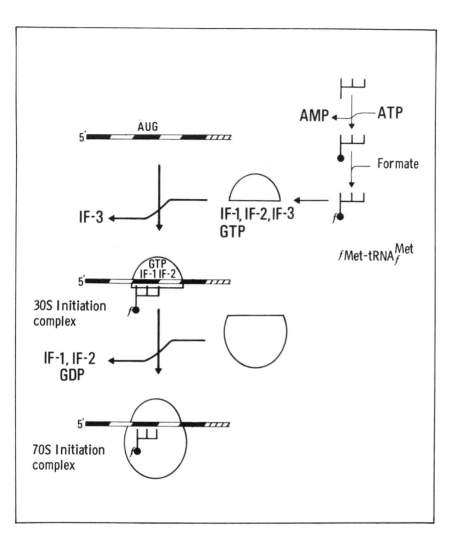

Figure 2.24: A Possible Sequence of Events Involved in Translational Initiation

It is still unclear whether the small subunit binds first message or the initiator tRNA, FMet-tRNA $_f^{Met}$. Note the formylase reaction prior to the formation of the 30S initiation complex (the structure of N-formylmethionine is given in Figure 2.23). The translational initiation factors IF-1 and IF-3 have been implicated in the maintenance of free ribosomal subunits and in recycling of IF-2; IF-2·GTP is necessary for stable fMet-tRNA $_f^{Met}$ binding (all three initiation factors are extra-ribosomal in nature and are equimolar, present at about one copy per 7 ribosomes). The scheme shows IF-3 leaving the 30S subunit on formation of the 30S pre-initiation complex, and IF-1 and IF-2 dissociating during junction with the 50S subunit. Recycling of IF-2 requires GTP hydrolysis. To date, only AUG (and GUG very rarely) have been found in mRNA initiator regions (see also Figure 2.2.2). Internal AUG, GUG and UUG codons have been used for 'restarts' — reinitiation of translation after premature termination at a nonsense codon.

S12 (molecular weight 13 600) and the 16S rRNA affect the species specificity of bacterial ribosomes supports the contention that the ability to discriminate between initiation signals on mRNA strands resides in the small subunit. Release of IF-2 upon GTP hydrolysis, after binding of the 50S subunit, primes the 70S initiation complex for polypeptide chain elongation.

(ii) *Elongation.* Two tRNA binding sites on the 70S ribosome, associated mainly with the large subunit, have been inferred from the studies using specific translational inhibitors: the *aminoacyl-tRNA acceptor* (or A) *site* and the *peptidyl-tRNA* (or P) *site.* Charged tRNAs normally enter the ribosomal complex via the A (substrate) site, whereas the growing polypeptide chain is carried in the P (product) site attached to a tRNA molecule (see below). There is evidence to suggest that the initiator tRNA, which carries a methionine residue with blocked N-terminus, resides in the P site of the 70S initiation complex. It is unique in this respect.

On this two-site model, subsequent aminoacyl-tRNAs enter the A site as a ternary AA-tRNA · EF-Tu · GTP complex (binding of EF-Tu to AA-tRNA requires a free N-terminus). The elongation factor EF-Tu (about ten molecules of this 44 000 daltons protein are present per ribosome) is required in conjunction with GTP for codon-anticodon pairing. Hydrolysis of GTP allows dissociation of the EF-Tu · GDP binary complex. EF-Tu · GTP is regenerated via the EF-Tu · EF-Ts intermediate in a reaction involving a second elongation factor, EF-Ts, with a molecular weight of about 30 000 (Figure 2.25); EF-Tu is *un*stable relative to the thermostable factor EF-Ts.) Binding of the first charged tRNA in the A site 'saturates' the ribosome since fMet-tRNA$_f^{Met}$ is also contained within the P site. The peptidyl transferase centre catalyses peptide bond formation between N-formylmethionine and the A-site residue. The reaction is restricted to the carboxy-group of methionine since it carries a blocked N-terminus. It involves cleavage of the carboxyl ester linkage between N-formylmethionine and its cognate tRNA, tRNA$_f^{Met}$, and fMet transfer to the A site, to form a dipeptidyl-tRNA species. A deaminoacylated donor tRNA (in this case tRNA$_f^{Met}$) remains in the P site (Figure 2.26). Binding of elongation factor EF-G (about 80 000 daltons in size) in the presence of GTP promotes displacement of this uncharged transfer RNA molecule by the newly formed dipeptidyl-tRNA. Translocation of the ribosome along the message by one codon unit in the 5'-3' direction ensures that the adjacent triplet lies in the A site. EF-G is released for recycling upon GTP hydrolysis.

Further elongation of the peptide chain occurs by repetition of this epicycle: binding of an AA-tRNA to the A site, peptide bond formation and translocation. At the beginning of each epicycle, the growing peptide chain is contained within the P site as peptidyl-tRNA while charged tRNAs enter into the A site. Two molecules of GTP[18] are consumed for every peptide bond formed. Polypeptide

18.   Three extraribosomal translation factors with ribosome-associated GTPase activity, IF-2, EF-Tu and EF-G, interact with L7/L12. Note that in the case of A-site binding and translocation, GTP hydrolysis is subsequent to the mechanical process and allows release of the carrier protein (EF-Tu and EF-G, respectively).

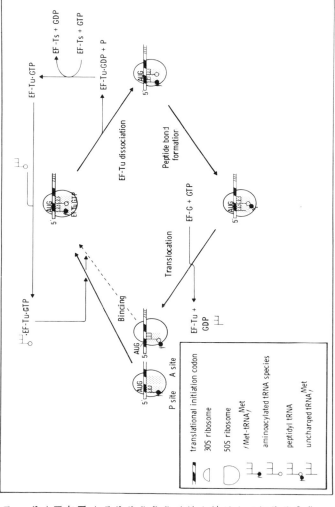

Figure 2.25: Polypeptide Chain Elongation

tRNAs other than the initiator species enter the ribosome as an AA-tRNA·EF-Tu·GTP ternary complex. The actual transfer of the AA-tRNA to the acceptor (A) site on the ribosome is accompanied by GTP hydrolysis and release of EF-Tu·GDP. EF-Tu·GTP is reformed in a reaction involving an EF-Tu·EF-Ts intermediate. Peptide bond formation is catalysed by the peptidyl transferase centre on the 50S ribosomal subunit (see Figure 2.26). Translocation involves the movement of peptidyl-tRNA from the A site to the product (P) site and movement of the ribosome through 3 ribonucleotides (in the 5' → 3' direction) such that the adjacent triplet appears in the A site; this reaction is catalysed by the translational elongation factor EF-G. Release of the deacylated tRNA initially present in the P site and GTP hydrolysis accompany the reaction. One turn of this epicycle (and hydrolysis of two molecules of GTP) is necessary for the polymerisation of each amino acid

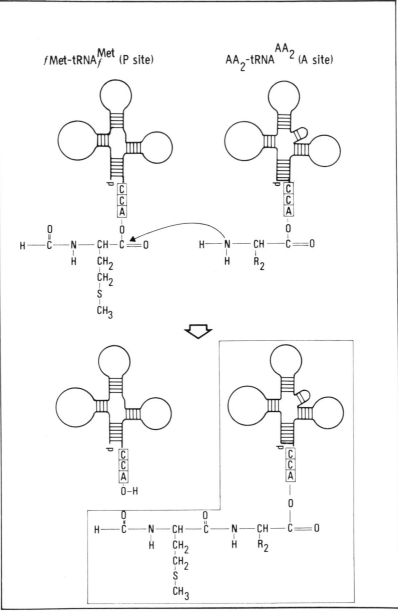

Figure 2.26: Peptide Bond Formation

The peptidyl transferase centre sited on the 50S ribosomal subunit is responsible for this reaction (as well as hydrolysis of the polypeptide chain away from the peptidyl-tRNA at the termination stage in translation; see Figure 2.27). The anticodons of tRNAs in both the A and P sites appear to be close together, as would be necessary for peptide bond formation. Simultaneous codon-anticodon pairing at both the A and P sites would require the mRNA to kink between the two adjacent codons. There is support for such pairing chain elongation continues at a rate of about 16-20 amino acids per second until an in-phase termination codon is encountered in the A site (or, presumably, to the end of the message if no such codon is present).

(iii) *Termination.* The bacterial genome does not normally code for tRNAs with anticodons complementary to chain termination triplets (but see Nonsense Suppression, Section 3.3.2). Thus, no amino acid is inserted into the nascent polypeptide chain at these sites. The presence of an in-phase termination codon triggers, instead, polypeptide chain termination. Though a single terminator is, apparently, sufficient for translational termination, 3'-distal mRNA sequences infrequently have two terminators in tandem. The dependence of nonsense suppression upon reading context suggests that other nucleic acid sequences — presumably adjacent to the terminators — might influence natural termination (a possible interaction between chain termination codons and the terminal 3'-sequence of 16S rRNA is discussed in Section 3.3.2). In addition, three cytosol proteins, termed release factors, are required. The ribosomal proteins S3, S9, S11 and L7/L12, and the 3'-terminus of 16S tRNA, have been implicated in RF binding. RF-1 and RF-2 (molecular weights 44 000 and 47 000, respectively) are codon-specific since the former responds to UAA and UAG codons while the latter functions at UAA or UGA. RF-3 and GTP promote their action.

The fate of the termination complex, mRNA · ribosome · peptidyl-tRNA · RF GTP, is unclear. The presence of a termination codon, and RF binding, signal some modification in the peptidyl transferase centre on the 50S subunit that results in hydrolysis of the terminal aminoacyl-tRNA ester bond (rather than the catalysis of peptide bond formation, Figures 2.25 and 2.27). This allows release of the nascent polypeptide chain[19] and, presumably, the 70S ribosome; RF release is possibly aided by GTP hydrolysis. Little is known about the sequence of events leading to ribosome dissociation.

(b) Post-Translational Modification. Translation *per se* is by no means the final stage of gene expression. Polypeptide folding and subunit association is, clearly, important for function. The tertiary structure may be formed partially during the translation process, itself, though renaturation studies indicate that folding of protein species into an active configuration does not require participation of the ribosomal complex nor other translation factors (that is, higher order structures are dictated by the amino acid sequence, alone). Conformational change at the post-translational stage is also important in the regulation of allosteric proteins (see *lac* Repressor, Section 9.1.2 (a)). In addition, once synthesised, prokaryotic gene products may be processed. Such reactions are, on the whole, essentially irreversible, involving covalent modification of residues or peptide bond cleavage. Processing may be necessary for function, or part of the cellular

---

19. In natural termination, the completed protein is released into the cytosol. If a ribosome encounters a termination codon prior to the normal terminator (because, say, of a mutational event) premature termination occurs with the result that only part of the gene product is synthesised. The effect of nonsense mutations on gene expression and their suppression is discussed in Chapter 3.

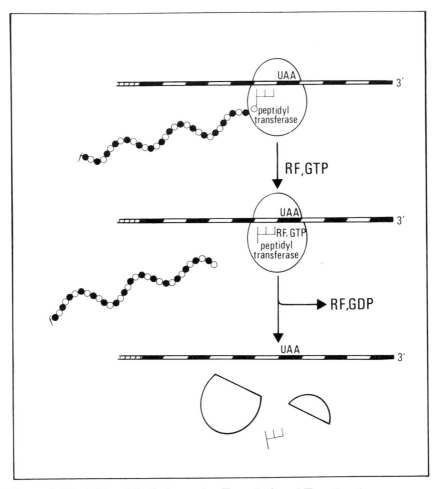

Figure 2.27: A Possible Scheme for Translational Termination
The actual order of events is unclear. Note that there are three release factors, RF-1, RF-2 and RF-3.
While RF-1 is active on UAR, RF-2 acts on URA (R is a purine base); RF-3 stimulates their action.
UAA and UGA appear to be used with equal frequency (and more often than UAG); tandem
termination triplets are rarely found

scavenging system that removes non-functional or lethal polypeptides.

Covalent modification of certain residues occurs after their incorporation into
the nascent polypeptide chain. Consider, for example, the removal of the amino-
terminal formyl group or N-formylmethionine, itself. Though prokaryotic
protein biosynthesis is initiated by fMet-tRNA$_f^{Met}$, many bacterial proteins lack a
blocked N-terminus (β-galactosidase, the *lacZ* gene product, for example,
begins with threonine though the first codon is AUG). This modification must,
therefore, occur at the post-translational stage either subsequent to initiation or
on completion of the polypeptide chain. Whereas the amino residues, aspara-

gine and glutamine, may spontaneously deamidate, acetylation of the ribosomal protein L12 (to give the L7 form) apparently occurs by a bacterial-coded system (the essential role of methylation in bacterial chemotaxis is another case in point). Covalent modification may also play a role in the control of gene expression, as exemplified by adenylation of the α subunit of RNA polymerase upon bacteriophage T4 infection.

Peptide-bond cleavage, the second major form of post-translational modification, is readily apparent in *E. coli:* approximately 1-2 per cent of cellular protein is degraded per hour during normal bacterial growth (possibly triggered by an endoproteolytic event). Large, acidic proteins tend to be removed more rapidly. The marked increase in the degradative rate upon entrance into stationary phase (or during amino acid starvation) suggests an adaptive response that compensates for poor growth conditions by supplying a pool of amino acids for essential cellular proteins. The production of guanosine tetraphosphate (also called 'magic spot I', MSI), pp G pp, mediated by the presence of uncharged tRNAs at the A site, has been implicated in the control of bacterial intracellular protein catabolism (as well as stable RNA accummulation, Section 9.3.2 (b)). Proteolysis acts not only to remove 'unwanted' proteins but also to regulate gene expression. Thus, proteolytic cleavage of the phage *lambda* repressor (mediated by the *recA* gene product, Sections 3.3.2 (a) and 6.2.1 (ai)) allows expression of genes required for phage production and lysis (Section 10.2.3). Moreover, removal of crucial gene products required for continued transcription of specific cistrons, further modulates gene expression (see *lambda N* gene, Section 10.2.3 (b)). Finally, protein cleavage plays a key morphogenetic role in the assembly of certain bacteriophages.

(c) Control. Initiation is probably the rate-limiting step in translation. The polypeptide chain initiation frequency (4-20 per minute) is low compared with the elongation rate (16-20 amino acids per second). The latter is invariant, at least with respect to growth conditions, whereas there is some evidence that the initiation rate is message dependent. Moreover, some protein factors in translational initiation may be present in limiting amounts at particular growth rates. The necessity of both specific initiation sequences on messenger RNA and protein factors might suggest the possibility of control of protein biosynthesis by variation in these requirements in a manner analogous to transcriptional regulation. There are, however, few clear instances of translational regulation of bacterial messages (but see Section 9.3.2 (c)). Messenger biogenesis (and its degradation) is the main form of prokaryotic gene control (particularly of *E. coli* cistrons) though there is now thought to be a link betwen RNA and protein production (Section 8.1.2 (a)).

Some phages clearly regulate their expression at the translational level. This control is mediated either by protein factors or by RNA secondary structure. In the first case, phage proteins act as translational repressors. For example, the coat protein of the RNA phage R17 inhibits production of the viral replicase, apparently by maintaining the initiator region for this cistron in a secondary structure unfavourable for ribosomal initiation (note that this only happens when coat protein accumulates and when no further replicase is required). There is now a large body of evidence indicating that the RNA structure, itself, is involved in RNA phage translation. The fact that only certain nonsense mutations — a genetic change that gives rise to a chain termination triplet (Section 3.1.1 (c)) — in the R17 coat protein cistron exert polar effects on expression of the distal replicase protein may be explained in terms of the

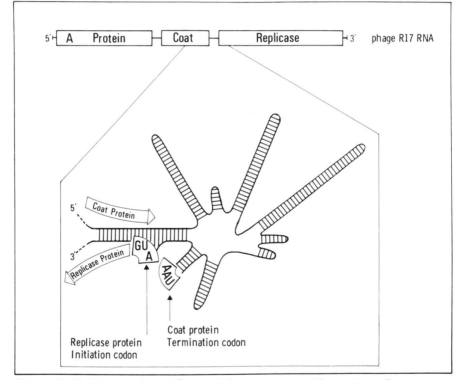

Figure 2.28: Translational Control; Importance of Secondary Structure
R17 is an RNA phage, its genetic material consisting of RNA rather than double-or single-stranded DNA (see Section 5.1.1 (b)). The viral genome, about 3600 ribonucleotides in length, encodes four separate proteins, the so-called maturation or A protein, the coat protein, replicase and lysis protein. (The latter, encoded in a+1 frame in the coat and replicase genes, is not shown for clarity. The replicase is responsible for production of the single-stranded RNA molecule during infection; the replicase complex consists of the phage-coded protein, and three bacterial proteins involved in translation: EF-Tu, EF-Ts and the ribosomal protein, S1). This RNA molecule acts both as a template for phage reproduction and as a long polycistronic message for translation. Secondary structure in RNA phage genomes appears to be important in regulating template function. Thus, in the 'flower' model, translation into the flower, leading to melting of the double-stranded region encoding the coat protein, is necessary to uncover the translational initiation region for the replicase. Mutations, such as nonsense lesions, which prevent complete coat protein production also stop replicase synthesis. After predicted secondary structure of the closely-related RNA phage MS2 by Min Jou *et al.* (Min Jou, W., Haegeman, G., Ysebaert, M., and Fiers, W. (1972) *Nature, 237*, 82-88)

'flower' model of R17 RNA secondary structure (Figure 2.28). Translation into the 'flower' is necessary to destroy a helical region in order to allow binding of the ribosome at the replicase initiatior region. Nonsense mutations that terminate translation in the coat protein gene prior to this helical region, therefore, impede replicase production (see also Polarity and its Suppression, Section 8.1.2 (a)). In conclusion, both RNA secondary structure as well as phage proteins allow a temporal sequence of viral protein synthesis. Moreover, it now appears that the three-dimensional conformations of both phage and bacterial messages are utilised to prevent ribosome recognition of non-initiator AUG, GUG or UUG codons.

(d) *Antibiotic Inhibitors of Translation.* Numerous antibiotics interfere with the translation process. The translation complex, particularly the ribosome, is a prime target for these antibacterial agents. Elucidation of the mechanism of protein biogenesis has been greatly helped by antibiotics that block specific steps in this process (though some substances elicit pleiotropic responses) and has allowed an understanding of their mode of action. The majority of antibiotics inhibit polypeptide chain elongation. In the main, the large subunit is the target of these inhibitors (for example, puromycin and choramphenicol), though both the small subunit (streptomycin) as well as translation factors (fusidic acid) are also susceptible.

Puromycin is similar in structure to, and presumably competes with, the aminoacyl-adenyl terminus of aminoacyl-tRNA (Figure 2.29). It is incorporated into the growing polypeptide chain (linked by its amino-side group to the carboxy-terminus). However, peptidyl-puromycin, unlike peptidyl-tRNA, does not contain an equivalent hydrolysable ester bond nor can it bind tightly to the ribosome. Thus, once formed, the transpeptidation process is inhibited, and nascent peptides, in which the C-terminus is blocked by puromycin, are released; they tend to be degraded by endogenous proteases. The mode of action of puromycin has allowed an operational definition of the P site — it is the puromycin-sensitive target since only donor peptidyl-tRNA is reactive. The broad-spectrum antibiotic chloramphenicol (Figure 2.29), on the other hand, acts at the peptidyl transferase centre to prevent peptide bond formation. Certain bacterial plasmids code for an enzyme that inactivates this antibiotic, thus rendering resistant strains that carry these small supernumerary chromosomes (Chapter 4).

Antibiotics of the aminoglycoside group are diverse in their action. Streptomycin, the most extensively investigated inhibitor, binds to the 30S ribosomal protein S12. This prevents chain initiation and elongation (possibly by interference with aminoacyl-tRNA binding); it also causes misreading at low drug concentrations, a property independent of the bactericidal function (see below, and Section 3.3.3). Streptomycin binding causes release of peptidyl-tRNA and concomitant polysome breakdown. The fact that mutations conferring streptomycin resistance result in a modified S12 protein and reduced chain growth rate supports the role of this r-protein in polypeptide chain elongation.[20] It has been suggested that the difference in the action of streptomycin at high and low concentrations depends upon the frequency with which a streptomycin molecule encounters an initiating or elongating ribosome, the former event being responsible for the dominant bactericidal effect, i.e. ribosomes to which streptomycin has bound irreversibly form unstable initiation complexes and, thus, prevent streptomycin-free ribosomes from functioning even after removal of the drug. This is in keeping with other data that implicate S12 in polypeptide chain initiation (as well as elongation).

Fusidic acid prevents re-utilisation of the elongation factor EF-G by 'freezing' the binary EF-G · GDP complex on the ribosome: the translocation step is, thus, inhibited. The site of action of this steroid antibiotic has been demonstrated by the isolation of resistant mutants that code for an altered elongation factor.

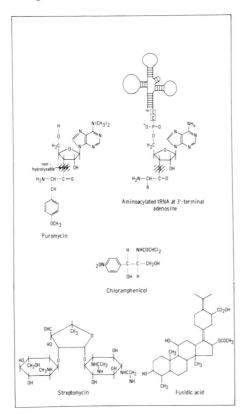

Figure 2.29: Structure of Some Antibiotics that Inhibit Translation

20.   These mutations are chromosomal lesions. There is also plasmid-mediated resistance (see Section 4.2.1) which functions through a detoxification system, analogous to the plasmid-coded inactivation of chloramphenicol.

Bibliography

Crick, F.H.C. (1970) 'Central dogma of molecular biology', *Nature, 227,* 561-563.
Jacob, F. and Monod, J. (1961) 'Genetic regulatory mechanisms in the synthesis of proteins', *J. Mol. Biol., 3,* 318-356.

2.1 DNA Transcription

Chamberlin, M.J. (1976) 'RNA polymerase — an overview', in R. Losick and M. Chamberlin (eds.), *RNA Polymerase* (Cold Spring Harbor Laboratory, New York), pp. 17-67.
Lathe, R. (1978) 'RNA polymerase in *Escherichia coli*', *Curr. Topics Microbiol. Immunol., 83,* 37-92.

2.1.1 The Elements of Transcription

Bautz, E.K.F. (1976) 'Bacteriophage-induced DNA-dependent RNA polymerase', in R. Losick and M. Chamberlin (eds.), *RNA Polymerase* (Cold Spring Harbor Laboratory, New York), pp. 273-284.
Burgess, R.R., Travers, A.A., Dunn, J.J. and Bautz, E.K.F. (1969) 'Factor stimulating transcription by RNA polymerase', *Nature, 221,* 43-46.
Burgess, R.R. (1976) 'Purification and physical properties of *E. coli* RNA polymerase', in R. Losick and M. Chamberlin (eds.), *RNA Polymerase* (Cold Spring Harbor Laboratory, New York), pp. 69-100.
Krakow, J.S., Rhodes, G. and Jovin, T.M. (1976) 'RNA polymerase: Catalytic mechanisms and inhibitors', in R. Losick and M. Chamberlin (eds.), *RNA Polymerase* (Cold Spring Harbor Laboratory, New York), pp. 127-157.
Yura, T. and Ishihama, A. (1979) 'Genetics of bacterial RNA polymerases', *Ann. Rev. Genet., 13,* 59-97.
Zillig, W., Palm, P. and Heil, A. (1976) 'Function and reassembly of subunits of DNA-dependent RNA polymerase', in R. Losick and M. Chamberlin (eds.), *RNA Polymerase* (Cold Spring Harbor Laboratory, New York), pp. 101-125.

2.1.2 The Mechanism of Transcription

(a) *The transcription cycle*

Adhya, S. and Gottesman, M. (1978) 'Control of transcription termination', *Ann. Rev. Biochem., 47,* 967-996.
Chamberlin, M.J. (1976) 'The selectivity of transcription', *Ann. Rev. Biochem., 43,* 721-775.
Doi, R.H. (1977) 'Role of ribonucleic acid polymerase in gene selection in prokaryotes', *Bacteriol. Rev., 41,* 568-594.
Pribnow, D. (1979) 'Genetic control signals in DNA', in R.F. Goldberger (ed.), *Biological Regulation and Development I: Gene Expression* (Plenum Press, New York), pp. 219-277.
Rosenberg, M. and Court, D. (1979) 'Regulatory sequences involved in the promotion and termination of RNA transcription', *Ann. Rev. Genet., 13,* 319-353.
Siebenlist, S., Simpson, R.B. and Gilbert, W. (1980) '*E. coli* RNA polymerase interacts homologously with two different promoters', *Cell, 20,* 269-281.

(b) *Control*

Apirion, D. (1973) 'Degradation of RNA in *Escherichia coli:* A hypothesis', *Molec. Gen. Genet., 122,* 313-322.
Gallant, J.A. (1979) 'Stringent control in *E. coli*', *Ann. Rev. Genet., 13,* 393-415.
Maaløe, O. (1979) 'Regulation of the protein-synthesising machinery — Ribosomes, tRNA, factors, and so on', in R.F. Goldberger (ed.), *Biological Regulation and Development, I: Gene Expression* (Plenum Press, New York), pp. 487-542.
Nierlich, D.P. (1978) 'Regulation of bacterial growth, RNA and protein synthesis', *Ann. Rev. Microbiol., 32,* 393-432.
Perry, R.P. (1976) 'Processing of RNA', *Ann. Rev. Biochem., 45,* 605-629.

(c) *Antibiotic inhibitors of transcription*

Goldberg, I.H. and Friedman, P.A. (1971) 'Antibiotics and nucleic acids', *Ann. Rev. Biochem., 40,*

775-810.

Kersten, H. and Kersten, W. (1974) *Inhibitors of Nucleic Acid Synthesis: Biophysical and Biochemical Aspects* (Springer-Verlag, New York).

Sarin, P.S. and Gallo, R.C. (eds.) (1980) *Inhibitors of DNA and RNA Polymerases* (Pergamon Press, New York).

Suhadolnik, R.J. (1979) 'Naturally occurring nucleoside and nucleotide antibiotics', *Prog. Nucl. Acid. Res. Molec. Biol., 22,* 193-291.

Werhli, W. (1977) 'Ansamycins: Chemistry, biosynthesis and biological activity', *Topics Curr. Chem., 72,* 21-49.

2.2 RNA Translation

2.2.1 The Elements of Translation

Chambliss, G., Craven, G.R., Davies, J., Davis, K., Kahan, L. and Nomura, M. (eds.) (1980) *Ribosomes: Structure, Function and Genetics* (University Park Press, Baltimore).

(a) *The genetic code*

Grantham, R., Gautier, C., Gouy, M., Jacobzone, M. and Mercier, R. (1981) 'Codon catalog usage is a genome strategy modulated for gene expressivity', *Nucl. Acids Res., 9,* r43-r74. (This list is updated periodically.)

Jukes, T.H. (1978) 'The amino acid code', *Adv. Enzymol., 47,* 375-432.

Umbarger, H.E. (1978) 'Amino acid biosynthesis and its regulation', *Ann. Rev. Biochem., 47,* 533-606.

Yčas, M. (1969) *The Biological Code* (North-Holland, Amsterdam).

(b) *Transfer RNA*

Altman, S. (1978) 'Biosynthesis of tRNA' in S. Altman (ed.), *Transfer RNA* (The MIT Press, Cambridge, MA), pp. 49-77.

Barrell, B.G. and Clark, B.F.C. (1974) *Handbook of Nucleic Acid Sequences* (Joynson-Bruvvers, Oxford).

Celis, J.E. (1980) 'Collection of mutant tRNA sequences', *Nucl. Acids Res., 8,* r23-r29.

Clark, B.F.C. (1979) 'Structure and function of tRNA' in J.E. Celis and J.D. Smith (eds.), *Nonsense Mutations and tRNA Suppressors* (Academic Press, London), pp. 1-46.

Crick, F.H.C. (1966) 'Codon-anticodon pairing: The wobble hypothesis', *J. Mol. Biol., 19,* 548-555.

Feldman, M. Ya. (1978) 'Minor components in transfer RNA: the location-function relationship', *Prog. Biophys. Molec. Biol., 32,* 83-102.

Gauss, D.H. and Sprinzl, M. (1981) 'Compilation of tRNA sequences', *Nucl. Acids Res., 9,* r1-r23. (An up-to-date collection of tRNA sequences is published each year by NAR.)

Goddard, J.P. (1977) 'The structures and functions of transfer RNA', *Prog. Biophys. Molec. Biol., 32,* 233-308.

Kim, S.-H. (1978) 'Three-dimensional structure of transfer RNA and its functional implications', in A. Meister (ed.), *Advances in Enzymology and Related Areas of Molecular Biology, 46* (Wiley, New York), pp. 279-315.

Nishimura, S. (1978) 'Modified nucleosides and isoaccepting tRNA', in S. Altman (ed.), *Transfer RNA* (The MIT Press, Cambridge, MA), pp. 168-195.

Schimmel, P.R. and Söll, D. (1979) 'Aminoacyl-tRNA synthetases: General features and recognition of transfer RNAs', *Ann. Rev. Biochem., 48,* 601-648.

(c) *The ribosome*

Brimacombe, R., Stöffler, G. and Wittmann, H.G. (1978) 'Ribosome structure', *Ann. Rev. Biochem., 47,* 217-249.

Brosius, J., Palmer, M.L., Kennedy, P.J. and Noller, H.F. (1978) 'Complete nucleotide sequence of a 16S ribosomal gene from *Escherichia coli'*, *Proc. Nat. Acad. Sci. USA, 75,* 4801-4805.

Brosius, J., Dull, T.J. and Noller, H.F. (1980) 'Complete nucleotide sequence of a 28S ribosomal RNA gene from *Escherichia coli'*, *Proc. Nat. Acad. Sci. USA, 77,* 201-204.

Erdman, V.A. (1976) 'Structure and function of 5S and 5.8S RNA', *Prog. Nucl. Acid Res. Molec. Biol., 18,* 45-90.

Fellner, P. (1974) 'Structure of the 16S and 23S ribosomal RNAs', in M. Nomura, A. Tissières and P. Lengyel (eds.), *Ribosomes* (Cold Spring Harbor Laboratory, New York), pp. 169-191.

Stöffler, G. and Wittmann, H.G. (1977) 'Primary structure and three-dimensional arrangement of proteins within the *Escherichia coli* ribosome', in H. Weissbach and S. Pestka (eds.), *Molecular Mechanisms of Protein Biosynthesis* (Academic Press, New York), pp. 117-202.

2.2.2 The Mechanism of Translation

(a) The ribosome cycle

Bermek, E. (1978) 'Mechanisms in polypeptide chain elongation on ribosomes', *Prog. Nucl. Acid Res. Molec. Biol., 21*, 63-100.

Caskey, C.T. (1977) 'Peptide chain termination', in H. Weissbach and S. Pestka (eds.), *Molecular Mechanisms of Protein Biosynthesis* (Academic Press, New York), pp. 443-465.

Grunberg-Manago, M., Buckingham, R.H., Cooperman, B.S. and Hershey, J.W.B. (1978) 'Structure and function of the translation machinery', *Symp. Soc. Gen. Microbiol., 28*, 27-110.

Pongs, O. (1978) 'Transfer RNA function in protein synthesis: Ribosome (A-sites and P-sites) and mRNA interactions' in S. Altman (ed.), *Transfer RNA* (The MIT Press, Cambridge, MA), pp. 78-104.

Steitz, J.A. (1979) 'Genetic signals and nucleotide sequences in messenger RNA', in R.F. Goldberger (ed.), *Biological Regulation and Development, I: Gene Expression* (Plenum Press, New York), pp. 349-399.

(b) Post-translational modification

Goldberg, A.L. and Dice, J.F. (1974) 'Intracellular protein degradation in mammalian and bacterial cells', *Ann. Rev. Biochem., 43*, 835-869.

Goldberg, A.L. and St. John, A.C. (1976) 'Intracellular protein degradation in mammalian and bacterial cells: Part 2', *Ann. Rev. Biochem., 45*, 747-803.

Hershko, A. and Fry, M. (1975) 'Post-translational cleavage of polypeptide chains: Role in assembly', *Ann. Rev. Biochem., 44*, 775-797.

Mount, D.W. (1980) 'The genetics of protein degradation in bacteria', *Ann. Rev. Genet., 14*, 279-319.

(c) Control

Lodish, H.F. (1976) 'Translational control of protein synthesis', *Ann. Rev. Biochem., 45*, 39-72.

Weissman, C. (1974) 'The making of a phage', *FEBS Letts., 40S*, 10-18.

(d) Antibiotic inhibitors of translation

Pestka, S. (1977) 'Inhibitors of protein biosynthesis', in H. Weissbach and S. Pestka (eds.), *Molecular Mechanisms of Protein Biosynthesis* (Academic Press, New York), pp. 467-553.

Suhadolnik, R.J. (1979) 'Naturally occurring nucleoside and nucleotide antibiotics', *Prog. Nucl. Acid Res. Molec. Biol., 22*, 193-291.

Vázquez, D. (1979) *Inhibitors of Protein Biosynthesis* (Springer-Verlag, Berlin).

3 Mutation

The genetic material of an organism (be it DNA or RNA[1]) may be damaged during growth by factors in its external or internal environment. A heritable change that permanently affects the chromosome, termed a *mutation*, results in an altered, *mutant* organism. Mutations in a particular gene give rise to a number of different *alleles* of that gene; these homologues may differ in one or more nucleotide pairs. The selective advantage that any one allele imparts to the mutant strain allows its preferential survival. Non-lethal chromosomal lesions (see below) are the source of genetic variation and, thus, provide the basis for evolution.

What is the effect of a mutational alteration? A genetic lesion may inactivate a gene product, or modify its activity to perform a slightly different task: the creation of an entirely new property is a rare event. Loss of function would be expected to occur more frequently since it could result from a number of different lesions (in one or several genes). Change of function, on the other hand, requires a genetic alteration that modifies a specific property without the concomitant gene inactivation.

Lesions that inactivate essential functions (the replicative apparatus, for example, or processes involved in gene expression) are *lethal*. Cellular death may be avoided if a back-up mechanism or second gene copy exists. In contrast, *non-lethal* mutations affect dispensable gene products — dispensable because their absence either can be compensated for externally or does not affect growth. *Auxotrophic* lesions, a major class of non-lethal mutations, are responsible for changed growth requirements. Thus, the mutant organism, termed an *auxotroph*, requires extra nutrients for survival, while the *prototrophic* wild-type strain is maintained by an aqueous minimal medium consisting of only inorganic salts and glucose (Section 1.3.2 (a)). In addition, rather than causing auxotrophy, non-lethal mutations may impart the ability to survive in the presence of external bacteriostatic or bactericidal factors such as antibiotics *(drug resistance)* and bacterial viruses *(phage resistance;* not that all antibiotics are bacteriostatic, nor all phages bactericidal).

The ability to isolate lesions in specific genes, to locate (map) the site of these genetic defects and to determine their effect upon cellular growth, is one

---

1. Certain minute bacterial viruses contain RNA rather than DNA as their genetic material (Chapter 5). Lesions affecting expression of these phages are, therefore, at the level of RNA and not DNA.

This chapter concentrates mainly on the bacterium *E. coli*. The terms, concepts and many of the mutagenic techniques apply equally to other prokaryotes, including bacteriophages.

cornerstone of molecular biology. The role of a gene product in a metabolic pathway, for example, may be readily determined by analysis of mutant strains that fail to express its function (Section 7.1.2(d)). Most importantly, mutant isolation allows the study of gene structure and function (see Chapter 7). Biochemical and genetic analysis of mutant strains is, therefore, a powerful technique for the study of cellular processes at the molecular level.

3.1 Mutation Classification

Mutations may have gross effects, altering extensive portions of the organism's genome. Chromosomal rearrangements such as inversions, duplications or deletions are examples (Figure 3.1). A change that extends over several base pairs

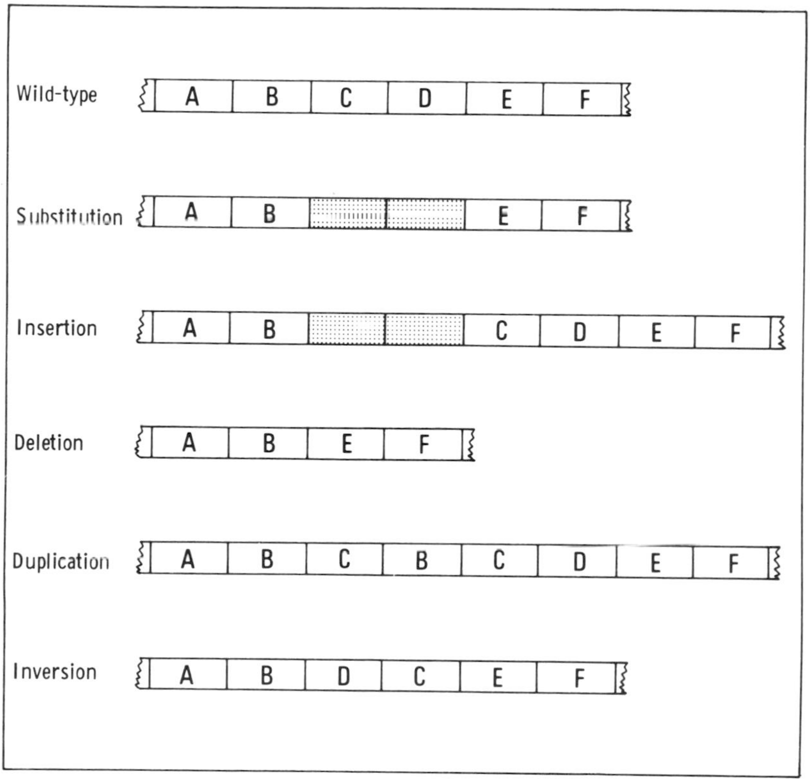

Figure 3.1: Chromosome Rearrangements

The top sequence represents a portion of the wild-type chromosome from which the other sequences are derived. Whereas in a substitution multisite mutation, one region of DNA is replaced by another (shown as a *stippled* sequence), an insertion involves the integration of a novel piece of DNA, which increases the overall size of the genetic element. A segment of nucleic acid is lost in a deletion macrolesion, bringing adjacent markers closer together. Deletions are defined as mutations that fail to recombine with two or more point mutations. In an inversion, a segment of the chromosome is inverted in relation to the neighbouring sequences is termed *multisite*.

A lesion that affects just one (or only a very few) nucleotides is referred to as a *point* mutation. Thus, a *base substitution*, the replacement of one base pair by another without an alteration in the total number of nucleotide pairs, is classified as a point mutation. Similarly, the addition or deletion of one or two base pairs, causing a *frameshift* mutation, also fits into this category. The term *transition* has been used to describe a single base substitution that replaces one base with another of the same type, i.e. a purine base with another purine, or a pyrimidine base with another pyrimidine (Figure 3.2). A *transversion*, on the other hand, is the replacement of a purine base by a pyrimidine (or *vice versa*).

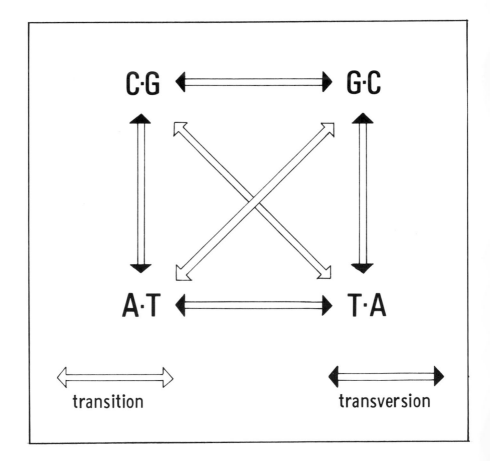

Figure 3.2: Base Substitutions: Transitions and Transversions
The terms transition and transversion were first introduced by Freese (Freese, E. (1959) *Brookhaven Symp., 12,* 63-73)

3.1.1 Types of Point Mutation

A point mutation may occur within DNA that is expressed (either as RNA or protein) or within the regulatory sequences responsible for controlling its transcription or translation. The level of expression but not the sense is altered if a control site carries a lesion. That is, in the presence of a *regulatory mutation*, the same gene product is synthesised albeit at a changed rate. On the other hand, a genetic change within a structural gene affects the function of its product. When the gene that carries the point mutation encodes a non-translatable RNA species, a single ribonucleotide change ensues. In the case of a gene that codes for a protein product, point mutations may be classified into four main types (Figure 3.3 and below).

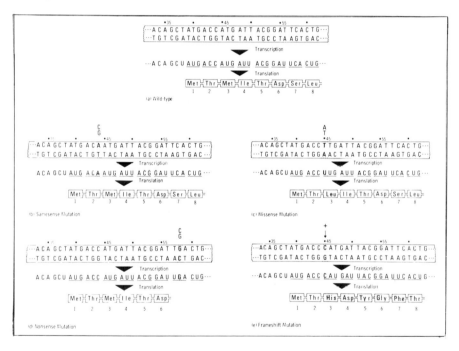

Figure 3.3: Types of Point Mutations in a Structural Gene
The N-terminal region of the *lacZ* gene (Section 9.1) is shown. After Smith-Keary (Smith-Keary, P.F. (1975) *Genetic Structure and Function* (Macmillan, London))

(a) Samesense Mutation. Degeneracy in the genetic code (Section 2.2.1 (a)) permits certain base pair changes to occur without a concomitant amino acid alteration. Such mutations are *cryptic* since they are not readily detectable by normal genetic techniques (the isolation of the DNA region concerned, Section 7.3, and nucleic acid sequencing will, of course, indicate the presence of an altered base).

(b) Missense Mutation. A different amino acid (from wild) is encoded in this case. The mutant polypeptide may be *pseudo wild-type* in character, functioning nearly or as well as the wild-type protein (when the altered polypeptide has activity sufficiently similar to the original wild-type protein, the *neutral* amino acid change will remain undetected: the mutation will be cryptic). Alternatively, it may be partially or totally inactive.

(c) Nonsense Mutation. The base substitution results in the formation of a codon, UAG, *amber*, or UAA, *ochre*, or UGA (Figure 3.4), which is not translated but rather codes for termination of polypeptide biosynthesis; the triplet has no amino acid sense (Section 2.2.1 (a)). Chain-terminating mutations are, in fact, defined by their ability to be suppressed by characterised nonsense suppressors (discussed in Section 3.3.2). Termination occurs prematurely, to produce a truncated, generally inactive polypeptide lacking the normal carboxy-terminus (in the absence of a functional nonsense suppressor, Figure 3.3). A nonsense mutation in a polycistronic operon may affect expression not only of the cistron in which the terminator resides but also of promoter-distal genes. Such lesions are said to be *polar* (Figure 3.5). There are gradients of polarity within a particular gene, though early (promoter-proximal) mutations tend, in general, to be most polar. Translational termination, mediated by nonsense codons, is now thought to induce premature termination of transcription and thus inhibit the production of distal mRNA (this model is discussed in Section 8.1.2 (a)).

(d) Frameshift Mutation. The addition or deletion of one or two base pairs alters the translational reading-frame of the messenger RNA carrying the frameshift mutation (Figure 3.3). This is a consequence of the code being read in blocks of three nucleotides, codons, initiated at a fixed starting point (Section 2.2.1 (a)). The addition of one base-pair (+1 frameshift), or the removal of two, causes the reading-frame to move one base backward. Conversely, the removal of one base pair (or the addition of two) results in the reading-frame shifting one base forward (−1 frameshift). If the number of base-pairs altered is a multiple of three (equivalent to one or more codons), no shift in reading-frame occurs. Rather, one or more amino acid(s) are inserted or deleted at the mutant site (discussed further in relation to frameshift suppression, Section 3.3.2). The amino acid sequence preceding a frameshift lesion is unchanged but all residues past this site are altered, even though the genetic information is retained. Alteration of the reading-frame will frequently 'uncover' nonsense codons that terminate the out-of-phase product. Accordingly, frameshift mutations are also often polar (cf. Nonsense Mutation, above).

3.1.2 Conditional Mutants

The phenotype of a *conditional* mutant is expressed only when the organism is grown under a particular set of conditions. *Temperature-sensitive (Ts)* mutants

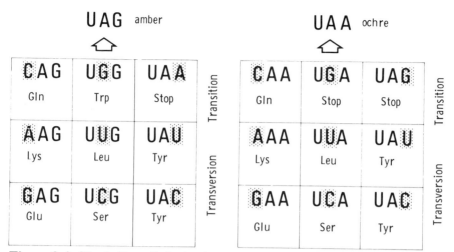

Figure 3.4: Single Base Substitutions that Give Rise to Amber and Ochre Mutations

Stop codons are not present internally in a structural gene (in the correct reading frame, that is). There are, however, 8 amber- and 7 ochre-mutable codons (8 and 7 sense codons, respectively) that give rise to an amber or ochre mutation in a single base change

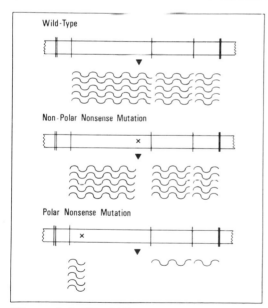

Figure 3.5: The Effect of Non-Polar and Polar Nonsense Mutations on Distal Gene Expression

Nonsense lesions cause premature termination of translation (and the production of truncated polypeptides). Such mutations, when present in a gene that is part of a polycistronic operon, may also reduce/abolish expression of distal genes of the same operon. Polar nonsense mutations lie generally in the N-terminal region (but see Gradients of Polarity in the *lacZYA* Polycistronic Operon, Figure 8.6) and strains carrying nonsense lesions have been most commonly studied. Ts strains grow normally at the low, *permissive temperature* (30-33°C), but exhibit the mutant phenotype at the elevated, *restrictive* (or *non-permissive*) temperature (39-43°C); *cold-sensitive* conditional mutants, which are unable to grow at 25° but survive at 37°, have also been studied. A Ts auxotroph, a *conditional non-lethal* mutant, is unable to grow on minimal medium at the *non-permissive* temperature unless the auxotrophic requirement is satisfied (tryptophan in the case of a *trp* (Ts) strain).

The Ts phenotype is thought to arise from unfolding and consequent inactivation of the mutant polypeptide chain at the restrictive temperature. An amino acid replacement that alters the gene product's conformational stability is, presumably, responsible for this conditional response. The isolation of Ts strains with lesions at a specific locus was initially taken as evidence that the locus encodes a protein product. However, mutations in non-translatable genes (such as those coding for tRNAs) may also result in a Ts phenotype; at high temperature the mature functional tRNA species is not formed due to lack of processing of the mutant precursor (Section 9.3.1 (b)). The existence of a Ts mutant *per se* is, thus, not sufficient proof for a structural gene.

Clearer evidence for a translatable locus is afforded by the isolation of nonsense mutations at this site, and identification of the mutation-specific nonsense fragment. These lesions also result in a conditional phenotype since premature termination of translation can be reduced by the presence of a suitable suppressor (see Figure 3.22). The latter can be introduced on a plasmid (Section 4.1.3 (b)) or by transduction (Section 5.2).

Conditional mutants have proved extremely useful in the analysis of essential genes. Lesions at these loci are normally lethal (unless a back-up mechanism or second, functional gene copy is also present), whereas the viability of a Ts *conditional-lethal* mutant is maintained at the permissive temperature; the existence of a lethal mutation in an indispensable gene is essentially cryptic under these non-restrictive conditions. (The same procedures may be applied to chain-terminating mutations in the presence of a temperature-sensitive nonsense suppressor.) Thus, the lesion can be transferred and mapped by standard bacterial genetic techniques (Section 7.1). Moreover, the functional role of an essential gene product can be determined by studying the physiological changes that ensue upon its inactivation at the elevated temperature. Finally, identification of the mutant protein may lead to the isolation of the essential gene product, itself. Nonsense mutants are particularly amenable to this type of analysis because of the difference in size between the wild-type and foreshortened polypeptides (Section 3.4).

3.2 Mutagenesis

The nature of the genetic material renders it intrinsically susceptible to chemical change. An event that would occur with negligible probability in a small molecule has a high chance of incidence in the DNA macromolecule (the bacterial genome, itself, has over one hundred million covalent bonds). Moreover, the existence of only a very limited number of copies of the genetic material makes damage to one a far more significant event, particularly because these copies are required for their own replication: other macromolecules are synthesised from a separate template and, thus, a structural modification does not, normally, affect their *de novo* biosynthesis.

The study of mutagenesis has been greatly facilitated by the advent of well-characterised prokaryotic systems. Bacteria and their bacteriophages are readily manipulated because of their short generation time. Moreover, powerful selection techniques allow the analysis of rare molecular events (see Section 3.2.2(b)). In addition, the expression of recessive traits is made possible by their haploid status. The accurate location of lesions by fine genetic mapping (Section 7.1) and the isolation and cloning of pure DNA (Section 7.3) allows an understanding of the molecular processes responsible for mutagenesis.

3.2.1 The Molecular Basis of Mutagenesis

Genetic defects may be split, conveniently, into two main categories, spontaneous and induced (Table 3.1). *Spontaneous mutations* occur 'naturally' at low frequency and are due to internal factors such as DNA replication errors or misrepair of DNA damage. The environment, itself, is unlikely to be mutagen free. Background radiation, for example, may give rise to genetic lesions. Moreover, visible light is known to be mutagenic. A rigorous definition for spontaneous mutation must, therefore, exclude externally-mediated events and concentrate, instead, upon genetic defects caused by intracellular perturbations that originate from within the bacterium.

Induced mutations stem from external factors, for instance irradiation, chemical mutagens or heat (Table 3.1). The increase in mutation frequency brought about by these agents is advantageous for the isolation of mutational defects that occur rarely as spontaneous events. Moreover, the type of lesion induced, the *mutagenic spectrum*, is mutagen dependent (Table 3.1). In general, base mispairing or misrepair are responsible for the majority of mutagenic events.

(a) Spontaneous Mutation. A number of *mispairing* schemes have been postulated, in addition to base tautomerisation, that fit into the standard dimensions of the DNA double helix (pairing in which N-glycosyl bonds are only approximately similar to those in normal purine-pyrimidine interactions have also been proposed). The spontaneous occurrence of all possible base pair substitutions (albeit at low frequency, see below) supports these mispairing schemes.

The bacterial and bacteriophage replicative machinery is very accurate. The $3' \rightarrow 5'$ exonuclease activity associated with prokaryotic DNA polymerases has a 'proof-reading' function, excising newly incorporated $3'$-terminal nucleotides

Table 3.1 Mutagen Action

Mutagen	Specificity	Mechanism [a]
Spontaneous	substitution	mispairing
	frameshift	slipping
	multisite	recombination
	all types	misrepair
UV radiation	all types	misrepair
Base analogues		
5-bromouracil	A·T ⟷ G·C [b]	mispairing
2-aminopurine	A·T ⟷ G·C [c]	mispairing
Base modifiers		
nitrous acid	A·T ⟷ G·C	mispairing
hydroxylamine	G·C ⟶ A·T	mispairing
alkylating agents	mainly transitions [d]	mispairing
	all types	misrepair
Intercalators	frameshift	slipping
	all types [e]	misrepair

a   Mispairing, non-standard base pairing, may arise spontaneously or through the presence of nucleotide derivatives.
Slipping refers to imperfect pairing between complementary strands due to base sequence redundancy. All types of mutations may be induced indirectly by faulty repair mechanisms, misrepair.
b   G · C → A · T is favoured.
c   A · T → G · C transitions occur 10-20 fold more frequently.
d   EMS and MNNG are highly specific for G · C → A · T transitions (though other mutations are induced by error-prone repair).
e   The wide range of lesions induced by ICR compounds may stem from their alkylating side-chain.

that are incorrectly paired with the template (discussed in Section 6.6.1 (a)). The copy-editing system, in conjunction with base-pair selection specificity, reduces errors to about one per 10^{10} nucleotide pair replication (estimated from the mutation frequency for a number of bacterial genes, about 10^{-8} per cell per generation). This is equivalent to a net spontaneous error rate in *E. coli* (genome size 3. 8×10^6 base pairs) of about 4×10^{-4} per genome replication. Only a fraction of genetic changes are likely to be deleterious, the majority may be silent. Spontaneous mutations are, therefore, at best rare events.

The presence of *mutator* alleles of DNA polymerases, such as *E. coli* polymerase I and III (the *polA* and *polC* gene products, respectively), and of other genes involved with DNA synthesis, for instance the DNA unwinding protein of bacteriophage T4 (gene 32 product), increases mutation rates by as much as 100-fold. In addition, *E. coli* strains carrying mutator alleles (designated *mut*⁻) of genes other than those known to be involved in replication show an increase of 100- to 100 000-fold in the spontaneous mutation rate. There is evidence suggesting that most *mut*⁺ gene products are non-essential. The mutator *mutT*⁻, unlike others, is highly specific, causing only unidirectional A · T → C · G transversions; the site of action of this gene product may be the replication fork. *Anti-mutator* DNA polymerase mutants have also been isolated that show reduced mutation rates. The increase in nuclease to polymerase ratios of some anti-mutator DNA polymerases (and the decrease for the corresponding mutator enzymes) support a model in which the DNA polymerase-dependent mutation rate is governed by the efficiency of copy editing. Nevertheless, base selection is, clearly, also important in maintaining the high fidelity of DNA replication.

Direct mispairing would be expected to predominantly generate base substitutions. Frameshift mutations may also occur spontaneously, possibly by slipped mispairing during DNA replication or recombination (see Streisinger model, Figure 3.13). Moreover, genetic lesions (including mutagen-induced lesions) can be produced indirectly by induction of an *error-prone DNA repair* system that has a mutator effect (see below). A wide variety of mutational types are produced by *misrepair*.

Chromosomal rearrangements such as duplications and deletions (see Figure 3.1) also arise spontaneously. Tandem duplications occur at a frequency of about 10^{-3} to 10^{-5} in *E. coli*; duplications of as much as 22 per cent of the bacterial chromosome have been reported. The rate is markedly reduced in recombination-deficient *(recA⁻)* strains. Moreover, duplications are readily lost by *recA*-mediated recombination. Intrachromosomal recombination between similar or identical sequences (possibly between two sister arms of a replicating chromosome) is likely to be the cause of these events.

Loss of DNA from a certain region, as compared with its duplication, is, generally, less frequent. These multisite events occur in the absence of a number of functions concerned with DNA metabolism, including the *recA* product. The mechanism of this illegitimate recombination process has yet to be elucidated. The location of short repeated sequences at the endpoints of certain deletions suggests that intramolecular (or intermolecular) recombination between non-tandem direct repeats is one pathway for spontaneous deletion formation (discussed in Section 6.2.2 (b)).

(b) Induced Mutation. Lesions may be induced by a variety of external perturbations. Electromagnetic radiation and a number of chemical mutagens are considered below.

(i) *Radiation.* Ultraviolet (UV) irradiation at 254 nm, the wavelength absorbed by most common purine and pyrimidine bases, is convenient for the induction of genetic lesions. Higher energy sources that have greater penetration properties, for instance ionising X-rays and gamma rays, are not required with prokaryotes since the bacterial membrane is only slightly refractory to the weakly penetrating UV radiation. Pyrimidine dimers, most commonly thymine-thymine *diadducts*, are the primary UV photoproduct (Figure 3.6). A standard germicidal lamp can be used; about 6 dimers per bacterial genome are induced in a wild-type strain by a UV fluence of $0.1\mu J/mm^2$.

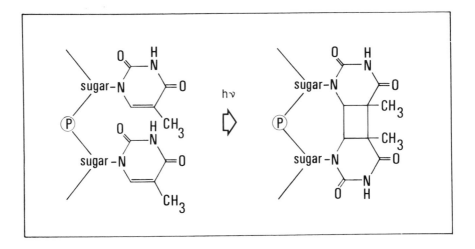

Figure 3.6: Structure of a Thymine Photodimer

Bases in pyrimidine dimers are unable to hydrogen bond. Repair of UV damage in DNA of surviving bacteria is carried out efficiently by relatively 'error-proof' mechanisms: photoreactivation, excision repair and daughter-strand gap repair (see DNA Repair, Section 6.3). Whereas 40 per cent of the cell population survive the presence of 3000 dimers (generated by a fluence of $50\mu J/mm^2$), one or two diadducts are lethal in an organism lacking any suitable repair system (Figure 3.7). The *E. coli uvr* (*UV* resistance) genes are involved in excision repair of UV dimers in UV-irradiated bacterial or bacteriophage DNA. Thus, genetic lesions at the *uvr* loci result in increased sensitivity to UV (Figure 3.7). Moreover, the induced mutation frequency is several orders of magnitude greater than in an *uvr*+ bacterium.

The fact that certain recombination-deficient derivatives of the UV-sensitive mutants are less, rather than more sensitive to the mutagenic effect of irradiation, might, at first sight, appear contradictory (these *recA⁻ uvrA⁻* double mutants are, of course, more sensitive to the killing action of UV irradiation but the frequency of induced mutations is no greater than the spontaneous rate in the absence of irradiation). This observation, in fact, led to the realisation that the mutagenic event is not the consequence of an inactive excision repair pathway (in the *uvr⁻* background). Rather, inaccurate post-replication repair of unexcised pyrimidine dimers, in cells that survive despite the presence of the diadduct, is generally the cause. This postulated error-prone bacterial system, responsible for both bacterial and phage UV mutagenesis, has been called *SOS repair* since it is apparently inducible by DNA damage or by perturbations to the synthesis of DNA, to presumably promote survival of the organism. Expression of the inducible functions associated with the SOS system requires the *recA⁺* and *lexA⁺* gene products; the former's role in UV mutagenesis appears to be

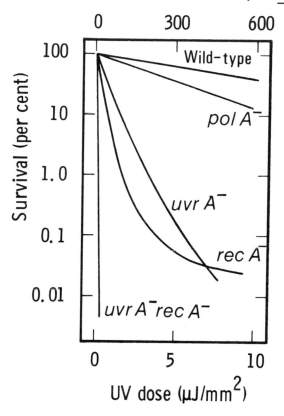

Figure 3.7: The Sensitivity of Bacterial Strains to UV Irradiation
Note the log scale for per cent survival. *polA⁻* and *uvr⁻* strains are deficient in excision repair and
recA⁻ mutants are unable to perform daughter-strand gap repair. Clearly, *uvr⁻ recA⁻* double
mutants lack both repair pathways (see DNA repair, Section 6.3). After Hanawalt (Hanawalt, P.C.
(1972) *Endeavour, 31,* 83-87)

independent of its normal function in general recombination (see Sections 6.2.1
and 6.3.2 (b)).[2]

In conclusion, pyrimidine dimers that are not eliminated by error-free repair
systems, block DNA replication and, thus, cause gaps in the complementary
strand. Generation of these gaps may be avoided by induction of the SOS repair
system. However, it is this DNA synthesis opposite photodimers that apparently
leads to misincorporation (presumably because of reduced base pairing specifi-

2. A number of functions have been associated with the SOS response that include error-prone
DNA repair (Section 6.2.3 (b)), cell division delay, and prophage induction (Section 5.1.2 (c)).
They have in common a dependence on the *recA⁺* and *lexA⁺* gene products (Section 6.2.1 (a)) for
expression, after generation of the SOS regulatory signal.

city) and results in a high proportion of tandem, double base pair changes. In addition, UV and other agents that trigger SOS misrepair (for example, mitomycin C or thymine starvation) induce a wide spectrum of mutagenic events, including base substitutions (both transitions and transversions), frameshift mutations and deletions. Abnormal induction of the SOS error-prone system in undamaged cells generates genetic lesions and, hence, leads to an increase in the 'spontaneous' mutation rate.

(ii) *Chemical Mutagens.* Many compounds originally thought to be harmless are now known to be mutagens, causing chromosomal damage to both prokaryotes and higher organisms (Table 3.2). The development of rapid, relatively inexpensive screening systems, such as the Ames *Salmonella* mutation

Table 3.2 Some Common Carcinogenic Sources

Source	Component	Action [a] Mutagenic [b]	Carcinogenic
Native agents			
fungi	aflatoxin B_1	+	+
cycad nuts	cyasin	+	+
sassafras oil	safrole	+	+
tobacco tar	various	+	+
food pyrolysis products	various	+	+
food amines and nitrites	N-nitroso compounds	+	+
Synthetic agents			
hair dyes	various	+	+
antibiotics	various	+	+
pesticides	various	+	+
flame retardants	various	+	+
disinfectants	sodium hypochlorite	+	not known
aerosol propellants	vinyl chloride	+	+
insulation	asbestos	-	+

a   Where several components are active, there is not necessarily a correlation between mutagenicity and carcinogenicity in each case.
b   Most substances require metabolic activation for mutagenicity in bacterial tests (see text).

test (Figure 3.8), has led to a realisation of the widespread occurrence of mutagenic compounds, both natural and man-made. These chemical mutagens can be divided into three main groups., base analogues, compounds that alter the chemical structure of DNA, and DNA binding/intercalating agents.

Base analogues: 5-bromouracil (5-BU) and 2-aminopurine (2-AP), two analogues investigated early, have structures similar to the common bases thymine and adenine, respectively (Figures 3.9 and 10). Some replacement of the

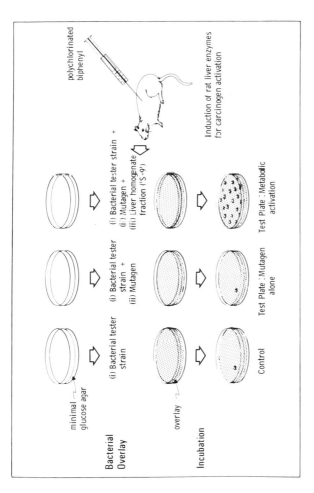

Figure 3.8: Bacterial Mutation Test

Well-characterised auxotrophic mutants of *S. typhimurium* have been used in the Ames test. The number of revertants able to grow in the absence of nutrient requirements is determined after culturing a lawn of mutant cells on agar minimal medium (Section 1.3.2 (a)) in the presence or absence of the test compound. Ingeniously, since many mutagens are apparently activated within the mammalian body, a rat liver fraction (in which carcinogen-activating enzymes have been induced with polychlorinated biphenyl) and the mutagen under test are applied to the cells in an overlay (see, for example, Figure 5.8). The rat liver fract on is no:, itself, mutagenic. A potential mutagen — a compound that produces reverse mutations — is rapidly detected in this economic system. About 90 per cent of compounds known to have carcinogenic effects on animals were mutagens in the Ames test (see Table 3.2). Similarly, about 86 per cent of chemicals believed to be non-carcinogenic by animal tests proved not to be mutagenic. Although the bacterial mutation test is not entirely accurate — some mutagens escape the test while certain compounds appear to be mutagenic *in vitro* but not *in vivo* — animal tests require long-term and unpleasant experimentation to obtain proof about a putative carcinogen

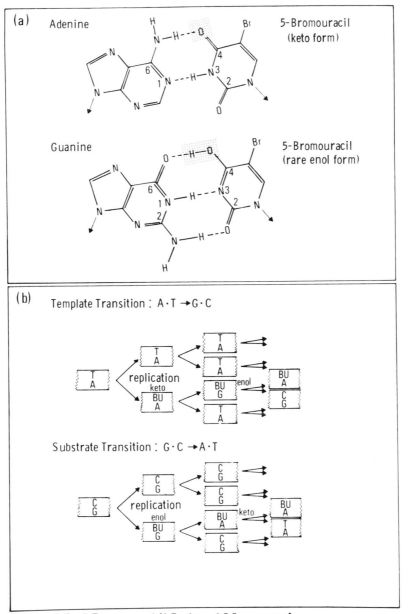

Figure 3.9: 5-Bromouracil-Induced Mutagenesis

The thymine analogue 5-bromouracil *(BU)* pairs with adenine in the normal keto form but with guanine in the rare enol form (a). Base tautomerisation, thus, leads to mispairing (b). When BU incorporated into a DNA chain in the keto state subsequently interacts in the enol state, an A·T → G·C transition is generated: conversely, a G · C → A · T transition occurs if BU in a BU · G pair interacts in the keto form during replication. G·C → A·T transitions occur more frequently with this analogue

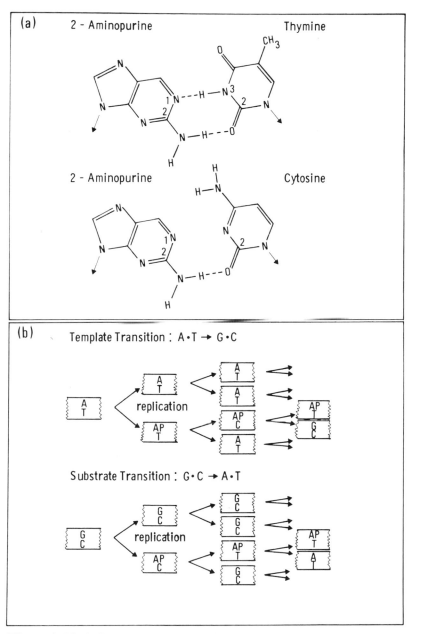

Figure 3.10: 2-Aminopurine-Induced Mutagenesis

The adenine analogue 2-aminopurine *(AP)* can pair with both thymine and cytosine, albeit with one hydrogen bond in the latter case (a). AP, like 5-bromouracil (Figure 3.9), can, therefore, generate A · T → G · C transitions (b), depending upon whether AP incorporated as T · AP interacts with cytosine (A · T → G · C) or whether a C · AP base pair allows T · AP interaction (G · C → A · T). A · T → G · C transitions are favoured naturally occurring bases takes place during cellular growth in the presence of an analogue. However, it is not the presence of the changed base *per se* that is the cause of the mutagenic event. Introduction of the base analogue results in mispairing during replication at the site of the modified base, owing to the property of certain of these analogues to tautomerise far more frequently than the natural bases and the fact that different tautomers can pair with different bases. Thus, whereas 5-BU in the normal keto state pairs with adenine, the enol form pairs with guanine (in the case of the adenine analogue 2-AP, the normal form can pair with cytosine, albeit with a single hydrogen bond; Figures 3.9 and 3.10). In summary, the highly specific pairing of normal bases, essential for accurate DNA replication (Section 6.1.2), is overriden in the presence of base analogues.

Mispairing may occur between an analogue present in the DNA template and the progeny strand *(template transition):* it may also occur between a normal base in the template and an incoming analogue *(substrate transition)*. In both situations, the base pair change is induced by faulty replication due to the pairing characteristics of these unusual bases (Figures 3.9 and 10). Both analogues induce A · T ↔ G · C *two-way transitions* (the forward reaction occurs preferentially with 2-AP, the backward with 5-BU). In general, therefore, mutations induced by base analogues (and nitrous acid, see below) are also revertible by these agents.

Compounds that alter the chemical structure of DNA: certain mutagens, such as nitrous acid (HNO_2) and hydroxylamine ($HONH_2$), act directly on template DNA, modifying specific bases *in situ* rather than by causing mispairing during replication (alkylating agents, which act both directly and indirectly, are discussed below). These modified bases have changed pairing characteristics (as compared with the original nucleotides). Thus, daughter strands carry incorrect bases through standard base pairing between these modified bases and substrate deoxyribonucleotides (Figure 3.11).

The potent mutagen, nitrous acid, acts directly on DNA templates to produce A · T ↔ G · C two-way transitions, which are revertible by both HNO_2 and base analogues (Figure 3.11). Oxidative deamination of amino-bases converts adenine and cytosine to hypoxanthine and uracil, respectively.[3] These HNO_2-modified bases pair with cytosine and adenine instead of thymine and guanine and, hence, effectively generate T · G and G · T mispairs (guanine is similarly converted to xanthine, which has unchanged base-pairing characteristics, albeit with two rather than three hydrogen bonds).

Hydroxylamine is a very useful mutagenic agent since it specifically induces G · C → A · T *one-way transitions* (Figure 3.11). $HONH_2$-generated mutations can, therefore, not be reverted by the same mutagen. This chemical reagent reacts preferentially with cytosine (or 5-hydroxymethylcytosine),

---

3.   Of the four naturally occurring DNA bases, cytosine is most susceptible to heat-induced degradation. Spontaneous deamination of adenine to hypoxanthine also occurs at a slow but significant rate. There are specific enzymes, DNA glycosylases, that remove uracil and hypoxanthine (and 3-alkyl adenine derivatives) *in toto* from DNA chains, permitting DNA repair (see Section 6.3).

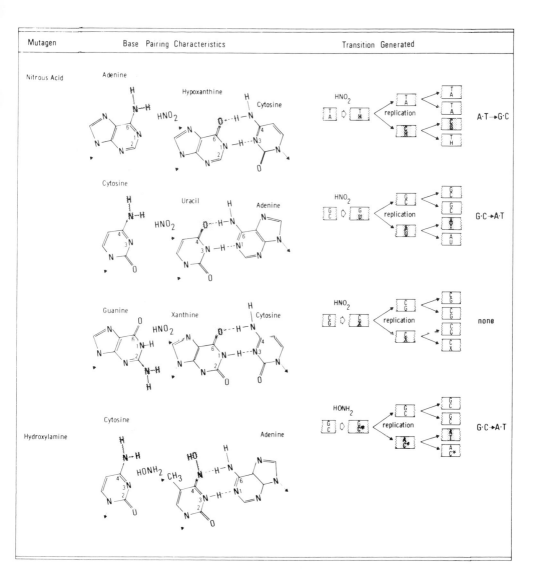

Figure 3.11: Mutagenesis by Nitrous Acid and Hydroxylamine

Whereas base analogues are not mutagenic *per se* (they are not detrimental to an organism until mispairing, Figures 3.9 and 3.10), compounds that alter DNA structure act directly. That is, transcription and translation of a gene carrying a chemically altered base leads to mutant gene product irrespective of whether replication has taken place. Moreover, the presence of a modified base with pairing characteristics different from the wild-type residue generates transitional mutations. Nitrous acid (HNO₂) gives rise to G · C ↔ A · T transitions, depending upon which base (adenosine or cytosine) is deaminated; although guanine residues are converted to xanthine, they can still pair with cytosine, albeit with two hydrogen bonds. Hydroxylamine (HONH₂), on the other hand, produces G · C → A · T one-way transitions. HONH₂-induced mutations are, consequently, not revertible by the same mutagen modifying the base *in situ*, such that it, apparently, now attracts adenine.

Alkylating agents, the largest class of mutagens (extensively used in industry — for other purposes!), alkylate most, if not all, susceptible sites of the normal DNA bases: ring nitrogens, amino groups, phenoxy groups and ring carbons. These substances induce mutations both directly through mispairing, and indirectly by error-prone repair ((i) above and Section 6.3.2 (b)). They, therefore, give rise to a plethora of genetic defects. Methylmethane sulphonate (MMS), ethylmethane sulphonate (EMS) and N-methyl-N'-nitro-N-nitro-soguanidine (MNNG) are mutagens commonly applied to prokaryotic systems (Figure 3.12) Alkylation of guanine O^6 and thymine O^4 has been implicated in

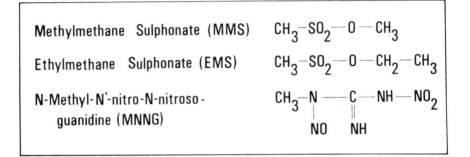

Figure 3.12: The Structure of Some Alkylating Mutagens

directly induced alkylation mutagenesis (rather than 7-alkyl-guanine and 3-methyl-adenine, the major products). The weakly mutagenic character of MMS may be attributable to inefficient methylation of guanine O^6. Certainly, EMS-induced mispairing appears to occur at high frequency, while most mutations induced by MMS are the consequence of error-prone repair *(indirect mutagenesis)*. The high preference for $G \cdot C \rightarrow A \cdot T$ transitions after EMS and MNNG mutagenesis supports formation of O-6-alkyl guanine as the primary mutagenic event.[4] MNNG acts predominantly at the replication fork, causing closely spaced multiple mutations (single-stranded DNA in the region of the fork may be the favoured target). This property of MNNG has been employed for *localised co-mutagenesis* in which selection of a genetic lesion at one locus allows the isolation of strains that carry separate mutations close to this marker.

DNA-Binding Compounds: The ICR frameshift mutagens are acridine-like derivatives (Figure 3.13), planar heterocyclic compounds that carry alkylating side-chains necessary for penetration of the bacterial membrane (acridine is, itself, highly mutagenic for bacteriophages such as T4).[5] These compounds are

4.   The alkyl group of which is situated in the major groove.
5.   Base-pair substitutions, presumably caused by the alkylating side-chains, are also induced by the ICR compounds.

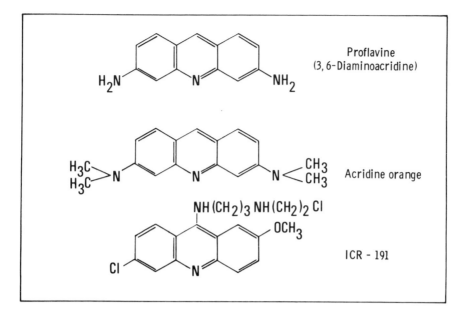

Figure 3.13: The Chemical Structure of Some Frameshift Mutagens
The term ICR for certain mutagens comes from the fact that these compounds were synthesised by Creech and his colleagues at the Institute for Cancer Research

thought to act by intercalation between stacked nucleotide bases. The Streisinger group have proposed a model to explain the role of *intercalating agents* in frameshift mutagenesis. This involves stabilisation of the imperfect pairing that arises when one DNA strand 'slips' relative to the complementary one (Figure 3.14). This model is strongly supported by the observed base sequence redundancy (monotonous runs of G · C pairs) at the site of many frameshift mutations (generating mutational 'hotspots', see below). The different responses of *E. coli* and bacteriophage T4 to intercalating agents may, thus, be due to their different base compositions (G+C content: 51 per cent *E. coli* K12, 34 per cent T4). However, the model does not, as yet, explain the lack of correlation between intercalation ability and mutagenicity. Perhaps these agents stabilise looped-out bases (extrahelical stacking) rather than intercalate into the DNA duplex.

(c) Mutational Hotspots. Mutagenesis is not entirely a random event: some sites are more susceptible to mutation than others.[6] Sequences neighbouring mutant sites are, therefore, likely to be responsible for 'hotspot' formation (Figure 3.15).

6.   Of course, not all lesions are deleterious, resulting in a mutant phenotype. Some polypeptides have domains that are less sensitive to amino acid replacement. The 'silent' region of a structural gene that encodes this domain might appear mutation free, thereby giving rise to artificial clustering of those lesions that do cause a detectable phenotype.

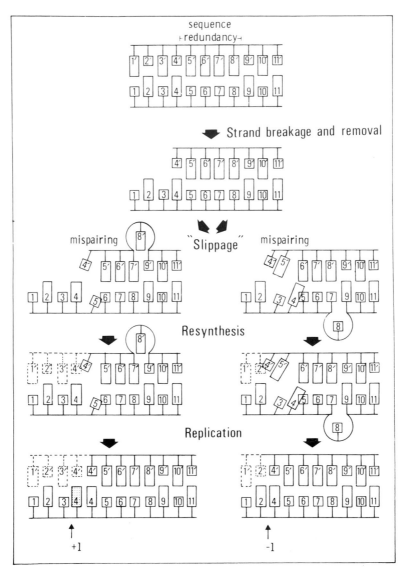

Figure 3.14: The Streisinger Model for the Production of Frameshift Mutations

It is postulated that, after random strand breakage and removal, 'slippage' of the two DNA strands occurs due to mispairing adjacent to the redundant regions (the base pairs $5 \cdot 5'$, $6 \cdot 6'$, $7 \cdot 7'$ and $8 \cdot 8'$ represent a run of, say, $G \cdot Cs$ and $6 \cdot 5'$, $7 \cdot 6'$ and $8 \cdot 7'$ etc. the 'slipped' $G \cdot C$ region). The frameshift is created, following gap-filling (resynthesis) of the damaged strands *(dotted lines)*, when this completed strand is used as a template for DNA replication. Whereas replication on a resynthesised strand carrying a looped-out chain gives rise to a $+ 1$ insertion (represented as a *stippled* base), the inverse situation results in a -1 deletion. After Streisinger *et al.* (Streisinger, G., Okada, Y., Emrich, J., Newton, J., Tsugita, A., Terzaghi, E. and Inouye, M. (1966) *Cold Spring Harbor Symp. Quant. Biol., 31,* 77-84) and Roth (Roth, J.R. (1974) *Ann. Rev. Genet., 8,* 319-346)

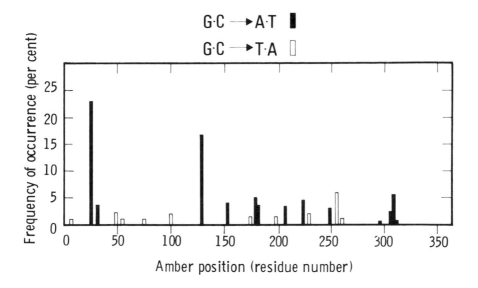

Figure 3.15: Base Substitution Hotspots in the *lacI* Gene

Distribution of 221 amber mutations found spontaneously in the *lac* repressor gene (Section 9.1) of *E. coli*. Only events that stem from G · C → A · T transitions (75.6 per cent) and G · C → T · A transversions (16.7 per cent) are shown since they represent the vast majority of the mutant collection (the transversions A · T → T · A, A · T → C · G and G · C → C · G occur at a frequency of 1.8, 5 and 0.9 per cent, respectively). Two hotspots dominate the amber spectrum, both involving G · C → A · T transitions, at positions 26 and 131 (23.1 and 16.7 per cent). Sequence analysis has shown that both these hotspots occur at the second cytosine in CCAGG (see text). Adapted from Coulondre *et al.* (Coulondre, C., Miller, J.H., Farabaugh, P.J. and Gilbert, W. (1978), *Nature, 274*, 775-780)

The Streisinger model postulates frameshifts at monotonous runs or regions of alternating doublets (both of which allow slipping and mispairing). Hotspots in deletion endpoints have been mentioned above (see also Section 6.2.2 (b)).

Recently, Miller and co-workers have proposed an attractive hypothesis for the molecular basis of spontaneous base substitutions that involves 5-methyl-cytosine residues (indicated with an asterisk) in the sequence,

$$
\begin{array}{c}
* \\
5'\ \text{C C A G G} \\
\text{G G T C C}\ 5' \\
*
\end{array}
$$

The *E. coli dcm mec* gene product, which is responsible for this modification, protects against cutting by the endogenous RII restriction endonuclease (Section 7.3.3). Spontaneous deamination of 5meC, yielding thymine (5-methyl uracil), is suggested to be the major cause of spontaneous G · C → A · T transitions, whereas uracil residues formed by spontaneous deamination of unmethylated cytosine residues can be rapidly excised by the host-coded enzyme, uracil-DNA glycosylase 9the *ung* gene product, Section 6.3.1 (a)). Thus, there is a system in *E. coli* for the correction of deaminated cytosine residues but not for 5-methylcytosine, which is, itself, the product of a protection process.

3.2.2 The Application of Mutagenesis to Bacterial Systems

(a) Isolation of Independent Mutants that Carry Single Lesions. Many of the mutant cells present in a single culture will be identical since inoculation of the growth medium with a mutagen-treated cell suspension containing a few mutants results in the multiplication of mutant (and wild-type) cells. In some cases, the mutant allele imparts a selective advantage (see below). Even mutant isolates obtained from uninduced cultures are not necessarily independent. When the spontaneous lesion occurs early during growth, the number of mutant descendants is amplified by successive rounds of cellular divisions. Only a mutational event that takes place just prior to plating will generate a single unique mutant organism (Figure 3.16).

Independent mutant isolates are obtained, in practice, by inoculating a series of culture vessels with a small number of cells (about 10^3 or less) or a number of different colonies (see legend to Figure 3.16). Routinely, only one mutant is retained from each independent culture. Even these strains, however, need not necessarily carry different mutant alleles since a non-random distribution of genetic lesions can result from hotspots.

Spontaneous mutations are rare (Section 3.2.1 (a)). Perhaps one in 10^6 to one in 10^{10} bacteria carry genetic defects. The frequency of multiple-point mutations that occur naturally (in the absence of external factors) is, therefore, so rare as to be normally undetectable. However, this is not the case when lesions are artificially induced. The importance of mutagens lies in their very property of increasing the mutation frequency. In addition, MNNG specifically induces closely-spaced multiple defects.

Clearly, conditions that cause excessive mutagenesis must be avoided. Thus, after mutagenesis, the DNA region of interest is, commonly, transferred to another, unmutagenised strain (by Conjugation or Transduction, Sections 4.1.3 and 5.2). This avoids the complexity created by the presence of multiple genetic lesions, and allows analysis of the mutant phenotype in a 'clean' background.

A useful technique, *localised mutagenesis*, involves the mutagenesis of a bacteriophage that carries a small region of the bacterial chromosome (Transducing Phage, Section 5.2). Subsequent transfer to an unmutagenised host allows the induction of genetic defects restricted to a small, localised region of the bacterial chromosome. *In vitro* techniques (Section 7.3.3) have increased the scope and precision of such approaches.

(b) Mutant Selection. Strains carrying genetic defects may be obtained either spontaneously or by induction (Section 3.2.1). Mutants *resistant* to particular

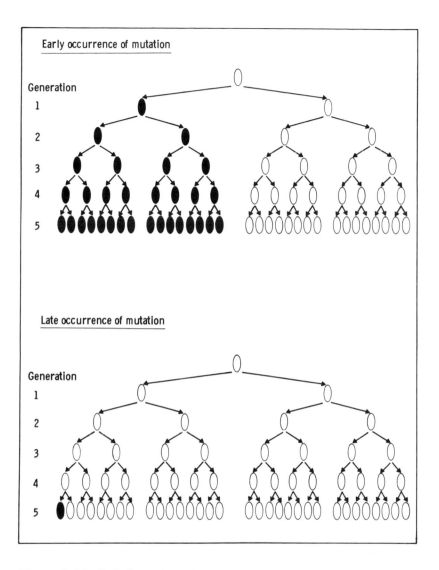

Figure 3.16: Cell Growth and Mutation

Schematic representation of the appearance of a mutant organism *(dark cell)* and its proliferation. When the mutation occurs within the first generation, one half ($^{16}\!/_{32}$) of the sibs are altered, whereas a lesion in the fifth generation results in only $^1\!/_{32}$ daughter cells being mutant. A small inoculum (say, one millionth of a master culture of 10^9 cells/ml) reduces the likelihood that any mutants pre-existing in the master culture are subsequently introduced into a single culture vessel (the use of single-cell inoculums would, obviously, make such an event even less probable!). A wide fluctuation in the number of bacteria carrying particular mutant markers is expected since the proportion in an overnight culture is dependent upon the generation in which the genetic defect arises. Luria and Delbruck (1943) first employed the fluctuation test to demonstrate that bacterial mutations arise spontaneously rather than from Lamarckian adaptation; replica-plating (Section 3.2.2 (bi)) experiments of Lederberg and Lederberg (1952) supported these conclusions factors, for instance to bacteriophages or antibiotics, can be readily isolated for they represent the sole survivors from among the population of cells subjected to this selective treatment. In many cases, however, there is no straightforward *positive selection*. Mutation to auxotrophy, say, results in the inability of the organism to grow on minimal medium (other supplement(s) are required). Survivors that grow on minimal plates are, therefore, not auxotrophic mutants. True, some of the bacterial cells growing on nutrient plates may be auxotrophic, but how does one screen several million bacteria for a single mutant? A number of techniques have been developed to surmount this problem.[7]

(i) *Replica-plating*. Colonies gridded onto an agar plate or well-separated colonies (normally less than 1000, though many more colonies can be screened when there is a positive selection) can be transferred by *replica-plating* to a number of test plates. The colonies are imprinted, initially, onto a small sterile sheet of velvet from the *masterplate*; this imprint can then be transferred to other plates (Figure 3.17). The technique allows rapid screening of a large number of isolates (with, say, 1000 colonies per plate, only 1000 plates are required to screen one million isolates!)

(ii) *Penicillin Enrichment*. The antibiotic penicillin (Figure 3.18) irreversibly blocks cross-linking of the peptidoglycan layer, particularly at the septal area where major growth occurs. Thus, in the presence of the antibiotic, growing bacteria suffer osmotic lysis (see Figure 1.17, Section 1.2.1). However, mutants that are unable to grow are unaffected since only growing cells synthesise their cell wall constituents (penicillin is innocuous to a Trp$^-$ auxotroph, for example, in minimal medium that lacks the amino acid). It is, of course, important to allow growth of the cell population prior to antibiotic treatment to permit expression of the mutant phenotype, that is, to avoid *phenotypic lag*. Consider, for example, a mutation that imparts a Trp$^-$ phenotype. Even after inactivation of say, *trpA* (Section 9.2), some functional tryptophan synthetase remains in the cytoplasm. Thus, growth is necessary after mutagenesis for dilution of 'old' enzyme. Recovery of a mutant strain also demands that the genetic lesion does not have an irreversible *(bactericidal)* effect on bacterial growth. After penicillin treatment, survivors are plated-out under conditions at which the specific mutant strain grows (minimal medium supplemented with tryptophan for the Trp$^-$ auxotroph). This technique is only semi-selective: it is more a purification step

---

7.   'Suicide' techniques have also been employed. This involves the use of substances whose metabolism or assimilation is lethal to cellular growth. Derivatives of lactose, for example, that when cleaved by β-galactosidase (Section 9.1) give rise to a toxic product allow the isolation of mutants defective in the catabolic enzyme. Similarly, tritium suicide has been used to obtain temperature-sensitive strains defective in DNA synthesis at the non-permissive temperature. In this procedure, cells carrying a *dna*-like lesion (Section 6.1) are unable to incorporate the ^{3}H-thymidine into their DNA at high temperature and are, thus, saved from the lethal effects of β particles emitted upon long-term storage in the cold.

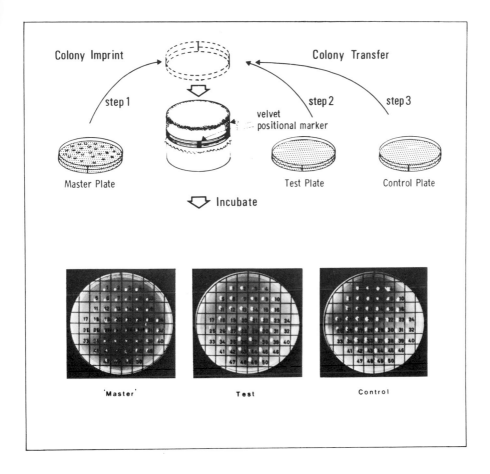

Figure 3.17: Replica-Plating

Colonies are transferred from the master plate to velvet *(step 1)* and from there to the test *(step 2)* and control plates *(step 3)*. The latter is included to ensure that the imprinting and transfer process has not resulted in artifactual loss of colonies. Thus, the composition of the control plate (and its incubation conditions) is identical to that of the master. Colonies that grow on these plates but not on the test plate are putative mutant species (see grid positions 3, 4, 7, 15, 19, 20, 24, 26, 29, 35, 38, 44 and 48)

that enriches about 10-fold for non-growing auxotrophic (or conditional) mutants.

3.3 Suppression

The molecular processes responsible for changes in the genetic material that lead to point mutations are normally reversible. Thus, not only can strains be isolated carrying gene defects, but also *revertants* that have the original wild

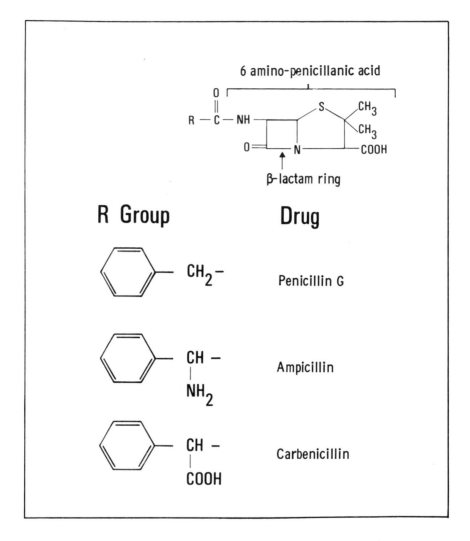

Figure 3.18: The Structure of Some Penicillins

phenotype restored. Many different changes within one gene (or, perhaps, several genes) can, clearly, give rise to a particular mutant phenotype. On the other hand, reversion might, at first, appear to involve a specific change at the site of the initial lesion to restore the exact, original nucleotide sequence. The fact that the frequencies of the *forward* and *reverse* mutations are not necessarily the same is in agreement with this (but see below). The reversion rate is, presumably, also dictated by the molecular events responsible for the primary genetic lesion.

As an alternative to restoring the original base sequence, an independent mutation may occur at a site distinct from the first lesion and wholly or partially

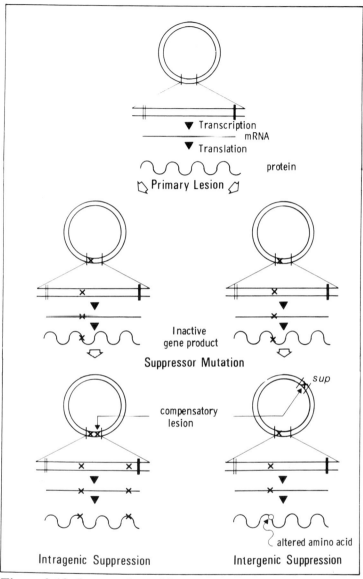

Transcription
mRNA
Translation
protein
Primary Lesion

Inactive
gene product

Suppressor Mutation

compensatory
lesion

sup

altered amino acid

Intragenic Suppression

Intergenic Suppression

Figure 3.19: Intragenic and Intergenic Suppression

The presence of a lesion in a structural gene results in an altered gene product — the polypeptide encoded contains an altered amino acid (only one residue is affected in the case of a base substitution other than a nonsense lesion). A second mutational event may suppress the effect of this primary lesion, either through the occurrence of a compensatory lesion within the original gene *(intragenic suppression)* or in another, unlinked locus *(intergenic suppression)*. In the former, the suppressed, mutant polypeptide contains at least two altered residues. In both cases, the suppressed, mutant organism carries two distinct mutations. The present example shows intragenic suppression of a base substitution and informational (intergenic) suppression of a missense mutation (see Table 3.3 and Figures 3.20-3.23) restore gene function. These *second-site revertants* carry, therefore, both the initial base change and a second *suppressor* mutation. The latter suppresses the phenotypic expression of the primary mutation. This phenomenon has been termed *intragenic* or *intergenic suppression*, depending upon whether the second, suppressor mutation lies within the gene carrying the initial lesion or in another region of the genome (Figure 3.19 and Table 3.3).

Table 3.3 Suppressor Classification

	Suppressor Type		
Point Mutation	Intragenic	Intergenic	
		Indirect	Direct
			(informational)
missense	+ a	+	+ c
nonsense	-	+	+
frameshift	+ b	+	+ d

a   Intragenic suppression of missense mutations is, in general, a rare event and depends upon the nature of the gene product.
b   Generally requires a frameshift mutagen (Section 3.2.1 (bii)).
c   Only a few examples of missense informational suppressors are known; they are highly specific owing to the nature of the process (see text).
d   Characterisation of informational frameshift suppressors *(suf)* has, to date, been restricted to *S. typhimuruum*; few examples are known.

3.3.1 Intragenic Suppression

A wild (or pseudo-wild) phenotype derived from second-site reversion may stem from the introduction of a second mutation into the gene harbouring the primary mutational site. When the primary lesion is a point mutation within a structural gene, second-site revertants will carry at least two altered residues: one that arises from the first genetic change, and a second suppressing the defect in gene function due to the first. A rationale for this type of intragenic suppression can be proposed by equating the original amino acid replacement with a conformational change in the gene product. A second replacement elsewhere in the polypeptide chain would then allow restoration (or partial restoration) of the wild-type conformation. Clearly, such second, suppressor events are restricted to those genetic changes that recreate amino acid interactions essential for the protein's secondary, tertiary and quaternary structure.

A second type of intragenic suppression is concerned with frameshift mutations. Introduction or deletion of multiples of one or two base pairs (but not three) alters the translational reading-frame, perhaps even uncovers nonsense codons (Section 3.1.1 (d)). However, the reading-frame is unaffected when the

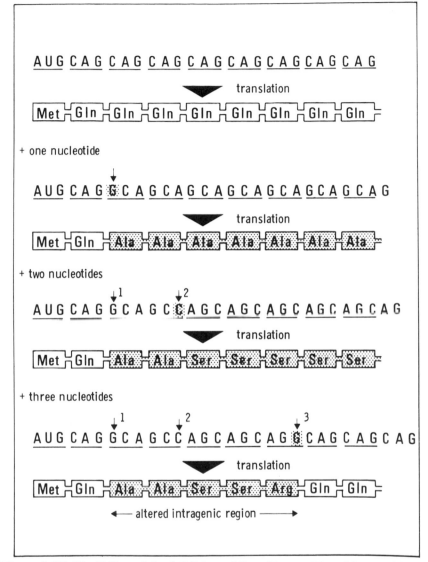

Figure 3.20: The Effect of the Addition of One, Two or Three Nucleotides on the Translational Reading-Frame

When additional bases are present in a structural gene (a +1 or +2 frameshift mutation), the reading-frame is changed, resulting in the formation of a drastically altered gene product: all residues are changed *(stippled)* downstream from the site of the frameshift(s). The addition (or deletion) of three closely-spaced but separate bases (or the combination of, say, one '+' and one '−' mutation) restores the reading-frame with only the region between these mutations still being translated incorrectly. The suppressed polypeptide is likely to be less active than wild-type (pseudo-wild phenotype) because of the presence of the remaining misread region. Note that in such intragenic suppression of frameshift mutations, the suppressed gene carries multiple lesions (only the transcript sequence is shown above); compare with informational (intergenic) frameshift suppression (Figure 3.23) number of base pairs changed is three owing to the triplet nature of the genetic code (Section 2.2.1 (a)). Rather, one (or more) amino acids are inserted or deleted. The presence of two closely-spaced frameshift mutations of opposite sign (or three of the same sign) can, therefore, be mutually corrective. The reading-frame is restored after the second alteration (or third in the case of mutations of the same type). Only the region between the mutations (stippled in Figure 3.20) is read incorrectly.[8] Thus, the first and second lesion must be closely spaced to reduce to a minimum this remaining out-of-phase intragenic region.

3.3.2 Intergenic Suppression

Intergenic suppressors can be divided into two classes, *indirect* or *direct* (see Table 3.3). The former bypass the effect of the original mutation without restoring either a functional gene or gene product. Thus, a structural gene mutation that gives rise to an altered, less active enzyme may be compensated for by activation of an alternative pathway or by a regulatory mutation that, say, increases the rate of transcription of the mutant allele.

Direct *informational suppressors* function at the translation level through mutant tRNA species. *Missense suppressors* of this class allow the introduction of the wild-type (or another suitable) amino acid into the growing polypeptide chain at a site where a mutant, non-functional residue is encoded (Figure 3.21). In *nonsense suppression*, the introduction of an amino acid prevents premature termination of translation (Figue 3.22). Finally, *frameshift suppressors* introduce residues at codons carrying base insertions or deletions (Figure 3.23). In each case, a single alteration in the anticodon[9] of a tRNA molecule is usually responsible for this reduction in the fidelity of translation (such intergenic suppressor mutations define, therefore, a particular tRNA gene). The suppressor mutation changes the codon-recognition properties of the suppressor tRNA species without, normally, affecting its charging characteristics (but see Table 3.4).

The altered anticodon of a missense suppressor tRNA allows pairing with non-cognate triplets. A residue different from that specified by the mutant codon is, thus, inserted into the polypeptide chain at the missense site (Figure 3.21). Similarly, the modified anticodon of a nonsense suppressor tRNA allows interaction with a chain-terminating codon where previously no such pairing occurred (Figure 3.22). Frameshift suppressors have a different number of ribonucleotides in the anticodon which restores, presumably, the correct reading-frame (Figure 3.23). In summary, informational suppression is mediated by mutant tRNAs and involves the insertion of a competent amino acid at the

8.   In fact, nucleic acid sequences involved in frameshift mutations have been inferred by comparison with amino acid sequences of second-site revertant proteins — the sequence encoded by the altered intragenic region — with that from the wild-type protein. Such studies provided an early *in vivo* confirmation of the genetic code (and its direction of reading) as determined *in vitro*.
9.   There are exceptions to this rule: the chain terminator UGA is, for example, suppressed by a mutant tRNA[Trp] that has a change only in the dihydrouridine stem, G 24 to A (Table 3.4).

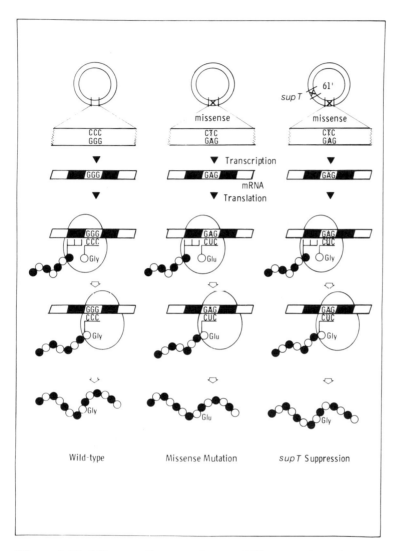

Figure 3.21: Missense Suppression: *supT* **Suppression of a Missense Lesion**
The presence of a missense mutation in a structural gene gives rise, upon translation, to a protein
product carrying a single amino acid substitution. In the present example, a $G \cdot C \rightarrow A \cdot T$ transition
results in the insertion of Glu rather than Gly (the wild-type situation is shown in the *left-hand panel*,
the missense case in the *centre*). Intergenic suppression *(right panel)* may alleviate the effect of a
missense mutation by virtue of a mutant tRNA $_I^{Gly}$ species misreading the 5'GAG 3' (Glu) codon
with its altered anticodon, 3'CUC5' (the *supT* (Su_{GAG}) allele of *glyU* maps at 61 min). The
suppressed organism carries, therefore, two separate mutations: the primary lesion and one in a
tRNA gene. Missense suppressors are allele-specific. That is, a particular mutant tRNA will
suppress only a particular missense lesion in one structural gene. *supT*, for instance, suppresses the
trpA461 (GAG) allele and inserts Gly rather than Glu. Missense suppressors are generally
inefficient due to a low level of aminoacylation. Strong suppression would, in fact, be a lethal event
since it would result in amino acid substitution in both non-essential and essential proteins

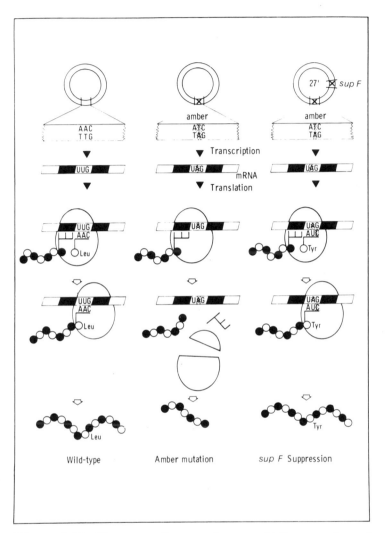

Figure 3.22: Nonsense Suppression: *supF* **Suppression of a Nonsense Lesion**

When a nonsense lesion occurs within a structural gene, there is premature termination of translation and the production of only a truncated polypeptide. Nonsense mutations, therefore, almost always give rise to non-functional gene products. Termination, in the above example, takes place at an amber mutation in the eighth codon due to an A · T → T · A transversion in the Leu triplet, UUG (the wild-type situation is shown in the *left panel*, the amber case in the *centre*). A second mutation in a tRNA gene may suppress premature termination of translation *(right-hand panel)* thanks to the synthesis of a mutant tRNA species carrying an altered anticodon (but see *sup9*, Table 3.4). An amber anticodon is sufficient to allow translational readthrough past an amber mutation but the function of the suppressed polypeptide depends upon whether the amino acid inserted (and, hence, the nature of the suppressor tRNA species) is suitable for that site and that protein product. Thus, the *supF* derivative of tRNA$_1^{Tyr}$ always inserts tyrosine at amber sites (the *supF* (Su3) allele of *tyrT* maps at 27 min, Table 3.4/Figure 3.25)

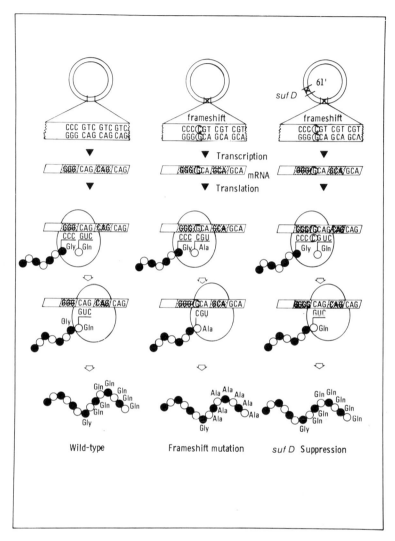

Figure 3.23: Frameshift Suppression: *sufD* Suppression of a Frameshift Lesion

The addition or deletion of one or two base pairs to a DNA sequence encoding a protein product has the property of totally altering the translational reading-frame since the code is read in blocks of three nucleotides without signals punctuating these codons (see Figure 3.20). The frameshift polypeptide synthesised *(centre panel)* is quite distinct from the wild-type sequence *(left-hand panel)*, all amino acids incorporated past the mutant site being different. The scheme shows the addition of a G·C base pair to a monotonous run of G·Cs as sequence redundancy has been implicated in frameshift mutagenesis (see Figure 3.14). The reading-frame may be restored if the tetra-nucleotide sequence generated by base insertion is translated as one codon. Informational suppressors that accomplish this task have been identified in *S. typhimurium (right panel)*. The *sufD* derivative of tRNA$_1^{Gly}$ (located at 61 min on the *S. typhimurium* genetic map) contains an additional cytidine residue in its anticodon (3'CCCC5' for 5' GGGG3') mutant site.

The act of insertion is not, itself, necessarily sufficient for the production of a functional gene product. The inserted residue must be acceptable for that particular position in the polypeptide chain. Thus, the nonsense suppressor, *supF* (Table 3.4), partially suppresses premature translational termination by allowing the extension of a proportion of polypeptide molecules past amber codons. However, only those polypeptides that function with the aromatic amino acid tyrosine at the amber site will have biological activity. Nonsense suppression by an unsuitable mutant tRNA results in the synthesis of a full-sized but inactive molecule (which may be subsequently degraded by endogenous proteases, Section 2.2.2 (b)).

Informational suppressors tend to be highly specific since their mechanism of action demands a fidelity of anticodon-codon interaction similar to that of normal translation. Clearly, then, missense suppressors do not function on nonsense mutations (and *vice versa*). Moreover, there is specificity within each class. Thus, amber suppressors only inhibit premature translational termination at amber codons, and UGA suppressors interact specifically at UGA codons (but see Table 3.6). However, wobble in the third base (Section 2.2.1 (b)) allows ochre suppressor tRNAs to act both on ochre and amber lesions, albeit weakly (Figure 3.24). This general specificity makes informational suppressors, particularly nonsense suppressors, useful genetic probes (Section 3.1.2).

Misreading engendered by informational suppression is not restricted to the primary mutant mRNA. Any codon recognised by the modified anticodon may be translated incorrectly. Moreover, if the bacterial chromosome carries only one copy of a specific tRNA gene, mutation to suppressor phenotype results in lack of translation of the codon recognised by the wild-type tRNA species. Such events could, of course, be deleterious to the cell.[10] In general, however, suppressor mutations are not lethal. Firstly, codons that might be misread incorrectly by a suppressor tRNA may be 'misread' correctly by isoaccepting tRNA species due to wobble in the third base (Section 2.2.1(b)). Secondly, more than one copy of each tRNA gene may exist (Figure 3.25): there are three tRNA^{tyr} genes, for example, of which two code for identical species and the third is almost identical (Section 2.2.1(b)). Finally, informational suppressors are not 100 per cent efficient (Table 3.4 and below).

Internal termination codons that have arisen from nonsense mutations will be recognised, presumably, by the translational machinery as normal translational stop sites. Aberrant readthrough by nonsense suppressor tRNAs must, therefore, compete with the action of release factors and other components of the prokaryotic termination system (note that both R1 and R2 are specific for UAA,

10. The *E. coli.* chromosome possesses a single copy of the tRNA^{Trp} gene *(trpT)*. Nonsense suppressor mutations in *trpT* are, therefore, lethal unless a copy of the wild-type gene is also present (say, on a F-prime plasmid, Section 4.1.3 (b)). The mutant *trpT⁻* (Su⁺) gene encodes a tRNA with the anticodon CUA rather than CCA. Strangely, this change results in both nonsense suppression and mischarging by glutamine, such that Gln and Trp are inserted at a frequency of 9 : 1. Analysis of such mutant tRNAs has, in fact, given insight into structure-function relationships.

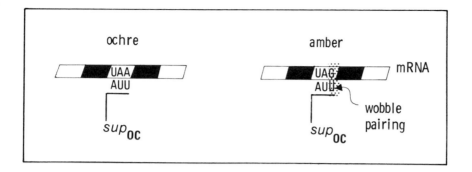

Figure 3.24: Ochre Suppression of Ochre and Amber Mutations by Virtue of Wobble

In general, nonsense suppression is a specific process — as would be expected since it is mediated by codon-anticodon pairing — such that amber suppressors act only on amber mutations and UGA suppressors on UGA lesions (but see Table 3.6). Ochre suppressors, on the other hand, interact both with ochre and amber mutations, albeit inefficiently (Table 3.4). This property may be ascribed to 5'U in the ochre anticodon pairing with either A or G (see Wobble Pairing, Table 2.6). Ochre mutations are characterised by their ability to be suppressed only by ochre suppressors

whereas only a single release factor is involved in termination at the other two terminators, Section 2.2.2 (aiii)). It is, consequently, not surprising that these nonsense suppressors are inefficient; *supF*, for example, although a particularly strong suppressor, generally allows only six out of ten initiated polypeptide chains to be elongated past an amber codon (while four out of ten terminate at the nonsense mutation, Table 3.4). In addition, a given suppressor's efficiency is dependent upon *reading context* — the nucleotide sequence neighbouring the termination codon. Amber and UGA suppressors are, in general, most efficient. This might be explained in terms of an interaction between the nonsense codon on the mRNA strand and the 3'-terminal sequence of 16S rRNA (Figure 3.26). The relative efficiency of nonsense suppressors would then be inversely proportional to the strength of binding of these two RNAs.

This pleasingly simple story may, at the best, apply only to *E. coli* and a few other species, since there is a wide variation in the 3'-terminal sequence. The lack of conservation of the 3'-terminal UUA-OH triplet in prokaryotic 16S rRNAs suggests that if translational termination (like initiation) proceeds through an mRNA · rRNA interaction, it may not involve the terminal 16S sequence itself. The 3'-terminus, or at least the first 49 nucleotides, is, however, required for RF binding (Section 2.2.2 (aiii)).

Are ochre codons the 'natural' translational terminators? Although this hypothesis is supported by the inefficiency of ochre suppression, sequence analysis has indicated otherwise. Rather, UGA and UAA are used equally frequently (as well as UAG in fewer instances), and tandem terminators define the translational end of a few mRNA sequences. Moreover, the importance of the reading context in nonsense suppression suggests that the chain terminating machinery may recognise a sequence of nucleotides adjacent to the natural terminator(s).

3.3.3 Phenotypic Suppression: The Role of the Ribosome in Misreading

Informational suppression functions through mutant tRNA species at the level of translation. The ribosome is by no means an inactive partner with respect to the fidelity of translation. In fact, an intrinsic low-level misreading appears to be built into the ribosomal matrix. Three ribosomal proteins of the small 30S subunit are associated with this 'natural' infidelity, the S12, S5 and S4 proteins (encoded by *rpsL, rpsE* and *rpsD*, respectively; these genes were originally denoted by *strA, spcA* and *ram* owing to the phenotype of mutants carrying lesions at these two loci — *str*eptomycin resistance, *spc*tinomycin resistance and

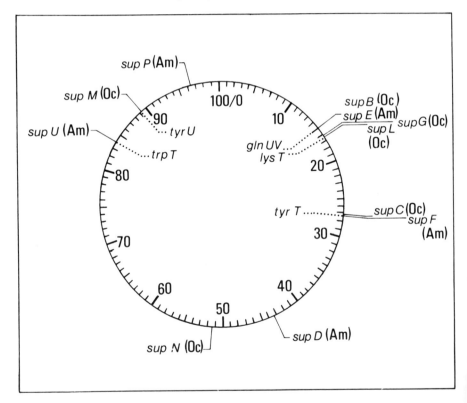

Figure 3.25: *E. coli* **Map Showing the Location of Some tRNA Suppressor Genes**
Suppressor alleles *(external)* and, where known, their equivalent tRNA gene *(internal)* are shown. Thus, *supF* is a mutant allele of *tyrT* and *supU* maps at *trpT* (Table 3.4). Compare with the genetic map for the genes of the translational apparatus (Figure 2.20). Map locations from Bachmann and Low (Bachmann, B.J. and Low, K.B. (1980) *Microbiol. Rev., 44,* 1-56)

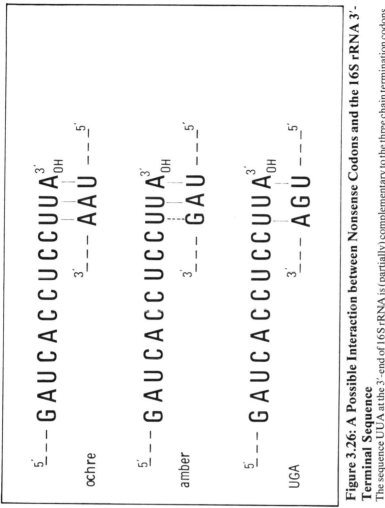

Figure 3.26: A Possible Interaction between Nonsense Codons and the 16S rRNA 3'-Terminal Sequence

The sequence UUA at the 3'-end of 16S rRNA is (partially) complementary to the three chain termination codons, UAA (ochre), UAG (amber) and UGA. Whereas the ochre codon can pair with all three bases in a standard manner *(closed lines)*, UAG interaction would require wobble pairing of 3'G *(dotted line)*; only pairing of two bases is possible with UGA. Since there appears to be an inverse relationship between the strength of chain termination codon-rRNA interaction and suppressor efficiency (ochre suppressors are least efficient, Table 3.4), the 3'-terminal rRNA sequence may play a role in translational termination. Despite the fact that the UUA triplet is not always conserved, the terminal region of 16S rRNA has been implicated in the termination process in *E. coli*. It is certainly involved in translational initiation (Figure 2.22)

Table 3.4ᵃ Nonsense Suppressors

Suppressor[b]	Map Location (min)	tRNA	Codon(s) Suppressed	Anticodon[c] Suppressor	Anticodon[c] Wild-type	Amino Acid Inserted	Efficiency[d]
Amber							
supD	43	nk	5'UAG	nk	nk	Ser	50%
supE	15	Gln2(glnU)	5'UAG	3'AUC	3'GUC	Gln	22%
supF	27	Tyr1(tyrT)	5'UAG	3'AUC	3'AUG	Tyr	68%
supP	96	nk	5'UAG	nk	nk	Leu	55%
supU	84	Trp (trpT)	5'UAG	3'AUC	3'ACC	Gln/Trp[e]	46%
Ochre							
supB	15	Gln1(glnV)	5'UAA/G	3'AUN	3'GUN	Gln	5%
supC	27	Tyr1(tyrT)	5'UAA/G	3'AUU	3'AUG	Tyr	1%
supG	16	nk	5'UAA/G	nk	nk	Lys	3%
UGA							
sup9	84	Trp (trpT)	5'UGA	G24 ⟶ A[f]		Trp	50%

a   Abbreviation: nk, not known.
b   A rather varied nomenclature has been applied to wild-type and suppressor genes In general, *sup⁻* (the negative sign is left out in the above table) refers to the mutant suppressor allele and Su⁺ indicates the resultant suppressor phenotype (Su1, 2, 3, 6, 7, B, C, G and 9 for the suppressors listed). Thus, a wild-type strain with normal *sup⁺* tRNA genes is phenotypically Su⁻.
c   The anticodon base that is changed to allow pairing with a nonsense triplet is underlined. Although the anticodons of some suppressors have not been determined, they are expected to be complementary to the codons with which they interact. (Q is queuosine, a guanosine derivative; N is a 2-thiouridine derivative.)
d   Suppressor efficiency is determined by measuring the expression of a gene carrying a known nonsense mutation. In many instances, enzymatic activity rather than absolute expression is monitored, giving a value which depends upon both suppression of premature translational termination and the suitability of the amino acid inserted (both of which are dictated by reading context).
e   *supU* (*trpT*$_{UAG}$) inserts both glutamine and tryptophan (albeit in the ratio 9 : 1) at amber sites. Since the mutant tRNA carrys only a single base change, an alteration moreover in the anticodon, the insertion of glutamine by *s-apU* indicates that the anticodon is important in the charging reaction.
f   *sup9* (*trpT*$_{UGA}$) is altered in the D stem (Section 2.2.1 (b)) and still carries the normal 3′ACC anticodon, suggesting a role for the D region in anticodon–codon pairing. Note that even the wild-type Trp tRNA species translates UGA at low efficiency, giving rise to 'leaky' UGA lesions. Only one copy of the *trpT* gene exists and, thus, mutations at this locus are generally lethal unless a second, functional copy is present. The *sup9* derivative is an exception since it still translates Trp codons.

*r*ibosomal *am*biguity).

The *rpsL* gene product, ribosomal protein S12 (about 15 000 daltons), appears to be responsible for the *intrinsic infidelity* since certain mutations to streptomycin-resistance (Str-R phenotype), which result in an altered S12 protein, *restrict* this misreading. Nonsense suppressors, for example, frequently show reduced efficiency of suppression in the restrictive *rpsL* (Str-R) background. This restriction can be counteracted, in certain instances, by the addition of streptomycin to the growth medium. Thus, some strains exhibit a *conditional streptomycin dependent* phenotype, Str-D (or CSD), since they require the antibiotic to increase the translational ambiguity. Gorini and co-workers have termed this streptomycin effect *phenotypic suppression* in analogy to informational suppression; the latter is caused by a genetic change in the translational apparatus whereas phenotypic suppression requires a modification of the cellular environment (for example, the presence or absence of streptomycin in the growth medium[11]).

The S4 and S5 ribosomal proteins encoded by *rpsD* and *rpsE* seem, on the other hand, to be required for accurate translation. This involvement in translational fidelity (rather than ambiguity) is suggested by the *ribosomal ambiguity mutant (ram)* class which shows increased misreading (mimicking the action of streptomycin, itself), associated with lesions in the *rpsD* and *rpsE* loci. In addition, the S4 (molecular weight 23 000) and S12 ribosomal proteins apparently interact since there is a near-normal degree of mistranslation in some *rpsL* (Str-R) *rpsD* (Ram) double mutants (Table 3.5).

The natural infidelity of the ribosomal matrix may be important in allowing wobble mispairing (Section 2.2.1 (b)). Certainly, an appreciable amount of ambiguity, intrinsic to codon-anticodon pairing, exists (Table 3.6). Read-through across termination codons, producing functional products, also occurs at a low level in the absence of nonsense suppressors (it may be difficult, therefore, to distinguish between inefficient nonsense suppression and natural 'leakiness'). This translational 'leakiness' can, therefore, be regarded as weak informational suppression since it is mediated through the same molecular process, namely, codon-anticodon pairing.

3.4 Summary: Mutation Identification

Four types of point mutations can arise from a single base change in a codon: samesense, missense, nonsense or frameshift (Section 3.1.1). The triplet formed

11.   The analogue 5-fluorouracil (5-FU) and the unusual nucleotide ppGpp also mediate in phenotypic suppression. Substitution of amber codons with (5-FU) AG and their occasional mispairing as CAG (i.e. a non-heritable U ↔ C transition) results in the translation of some amber terminators as glutamine. The functionality of 5-FU suppressed proteins (as for informational suppression) depends upon whether the amino acid 'replacement', glutamine, is suitable at the terminator sites. Guanosine tetraphosphate acts at the ribosomal level to prevent misreading (Section 9.2.3 (b)).

Table 3.5 Interaction of the Ribosomal Proteins S12 *(rpsL)* and S4 *(rpsD)*

Mutant Allele		Degree of Misreading
rpsL	rpsD	
str A$^+$	ram$^+$	intrinsic low level
str A$^-$	ram$^+$	restriction
str A$^+$	ram$^-$	increased ambiguity
str A$^-$	ram$^-$	near-normal (as str A$^+$ ram$^+$)

Table 3.6 Intrinsic Ambiguity in Codon-Anticodon Pairing as Shown by the Range of Nonsense Suppression[a]

Suppressor	Nonsense Suppression [b]		
	UAG	UAA	UGA
sup$_{UAG}$	+++	+	-
sup$_{UAA}$	+++	++	+
sup$_{UGA}$	-	+	+++

Suppression of UAG by *sup*$_{UAA}$ is explained by wobble (Section 2.2.1 (b)). In the other cases, misreading occurs either at the centre nucleotide or in a 'non-standard' manner at the third base. Streptomycin increases the level of suppression but does not change the misreading specificity; *ram* alleles have the same effect, whereas *strA* decreases suppression.

a   After Strigini and Brickman (1973) (*J. Mol. Biol.,* 75, 659-672).
b   The observed level of suppression increases in the order '-', '+' and '+++'.

in such an event (there are nine possible permutations; Figure 3.27) is dependent upon whether the mutation is a transition or transversion, and which of the three bases is altered. The probability of any one event occurring and, thus, the incidence of one particular mutant codon, is not necessarily equal due to the apparent lack of randomness of mutagenesis (Section 3.2.1 (c)).

The type of lesion formed in a mutagenic event (spontaneous or induced) may be determined by genetic techniques (Table 3.7). Point and multisite mutations are readily distinguished by reversion and analysis. Clearly, a chromosomal rearrangement that includes removal of a region of DNA cannot be repaired to the original state — there is no template upon which to reform these missing sequences. Deletions may, of course, inactivate more than one gene. In addition,

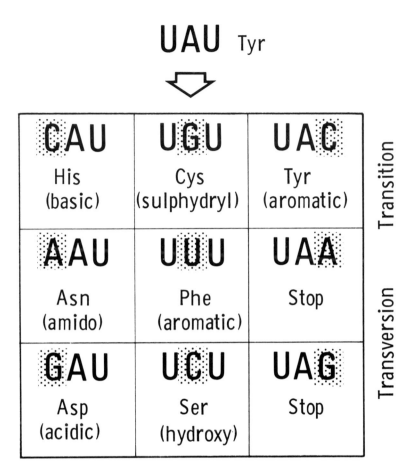

Figure 3.27: The Nine Possible Codons Arising from a Single Base Substitution in the Tyrosine Triplet, UAU

There are a total of 549 (61×9) possible substitutions in the sense codons, of which about one quarter fail to produce amino acid substitutions owing to codon degeneracy (see Degeneracy of the Genetic Code, Table 2.4). These are samesense mutations (Figure 3.3). Thus, in the case of the aromatic amino acid tyrosine, one replacement in the UAU triplet gives rise to the second tyrosine codon, UAC. That is, the tyrosine codons UAY (where Y is a pyrimidine base) are insensitive to transitional mutations in the third position. Note the charge changes, Tyr → His(+1) and Tyr → Asp(−1), which are likely to alter the protein's isoelectric point (see also Table 3.7)

these chromosomal arrangements affect recombination of adjacent markers and are characterised by their failure to recombine with two or more point mutations within regions over which they extend (see Section 7.1). Point mutations, on the other hand, arise from single base changes and are, consequently, revertible: an error in replication, for instance, can both create point lesions as well as restore DNA sequences. Suppression analysis, in particular nonsense suppression, is a useful technique for mutation identification (Table 3.7).

Table 3.7 Characterisation of Structural Gene Lesions

Mutation	Reversion	Type of Analysis		Characteristics of Gene Products	
		Suppression [a]			
		Intragenic	Intergenic (informational)	Size	Charge
missense	spontaneous	+	+	wild-type	wild-type/different [c]
nonsense	spontaneous	-	+	fragment	different [d]
frameshift	f/s mutagen [b]	+	+	different/wild-type [e]	different/wild-type [e]
deletion	-	-	-	different [f]	different [f]

a The specificity and low number of missense (and frameshift) informational suppressors restricts the number of different lesions identifiable by this technique (Table 3.3). Informational suppression, however, is particularly useful for verifying nonsense mutations (Table 3.4).

b Some frameshifts can revert spontaneously as well as being induced to revert, whereas others are not revertible by ICR compounds.

c The observable charge change depends upon both the amino acid replacement and the size of the polypeptide.

d A nonsense fragment would be expected to have a different charge-to-mass ratio owing to its smaller size.

e Frameshift mutations can give rise to a fragment ('uncovering' of a chain termination codon) or a larger gene product ('covering' of a natural terminator), with varying degrees of charge changes.

f The extent of alteration is, of course, dictated by the amount of DNA removed.

Each type of lesion has a characteristic effect upon the mutant protein product (Table 3.7). Molecular weight and charge properties are readily analysed by electrophoresis in semi-solid medium. Chain-termination mutations give rise to truncated polypeptides. The shorter the protein product, the faster it migrates when electrophoresis takes place in polyacrylamide gels in the presence of sodium dodecylsulphate, a process that separates polypeptide chains according to size (see Section 7.2.1). Under non-denaturing conditions, mobility is dependent upon both charge and size. Moreover, the pore size of the semi-solid medium may be adjusted to maximise the charge effect. Missense mutations (or rather their product) may be detected in this type of electrophoretic system since mobility is sensitive to an amino acid substitution, particularly a charge change. Frameshift and multisite mutations would be expected to cause gross alterations in protein structure. Clearly, standard genetic and biochemical techniques allow identification of mutation type. Mapping of genetic lesions — locating the site of genetic changes on the genome — is discussed in Chapter 7.

Bibliography

3.1 Mutation Classification

Hayes, W. (1968) *The Genetics of Bacteria and their Viruses*, 2nd edn (Blackwell, Oxford).

3.2 Mutagenesis

3.2.1 The Molecular Basis of Mutagenesis

(a) *Spontaneous mutation*

Anderson, R.P. and Roth, J.R. (1977) 'Tandem genetic duplications in phage and bacteria', *Ann. Rev. Microbiol., 31*, 473-505.

Cox, E.C. (1976) 'Bacterial mutator genes and the control of spontaneous mutation', *Ann. Rev. Genet., 10*, 135-156.

Starlinger, P. (1977) 'DNA rearrangements in prokaryotes', *Ann. Rev. Genet., 11*, 103-126.

Weisberg, R.A. and Adhya, S. (1977) 'Illegitimate recombination in bacteria and bacteriophage', *Ann. Rev. Genet., 11*, 451-473.

(b) *Induced mutation*

Ames, B.N., McCann, J. and Yamasaki, E. (1975) 'Methods for detecting carcinogens and mutagens with the *Salmonella*/mammalian-microsome mutagenicity test', *Mut. Res., 31*, 347-364.

Ames, B. and Hooper, K. (1978) 'Does carcinogenic potency correlate with mutagenic potency in the Ames assay?', *Nature, 274*, 19-22.

Bridges, B.A. (1976) 'Short term screening tests for carcinogens', *Nature, 261*, 195-200.

Drake, J.W. and Baltz, R.H. (1976) 'The biochemistry of mutagenesis', *Ann. Rev. Biochem., 45*, 11-37.

Nagao, M., Sugimura, T. and Matsushima, T. (1978) 'Environmental mutagens and carcinogens', *Ann. Rev. Genet., 12*, 117-159.

Roth, J.R. (1974) 'Frameshift mutations', *Ann. Rev. Genet., 8*, 319-346.

Streisinger, G., Okada, Y., Emrich, J., Newton, J., Tsugita, A., Terzaghi, E. and Inouye, M. (1966) 'Frameshift mutations and the genetic code', *Cold Spring Harbor Symp. Quant. Biol., 31*, 77-84.

Witkin, E.M. (1976) 'Ultraviolet mutagenesis and inducible DNA repair in *Escherichia coli*', *Bacteriol. Rev., 40*, 869-907.

Zimmermann, F.K. (1977) 'Genetic effects of nitrous acid', *Mut. Res., 39*, 127-148.

(c) *Mutational hotspots*

Coulondre, C., Miller, J.H., Farabaugh, P.J. and Gilbert, W. (1978) 'Molecular basis of base substitution hotspots in *Escherichia coli*', *Nature, 274*, 775-780.

Farabaugh, P.J. and Miller, J.H. (1978) 'Genetic studies of the *lac* repressor, VII: on the molecular nature of spontaneous hotspots in the *lacI* gene of *Escherichia coli*', *J. Mol. Biol., 126*, 847-863.

3.2.2 The Application of Mutagenesis to Bacterial Systems

Hopwood, D.A. (1970) 'The isolation of mutants', in J.R. Norris and D.W. Ribbons (eds.), *Methods in Microbiology, 3A* (Academic Press, London), pp. 363-433.

Lederberg, J. and Lederberg, E.M. (1952) 'Replica-plating and indirect selection of bacterial mutants', *J. Bacteriol., 63*, 399-406.

Luria, S.E. and Delbrück, M. (1943) 'Mutations of bacteria from virus sensitivity to virus resistance', *Genetics, 28*, 491-511.

Miller, J.H. (1972) *Experiments in Molecular Genetics* (Cold Spring Harbor Laboratory, New York).

Tomasz, A. (1979) 'The mechanism of the irreversible antimicrobial effects of penicillins: How the beta-lactam antiobiotics kill and lyse bacteria'. *Ann. Rev. Microbiol., 33*, 113-137.

3.3 Suppression

Crick, F.H.C., Barnett, L., Brenner, S. and Watts-Tobin, R.J. (1961) 'Triplet nature of the code', *Nature, 192*, 1227-1232.

Gorini, L. (1974) 'Streptomycin and misreading of the genetic code', in M. Nomura, A. Tissieres and P. Lengyel (eds.), *Ribosomes* (Cold Spring Harbor Laboratory, New York), pp. 791-803.

Hartman, P.E. and Roth, J.R. (1973) 'Mechanisms of suppression' *Adv. Genet., 17*, 1-105.

Hill, C.W. (1975) 'Informational suppression of missense mutations', *Cell, 6*, 419-427.

Ninio, J. (1974) 'A semi-quantitative treatment of missense and nonsense suppression in the *strA* and *ram* ribosomal mutants of *Escherichia coli*: Evaluation of some molecular parameters of translation *in vivo*', *J. Mol. Biol., 84*, 297-313.

Seege, D.A. and Söll, D.G. (1979) 'Suppression', in R.F. Goldberger (ed.), *Biological Regulation and Development I: Gene Expression* (Plenum Press, New York), pp. 433-485.

Smith, J.D. (1979) 'Suppressor tRNAs in prokaryotes', in J.E. Celis and J.D. Smith (eds.), *Nonsense Mutations and tRNA Suppressors* (Academic Press, London), pp. 109-125.

Part III
Gene Transfer

4 Plasmids

Some cells harbour small *extrachromosomal genetic elements* in addition to, and distinct from, the bacterial genome ('extrachromosomal' is used in this context to indicate that these elements are present in the cytoplasm and are not covalently linked to the chromosome). The term plasmid has been employed to describe those extrachromosomal genetic elements that are stably inherited.[1] Plasmids are distinct replicons — genetic units capable of replicating independently from the host chromosome — consisting of covalently-closed circular molecules of DNA. The replicated unit is 'trapped' within the daughter cell during septum formation, possibly through association with the cytoplasmic membrane (Section 6.1.2).

Whereas no genetic exchange normally exists between plasmid-free bacteria (or bacteria harbouring *non-conjugative* elements), plasmids may mediate in the transfer of genetic material from one organism (the *donor*) to another (the *recipient*) — a process termed *bacterial conjugation*. This property of *conjugative* plasmids, which allows both recombination and complementation analysis, has elevated the field of bacterial genetics (of *E. coli* and other bacterial species in which bacterial conjugation occurs) to a powerful science. In general, naturally occurring bacterial plasmids are dispensable but may be beneficial to the host cell. These genetic elements behave, therefore, rather like small supernumerary chromosomes that have gained the ability to transfer between cells. Moreover, in a manner analogous to their lethal counterparts — the bacterial viruses (Chapter 5) — certain plasmids may exist either autonomously in the cytoplasm or integrated into the bacterial chromosome. Such *episomes* are a subgroup. Thus, though all episomes are plasmids, some plasmids exist only in the extrachromosomal state.

What is the dividing line between a plasmid and bacteriophage? Plasmids are, in general, non-lethal replicons (at least to their host cells), whereas bacterial viruses are potentially lethal replicons encapsidated in a protein coat. The division, certainly, is not always clear; there is no strictly logical distinction between a non-viral extrachromosomal genetic element and a defective prophage (consider, for example, the lysogenic state of bacteriophage P1 or the plasmid derivative of *lambda, λdv*, Section 5.1.2 (c)). It is interesting to speculate

1.　Thus all plasmids are extrachromosomal genetic elements: the reverse is not necessarily true. A generalised transducing fragment, for example, is a genetic element that is not stably inherited unless recombined into the chromosome (Section 5.2.1).

whether phages, particularly temperate phages, are the evolutionary product of plasmids.

4.1 The F Plasmid

The first extrachromosomal genetic element identified in *E. coli* was the F plasmid (or F sex factor as it was originally called).[2] The F plasmid is a small covalently-closed circular molecule of DNA, approximately two per cent the size of the bacterial chromosome (62 Mdal, 94.5 kb). There is sufficient DNA to encode about 94 average-sized proteins; approximately one third of the genetic element (the *tra* region) is involved in conjugation (the genetic organisation of the F plasmid is considered in Section 10.1).

F is an episome since it replicates either independently of the bacterial chromosome, or as part of the host genome. The *autonomous* and *integrated* states are mutually exclusive.

4.1.1 Vegetative Replication

The F plasmid is stably maintained: it replicates as a single unit more or less in synchrony with the host genome and segregates into each of the two daughter cells at division. Initiation of replication is dependent upon host RNA and protein synthesis but evidently independent of the *dnaA* gene product (required for initiation of new rounds of chromosomal replication, Section 6.1.1 (b)). The host-mediated cytoplasmic factors work in conjunction with plasmid-specific initiation and/or repressor substances (see below). This *vegetative* form of replication (see F Transfer, below, and Section 6.1.3 (a)) of the cytoplasmic autonomous unit involves a unique site of initiation on the plasmid DNA, the origin *oriV*, and is likely to be in the Cairn's bidirectional model (Section 1.1.2 (b)).

The number of plasmid copies is accurately controlled at one to two per chromosome (although F replication is not directly coupled with chromosomal replication or cell division); *copy number* depends upon bacterial growth rate, increasing as generation time increases. Two main models have been proposed for *stringent* control of F plasmid replication (and for putative chromosomal regulatory system(s)), with the rate of replication being governed by the timing of initiation of new rounds of plasmid replication.

2.   The discovery that genetic material could be transferred in *E. coli* by cell-to-cell contact, mediated by the F plasmid, has led to a nomenclature steeped in sexual terms. Thus, the transfer process has been described as *sexual conjugation* (now called *bacterial conjugation*), the F plasmid is the *sex plasmid* (or *fertility factor*), and the cells involved are the *male (donor)* and *female (recipient)*. Even the popular press have had banner headlines proclaiming 'bacteria have sex'! However, as sexual reproduction in eukaryotes normally involves complete nuclear and cytoplasmic fusion of two haploid gametes (carrying homologous chromosomes) to form the diploid zygote, and since it does not result in a female being converted into a male, bacterial conjugation is not sexual in the true sense of the word. Certainly, the male and female cells are different, both morphologically and cytoplasmically. 'Factor' has now been replaced by the more specific term 'plasmid'.

Positive control: This model (based originally upon the replicon concept of Jacob, Brenner and Cuzin (1963)), postulates the existence of an effector of replication: a specific plasmid attachment site on the cytoplasmic membrane which duplicates during cellular growth, triggering the initiation of plasmid replication. Segregation occurs through membrane growth between attachment sites, and septum formation.

Negative control: On this hypothesis, a diffusible plasmid-coded repressor, produced after initiation, inhibits further initiation events (Pritchard, Barth and Collins (1969)). A new round of replication is started when the (putative) repressor is diluted below a critical concentration during subsequent cell growth.

The repressor model is currently favoured, particularly as it more readily explains plasmid incompatibility (Section 4.1.2(b)). However, a membrane complex is likely to be involved at some stage in replication, notably in the physical and symmetric distribution of replicated plasmids to the two daughter cells during division.

Though stably maintained under normal conditions, the cytoplasmic plasmid may be lost during bacterial growth. The *cured* cell will then lack phenotypic characters (see Section 4.1.2(a), below) associated with the F plasmid. Curing may occur spontaneously, caused presumably by some perturbation to the replicative machinery (Figure 4.1). Alternatively, acridine orange at pH 7.6 can be used to isolate cured derivatives. The precise mechanism of acridine orange curing is not known — the dye probably influences plasmid replication by preferentially inhibiting plasmid, relative to chromosomal, initiation. That the effect of this intercalator is specific to F-mediated replication is indicated by the resistance to curing of hybrid derivatives of this plasmid (F-Primes, see Section 4.1.3(b)) carrying the chromosomal origin of replication, *oriC* (replication is presumably initiated at *oriC* when F-directed replication is inhibited by acridine orange). The integrated state of F (Section 4.1.3(a)) can, therefore, be distinguished from the autonomous state for only the latter is readily susceptible to curing.

4.1.2 F Transfer

Previous discussion on F replication has been concerned with the maintenance of plasmid copy number. In this replication mode, vegetative DNA synthesis is necessary to avoid plasmid loss during cellular growth. The importance of F lies in the fact that it is a *conjugative* plasmid: it exhibits the property of DNA transmission between bacterial cells (both of F material and other nucleic acid sequences). This involves a mode of DNA production termed *transfer replication.*

(a) The Mechanism of Transfer. Cells that carry the F plasmid, F^+ bacteria, are morphologically distinct from their F^- counterparts. In addition to the large number of small 'common' pili frequently present, there are a few (typically, one to three for an exponential culture) long, thin proteinaceous appendages, termed

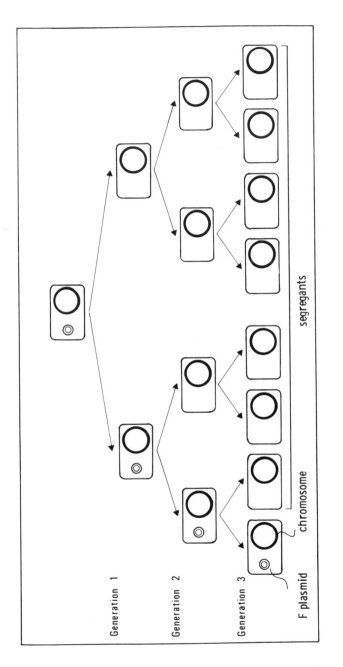

Figure 4.1: Segregation of a Non-Replicated Plasmid: Unilinear Inheritance

If, for some reason, an autonomous replicon fails to get copied, the unreplicated element will partition into one of the two daughter cells at division (though each sib will, of course, carry a copy of the bacterial chromosome). At each successive round, this process will be repeated — unless the segregated replication regains the ability to replicate correctly — and, thus, the proportion of cells carrying the element decreases exponentially ($\frac{1}{2}$, $\frac{1}{4}$, $\frac{1}{8}$ etc.)

F (or sex) *pili* (see Figure 1.14). These thread-like appendages (up to several micrometres in length and about 8 nm in diameter) are primarily made up of subunits of a phosphorylated glycoprotein, *F pilin* (about 11 000 daltons; *traA* or *traJ* product, see Section 10.1.1 (a)), forming a hollow cylindrical rod (about 2 nm inner diameter). F pili are encoded by the F plasmid and, thus, disappear concomitantly with plasmid loss; they are said to be *male-specific*. The F pilus, itself, is the adsorption site for male-specific bacteriophages (Section 5.1.2): filamentous DNA phages adsorb to the tip, whereas spherical RNA phages adsorb to the sides of the pilus (see Figure 1.14). An F^+ cell cured of its plasmid by, say, acridine orange treatment, becomes resistant to both types of male-specific virions. These phages can, therefore, be used as probes for the determination of pilus function.

Cell-cell interaction between the F^+ donor and F^- recipient occurs shortly after mixing donor and recipient cultures. Formation of the *mating-pair*[3] appears to be a random collision process, dependent upon cell density, temperature and viscosity of the mating medium. The tip of the pilus is involved in stable union formation, apparently interacting with a specific recognition site on the surface of the F^- cell (stable pair formation involves the *ompA* gene product, a major outer membrane protein of molecular weight 28 000 present in about 10^5 copies per cell). The pilus may contract or be resorbed upon interaction with this bacterial receptor. Though mating-pair formation is obligatory for gene transfer, transfer does not occur invariably upon this conjugal event. Moreover, it is still unclear whether DNA is actually transferred through the pilus (a *conjugation tube*) or whether this appendage just serves to bring the bacteria close together to form wall-to-wall contact upon pilus contraction. Certainly, the F pilus is required at an early stage in bacterial conjugation prior to DNA transfer. The less specific term *conjugation bridge* might, therefore, be preferable.

It should be emphasised that there is a natural barrier against the entrance of 'naked' DNA; the multi-layered rigid envelope. Thus, in some manner as yet undefined, fusion of the two cell envelopes is effected in bacterial conjugation to allow DNA passage. Polyelectrolyte penetration of the lipid bilayers — those of the outer and inner membrane — may require a protein channel. It is not known, in fact, whether the transferred strand is truly naked or has a 'pilot' protein bound at the 5'-terminus. Perhaps, adhesion points (Section 1.2.1) mediate in this transmembrane transport: the observed proximity of F pili to such sites of fusion suggests a role in analogy with their proposed function in viral DNA entrance (Section 5.1.2).

Transfer of F DNA in an $F^+ \times F^-$ *'cross'* is initiated from a unique site on the plasmid, *oriT* (the origin of transfer is distinct from the origin of vegetative replication), shortly after stable mating-pair formation (Figure 4.2). A specific

---

3.   There is now good evidence that mating mixtures form *mating aggregates* consisting of up to fifty cells, rather than only mating pairs. The latter term has been retained in the text to facilitate description of the transfer process.

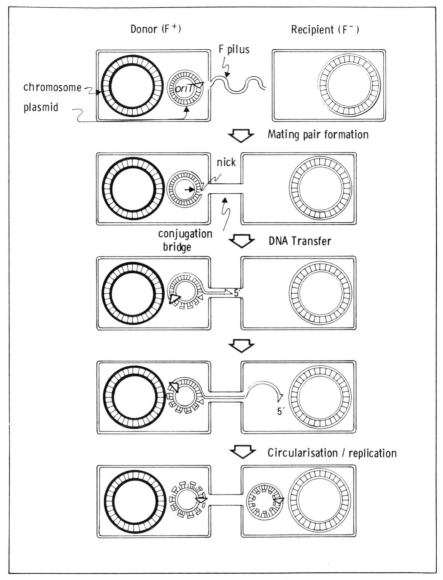

Figure 4.2: Possible Steps in F Plasmid Transfer

The process is initiated by cell-to-cell contact, mediated by the F-plasmid coded pilus on the donor cell. It is still unclear whether the F pilus supplies a tube through which single-stranded DNA is transferred across the two cell walls or whether some sort of additional bridge is set up between the cells once contact has been made. A nick at the origin of transfer site, *oriT*, supplies the 5'-terminus that invades the recipient cell (F DNA synthesis can but need not occur simultaneously on the intact covalently-closed circular molecule of single-stranded DNA, see also Hfr-Mediated Transfer of Chromosomal DNA, Figure 4.4). Transfer of a genetic element capable of autonomous replication requires production of the complementary molecule on the transferred strand (continuous DNA synthesis is shown by *dashed lines*), and circularisation. This rolling-circle mode of replication, transfer replication, is discussed in Section 6.1.3 (a) (see Figure 6.13) plasmid-coded endonuclease (the *traYZ* gene products), triggered by conjugal-pair formation, nicks the covalently-closed circular molecule at this site. Transfer is orientated and unidirectional: a single linear strand of F DNA enters the recipient, 5'-end first, leaving the complementary strand behind in the donor (a protein bound at the 5'-terminus could assist in DNA recircularisation after transfer). F transfer proceeds at a uniform rate of about 10^4 nucleotides per minute at $37°$ (approximately equal to the rate of bidirectional replication of the bacterial chromosome) and is, therefore, completed in about one to two minutes. Normally, in the case of an $F^+ \times F^-$ cross, both single strands are immediately used as templates for synthesis of double-stranded DNA by the bacterial DNA polymerase III (the 'transferred' strand is expected to be replicated discontinuously as the 5'-terminus is transferred first, Section 6.1.3 (a)). However, coupling of DNA transfer with replication in donor or recipient is not obligatory for DNA movement. How is the parental F plasmid replicated during transfer? The rolling-circle mode of replication is favoured (Section 6.1.3 (a)), the remaining intact strand acting as template for *de novo* DNA synthesis, primed by the 3'-terminus of the nicked strand. Transfer replication is distinct from vegetative plasmid (and chromosome) replication since it appears to proceed in the absence of the host *dnaB* gene product (a protein involved in the initiation and elongation stages of host DNA replication, Section 6.1.1 (b)).

The recipient cell can be seen to co-operate in the survival of the transferred nucleic acid as it allows the single strand of 'foreign' DNA to replicate and circularise in the presence of endogenous nucleases, and the stable inheritance of the mature double-stranded molecule (the conversion of the transferred strand to an element capable of autonomous maintenance has been called *repliconation*). Both plasmid and host-coded functions are required for this survival. The product of the conjugal event, if DNA replication and re-circularisation occurs, is the transfer of an intact covalently-closed circular F plasmid to the F^- recipient and the retention of an identical copy in the donor. Thus, the *transconjugant* — the recipient cell harbouring the transferred plasmid — becomes a further vehicle for F transmission.

(b) Barriers to Transfer. Transfer of the F plasmid to an F^- recipient might at first appear to be an inefficient process. Consider a mating mixture consisting of equal numbers of F^+ donors and F^- recipients: in the absence of specific transfer barriers, there will be an equal probability of transfer to another donor or to a recipient. In fact, the efficiency of conjugation approaches 100 per cent under optimal conditions. Transfer is normally restricted to F^- cells since bacteria that already harbour the plasmid are poor recipients. A further mechanism inhibits replication of plasmids that have escaped this barrier.

Entry (or Surface) Exclusion, Eex: Two F-coded gene products, the *traS* and *traT* proteins (18 and 25 kdal, respectively; the latter an outer membrane protein present in about 10^4-10^5 copies per cell) are responsible for reduced formation of stable mating-pairs between cells carrying homologous plasmids. This phenomenon (previously designated surface exclusion or Sex), which is independent of F pilus function, presumably operates through outer membrane effects. The barrier is partly overcome in stationary phase (starved) donor cultures. Expression of the entry exclusion effect is reduced in these phenotypically F⁻ (Eex⁻) cells, termed *F⁻ phenocopies*. This change is only temporary; the normal surface properties are regained upon subculture in fresh medium. *Phenocopy mating* has proved useful for the transfer of F and its derivatives (Section 4.1.3 (b)) into strains already carrying an F plasmid.

Incompatibility, Inc: Plasmids may be grouped according to their inability to co-exist stably in the cytoplasm of their bacterial host. Thus, the F plasmid is the prototype for the IncFI group (see also Section 4.2). In general, incompatible genetic elements of the same group are excluded symmetrically, either of the pair being stably maintained. This barrier occurs after a plasmid has entered a cell already harbouring a plasmid of the same group, and is thought to result from inhibition of plasmid replication (plasmids of the same Inc group, presumably, have similar mechanisms for replicative control, see below); segregation of one plasmid ensues.

These two barriers are distinct. Lack of establishment due to incompatibility results from failure of the transferred plasmid to replicate rather than transfer inhibition as in the case of the exclusion mechanism. Moreover, they are also independent, for plasmids of the same Inc group may exhibit different entry exclusion properties and *vice versa*.

According to the positive control model for F plasmid replication (Section 4.1.1, above), incompatible plasmids compete for a limited number of specific membrane maintenance sites, while compatible plasmids of different Inc groups initiate replication at different sites. However, this maintenance site (positive control) model was initially proposed before the widespread nature of bacterial plasmids was known (at least 26 Inc groups among the plasmids of enteric bacteria have been classified to date). It now appears unwieldy as control mediated through a bacterial effector, rather than a plasmid-coded factor, implies the pre-existence of numerous sites on the cytoplasmic membrane, a different one for each incompatibility group. The negative control model requires a common plasmid-encoded repressor molecule for control of incompatible plasmids. Introduction of the second plasmid increases the concentration of this controlling element and thereby ensures that the total number of copies of these two plasmids does not exceed the normal copy number of one of them alone; only one of the two plasmids segregates at cell division (discussed further in Sections 6.1.2 (c) and 6.1.3 (a) with respect to replication control mechnisms).

4.1.3 F-Mediated Transfer

In addition to its property of self-transfer, the F plasmid also mediates in the exchange of DNA sequences other than that of the plasmid, itself. This process of gene transfer using the F plasmid as a genetic vector is sometimes referred to as *F-duction* (or sexduction) in analogy with phage-mediated gene transfer (see Transduction, Section 5.2.2). It has allowed recombination and complementation analysis of bacterial genes, techniques that require a partial diploid status (discussed in Section 7.1.2). F-duction takes two forms: mobilisation of the whole bacterial chromosome (an Hfr $\times$ F$^-$ cross), or transfer of small regions of donor material (an F-prime $\times$ F$^-$ cross).

(a) F Integration: Hfr Formation. Mobilisation of donor DNA stems from insertion of the F plasmid into the chromosome. Bacteria that harbour this plasmid in the integrated, rather than cytoplasmic, state are said to be *Hfrs* since they give a high frequency of recombination of donor markers. Chromosome mobilisation in Hfr strains may be considered as F self-transfer, in which F and the bacterial chromosome form a *cointegrate* — the fusion of two complete replicons.

(i) *Insertion.* Insertion of F into the chromosome occurs at a frequency of about 10^{-5}-10^{-7} per generation. Campbell (1962) has proposed that integration involves interaction between the two, covalently-closed circular molecules of DNA — the F plasmid and the host chromosome — resulting in a reciprocal crossover event to form a single circular structure, in which the bacterial and episomal genes are contiguous. The inserted linear F DNA is then a circular permutation of the autonomous plasmid. The effect of splitting the plasmid DNA at the site of insertion is to move F genes in this region to the ends of the linear integrated form as well as to increase the distance between chromosomal markers lying on either side of the episome (see Figure 4.3). Neither F nor chromosomal DNA is lost in this *integrative recombination* step (see Section 6.2.2 (a)). Moreover, the process is reversible, leading to the reconstruction of the circular cytoplasmic genetic element (F Excision, see below). There is convincing support for this model, both with regard to F plasmid and bacteriophage *lambda* integration (Section 5.2.2).

Integration is not a random event. There appears to be a number of specific sites scattered around the chromosome (a review by Low (1972) reports 22, see Figure 4.8) at which F inserts in a fixed orientation, normally in non-essential regions (this may be compared with the secondary attachment sites of bacteriophage *lambda*, see Section 6.2.2 (ai)). These regions, presumably, show homology with some portion of the F plasmid (a rationale for this, in terms of *insertion sequences*, will be discussed in Sections 6.2.2 (aii) and 10.1.1 (c)). This type of site-specific integration seems to require, in general, a functional host recombination system (*recA*$^+$ gene product) suggesting that the F plasmid does not encode an integrase-like enzyme (cf. λint, Section 5.2.2) that acts in the absence of the *recA* product. (There are, however, reports of *recA*-independent Hfr formation, albeit at low frequency). Once integrated, the F replicon can 'drive' replication of the host chromosome under conditions in which the host *dnaA* gene product (required for initiation of DNA synthesis at the bacterial

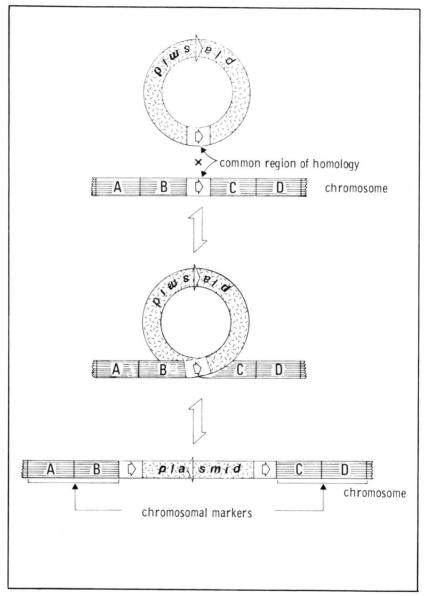

Figure 4.3: The Campbell Model for F Plasmid Insertion (and Removal)
Integrative recombination of F appears to be mediated, in the main, by small regions of homology *(arrows)* termed insertion sequences (see Section 10.1.1 (c)). The general recombination system is, however, responsible for the majority of insertion events. Integration increases the distance between chromosomal markers. The forward and reverse reactions, insertion/excision, are generally accurate processes but errors may occasionally occur (see Accurate and Aberrant Excision of F, Figure 4.10). This model applies to both plasmids and phage *lambda* (Section 5.2.1, Figure 5.14). After Campbell (Campbell, A.M. (1962) *Adv. Genet., 11,* 101-145)

origin of replication, *oriC*, Section 6.1.2 (ai)) is non-functional. This suppression of the mutant DnaA⁻ phenotype by plasmid integration has been termed *integrative suppression*.

(ii) *Hfr Transfer*. The integrated F plasmid is clearly not in a passive state; integrative suppression indicates that the replicon, itself, is active under certain conditions. In addition, the F transfer system is also functional, even though the plasmid is contained within the fifty-fold larger bacterial chromosome. The integrated plasmid is, however, no longer independently transmissible; transfer involves both F and bacterial DNA.

Mobilisation of the bacterial chromosome appears to be similar to F transfer (Section 4.1.2(a), above).[4] In an *Hfr* × *F⁻* cross, mating-pair formation, mediated by donor sex pili, is a necessary prerequisite for DNA transfer. Transfer is initiated at *oriT* on the integrated plasmid (arrowed in Figure 4.4), the 5′-end first entering the recipient. This orientated transfer, starting from a region within the integrated plasmid DNA, results in the early entry of a small part (only) of the F DNA, followed by the bacterial genes close to the leading side of *oriT*.

Transfer proceeds at a uniform rate of about one per cent of the chromosome per minute (i.e. about 10^4 nucleotides per minute at $37°$, the same rate as for F transfer, Section 4.1.2 (a)). The size of the bacterial genome indicates that a minimum time of one hundred minutes is required for transfer of the total chromosome. Consequently, the order of bacterial genes entering the recipient bacterium is temporally controlled; cistrons close to the leading side of *oriT* will enter early, those far from *oriT* will be transferred at significantly later times (transfer of the autonomous plasmid, on the other hand, requires only one or two minutes). Entry of the entire chromosome of the donor cell occurs very rarely (from about 10^{-4} of donor cells, except under special conditions such as immobilisation of the mating mixture on filters) since random shear of the linear single-stranded donor DNA terminates the process spontaneously, leaving a portion of DNA within the recipient's cytoplasm.

The presence of a donor fragment within the recipient creates, transiently, a *partial diploid* (or *merodiploid*) for the bacterial genes carried on the fragment. What is the fate of this 'ingested' DNA? As only part of the F plasmid is initially transferred due to splitting of the resident genetic element at *oriT* and since mobilisation rarely involves the entire chromosome, DNA that survives the transfer process is limited to the leading region of F and a portion of donor chromosome (whereas the autonomous genetic element is transferred *in toto*). This incomplete chromosome is likely to have lost the bacterial origin of replication (*oriC* at 83 min, see Section 6.1.2 (ai)) and will lack the region on the

---

4. An excess of Hfr cells has a killing effect on F⁻ but not F⁺ cells. The resident plasmid protects against this *lethal zygosis* both by surface exclusion (Section 4.1.2 (b)) and by a second line of defence, termed *ilz* (*i*mmunity to *l*ethal *z*ygosis). The fact that lethal zygosis is exhibited in Hfr but not F crosses suggests *ilz* is not in the leading F segment (see also Section 10.1.1 (a)).

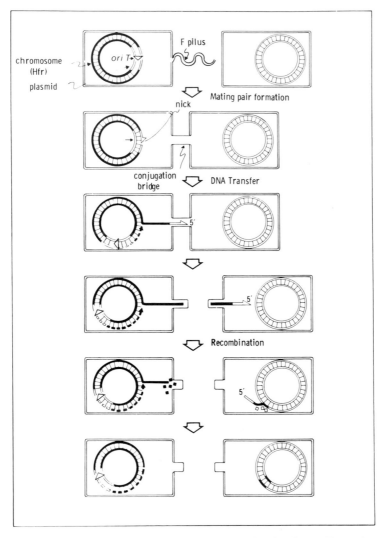

Figure 4.4: A Possible Mechanism for Hfr-Mediated Transfer of Chromosomal DNA

Hfr and F transfer (Figure 4.2) would seem to involve the same processes: cell-to-cell contact, strand breakage and single-strand transfer through some sort of conjugation bridge set up between the donor and recipient cells. However, survival of the Hfr strain requires concomitant DNA extension from the residual 3'-OH (continuous DNA synthesis is shown by *dashed lines;* see Rolling-Circle Mode of Replication, Section 6.1.3 (a)). A further difference between Hfr and F transfer is that only a portion of donor DNA, the 5'-leading strand, is transferred, owing to random rupture of the conjugation bridge prior to entrance of the entire bacterial chromosome (fifty-fold larger than the F plasmid, itself). Note that F loci necessary for transfer (Section 10.1.1) are last to enter the recipient cell. The majority of transferred fragments of DNA cannot be maintained as such (see Segregation of a Non-Replicated Plasmid, Figure 4.1) and require recombination with the recipient for preservation. This later process may be a single-strand or double-strand event (Figure 4.5)

F plasmid encoding the transfer functions (*tra* operon, Section 10.1.1 (a)). The fragment is unlikely to be circularised nor stably maintained and will be lost upon cell division. However, the chromosomal part of the donor strand may integrate into the recipient's chromosome by recombination with a homologous region.

Integration requires a functional host recombination system. Rec-mediated recombination possibly occurs through formation of a transient heteroduplex intermediate and subsequent ligation (Figure 4.5). Certainly, there is evidence for processing of heteroduplex joints. Alternatively, both strands of the recipient are replaced by a single donor strand, which can then be used as a template for synthesis of the complementary molecule. Both models, in which either one or both strands of the recipient is replaced by donor DNA, require some homology for pairing (though the first requires a greater extent). It is still unclear whether this recombination event occurs prior to replication of the transferred strand. Whatever the mechanism, part of the recipient's chromosome is replaced by the transferred donor fragment through breaking and rejoining. The donor fragment, once integrated, is replicated as part of the chromosome and is, therefore, stably maintained, passing to daughter cells in conjunction with the chromosome.

The frequency of transfer of bacterial genes close to the leading edge (i.e. early cistrons) approaches 100 per cent under optimal conditions. Once transferred, these sequences can integrate by recombination at an efficiency of about 0.5. The overall recombination frequency is, therefore, about 50 per cent. Strains that give *high frequency* of *recombination* of donor characters, albeit to a limited region of the donor chromosome, are termed *Hfr*. F$^+$ populations give low frequency of recombination (10^3-10^4-fold reduced) arising from the small proportion of Hfr cells (10^{-5}-10^{-7} per generation) formed through integrative recombination.

(iii) *The One Hundred Minute Map.* A particular Hfr strain transfers chromosomal markers in a characteristic and highly reproducible order, readily ascertained in practice. Hfr transfer is interrupted at various times by vigorous agitation, and the temporal order of inheritance by the F$^-$ population of Hfr genetic markers is determined on selective media (Figure 4.6). The time of entry of donor gene(s) into the recipient serves as a measure of genetic distance.[5] That is, in this novel situation, the temporal unit is used for measuring relative, molecular distances on the bacterial chromosome (Figure 4.7). Two assumptions are made for the translation of genetic distance into a time scale: firstly, rate of chromosome mobilisation is uniform upon formation of the stable mating-pair; and, secondly, recombination frequency per unit length is constant, independent of the region involved. These assumptions are upheld for the proximal third of the transferred chromosome (excluding the bacterial genes close to the leading F DNA).

5.   Time of entry mapping involves the determination of the entry of unselected marker(s) relative to a selected marker to allow for the time required for stable mating-pair formation and the onset of DNA transfer (see Section 7.1.2 (ai)).

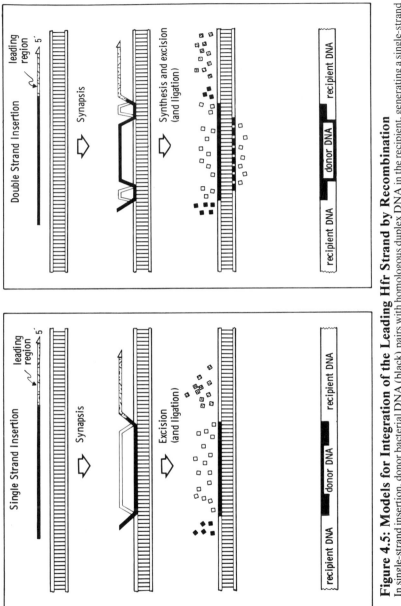

Figure 4.5: Models for Integration of the Leading Hfr Strand by Recombination

In single-strand insertion, donor bacterial DNA (black) pairs with homologous duplex DNA in the recipient, generating a single-strand loop (and tails) susceptible to nucleolytic attack. In the double-stranded event, flanking regions of donor material synapse and then the intervening portion of the recipient's duplex is removed, the transferred strand acting as template for repair synthesis (black dotted line). Note that replication of the product as a single-strand gives rise to both parental and recombinant molecules

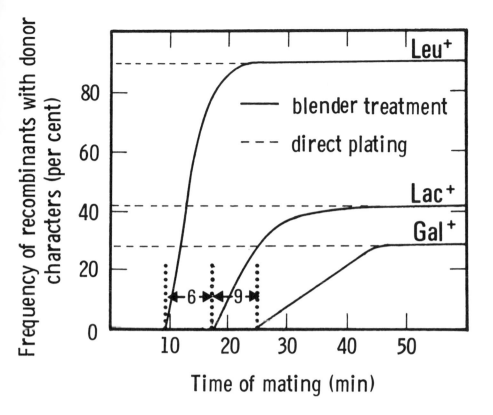

Figure 4.6: Temporal Order of Inheritance; The Results for *leu, lac* **and** *gal*
Aliquots of a liquid mating mixture were taken at various times and either plated directly on selective medium (medium lacking leucine, or containing lactose or galactose instead of glucose as sole carbon source) or treated initially in a Waring blender to disrupt all mating pairs. Recombinants were isolated by selecting for a distal locus and screening for unselected markers (*leu, lac* or *gal*). Time of entry mapping involves determination of the entry of unselected marker(s) relative to a selected one to allow for the period required for stable mating-pair formation and the onset of DNA transfer (Section 7.1.2 (ai)). Rate of transfer appears to be uniform though it is complicated by the apparent asynchrony of initiation in a population of Hfr bacteria that increases with distal markers (this can explain the slopes). The gradient of transmission (giving rise to the different plateau heights) appears to be due to spontaneous random breakage of the chromosome during its transfer from donor to recipient. The probability of random shear increases with time and, therefore, with genetic distance. The probability that any donor marker is incorporated by recombination (0.5) is reduced for markers close to the origin of transfer, presumably restricted by non-homologous F DNA. The experiment shown gives different intervals for *leu-lac* and *lac-gal* (9.5 to 6 min) than is now believed; the current values (6 and 9 min) are placed between the dotted lines (see also Figure 4.7). After Jacob and Wollman (Jacob, F. and Wollman, E.L. (1961) *Sexuality and the Genetics of Bacteria* (Academic Press, New York))

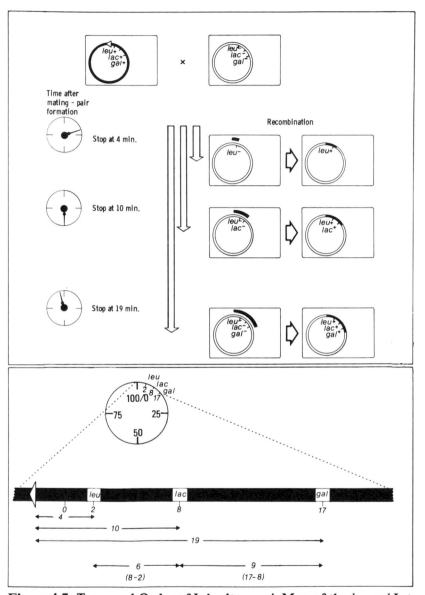

Figure 4.7: Temporal Order of Inheritance: A Map of the *leu-gal* Interval
The orientated nature, and uniform rate, of Hfr transfer (Figure 4.4) implies that the extent of DNA transferred is proportional to the time allowed for strand transfer (the time for mating-pair formation may be considerably longer than that for DNA transfer, itself). Thus, the time of entry of each donor gene into a F⁻ recipient provides a measure of genetic distance (assuming that recombination is constant for any region of the bacterial chromosome). In the present example, the Hfr strain (Hfr H, see Figures 4.8 and 4.9) transfers the *leu, lac* and *gal* loci in 4, 10 and 19 min, respectively, after the onset of the transfer process (a), placing these sites at 2, 8 and 17 min on the *E. coli* genetic map (b). Mapping is discussed in Section 7.1

The entire *E. coli* chromosome may be transferred in one hundred minutes at 37° under suitable conditions. This has led to the division of the circular genetic map into units of time, the total map covering one hundred minutes (see Figure 4.8, for example).

The direction of chromosome mobilisation is governed by the orientation of *oriT* and, hence, of the inserted plasmid. Moreover, the bacterial genes that are carried on the leading sequence (the 'early' genes) are dictated by the site of insertion (Figure 4.8). All loci on the bacterial map can be located in relation to

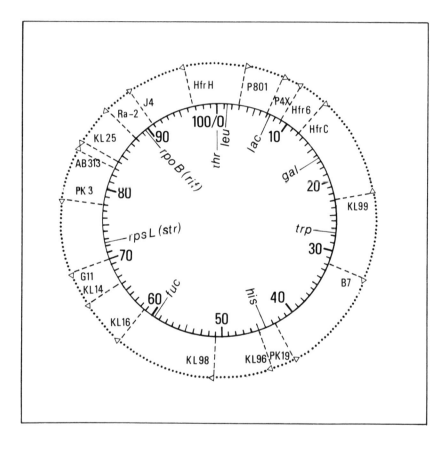

Figure 4.8: Sites of F Plasmid Insertion on the *E. coli* Chromosome
Open arrowheads indicate the sites of insertion and orientation of transfer (gene symbols are explained in Table 1.3/Figure 1.13). Thus, F in HfrH is located at about 98 min on the bacterial map, the Hfr strain transferring *thr-leu-lac* early (see Transfer of Different Chromosomal Markers by Different Hfrs, Figure 4.9). After Bachmann and Low (Bachmann, B.J. and Low, K.B. (1980) *Microbiol. Rev.*, *44*, 1-56)

one another by employing a group of Hfr strains, each transferring different, but overlapping, regions of the chromosome (Figure 4.9).

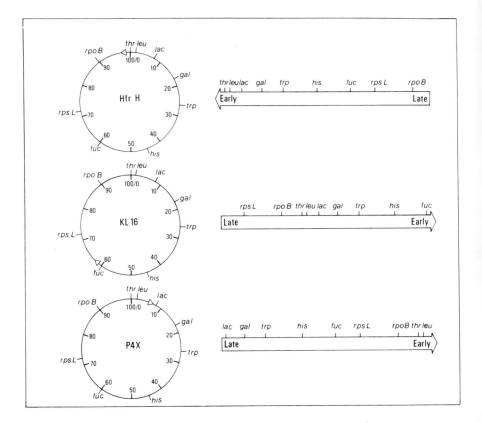

Figure 4.9: Transfer of Different Chromosomal Markers by Different Hfrs
Markers transferred 'early' and 'late' by HfrH, KL16 and P4X are shown to the right (gene symbols are explained in Table 1.3/Figure 1.13). Thus, whereas *thr* is one of the first genes to enter in an HfrH cross (with *rpoB* transferring at very low frequency), both *thr* and *rpoB* are on the leading end in P4X. Not only are the F plasmids at different sites in these different Hfrs but also in different orientations: 'anticlockwise' in HfrH and 'clockwise' in KL16 and P4X. The importance of such a collection of Hfr strains (see also Figure 4.8) is that genes transferring with low probability with one derivative enter at high frequency with another — every bacterial marker is available for analysis (see Genetic Analysis of Bacteria and their Viruses, Section 7.1)

(b) F Excision: F-Prime Formation. Integration of the F plasmid into the host chromosome is a reversible process. The excisive, reciprocal recombination event presumably occurs by genetic exchange between terminal homologous regions. Excision from the host chromosome restores the genome to its original structure and recreates the intact extrachromosomal genetic element in the process (Figure 4.10).

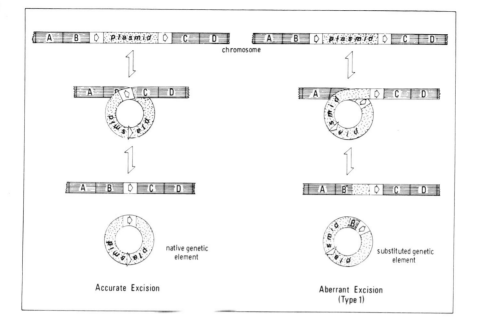

Figure 4.10: Accurate and Aberrant Excision of F

Accurate excision requires reciprocal recombination between the terminal repeats (arrows). If, however, synapsis occurs at other sites, a substituted plasmid may be generated. The scheme shows the formation of a type I F' plasmid that has lost the right-hand end of the integrated element but gained a portion of host gene, *B* (the notation *B'* and *'B* refers to two portions of this gene). F' plasmids carrying particular chromosomal markers may be conveniently isolated by crossing a suitable Hfr strain, for a short time, with a recombination deficient *(recA⁻)* recipient. This allows the stable and efficient recovery of substituted plasmids: proximal markers will not recombine with the recipient chromosome owing to the *recA⁻* background while terminal markers will not have transferred in the time allowed for mating. Thus, recipients that have received such markers are likely to contain substituted plasmids carrying these bacterial genes. For F-primes recovered from Hfr × F⁻ *(recA⁻)* crosses, HfrH gives *thr-leu-proA* derivatives at a frequency of 10⁻⁵ and KL98 gives F*his⁺* plasmids at a frequency of 10⁻⁷ (HfrH and KL98 are shown in Figure 4.8). See also Types of Substituted F Plasmids, Figure 4.11 (cf. Accurate and Aberrant Excision of Bacteriophage *Lambda*, Figure 5.16)

Occasionally, an error occurs, resulting in aberrant removal of the F plasmid to form a genetic element carrying both F and bacterial DNA (Figure 4.10); this hybrid is termed an *F-prime (F')* or *substituted F plasmid*. The event is, presumably, triggered by faulty synapsis between regions of partial homology rather than between the correct sequences — the terminal F sequences. Two types of aberrant genetic exchange are observed (Figure 4.11).

Exchange between a region on the bacterial chromosome and one within the integrated F plasmid (type I): The resultant substituted F' plasmid has lost some F DNA and gained genes from the bacterial chromosome. These bacterial genes are, in general, lost permanently from the chromosome of the Hfr strain.

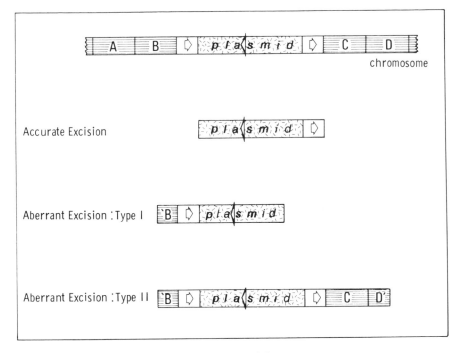

Figure 4.11: Types of Substituted Plasmids
Excision of an integrated plasmid may either be accurate, reforming the autonomous genetic element, or aberrant (Figure 4.10). Inexact excision gives rise to substituted plasmids carrying a portion of bacterial DNA. Type I plasmids have host DNA from only one side of the integrated element and lack some plasmid material while type II plasmids contain bacterial DNA neighbouring both sides of the insertion site (the notation '*B* and *D*' indicates distal and proximal portions of host genes *B* and *D*). The plasmids are shown as linear elements though they are, in fact, circular. The terms type I and type II are those introduced by Scaife (Scaife, J. (1967) *Ann. Rev. Microbiol., 21*, 601-638)

Moreover, F DNA missing from the F′ plasmid is left inserted in the chromosome. A direct consequence of the illegitimate exchange, therefore, is the reciprocal transfer of bacterial and plasmid genes (Figures 4.10 and 11).

Exchange between regions on the bacterial chromosome bordering the integrated plasmid (type II): The F′ plasmid generated consists of the entire extrachromosomal genetic unit, and carries, in addition, bacterial sequences located on either side of the F insertion. The bacterium loses out on both counts — no plasmid DNA remains, and the bacterial genes carried by this type II F-prime plasmid are also deleted (Figure 4.11).

Each Hfr can, consequently, give rise to a number of different substituted plasmids due to faulty exchange. The frequency of F′ formation varies with different Hfr strains (see legend to Figure 4.10). Type I events appeared, at first, to be more common. However, these early studies detected chromosomal DNA carried on the F prime by function: expression of particular bacterial sequences.

Analysis making use of restriction technology (Section 7.3.3) suggests, in fact, that type II hybrids occur most frequently, at least in the *lac-proC* interval (8-9 min). This recent observation is consonant with the nature of type I genetic elements.

Type I F-prime plasmids, formed by reciprocal recombination between 'pseudo' excision sites on the chromosome and within the integrated plasmid, are only stably maintained if the F regions required for vegetative replication have not been deleted from the hybrid. Moreover, loss of essential transfer genes (see Section 10.1.1(a)) will give rise to non-conjugative derivatives. Requirements for a functional F' plasmid clearly, therefore, impose constraints upon this type of faulty excision event.

These substituted plasmids (types I and II) are normally essential for survival of their primary bacterial host since acquisition of bacterial DNA results in its concomitant deletion from the host chromosome. The haploid status of the Hfr strain from which an F' plasmid is derived is, thus, usually unchanged. However, transfer of the F prime plasmid into an F$^-$ strain already carrying the bacterial genes present on the F-prime plasmid (an $F' \times F^-$ *cross*) creates a transconjugant diploid for the transferred region, a merodiploid (Figure 4.12). Obviously, the identity of the bacterial DNA sequences on the F' plasmid is governed by the site at which the plasmid first integrated (see Figue 4.7). The large number of Hfr strains available has allowed the isolation of F' plasmids that carry, in overlapping segments, the entire bacterial chromosome (Figure 4.13).

Homology between bacterial DNA on the F' plasmid and the chromosome in a merodiploid strain may lead, in a *recA*$^+$ background, to recombination of the plasmid-borne bacterial DNA into the recipient's chromosome (Figure 4.14); and, transfer of chromosomal markers to the F' plasmid in the case of a reciprocal recombination event. Alternatively, the F' plasmid can integrate as a complete entity into the chromosome (at a site independent of the F' bacterial sequences), resulting in *transposition* of the bacterial genes carried by the plasmid to this new site (Figure 4.14). Insertion may also occur through bacterial homology. *Fusions* between genes due to spontaneous loss of intervening sequences may be isolated from partial diploid strains since deletion of essential regions will not be lethal in such a background.

The ability to construct readily partial diploid strains (Figure 4.15), allowing genetic analysis previously applied only to conventional diploid organisms, has proved invaluable for standard genetic mapping (discussed in Section 7.1.2 (aii)) as well as for important manipulations such as gene transposition and gene fusion. Moreover, conjugation is not restricted to *E. coli:* intergenic transfer of bacterial genes can be effected, either directly or through intermediates, to some 40 bacterial genera.

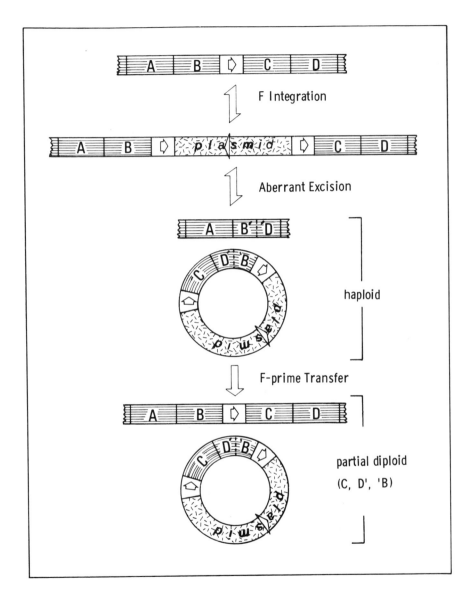

Figure 4.12: Creation of a Partial Diploid (Merodiploid) Status through F-Prime Transfer

The type II F-prime plasmid (see Figure 4.11) carrying the bacterial genes *C, D'* and *'B* (the notation *D'* and *'B* refers to the proximal and distal portions of these genes) creates a partial diploid — a strain diploid for *C, D'* and *'B* — upon transfer to a recipient carrying this region (see F-Prime Transfer: Counter-Selection against Donor and Recipient, Figure 4.15). Since the donor lacks this bacterial material, plasmid loss can be a lethal event (hence the single-headed arrow)

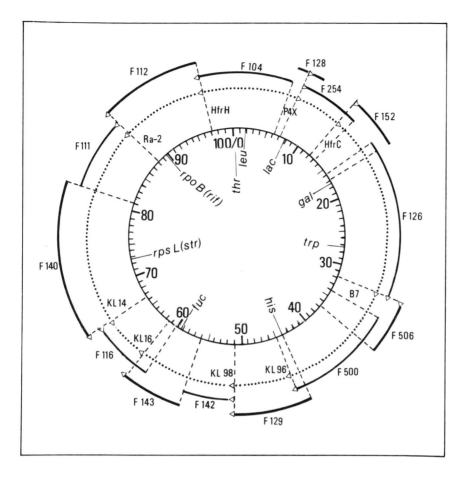

Figure 4.13: F-Prime Plasmids Covering the Bacterial Map

Substituted elements are shown externally; the extent of bacterial DNA present is indicated by *solid lines* (gene symbols are explained in Table 1.3/Figure 1.13). Hfrs responsible for these F-primes are also given (open arrowheads show the sites of F insertion and origins of transfer, see Figure 4.8). F104, for example, one of a number of F-primes obtained from HfrH, spans the interval 98-6 min (a total of 8 min, or 304 kb, of bacterial DNA). After Bachmann and Low (Bachmann, B.J. and Low, K.B. (1980) *Microbiol. Rev., 44*, 1-56)

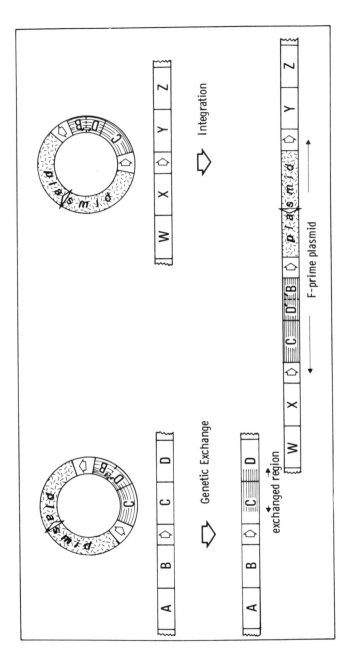

Figure 4.14: The Fate of the Transferred F-Prime Plasmid

When an F-prime enters an F⁻ recipient, it may either replicate autonomously in the bacterial cytoplasm or recombine with host DNA. Homology (*C*, *D'* and '*B*, cf. Figure 4.12) allows exchange of bacterial genetic material. Alternatively, the plasmid may integrate by recombination; the present example shows insertion through short stretches of homology (arrows) although it may also occur through the bacterial sequence carried on the substituted plasmid (the former case gives rise to gene transposition)

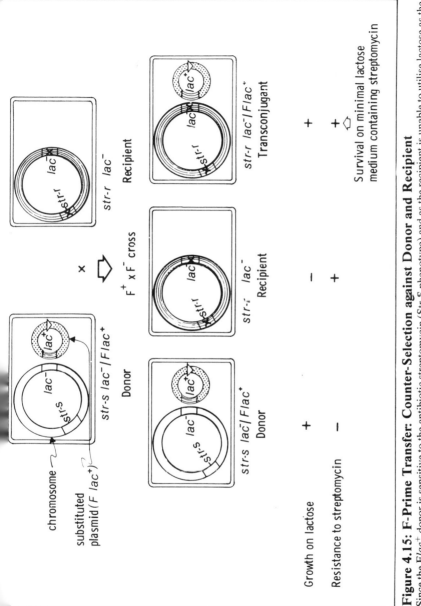

Figure 4.15: F-Prime Transfer: Counter-Selection against Donor and Recipient

Since the F*lac*+ donor is sensitive to the antibiotic streptomycin (Str-S phenotype) and as the recipient is unable to utilise lactose as the sole carbon source (Lac⁻ phenotype), only Lac⁺ Str-R transconjugants formed in a cross between these two strains are able to grow on lactose minimal medium containing streptomycin. Note that stable diploidy requires a recombination-deficient background (cf. Figure 4.14) and continued selection for the plasmid

4.2 A Guide to Naturally-Occurring Plasmids

The prototype of all stably inherited, extrachromosomal genetic elements, in terms of its discovery, is the F plasmid. Plasmids are, in fact, ubiquitous in nature. A rigorous search in *E. coli* and other bacteria over the last thirty years has led to the identification of a vast collection of these non-essential extra-chromosomal genetic elements, responsible for a number of different bacterial properties; they range in molecular weight from about five to one hundred Mdal (Table 4.1).[6] Many encode the genetic information for resistance to antibiotics, allowing both Gram-negative and Gram-positive bacteria harbouring these *R plasmids* to survive antibacterial agents. A second group, the *Col plasmids*, are responsible for the production of diffusible proteins, *colicins*, which are bactericidal to strains of *E. coli* (and of other enteric organisms) lacking the genetic elements. Some temperate bacteriophages or their derivatives may exist as stable cytoplasmic replicons or covalently associated with the bacterial chromosome (as discussed in Section 5.1.2 (c)) and can, therefore, also be included in this category.

Although F is transmissible, this is by no means a hard and fast rule for all extrachromosomal genetic elements. The naturally occurring *non-conjugative* class of plasmids neither encode donor pili nor transfer *per se:* they may, however, be *mobilised* by conjugative plasmids (see below). Even conjugative plasmids transfer, in general, at a markedly reduced rate (about 1000-fold) compared with F. This *repressed* state of conjugation is associated with little or no pilus production and may be overcome by the isolation of derepressed mutants *(drd⁻):* F is naturally derepressed. Transient high frequency transfer occurs immediately after conjugation, entry into a 'virgin' host resulting in physiological rather than genetic derepression. The presence of naturally occurring F-like[7] R plasmids (so called for their distinct homology with F DNA and the similarity betwen their encoded pili and that of F) may also repress F transfer when the two plasmids coexist in the same cell; this phenomenon is termed *fertility inhibition* (phenotype Fi⁺). The majority of Fi⁺ plasmids are F-like, coding for F-like pili (but in this case, stable mating-pair formation does not involve the *ompA* protein). There are, however, a number of different types of conjugative pili in addition to F and F-like, associated with Fi⁻ conjugative plasmids (Table 4.1).

Conjugative R and Col plasmids may mobilise both non-conjugative plasmids, and at low frequency, chromosomal markers. Whether the latter involves stable

---

6.   Numerous plasmids, in addition to R and bacteriocinogenic plasmids, have been found in bacterial species other than *E. coli*. They include elements that confer resistance to metal ions, reduced sensitivity to mutagens, and the ability to degrade certain organic compounds. They may also specify restriction-modification systems.

7.   Certain plasmids are conveniently grouped according to their similarity to F. These F-like plasmids, however, are clearly distinct from the prototype (see also Interactions between F and F-Like R Plasmids, Section 10.1.2).

Table 4.1 [a] **A Guide to Naturally-Occurring Plasmids** [b]

Plasmid Class	Representative Element			Determinants [c]			Copy Number [d]
	Name	Size (kb)	Inc.	Major Phenotype	Pilus Type	Self-Transmissible [e] (by conjugation)	
F	F	94.5	FI	-	F	+	1-2
R, Fi+ f	R1	94.5	FII	drug resistance	F-like	+	1-2
α, Fi-	R64	109.8	Iα	drug resistance	I-like g	+	1-2
Col, small	ColE1	6.4	- h	colicin	-	-	10-30
Col, large	ColIb-P9	99.1	Iα	colicin	I	+	1-2
phage, infective	PI	89.9	Y	prophage immunity i	-	-	1-2
phage, defective	λdv	6.4	-	-	-	-	50-100

a Abbreviations: Inc, incompatibility group, Fi+, fertility inhibition (with respect to F transfer).

b See Bukhari *et al.* (1977).

c Other genes are encoded by plasmids in addition to drug resistance, colicin production, pili and conjugative properties. These include functions necessary for replication and its control.

d Plasmids with high copy number generally exhibit a relaxed form of replication.

e Plasmids other than F generally transfer at a fraction the efficiency. Note that although ColE1 is not self-transmissible, it may be mobilised making use of the functions of a self-transmissible plasmid.

f Some elements with transfer systems apparently unrelated to that of F also inhibit F fertility.

g Fi- R plasmids encode I-like pili and other types.

h Relaxed plasmids such as ColE1 exhibit incompatibility; they have not as yet been assigned groups.

i Immune to superinfection (see Section 5.1.2 (c)).

integration into the bacterial genome rather than mobilisation via a less permanent interaction is still contentious.[8] Certainly, specific nucleotide sequences on both plasmid and chromosome (see Insertion Sequences, Section 6.2.2 (aii) and 10.1.1 (c)) may play an important role in plasmid-mediated chromosome mobilisation.

Stable inheritance requires a functional replicative machinery that controls plasmid replication, allowing accurate segregation during cell division. At first sight, this control might be expected to be mediated through some form of coupling of plasmid replication with that of the chromosome. Such synchrony is, apparently, not obligatory.[9] As a broad generalisation, *stringently* controlled plasmids are conjugative genetic elements, present as only a few copies per host genome: their counterparts, replicating under *relaxed* control, are small multi-copy non-conjugative plasmids usually present in between 10 and 30 copies per cell (ColEl is an example of the latter class, Table 4.1). Though these relaxed plasmids have, characteristically, an absolute requirement for the host DNA polymerase I (*polA* product, see Section 6.1.1 (a)), replication may proceed in the absence of host protein synthesis (it requires, however, RNA synthesis). As many as 3000 molecules per cell (copies apparently still containing their RNA primer) are synthesised under these conditions (brought about, for example, by the addition of chloramphenicol to the culture medium). Thus, though replication of the small multicopy plasmids is strictly controlled, it is 'relaxed' in so far as replication may be uncoupled from chromosomal DNA synthesis. (The replication characteristics of these two groups of plasmids are discussed in Section 6.1.3 (a)).

The presence of a functional control system in cells carrying stringent (and relaxed) plasmids is suggested by the inability of two homologous plasmids to be maintained within one cell (Section 4.1.2 (b)). Incompatibility is not limited to identical plasmids. In fact, this property has allowed stringent plasmids to be grouped into *incompatibility classes* — at least twenty-six have been identified to date — each class composed of plasmids that are mutually incompatible, but compatible with plasmids of other classes. Incompatibility may tell us more about a plasmid than just the presence of an active control system. Plasmids within an Inc group normally show extensive DNA homology as compared with genetic elements of different groups, and have similar transfer mechanisms (exemplified by pilus-type and host-range). The similarity of the F-like

8.   Plasmid integration may not be necessary for chromosome transfer. The small non-conjugative plasmids, for example, can be transferred using the transfer system of conjugative plasmids but without apparent association. Instability of some Hfrs may, therefore, be explained by transient integration; and, in some cases, R plasmid-mediated chromosome mobilisation has been found without the isolation of the equivalent *Hfr (R)* derivative. Chromosome transfer in *recA⁻* strains suggests even transient integration may not be required. The ability of a plasmid to exhibit integrative suppression and polarised transfer of chromosomal markers has been used as a test for plasmid integration (Section 4.1.3 (a)).
9.   Control of plasmid replication may depend upon the bacterial host. Certain R plasmids, for example, exist as multimers in *Proteus mirabilis*. Moreover, hybrid plasmids consisting of two fused replicons exhibit anomalous modes of replication.

plasmids, for example, has been attributed to a region coding for the transfer functions, the *tra* operon (Section 10.1.1 (a)). Variability in their resistance determinants may be explained by the ability of these genes to 'jump' (see Transposon, Section 6.2.2 (aii)).

4.2.1 R Plasmids

The increasing incidence of infections caused by antibiotic-resistance strains of bacteria has become a clinical problem of utmost importance. This rise in multiply-resistant micro-organisms (Figure 4.16) is not readily explained in terms of selection of spontaneously arising drug-resistant mutants: the probability of a single bacterium developing resistance to even two or three different drugs, when the mode of resistance is different in each case, is infinitesimally small (see legend to Figure 4.16). The presence of transmissible plasmids carrying drug-resistance determinants, *R plasmids* (originally termed *R-factors*), now allows us to understand the occurrence and spread of multiply-resistant bacteria.

R plasmids carry determinants for resistance against a large number of antibiotics and their derivatives. There is, not surprisingly, no general mechanism for plasmid-mediated antibiotic resistance (Table 4.2). Nevertheless, the types of resistance are clearly different from those resulting from spontaneous chromosomal mutations. Chromosomal-coded resistance is normally passive in its action (and recessive to the wild-type allele), with the mutation altering the target site of the antibiotic to allow the mutant gene product to function in the presence of the specific antibiotic (chromosomal-coded resistance to penicillin is an exception to this rule). The resistance determinant of an R plasmid usually encodes a novel gene product that abolishes drug function by either enzymatic detoxification (inactivation of the antibiotic) or by interference with drug transport across the cellular membrane (tetracycline resistance is a case in point, Table 4.2). Plasmid-mediated drug resistance may be partially non-specific, acting on both the antibiotic and its synthetic derivatives.

4.2.2 Col Plasmids

Many different strains, including Gram-positive bacteria, produce transportable proteins, *bacteriocins*, that are lethal to prokaryotes. The *colicins* (molecular weight 40 000-80 000) are specifically bactericidal to other enteric bacteria, the Enterobacteriacae, and presumably function in nature by modifying their natural environment, the intestinal flora. That colicins are distinct from antibiotics is indicated by their large size and the requirement for a suitable bacterial receptor for killing (see below).

Colicins investigated to date are all plasmid mediated. These genetic elements may be split into two main groups according to their size, the first group consisting of Col plasmids at most one tenth the size of the second (5 and 80 Mdal, respectively): ColE1 and ColIb are representatives of these two groups (see Table 4.1). In general, the small plasmids are non-conjugative, multicopy genetic elements, unable to transfer in the absence of a functional conjugative

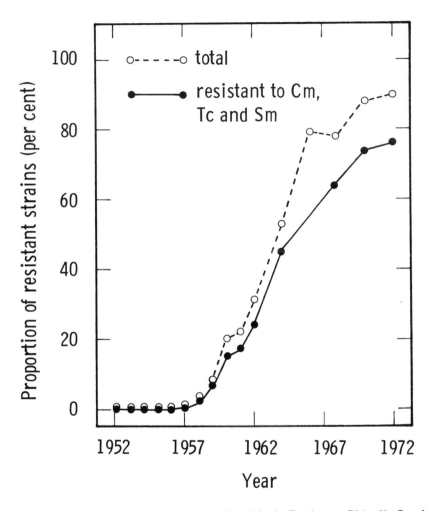

Figure 4.16: Incidence of Multiply Antibiotic-Resistant *Shigella* **Strains in Japan**

The 1960s saw a rapid rise in *Shigella* strains resistant to antibiotics. Most importantly, these strains survive the presence of more than one different antibiotic (abbreviations: Cm, chloramphenicol; Sm, streptomycin; Tc, tetracycline). Since mutations giving rise to antibiotic resistance occur spontaneously at a frequency of about 10^{-6}-10^{-10}, the frequency of resistance to two or three functionally different drugs would be expected to occur at about 10^{-12}-10^{-20} or 10^{-18}-10^{-30}. These figures are so small as to make the chance of multiple-resistant events occurring infinitesimal (if the events are unlinked). These and other bacterial strains are resistant to numerous antibiotics by virtue of the presence of drug-resistant plasmids — extrachromosomal genetic elements that encode resistance. Data from Mitsuhashi (Mitsuhashi, S. (1977), in S. Mitsuhashi (ed.) *R Factor Drug Resistance Plasmid* (University Park Press, Baltimore), pp. 3-24)

Table 4.2ᵃ A Comparison of Chromosomal and Plasmid-Mediated Antibiotic Resistanceᵇ

Antibiotic	Mode of Actionᶜ		Mechanism of Resistance	
	Process Inhibited	Target	Chromosomalᵈ	Plasmidᵉ
chloramphenicol	translation	50S ribosome subunit	nkᶠ	acetyltransferase
streptomycin	translation	30S ribosome subunit	$\underline{rpsL}$ (S12)	ATase,PTase
spectinomycin	translation	30S ribosome subunit	$\underline{rpsE}$ (S5)	ATase
tetracycline	translation	30S ribosome subunit	nk	Tet proteins
penicillin	cell wall synthesis	Murein crosslinking	$\underline{ampC}$ (β – lactamase)ᵍ	β – lactamase

a   Abbreviations: nk, not known; ATase, O-adenyltransferase; PTase, O-phosphotransferase.

b   Adapted from Davis and Smith (1978).

c   The mode of action of these antibiotics is discussed in Section 3.2.2 (b) (penicillin) and Section 2.2.2 (d).

d   Refers to the chromosomal genetic locus responsible for resistance (e.g. mutations at *rpsL* and *rpsE* are necessary for drug resistance).

e   These transferases act upon hydroxyl group(s) while β-lactamase cleaves the β-lactam ring; a number of different enzymes of both groups have been characterised to date. The production of Tet proteins (species induced by growth on tetracycline) appears to affect uptake of the antibiotic.

f   Chloramphenicol-resistant *B. subtilis* strains encode altered 50S ribosomal proteins.

g   The *E. coli* lactamase has restricted specificity as compared to R plasmid enzymes.

system. The majority of large Col plasmids, on the other hand, are transmissible and exhibit the stringent mode of replication. These conjugative Col plasmids have similarities with the transmissible F-like and I-like group of R plasmids, as indicated by the nature of the donor pilus and base sequence homology.

There are clear parallels between the regulation of colicinogenic plasmids and prophages, suggesting that similar mechanisms are involved: colicin synthesis is repressed in most cells harbouring Col plasmids, and induction by agents common to both systems (such as UV light, mitomycin and other agents known to induce SOS DNA repair functions, see Section 3.2.1 (bi)) appears to require a functional *recA* gene product.[10] Once synthesised, colicins are transported to the growth medium, though much of these antibacterial agents may, in fact, be bound to the outer membrane of the colicinogenic bacteria.

A single molecule of colicin is, apparently, sufficient to kill a sensitive cell. This bactericidal event requires at least two stages: receptor binding, and a subsequent, less well understood, energy-dependent step. After binding to cell-surface receptors, (certain) colicins are cleaved into fragments: the larger N-terminal portions possess receptor-binding activity while the C-terminal peptides have a killing function. The finding that certain bacteriophages and colicins have common receptors — strains selected for resistance to one lethal agent are *cross-resistant* to the other (and *vice versa*) — has supported a possible evolutionary relationship between these apparently diverse genetic entities. The 26 000 dalton *tsx* gene product, for example, is a receptor for both colicin K and bacteriophage T6 (and has been implicated in nucleoside uptake). Similarly, the E colicins (E1, E2 and E3) and the T5-like phage BF23 (as well as vitamin B_{12}) share a receptor specified by the *btuB* gene, a glycoprotein of 60 000 daltons.

The mode of action of colicins that seem primarily to affect the inner membrane is unclear. Energy-dependent processes such as transport and syntheses are inhibited, perhaps indirectly, by these proteins (often a single colicin gives apparently pleiotropic effects). Others, such as colicin E3, are now known to inhibit protein synthesis directly by endonucleolytic cleavage of the 16S rRNA of the 30 S ribosomal subunit at a specific target close to the 3′-end (49 nucleotides from the terminus).

Colicinogenic strains harbouring Col plasmids are protected from endogenous colicin by a plasmid-coded protein. The *immunity* engendered by this protein is highly specific. The colE3 *immunity protein* (about 11 000 daltons) functions by binding reversibly in 1 : 1 stoichiometric amounts to the E3 colicin (reversible binding is necessary for release of an active colicinogenic protein after transport to the outer membrane). Sensitive strains lacking this immunity may survive the

10. A low rate of colicin synthesis may occur in the absence of inducing agents. This is believed to result from the spontaneous induction of a small proportion of the colicinogenic bacterial population. Colicin production can be detected by plating the colicinogenic strain on agar medium and overlaying with a sensitive strain. Zones of inhibition or *'lacunae'* in the indicator lawn (analogous to the phage plaque, Section 5.1.3 (a)) result from colicin released from single cells upon spontaneous induction (and subsequent cellular death).

bactericidal effect by spontaneous mutations at a number of loci, resulting in either tolerance or resistance. Whereas the colicin receptor is non-functional in *resistant* derivatives,[11] *tolerant* strains bind the antibacterial agent without its lethal effect being realised. *tolA⁻* mutations, for example, which impart tolerance to colicins A, E2, E3 and K, are thought to affect the cytoplasmic membrane.

Bibliography

Broda, P. (1979) *Plasmids* (Freeman, Oxford).
Bukhari, A.I., Shapiro, J.A. and Adhya, S.L., (eds.) (1977) *DNA Insertion Elements, Plasmids and Episomes* (Cold Spring Harbor Laboratory, New York).
Lewin, B. (1977) *Gene Expression, 3: Plasmids and Phages* (John Wiley, New York).

4.1 The F Plasmid

Campbell, A.M. (1969) *Episomes* (Harper and Row, New York).
Jacob, F. and Wollman, E.L. (1961) *Sexuality and the Genetics of Bacteria* (Academic Press, New York).
Scaife, J. (1967) 'Episomes', *Ann. Rev. Microbiol., 21,* 601-638.

4.1.1 Vegetative Replication

Clark, A.J. and Warren, G.J. (1979) 'Conjugal transmisstion of plasmids', *Ann. Rev. Genet., 13,* 99-125.
Jacob, F., Brenner, S. and Cuzin, F. (1963) 'On the regulation of DNA replication in bacteria', *Cold Spring Harb. Symp. Quant. Biol., 28,* 329-347.
Pritchard, R.H., Barth, P.T. and Collins, J. (1969) 'Control of DNA synthesis in bacteria', *Symp. Soc. Gen. Microbiol., 19,* 263-298.
Rowbury, R.J. (1977) 'Bacterial plasmids with particular reference to their replication and transfer properties', *Prog. Biophys. Molec. Biol., 31,* 271-317.
Further references in Chapter 6.

4.1.2 F Transfer

Achtman, M. and Skurray, R. (1977) 'A redefinition of the mating phenomenon in bacteria', in J.L. Reissig (ed.), *Receptors and Recognition, B3: Microbial Interactions* (Chapman and Hall, London), pp. 233-279.
Bayer, M.E. (1979) 'The fusion sites between outer membrane and cytoplasmic membrane of bacteria: Their role in membrane assembly and virus infection', in M. Inouye (ed.), *Bacterial Outer Membranes: Biogenesis and Functions* (John Wiley, New York), pp. 167-202.
Manning, P.A. and Achtman, M. (1979) 'Cell-to-cell interactions in conjugating *Escherichia coli:* the involvement of the cell envelope', in M. Inouye (ed.), *Bacterial Outer Membranes: Biogenesis and Functions* (John Wiley, New York), pp. 409-447.
Reanney, D. (1976) 'Extrachromosomal elements as possible agents of adaptation and development', *Bacteriol. Rev., 40,* 552-590.
Tomoeda, M., Inuzuka, M. and Date, T. (1975) 'Bacterial sex pili', *Prog. Biophys. Molec. Biol., 30,* 23-56.

4.1.2 F-Mediated Transfer

(a) *F integration: Hfr formation*

Hayes, W. (1968) *The Genetics of Bacteria and their Viruses,* 2nd edn (Blackwell, Oxford).

---

11.   Colicins have been divided into two main groups based upon their action on bacterial mutants resistant to specific colicins, that is, the cross-resistance patterns of the mutants.

Holloway, B.W. (1979) 'Plasmids that mobilise bacterial chromosome', *Plasmid, 2,* 1-19.
Miller, J.H. (1972) *Experiments in Molecular Genetics* (Cold Spring Harbor Laboratory, New York).

(b) *F excision: F-prime formation*

Davidson, N., Deunier, R.C., Hu, S. and Ohtsubo, E. (1975) 'Electron microscope heteroduplex studies of sequence relations among plasmids of *Escherichia coli,* X: Deoxyribonucleic acid sequence organisation of F and F-primes, and their sequences involved in Hfr formation', in D. Schlessinger (ed.), *Microbiology 1974* (American Society for Microbiology, Washington), pp. 56-65.
Low, K.B. (1972) '*Escherichia coli* K-12 F-prime factors, old and new', *Bacteriol. Rev., 36,* 587-607.

4.2 A Guide to Naturally-Occurring Plasmids

Falkow, S. (1975) *Infectious Multiple Drug Resistance* (Pion, London).
Novick, R.P., Clowes, R.C., Cohen, S.N., Curtiss, R., Datta, N. and Falkow, S. (1976) 'Uniform nomenclature for bacterial plasmids: A proposal', *Bacteriol. Rev., 40,* 168-189.

4.2.1 R Plasmids

Chopra, I. and Howe, T.G.B. (1978) 'Bacterial resistance to the tetracyclines', *Microbiol. Rev., 42,* 707-724.
Davies, J. and Smith, D.I. (1978) 'Plasmid determined resistance to antimicrobial agents', *Ann. Rev. Microbiol., 32,* 469-518.
Hopwood, D.A. (1978) 'Extrachromosomally determined antibiotic production', *Ann. Rev. Microbiol., 32,* 373-392.
Mitsuhashi, S. (ed.) (1977) *R. Factor Drug Resistance Plasmid* (University Park Press, Baltimore).

4.2.2 Col Plasmids

Davies, J.K. and Reeves, P. (1975) 'Genetics of resistance to colicins in *Escherichia coli* K12: Cross-resistance among colicins of group B', *J. Bacteriol., 123,* 96-101.
Davies, J.K. and Reeves, P. (1975) 'Genetics of resistance to colicins in *Escherichia coli* K12: Cross-resistance among colicins of group A', *J. Bacteriol., 123,* 102-117.
Hardy, K.G. (1975) 'Colicinogeny and related phenomena', *Bacteriol. Rev., 39,* 464-515.
Holland, I.B. (1976) 'Colicin E3 and related bacteriocins: Penetration of the bacterial surface and mechanism of ribosomal inactivation', in P. Cuatrecasas (ed.), *Receptors and Recognition, B1: The Specificity and Action of Animal, Bacterial and Plant Toxins* (Chapman and Hall, London), pp. 99-127.
Reeves, P. (1972) *The Bacteriocins* (Chapman and Hall, London).

5 Bacteriophages

Viruses that grow on bacteria are termed *bacteriophages*; they contain no metabolic apparatus of their own and are unable to multiply in the absence of a sensitive host. *Coliphages* are bacteriophages whose host-range is restricted to the bacterium *Escherichia coli*. In many ways, bacterial viruses can be considered as transmissible extrachromosomal genetic elements (Chapter 4) that have gained the ability of elaborating a proteinaceous coat as well as killing their susceptible host. The complexity of bacteriophages varies greatly, approximately in proportion to the size of their genomes. Nevertheless, they are nothing more than a (lethal) replicon — a group of genes (present in the phage head either as DNA or RNA) able to replicate only in a permissive bacterial host. The protein coat acts as a more-or-less passive vehicle for transport of the viral genetic material. There are two main types of coliphages: *virulent* phages that multiply within the susceptible host, giving rise to progeny phage upon cell lysis and death; and *temperate* phages which can either enter this *lytic cycle*, or *lysogenise* the cell, undergoing controlled replication within the cell without lysis.

The impetus, sparked off by Max Delbruck and others in the 1940s, to the analysis of bacteriophages as model systems for biological research has led to an understanding of the structure and function of these viruses far exceeding early expectations. Coliphages *per se* are readily handled and do not require special containment facilities since they are non-pathogenic. They are now a fundamental weapon in the arsenal of the biologist wishing to probe gene expression at the level of the genetic material itself.

5.1 Bacteriophages and their Life-Cycles

There is much variation in the type of virus able to make use of *E. coli* functions (Table 5.1). Some appear to be unique (for example, phages P1, P2 and Mu): others bear marked similarities. There are, nevertheless, two major types of coliphages with respect to the response elicited upon infection. Whereas virulent phages kill their host cell, temperate species (or rather their nucleic acid) may persist within a bacterium without the inevitable lethal response (although the latter may also occur). In addition to this distinction, bacterial viruses may be placed in a small number of groups according to morphological characteristics, the nature (and size) of their genetic material and a dependence upon either a

Table 5.1[a] Comparative Phage Anatomy[b]

Phage	Class	Infective Cycle [c]	Virion Characteristics			Viral Genome		
			Head Morphology	Size (nm)	Tail (nm)	Size (kb)	Form	Ends
λ	lambdoid	temperate	isometric	60	135 x 15	48.6	ds,linear	cohesive (12 nucl.)
P1	-	temperate[f]	isometric	65	150 x 12	91.5	ds,linear [d]	redundant;permuted
P2	-	temperate	isometric	60	135 x 10	33	ds,linear	cohesive (19 nucl.)
Mu	-	temperate	isometric	54	100 x 18	38	ds,linear [d]	heterogenous host sequences
T4	T-even	virulent	anisometric	80 x 110	98 x 20	166	ds,linear [e]	redundant,permuted
T7	T-odd ♀	virulent	isometric	58	20 x 19	40	ds,linear	redundant (160 bp)
φx174	isometric	virulent [f]	isometric	25	-	5.4	ss,circular	-
M13	filamentous ♂	virulent	-	-	900 x 9	6.4	ss,circular	-
MS2	RNA phage ♂	virulent	isometric	26	-	3.6	ss,linear,RNA	-

a   Abbreviations: kb, 10^3 nucleotides; ♀, ♂, female (F$^-$)- or male (F$^+$)- specific.
b   Adapted from Bukhari *et al.* (1977) and Luria *et al.* (1978).
c   In the lysogenic mode, phages *lambda* and P2 integrate at specific chromosomal sites, designated $att^\lambda B$, $att^{P2}H$ and $att^{P2}II$ (at 17,43 and 86 min, respectively); Mu integrates at random. Phage P1 can either exist autonomously or inserted in the chromosome in the prophage state.
d   Although the temperate phages P1 and Mu are extremely different, they share a short region of DNA homology — the invertible sequences — responsible for their host-range specificity.
e   T4 DNA contains glucosylated 5-hydroxymethyl cytosine which enables the phage to escape host-controlled DNA restriction (See Section 7.3.3.).
f   P1 has a broad host range. φX174 infects *E. coli* strain C but is unable to adsorb to wild-type K12.

male or female host (F$^+$ or F$^-$, Chapter 4). Such grouping does not, however, necessarily imply that phages of the same group are identical.

The following discussion gives a general description of coliphages studied to date. Where possible, differences between (and within) groups are noted. Bacteriophages P1 and *lambda* are used as model systems for phage-mediated gene transfer in Section 5.2; regulation of phage *lambda* development is discussed in Section 10.2.

5.1.1 Anatomical Considerations

The basic phage structure is a protein head, the *capsid*, encapsidating the viral genome. In addition, some bacterial viruses have a protein tail through which phage – membrane attachment occurs, and the nucleic acid passes The bacteriophage genetic material may consist of linear double-stranded DNA, circular single-stranded DNA or linear single-stranded RNA, the size varying over two orders of magnitude (Table 5.1 and Figure 5.1). While the RNA phage MS2 and the small isometric phage φX174, for example, carry the genetic information for 3 and 11 proteins, respectively, approximately 135 genes have been mapped on the massive T4 genome (about 4 per cent the size of the bacterial chromosome).

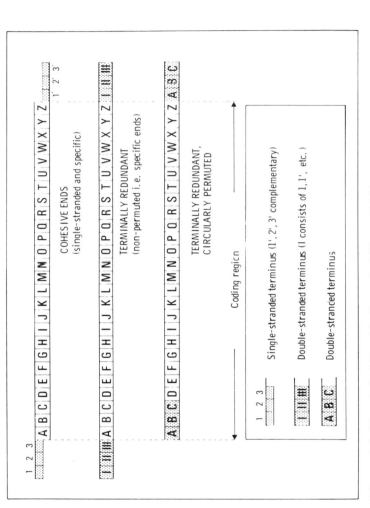

Figure 5.1: Termini of Double-Stranded Phage DNA
The coding material of double-stranded DNA viruses is frequently bracketed by specific sequences (Table 5.1). Bacteriophage *lambda*, for example, has cohesive ends, a complementary 12 nucleotide stretch that allows circularisation through end pairing (Figure 5.12). Phage T7, on the other hand, carries 160 bp redundant (repeated) sequences distinct from its coding region. The termini of bacteriophage T4 are also redundant but are not of a specific nature (instead of the sequence running *ABC → ZABC*, it could just as easily be *BCD → ZABCD, CDE → ZABCDE* etc.)

(a) Phage Morphology. The bacteriophage capsid is commonly icosahedral in structure,[1] appearing polyhedral (rather than semi-spherical) in electron micrographs, approximately 20-60 nm in diameter (Table 5.1 and Figure 5.2). Phage T4 has an elongated or *anisometric* capsid. The large DNA phages have long, narrow tails (150 × 10 nm) which may or may not be contractile; T7 is an exception in that its tail is particularly short. The minute phages carrying circular single-stranded DNA or linear single-stranded RNA lack a conventional tail. Whereas small DNA phages (φX174, G4, R, S13) have tail-like spikes on the icosahedral surface, necessary for infection, filamentous RNA phages (M13, fl, fd) appear 'headless', the nucleic acid being encapsidated in a long tail-like structure.

The phage capsid may also carry internal proteins, which either have a structural role, or are involved in regulation of the cell's metabolism. Other structural modifications include tail fibres (fine proteinaceous appendages attached to the tail, Figure 5.2) responsible for the host-range of the phage. The large virions are frequently more complex in structure. In fact, bacteriophages are a marvellous example of ordered complexity. The T-even phage, T4, for example, consists of more than fifty different proteins (encoded by about 40 per cent of the phage genome) assembled into a highly organised structure (Figure 5.2).

(b) The Viral Genome. There is some correlation between genome size and phage group, partially obscured by the early phage nomenclature based upon arbitrary laboratory designations. The T-odd phage, T5, is different both in growth characteristics and DNA size (T1, T3, T5 and T7 have molecular sizes of about 54, 36, 110 and 40 kb respectively.) In fact, T5 is more similar to the T-even phages in both these properties. T3 and T7 are members of a family of related phages but show considerable divergence even though they can form hybrid phage recombinants in mixed infection. T1 is unrelated to these and is unusual in its ability to survive extremes of dehydration. (Bacteriophage T1 can, therefore, spread rapidly through a laboratory and should be handled with special care.) The T-even group (T2, T4, T6) are of similar size and undergo recombination in mixed infection. The filamentous circular single-stranded DNA phage group, and the semi-spherical RNA viruses (MS2, f2, R17) are two coherent groups with respect to genome size and viral properties. φX174 and the other small isometric single-stranded circular DNA phages are similar and serologically related.

The form of the viral genome is not necessarily invariant (see Table 5.1). Once injected, linear double-stranded DNA frequently circularises ('the rule of the ring' at play again, Section 1.1.1 (c)). The intracellular form can be affected by

1.   There are two main types of regular viral capsids, helical and isometric (a semi-spherical closed shell that has identical linear dimensions along orthogonal axes), reflecting the two ways in which irregular subunits such as protein molecules can be arranged in stable, regular structures. Icosahedral symmetry is of the isometric type; it has five-fold, three-fold and two-fold axes of symmetry.

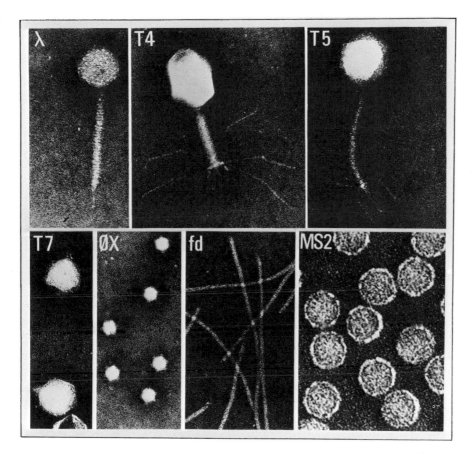

Figure 5.2: The Structure of Some Coliphages
Capsid sizes of λ, T4, T5, T7, φX174 and MS2 (Table 5.1) are 60, 80 × 100, 60, 58, 25 and 26 nm, respectively (i.e. approximately equal magnification apart from MS2 which is shown twice as large). MS2 and M13 can be seen attached to an F pilus in Figure 1.14; a schematic outline of *lambda*, showing the major structural features, is given in Figure 10.11. From Williams and Fisher (Williams, R.C. and Fisher, H.W. (1974) *An Electron Micrograph Atlas of Viruses* (Thomas Springfield)), reproduced with permission of the publishers and authors

the nature of the terminal DNA sequences (this aspect is described in greater detail in relation to bacteriophage *lambda*, Section 10.2.1 (c)). Moreover, replicative intermediates will also be present.

5.1.2 Phage Infection

The onset of phage infection occurs in three main steps: collision, adsorption (and irreversible binding), and injection of the phage nucleic acid (leaving, in some cases, a *phage ghost* — a virion devoid of its genome — attached to the outer membrane). Phage binding at the cellular surface not only activates release of nucleic acid but also may cause membrane alterations which facilitate its entry. Adsorption occurs, normally, at the cellular surface, though male-specific phages attach specifically to cellular appendages involved in bacterial conjugation (see F Pili, Section 4.1.2 (a)). There are site(s) on the bacterial surface that dictate, and are involved in, the specificity of phage attachment (Table 5.2). In the case of P1, T4, T7 and ϕX174, for example, the lipopolysaccharide has been implicated as the receptor substance. Bacteriophage *lambda*, on the other hand, requires a specific host-coded protein, the *lamB* gene product (mutations at this locus, thus, impart resistance to phage *lambda*). Metal ions, particularly divalent cations, are required for efficient binding (phages *lambda* and P1 have a specific requirement for Mg^{2+} and Ca^{2+}, respectively). Phage receptors may be essential cell components, thereby specifying more than one function. The *lamB* receptor, for example, is also required for uptake of the disaccharide maltose (Table 5.2). However, the involvement of one receptor in several functions does not necessarily imply a

Table 5.2[a] Some Phage Receptors[b]

Phage	Class	Receptor Gene	Receptor Map Location (min)	Characteristics Nature	Characteristics Gene Product Size	Characteristics Number of copies	Additional Components LPS	Additional Components tonB involvement	Multivalency Phage	Multivalency Colicin	Multivalency Other[c]
λ	lambdoid	lamB	91	protein	48,000	100,000	?	-	-	-	maltose
T6	T-even	tsx	9	protein	26,000	nk	+	+	-	K	nucleosides
BF23	T5-like	btuB	89	protein	60,000	200	- d	- e	-	E	B12
M13	filamentous♂	traJ	F-coded	F pilin	11,000	1-3 f	-	-	-	-	conjugation

a   Abbreviations: nk, not known; LPS, lipopolysaccharide; ♂, male-specific (with respect to F⁺ cells).
b   See Konisky (1979); Randall and Philipson (1980).
c   Refers to the uptake of the nutrients, maltose, nucleosides and vitamin B_{12}.
d   LPS is required for binding B_{12} *in vitro* but not phage BF23 or colicin E3.
e   *btuB*-dependent uptake of vitamin B12 requires a functional *tonB* product.
f   There are 1 to 3 copies of the F pilus per F⁺ cell (Section 4.1.2). Although male-specific phages, as a class, bind to this proteinacious appendage, different phages bind to different sites: filamentous phages such as M13 interact with the tip whereas the small RNA phages (for example, MS2) bind in large numbers to the side (see Section 1.2).

common mechanism. It has been suggested that some of these multispecific receptors (for example, the *tsx* gene product) are *diffusion pores* rather than receptors, particularly since there is no clear evidence for ligand binding. In contrast, the *btuB* product is clearly responsible for binding of vitamin B_{12}, the E colicins (Section 4.2.2) and the T5-like bacteriophage BF23. Moreover, pre-adsorption of B_{12} protects against these lethal agents.

A number of phage particles can adsorb onto one bacterium, particularly when the ratio of phage to bacteria is high. At high *multiplicity of infection (m.o.i.)*, the presence of a large number of phages can cause *lysis from without*

— disruption of the bacterial membrane and cell death. This process does not, apparently, require viral gene expression. High multiplicities of phage T6 are, in fact, an effective lysing agent.

(a) The Lytic Response. This homogeneous response of a bacterium to an infective phage is governed by the susceptibility of the host (the effect of phage mutations on infectivity is discussed in Section 10.2). *E. coli* mutations (either pre-existing — spontaneous — or induced) that affect the viral receptors impart *resistance* against infection. Alteration of multi-specific receptors affects adsorption of the different agents using the receptor (mutations in the *tonA* locus, for example, give rise to resistance against T1, T5 and the unrelated bacterio-phage, $\phi 80$).

Upon infection of a sensitive host, the phage nucleic acid is transferred (with little or no protein) into the cytoplasmic compartment, usually leaving the phage ghost still attached to its receptor; a slight shear force will remove the empty capsid, without affecting productive infection. Transmission is apparently unidirectional in the case of a linear viral genome. The mechanism by which phage DNA is transferred from the capsid through the multi-layered rigid envelope and into the cell is still unknown (see Section 4.1.2 (a)). It would seem that the tail supplies the delivery system Membrane adhesion sites (Section 1.2.1) have also been implicated in viral nucleic-acid injection.

Following entry of the viral genome into the susceptible host, the intracellular phage can go into the *lytic cycle* or *lysogenise* the host, depending upon the nature of the phage (virulent or temperate) and the physiological state of the bacterial host. The dividing line between virulence and temperance is not sharp; a single mutation in the genome of a temperate phage may lead to virulence. In the virulent (or *productive*) response, the bacterial metabolism becomes controlled by the phage genome (see Chapter 10). The phage genome is, in general, sequentially expressed, a process requiring both phage- and host-coded functions: structural proteins are synthesised, the genome replicated (making use of the host DNA as a material source) and encapsidated producing, finally, the next generation of mature phage particles. This results in death and lysis of the host cell, and release of phage progeny (filamentous phages, on the other hand, leave their host cell without the ensuing lysis or death). Entry of either virulent or temperate bacteriophages may give rise to the lytic response, whereas only temperate phage are able to generate the lysogenic response (see below and Section 5.2.2). Upon release, the newly synthesised infective phage particles can adsorb to, and infect, further sensitive cells and, thus, continue the lytic cycle (Figure 5.3).

The kinetics of intracellular phage production (Figure 5.4) indicate that there is an *eclipse* period after phage adsorption, lasting about half the life of the infected bacterium, during which no lysis occurs, and there are no infective phage particles present within the cell. The eclipse — the period of transient loss of infectivity — represents, therefore, the time required for synthesis of progeny

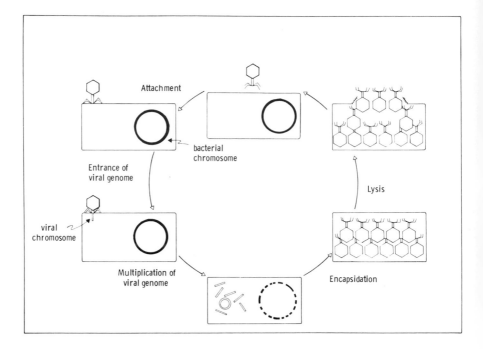

Figure 5.3: The Lytic Response
Phages bind to the outer membrane (see Some Phage Receptors, Table 5.2) and inject their nucleic acid. This latter event is responsible for control of the host's metabolism and the resultant phage multiplication and lysis. The process is cyclic in that released progeny phages can infect further susceptible host cells. See also the Lysogenic Response, Figures 5.5 and 5.6

phage (see Morphogenesis, below). The *latent period*, on the other hand, is the time required both for phage synthesis and release; this, consequently, includes the period for phage endolysin production and lysis from within. (Some phages lyse their host cells without a viral-encoded endolysin, e.g. T7.) Following the eclipse period, there is a gradual maturation of the non-infective intracellular phages, the *vegetative* form, and final release of progeny upon cell lysis (at the end of the latent period). The *burst size* of an infected bacterium, the average number of phages produced per infected cell, is given by the ratio of the final (the plateau value, Figure 5.4) to initial titre. Typical values are 50-200 infective particles (plaque-forming units, see (c), below) per bacterium (but 10^3 to 10^4 for the RNA phages).

Under conditions of multiple infection, a number of phages can both adsorb to, and infect, a single susceptible bacterial cell. Mixed infection by genetically similar bacteriophages (mutants or related), termed a *phage cross*, may give rise to mixed progeny, consisting of the infecting virions as well as phage recombinants (distinguished by their plating properties (see (c), below, and Section 7.1.3). On the other hand, unrelated phages (T1 and T2, for example) may show

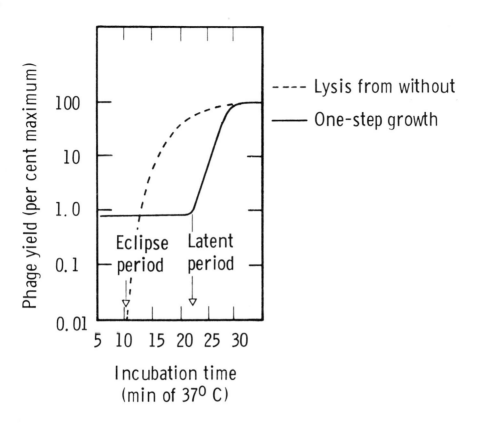

Figure 5.4: Intracellular Phage Kinetics

In one-step growth experiments, phages are incubated with host cells for a brief period to allow adsorption and then massively diluted, prior to further incubation, to prevent initiation of further infections by unadsorbed phages (and any newly released progeny viruses). Thus, phages liberated are those generated in a single life-cycle (Figure 5.3). Aliquots are taken at various times and assayed for extracellular phages. The plateau (solid line) spans the latent period, during which each single cell represents one infective centre. That is, when a single infected cell is plated out onto a suitable indicator lawn during the latent period it will produce only one plaque (see Some Phage Plaques, Figure 5.7). After rapid, semi-synchronous lysis (witness the steep slope), each infected cell produces some hundred progeny and, hence, several hundred plaques. Lysis from without, which may be achieved with chloroform, allows an estimation of the number of infective particles present in the cytoplasm prior to natural, phage-mediated lysis (from within). Liberated phages produced after a known period of incubation are asssayed on a suitable indicator strain (dotted lines). The length of the latent and eclipse periods depends upon temperature, bacteriophage and host cell (and its physiological state). The latent phases of T1, T4, and T5 are 13, 23 and 40 minutes, respectively, in nutrient medium at 37°; the T4 eclipse period is about 11 minutes under the same conditions. Bacteriophage T4 data from Doermann (Doermann, A.H. (1952) *J. Gen. Physiol., 35,* 645-656)

mutual exclusion, blocking production of infective particles. (In *superinfection exclusion*, prior infection modifies the cell such that it is no longer a suitable host for another or even the same phage type.)

(b) Morphogenesis. Encapsidation of complex double-stranded DNA phages follows a general pattern: production of the icosahedral protein shell, and synthesis of the chromosomal precursor with concomitant modification upon packaging. The order of events is not regulated by sequential gene expression. Rather, structural proteins produced in the latter half of the infection cycle interact in a specific manner. The protein *prehead* is first assembled, directed by 'scaffolding' protein(s) in a complex series of pathways, requiring, in general, both phage- and host-coded functions. The bacterial membrane appears to play an important role in assembly of some phages (*viz.* M13). Protein cleavage may also be involved in prehead morphogenesis; other host-coded functions include regulation of protein processing prior to cleavage (considered with respect to phage *lambda* in Section 10.2.1 (d)).

Phage DNA is packaged by entrance into the hollow capsid (rather than by construction of a protein shell around the viral genome). Once filled with the correct length or a complete genome, bacteriophage DNA is cut from its precursor to generate the mature, tightly wound, form. The general pattern among complex DNA phages is packaging of the monomeric viral genome from a linear concatamer (P2 is an exception in that circular monomeric DNA acts as substrate for packaging enzymes). For the T-even phages, the length of DNA packaged is dictated by the actual physical amount of DNA (this is the 'headful' model of Streisinger *et al.* (1967)). For phages such as T7 or *lambda*, cleavage is *site-specific* rather than *size-specific*; deletions, therefore, result in the encapsidation of a decreased amount of DNA. Heads of phages are larger than preheads. This increase in volume (the *lambda* head, for example, increases in diameter by about 20 per cent, corresponding to a 100 per cent increase in volume) is due to subunit rearrangement, apparently triggered by the presence of DNA for packaging. Packaging involves a massive increase in DNA concentration (certainly, *lambda* DNA is about 200 times more condensed than in the cytoplasm), the tightly packed DNA filling the head. Little is known about this remarkable aspect of bacteriophage morphogenesis.

Head assembly continues with modification of the collar (or base) to allow attachment of the protein tail, either synthesised in an independent pathway or formed *in situ* on the modified collar. Tail assembly, itself, involves a complex series of reactions (eleven gene products have been implicated in *lambda* tail formation and 21 for T4). Bacteriophage morphogenesis is, therefore, a highly organised process in which both phage- and bacterial-gene products act sequentially and at precise points in the assembly pathway.

(c) The Lysogenic Response. A bacteriophage, from the discussion above, would seem only to be able to multiply in its susceptible host to the detriment of the infected cell. This is true of virulent bacterial viruses (and virulent mutants, *vir*, of temperate phages). *Temperate* bacteriophages, on the other hand, are able to perpetuate their genetic material without the concomitant cellular death (and lysis) associated with virulence; the surviving bacterium (carrying a copy or

copies of the viral genome) is termed a *lysogen*. [The term lysogen implies that these cells retain the lysis proficiency of their symbiotic partner. Modification of bacterial phenotype by the presence of phage in the lysogenic state is called *phage conversion*.] Such a symbiotic relationship imparts *immunity* to the lysogenic cell against superinfection by genetically related phages (as well as protection against certain unrelated phages, see below). In short, infection by a temperate phage may give rise to either the lytic or lysogenic response. The molecular basis for this 'decision' is considered in Section 10.2 for *lambda*.

In lysogeny, the phage lethal functions are not expressed; the stable *prophage*, as this form of the carried phage is termed, remains within the cytoplasm, either integrated into and replicated in association with the host's chromosome (*lambda*, P2, Mu, for example) or existing as an autonomously replicating circular piece of DNA — a plasmid (*viz*. P1, *lambda* mutants such as λ*dv*).[2] The lysogenic cell maintains the prophage and at the same time is spared from death as the phage lytic functions are *repressed* in this prophage state by specific phage-coded regulatory elements (Figure 5.5). It is this blockage of lytic development, allowing cellular survival, that also imparts immunity to the lysogen against superinfection by the same phage, or phage of the same immunity type. Whereas mutation to resistance acts at the membrane level (inhibiting phage adsorption), superinfection immunity inhibits replication of the superinfecting phage genome; it is diluted out, *segregated* (see Figure 4.1), upon division of the immune lysogen. Immunity, therefore, is a second barrier against infection (see also Barriers to F Transfer, Section 4.1.2 (b)).

In addition to superinfection immunity, certain lysogens are protected from genetically unrelated bacterial viruses, apparently due to expression of other prophage functions. Thus, *lambda* lysogens do not plate the T4 *rII* (rapid lysis) mutants due to *rex* expression by the resident *lambda* prophage, and P2 lysogens inhibit phages such as λ*gam+ red+* (Section 10.2) that exhibit the Spi+ phenotype (*s*ensitive to *P2* *i*nterference).

Though stable, the lysogenic state is not necessarily permanent; the prophage may enter the lytic cycle upon phage *induction*. This event occurs spontaneously at low frequency (10^{-2}-10^{-5} per cell per generation).[3] Alternatively, it may be brought about by interference with the host cell's metabolism, particularly its DNA biosynthesis[4] (or by thermal induction of temperature-conditional phage

2. Since prophage replication does not require expression of all the genes of the productive cycle, some mutations that inhibit productive infection allow temperate phages to be perpetuated indefinitely in the prophage state as extrachromosomal genetic elements, rather than integrated into the bacterial chromosome. λ*dv* (defective *v*irulent) is a small fragment of the *lambda* chromosome (about 10-20 per cent the size of the wild-type genome) carrying part of the immunity region. It replicates in the cytoplasm without preventing bacterial growth; about fifty copies per bacterium are present (see also Section 10.2.4).

3. A characteristic of lysogenic cultures is the presence of phage particles in the medium. This is a consequence of spontaneous induction of a small proportion of lysogenic cells, which subsequently lyse, liberating phages into the culture medium. Survivors which still harbour their prophage will, of course, be immune to superinfection.

4. Not all temperate phages are artificially inducible (*viz*. P2).

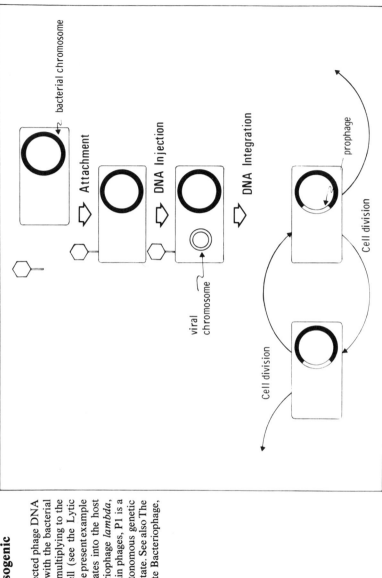

Figure 5.5: The Lysogenic Response

In the prophage state, injected phage DNA replicates in conjunction with the bacterial chromosome rather than multiplying to the detriment of the host cell (see the Lytic Response, Figure 5.3). The present example is for a phage that integrates into the host chromosome (say, bacteriophage *lambda*, Figure 5.14) though certain phages, P1 is a case in point, exist as autonomous genetic elements in the prophage state. See also The Life-Cycle of a Temperate Bacteriophage, Figure 5.6

repressor mutants). Once in the lytic phase, phage synthesis, maturation and release occur (and, of course, bacterial lysis, Figure 5.6). This process might be regarded in anthropomorphic terms as an 'escape mechanism' for the resident prophage from a doomed cell, liberating progeny virions to further infect viable cells. The prophage is clearly an innocuous genetic element that can impart a selective advantage to its host. It has, however, the potential to be lethal.

5.1.3 Phage Methodology

After addition of a small number of phages (say, less than 100) to a nutrient agar plate spread with sensitive bacteria, such that the host cells are present in marked excess, small circular clearings, or *plaques*, appear in the bacterial lawn upon incubation. These represent the infection of one cell by a single bacteriophage, followed by diffusion of the progeny and lysis of cells in the vicinity of the initial *infective centre*. (Addition of an excess of phage particles results in overlapping of the plaques, giving the appearance of a clear plate (see Confluent Lysis, Figure 5.7).) The rate of diffusion and, therefore, the plaque size, is dictated by the phage dimensions and the period required for phage production and release. Plaque morphology is generally characteristic of a given phage. Virulent bacteriophages give *clear* plaques, whereas plaques of temperate virions are *turbid* due to growth of lysogenic cells, immune to superinfection. Plaque size is, clearly, also phage-related (T-even plaques, for example, are small, T7 plaques large, Figure 5.7). Moreover, changes in the host cell or plating conditions can give rise to profound morphological differences.

(a) Titration: The Plaque Assay. The number of virulent phages in a population may be titred using its intrinsic all-or-nothing killing property: infection of a sensitive bacterium is either lethal or not. When lethal, lysis of a single cell, upon infection by a single bacteriophage, is amplified by diffusion of progeny phages. The plaque formed on solid medium is the culmination of this — a microscopic process magnified to the macroscopic level, involving the killing and lysis of several million bacteria.

In practice (Figure 5.8), phage dilutions are added to an excess of susceptible host. Phage aliquots are either spotted onto the cells spread in soft-agar *(top-agar)* across the surface of a nutrient agar plate or, better, phages and cells are mixed together to allow pre-adsorption, prior to immobolisation in top-agar.

Upon multiple adsorption (m.o.i.$>$1), sensitive cells are killed by phages irrespective of the proportion of inactive viral particles present in the population — there is *confluent lysis*, the region where the phage was added being clear of bacterial growth (apart from spontaneous phage-resistant bacterial mutants and lysogenic survivors). The true number of *infective particles* is, thus, hidden. However, when a small number of phages are used, only active particles produce *infective centres*. The number of *plaque forming units (p.f.u.)*, expressed in units per ml, is generally directly proportional to the phage concentration (inversely proportional to the dilution factor). This is, clearly, not an absolute assay: only

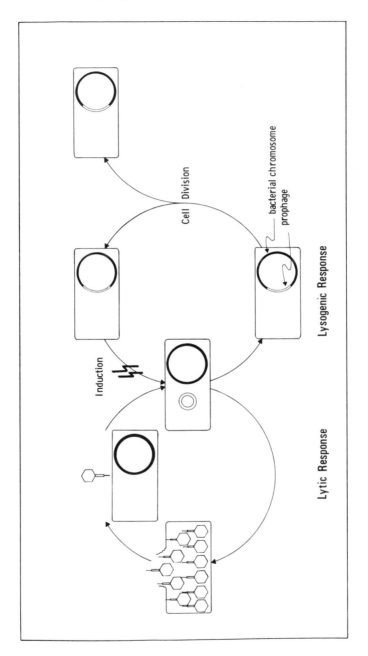

Figure 5.6: The Life-Cycle of a Temperate Bacteriophage
A temperate bacterial virus can either enter the lytic cycle upon infection or lysogenise its host (see also Figures 5.3 and 5.5). Interference with the host cell's metabolism, in particular DNA synthesis, induces the phage DNA out of the prophage state (in which the viral DNA may either be integrated into the bacterial chromosome or present as a stably inherited, autonomous genetic element)

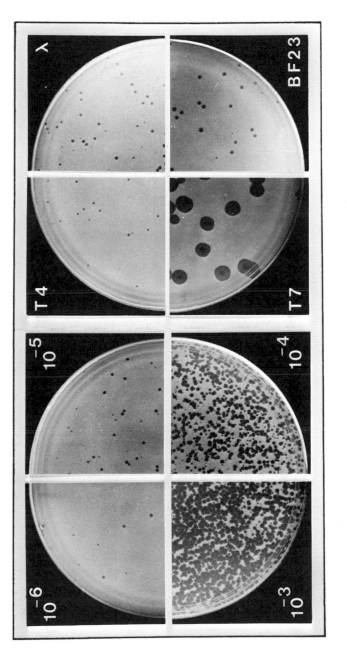

Figure 5.7: Some Phage Plaques

Composite for different phage dilutions (left) and different phages (right). At low phage number (high dilutions), the titre can be readily determined (25 *lambda* plaques in one quarter of a plate from 0.1 ml of a 10^{-5} dilution would be equivalent to $25 \times 4 \times 10 \times 10^5$ or 10^8 p.f.u. per ml — an appallingly poor titre!). As the number of infective particles increases (low dilution), individual plaques begin to overlap, resulting ultimately in confluent lysis. Plaque size of the 4 wild-type phages is in the order T4, λ, BF23 and T7 (under the present culture conditions, see text)

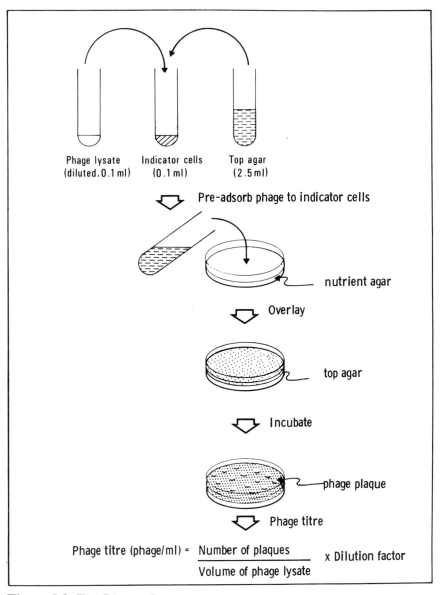

Phage lysate · Indicator cells · Top agar
(diluted, 0.1 ml) · (0.1 ml) · (2.5 ml)

Pre-adsorb phage to indicator cells

nutrient agar

Overlay

top agar

Incubate

phage plaque

Phage titre

$$\text{Phage titre (phage/ml)} = \frac{\text{Number of plaques}}{\text{Volume of phage lysate}} \times \text{Dilution factor}$$

Figure 5.8: The Plaque Assay

Phages and indicator (host cells) are spread onto nutrient agar in a molten overlay of top agar — a low-concentration (soft) agar solution. Each infective phage kills one cell, the liberated progeny particles acting on bacteria in their immediate environment. Each infective phage, therefore, produces an infective centre or plaque due to rounds of infection and lysis, and diffusion of progeny (see Some Phage Plaques, Figure 5.7). The number of viable phages (but not the total number of phage particles, see text) can be determined from the number of plaques in a given volume of diluted phage stock (Figure 5.9) active particles are determined. There is, nevertheless, generally good agreement between viral titres determined as p.f.u.s and the actual number present. (Note that there is a finite probability — extremely low when the m.o.i. is small — that more than one phage may infect a cell, even though only one phage is normally required for the lethal event.) The *efficiency of plating* (EOP), the proportion of total phage particles (enumerated by electron microscopy, say) able to give infective centres, lies between 40 and 100 per cent (RNA phages show lower values). The term EOP has also been used when comparing the plaque-forming abilities of phages on different hosts or under different plating conditions.

(b) Production of Phage Lines. A plaque represents bacteriophage multiplication resulting from the infection of one cell by a single active phage particle. The phages within the cleared area of a single plaque are identical in the absence of a mutational event occurring during phage multiplication. The phage plaque is, therefore, the viral equivalent of a bacterial colony: a group of identical organisms. *Lysates*, stocks of particular phage lines, are prepared starting with a single 'fresh' plaque. The *plate lysate* method (Figure 5.9) involves pre-adsorption of 'healthy' sensitive host cells with these plaque bacteriophages at an m.o.i. sufficient to give confluent lysis (higher m.o.i.s may reduce the yield), followed by 'harvesting' the soft-agar layer containing the liberated phages from the nutrient plate. An initial investment of about 10^6 p.f.u. can lead to a 10^5-fold return in a matter of hours — a lysate containing up to 10^{11} p.f.u.

5.2 Bacteriophage-Mediated Gene Transfer

Bacteriophages are 'vectors' of genetic information. Those that are virulent lyse and kill their hosts upon transfer of their genome. Temperate phages, on the other hand, are able to transfer their DNA without the ensuing cellular catastrophe. In either case, however, these transfer processes as described only involve phage genes. The property of certain phages, mainly temperate species (or their virulent derivatives), to act as vectors for the transfer of small segments of bacterial genetic material from one cell to another has proved to be of paramount importance in the analysis of gene structure at the molecular level. Phage-mediated gene transfer, *transduction*,[5] involves either movement of random regions of DNA *(generalised transduction)* or transfer of a particular limited portion of the bacterial chromosome *(specialised* or *restricted transduction)*; some phages can participate in both modes. Specific examples of these two types of transduction are described below.

5. The term 'transduction' was first applied by Zinder and Lederberg (1952) for transfer of genetic material by a phage vector (initially investigated in *S. typhimurium*). Specialised transduction refers strictly to the transfer of bacterial DNA (Morse *et al.* (1956)) though it is now used for DNA in general.

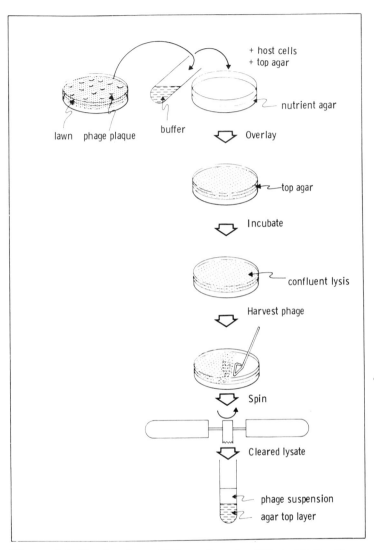

Figure 5.9: Production of Phage Lysates (Plate Method)

Phages within a fresh plaque (Figure 5.7) are commonly retrieved by 'stabbing' with a sterile thin wire or toothpick, or better, punching out the agar with a hollow glass tube; they are transferred to buffer containing chloroform to kill any residual, unlysed indicator cells. Host cells and phages — sufficient infective particles to give confluent lysis (say, m.o.i. $\simeq 10^{-2}$) — are pre-adsorbed and spread onto a fresh, moist nutrient agar plate in top agar (see Figure 5.8). After incubation (in the case of temperate phages, plates are left for the minimal time possible to avoid lysogeny), progeny phages are harvested by scraping off the top agar layer with a glass loop. Nutrient broth and chloroform (to lyse any intact cells) are added and cell debris and agar are removed by centrifugation. The clear supernatant contains the liberated phage particles (it may be concentrated by further centrifugal steps, Section 7.3.1)

5.2.1 Generalised Transduction

Growth of the large, DNA-containing phage P1 on *E. coli* results in the occasional encapsidation of a small amount of bacterial DNA in the place of the normal phage genome. Since erroneous packaging occurs at random, a small proportion of phage particles in a P1 lysate carry different regions of the host chromosome (the majority have P1 DNA). The bacterial genome is, therefore, represented in this phage population as a random distribution of short linear DNA fragments. The maximum size of a transducing fragment can be calculated from the size of the phage genome it replaces; transducing particles carry little, if any, phage DNA. The P1 genome (and, hence, the maximum size of a transducing fragment) is 91.5 kb (60 Mdal, Table 5.1), approximately 2.4 per cent the size of the host chromosome.

This phenomenon of encapsidation of foreign DNA is somewhat analogous to *phenotypic mixing*, in which mixed infection by two different phages can result in the normal phage complement being replaced by heterologous phage DNA; the alien viral DNA obtains the host-range of the capsid packaging it ('a virus changing its coat', Novick (1966)). Phenotypic mixing is not a genetic change since second-generation viral DNA is correctly encapsidated. As generalised transduction is restricted largely to temperate phages in *E. coli*,[6] the lysogenic property of the phage may be important. It is, nevertheless, difficult to build a model on this basis since P1 lysogeny does not generally appear to involve phage integration and a virulent mutant of this phage, P1 *vir*, still acts as a generalised transducing particle. Breakdown of bacterial DNA upon phage infection is another factor of importance (transducing particles incorporate only those fragments of bacterial DNA present at the time of infection; chromosomal replication is not involved). Clearly, formation of the tranducing particle is helped by some degradation of host DNA (and weak packaging specificity) but extensive turnover is likely to be detrimental.

The protein phage capsid acts as the vector in the transfer of random fragments of bacterial DNA. Phage P1 has also been used for the transfer of extra-chromosomal genetic elements (present in the cytoplasm at the time of infection); the limited capacity of the P1 capsid can give rise to *transductional shortening* of such genetic elements. Replacement of the viral complement by host material ensures gene transfer without its associated lethality.

Generalised transduction involves integration of part or all of one fragment of the donor chromosome into the recipient's genome. General recombination between homologous regions of DNA on the tranducing fragment and chromosome is responsible for integration of the transferred DNA. Transductants formed by recombination do not normally carry phage DNA (especially when precautions against P1 lysogeny are taken — by use of P1 *vir*, for example). The

---

6.   *Lambda* preheads may also package foreign DNA in special circumstances and can thus be used as a generalised transducing phage. Generalised transduction has also been observed with the virulent phage T1 and with a multiple mutant of T4.

proportion of transducing particles has been estimated to comprise about 0.3 per cent of the total phage population. The probability that a bacterial gene will be carried in a phage particle can be calculated from the frequency of transducing particles (0.003) and the fraction of the bacterial chromosome contained in such a particle (0.024), making the assumptions that all regions of the bacterial genome have an equal chance of incorporation, and that the chance of recombination is the same for any region. This probability is about 7.2×10^{-5}.

A proportion of both normal and transducing particles carries a shorter DNA length, apparently associated with a smaller phage head (circular permutation of the phage DNA allows *in situ* complementation of the small defective 'normal' virions, see Section 7.1.1 (a)). The diminished length of the transducing phage results in reduced frequency of the joint transfer of adjacent bacterial genes (Cotransduction, see below). Moreover, markers at either end of a transducing fragment are likely to synapse with homologous regions on the chromosome and exhibit reduced recombination frequencies. The effective size of a transducing fragment is, therefore, likely to be somewhat less than 2.4 per cent the size of the bacterial genome.

In some instances, particularly when the recipient cell lacks a functional recombination system, a transferred fragment is not integrated. The resulting *abortive transductant* carries a fragment of DNA which cannot replicate; it is, therefore, inherited in a unilinear manner (see Figure 4.1). Gene products encoded in the abortive fragment may be passed *cytoplasmically* to daughter cells, even though only one bacterium in a clone actually harbours the plasmid. Minute colonies may, thus, be formed from abortive transductants.

Delay in expression of integrated DNA may also give rise to micro-colonies. *Phenotypic lag* occurs noticeably when the gene transferred is recessive to the allele present in the recipient strain. Expression of a transduced recessive allele requires prior segregation of the dominant phenotype. Streptomycin resistance, for example, when caused by a genetic lesion in the gene for the S12 r-protein is *recessive* to the wild-type allele (a partial diploid strain carrying both resistant and sensitive alleles is drug sensitive due to the formation of unstable initiation complexes, Section 2.2.2 (d)). Therefore, entrance of a Str-R allele by P1 transduction only gives rise to resistant derivatives upon cytoplasmic dilution of existing sensitive ribosomes.

Phage P1 has proved an extremely useful tool in strain construction. Growth of P1 on one mutant *(propagation step)* allows transfer of a particular gene, or group of linked genes, to the bacterium of interest (*transduction step*, Figure 5.10). Subsequent propagation on a different host supplies a new gene pool for transduction. The technique is limited only by the necessity of selecting for the presence of the transduced region, and by the length of the DNA transferred (in any one transductional event). However, markers that are not readily selected may be transferred by selecting for gene(s) close to the region of interest (as long as they are not separated by a distance greater than 91.5 kb, Figure 5.11). This technique of joint transfer of genes on the same transducing fragment, termed *P1*

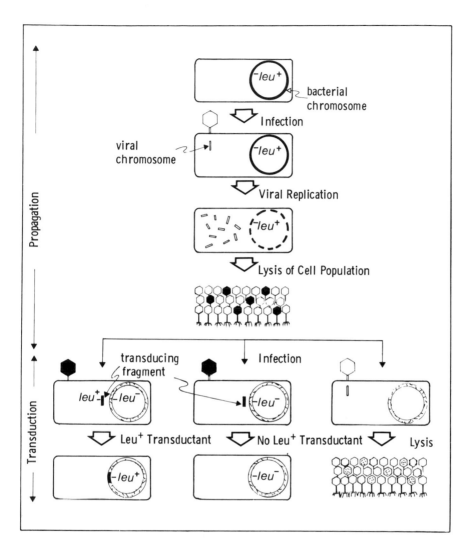

Figure 5.10: Stages in Generalised Transduction

In the propagation step, P1 is grown (cf. Figure 5.9) on a strain with the relevant genetic make-up (*leu*+ in the present example). Only a small proportion, about 10^{-5} of progeny particles, 'trap' this region during faulty encapsidation. Thus, in the transduction step, only a small proportion of cells (in this case, *leu*− derivatives) are infected with the specific bacterial sequence *(leu+)*. Whereas infection by P1, itself, results in cellular lysis and death (bottom, right), the injection of bacterial DNA avoids this lethal event. However, only those cells infected with *leu*+ DNA can give rise to *leu*+ transductants by recombination (Figure 5.11)

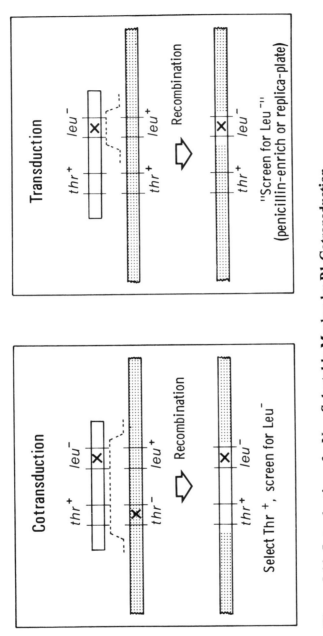

Figure 5.11: Introduction of a Non-Selectable Marker by P1 Cotransduction

Whereas it is eminently feasible to select for Leu⁺ transductants (only Leu⁻ revertants, or revertants, of a Leu⁻ auxotroph survive in the absence of leucine), introduction of a *leu*⁻ allele into a *leu*⁺ strain has to be done in a more roundabout fashion (cf. Mutant Selection, Section 3.2.2 (b)), i.e. since the nearby marker *thr* is cotransducible with *leu* (the distance separating these two sites being less than 91.5 kb), it is possible to obtain Leu⁻ recombinants by screening among Thr⁺ transductants *cotransduction*, has been used extensively in detailed mapping (see Section 7.1.2 (bi)). It also lends itself to localised mutagenesis (Section 3.2.2 (a)). Moreover, since most bacterial genes are continuously linked by P1 cotransduction, large regions of DNA can effectively be introduced.

Whereas P1 cannot, in general, be used for gene enrichment,[7] bacteriophage *lambda*, a specialised transducing particle, can. This property of *lambda* is discussed below.

5.2.2 Specialised Transduction

The head of the temperate coliphage *lambda* contains a linear double-stranded DNA molecule of 32 Mdal (approximately 48.6 kb, Table 5.1). At both ends of the phage genome there are single-strand GC-rich regions, twelve nucleotides long; these base sequences are complementary (Figure 5.12). *Lambda* virions adsorb to a specific receptor (the *lamB* gene product, Section 5.1.2) upon phage infection. The linear DNA is injected through the flexible phage tail into the sensitive bacterium, and rapidly circularises in the cytoplasm by covalent end-joining of its two single-stranded cohesive ends, m and m' *(cos)*, under the action of host DNA ligase (Figure 5.12).

In the lysogenic state, phage *lambda* DNA is inserted at a specific locus, the *lambda attachment site (att*$^\lambda$*R)* on the *E. coli* chromosome between the bacterial *gal* and *bio* genes, in one particular orientation (Figure 5.13). Integration of the circular phage genome occurs through reciprocal recombination between homologous regions which form parts of the chromosomal and phage attachment sites (*att*$^\lambda$*B* and *att*$^\lambda$*P*, respectively). There is an asymmetric, AT-rich common *core* region, *O*, of 15 nucleotide pairs.

$$5' \quad G\ C\ T\ T\ T\ T\ T\ T\ A\ T\ A\ C\ T\ A\ A$$
$$C\ G\ A\ A\ A\ A\ A\ A\ T\ A\ T\ G\ A\ T\ T \quad 5'$$

(the components of these two sites are designated BOB' and POP' where BB' and PP' refer to the different sequences bordering the common core in the bacterial and viral *att* sites, Figure 5.14). The circular vegetative viral chromosome splits within the phage attachment sequence to form a linear *prophage*, separated from bacterial sequences by the two *hybrid* (recombinant) attachment sites, termed *attL* and *attR* (consisting of the elements BOP' and POB', respectively). Splitting at *att*$^\lambda$*P*, rather than at *cos* (mm'), results in the prophage having a gene order different from that of the infective phage. Moreover, certain genes adjacent on the vegetative genome — a covalent circular molecule of DNA — are separate on the prophage. Thus, the prophage map is a *circular permutation* of the vegetative phage map (Figure 5.15). Integration of the circular phage genome into the continuity of the bacterial chromosome increases the length of

---

7.   There are derivatives of phage P1 that carry stably DNA other than the viral genome. P1 *cml*, for example, carries the genes for resistance to the antibiotic chloramphenicol integrated into the viral genome (the origin of these special derivatives is discussed in Section 6.2.2 (aii)).

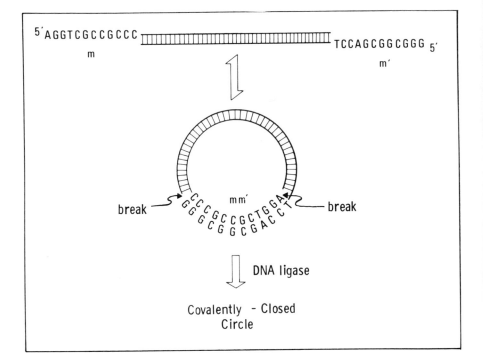

Figure 5.12: Circularisation of the Bacteriophage *Lambda* Genome
The two single-stranded termini, m and m' (12 nucleotides long), are complementary, thereby allowing specific circularisation through end pairing. Ligation (Section 6.1.1 (b)) completes the process

DNA (and decreases the linkage) between bacterial genes bounding to the attachment site. The increase is, clearly, directly proportional to the size of the phage genome inserted.

This form of lysogeny shows that bacterial and viral DNA can undergo recombination. The integrative event is, in fact, an example of site-specific recombination since insertion occurs at a unique site and in the absence of host recombination functions (but note that there are *secondary sites*, 'masked' by *att*$^\lambda$*B*, in which *lambda* may integrate at low frequency, Section 6.2.2 (ai)). It is presumably primarily through the common core region O that the *lambda int* gene product (molecular weight 40 000) catalyses site-specific recombination. Integrative recombination between the circular phage and bacterial genome is analogous to insertion of the F plasmid at specific sites on the bacterial chromosome (Campbell initially proposed this model for insertion of covalently

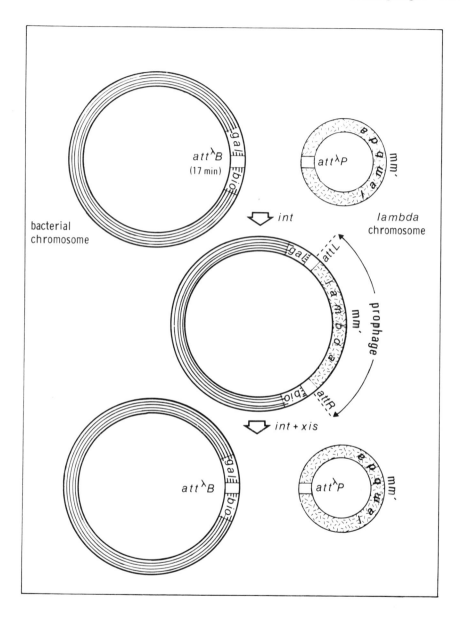

Figure 5.13: The Vegetative and Prophage States of Bacteriophage *Lambda*
Integration of the vegetative form occurs through specific sites on the two circular elements, *att*$^\lambda$*B*
and *att*$^\lambda$*P*. No genetic material is lost in the process, the resultant cointegrate consisting of the entire
phage and bacterial chromosomes (see The *Lambda* Prophage Map is a Circular Permutation of the
Vegetative Map, Figure 5.15). Whereas the forward reaction is catalysed by the phage-coded *int*
gene product (and certain bacterial proteins), excision requires both *lambda int* and *xis* functions

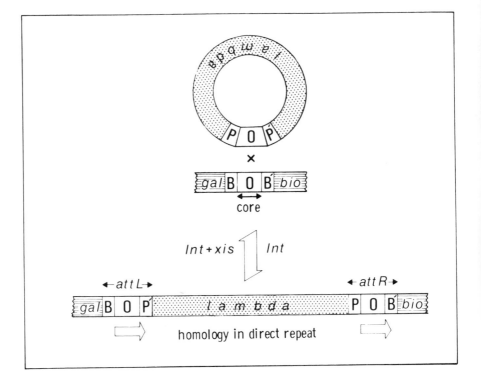

Figure 5.14: Integration of Bacteriophage *Lambda* **at** *att^λP*

Integrative recombination occurs at the phage and bacterial sites (*att^λP* and *att^λB*, respectively) through the common region of homology, O. Splitting at O generates two hybrid attachment sites, *attL* and *attR*, flanking prophage *lambda* and, thus, a non-tandem direct repeat of O (see The *Lambda* Prophage Map is a Circular Permutation of the Vegetative Map, Figure 5.15). The figure is based upon the Campbell Model (cf. the Campbell Model for F Plasmid Insertion, Figure 4.3)

closed circular molecules of DNA into the bacterial chromosome, see Section 4.1.3 (a)). Note that the reverse reaction, phage excision, requires the *lambda xis* protein as well as the *int* gene product (see below).

Under normal conditions of infection (and after superinfection), lysogens are formed that carry more than one copy of the phage genome. These *polylysogens* may represent as much as one third of lysogenised cells. In the case of normal infection, a single integration event appears to be responsible (presumably involving multimeric circles of *lambda*, formed by recombination, see Section 10.2.1 (c)). In superinfection (namely, a single lysogen infected with a second phage), polylysogens appear to arise from *int*-mediated integration at one of the resident hybrid attachment sites, *attL* or *attR:* they may also occur through normal recombination via *lambda* homology (such a dilysogen is shown in

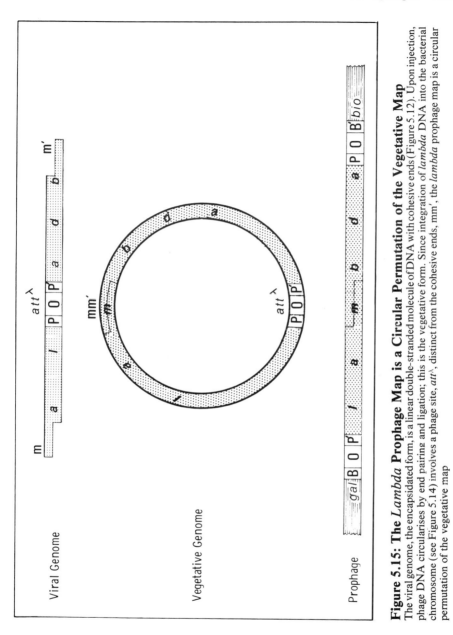

Figure 5.15: The *Lambda* Prophage Map is a Circular Permutation of the Vegetative Map

The viral genome, the encapsidated form, is a linear double-stranded molecule of DNA with cohesive ends (Figure 5.12). Upon injection, phage DNA circularises by end pairing and ligation; this is the vegetative form. Since integration of *lambda* DNA into the bacterial chromosome (see Figure 5.14) involves a phage site, *att*^λ, distinct from the cohesive ends, mm', the *lambda* prophage map is a circular permutation of the vegetative map

Figure 5.18). Polylysogens are unstable, the tandem repeated *lambda* sequences acting as substrate for the general recombination system (*recA*$^+$ dilysogens, for example, give rise to monolysogens at a frequency of 10^{-2}-10^{-3} per cell generation).

Phage-coded genes allow both accurate integration into, and excision from, the chromosomal attachment site (see Figure 5.13).[8] Occasionally, however, an error in excision occurs during recircularisation (at a frequency of about 10^{-5}). Rather than specific recombination between terminal attachment sites, breakage and joining takes place within non-homologous DNA. This illegitimate Rec-independent recombination event appears to involve the same molecular processes that generate deletion mutants and F′ plasmids (see Section 6.2.2(b)). The majority of specialised transducing phages appear to be formed at or after (spontaneous) phage induction though if phage genes are involved, these do not include *int* or *xis*. Improper looping-out produces a hybrid *specialised transducing phage* carrying some of the bacterial DNA bordering the chromosomal attachment site (Figure 5.16). Since the initial integration event is site-specific, transducing phages carry only a particular region of the bacterial chromosome (genes on either the *gal* or *bio* side of *att*$^\lambda$*B*). Aberrant excision often results in concomitant loss of phage DNA: only a limited amount of DNA may be packaged in the *lambda* capsid (about 10 per cent or 4.9 kb more than the normal *lambda* complement of DNA, and not less than 73 per cent). The amount of foreign DNA which can be carried by a *lambda* transducing phage is commonly augmented by making use of small *lambda* derivatives carrying deletions that remove non-essential functions. The b2 deletion, for example, allows an extra 12.1 per cent (5.9 kb).

The extent of bacterial DNA included and phage DNA lost varies from one transducing particle to another. Genes lost from the circularised bacteriophage genome are normally those in the prophage structure that are most distal to the bacterial genes included. Thus, *bio* transducing phages may lack part of the regulatory region, whereas essential morphological genes are excluded in those phages that carry the bacterial *gal* genes (see Section 10.2.1 for the organisation of the *lambda* genome). When this lost phage DNA encodes essential phage functions, the resulting specialised transducing phage is *defective*, unable to propagate on susceptible cells. These defective phages normally grow when co-infection takes place in the presence of a wild-type *helper* phage, supplying the missing essential functions. Gal transducing phages, λ*gal*, are an example of this defective class: λ*pbio* phages are not normally defective since only non-

8.   Aberrant excision may leave part of the prophage behind in the bacterial chromosome. Continued survival of the bacterium necessitates deletion (or inactivation) of lethal phage functions. Such survivors usually lack immunity to reinfection. These strains are no longer lysogenic since they lack the potential for spontaneous production of phage particles. This type of defective prophage is termed *cryptic* — its presence is not detectable by immunity and the resident mutant prophage appears functionally inactive.
      Several authors have established the existence of cryptic prophage(s) integrated into the *E. coli* chromosome. The Rac prophage, for example, is lambdoid in type (see Section 6.2.1 (aii)).

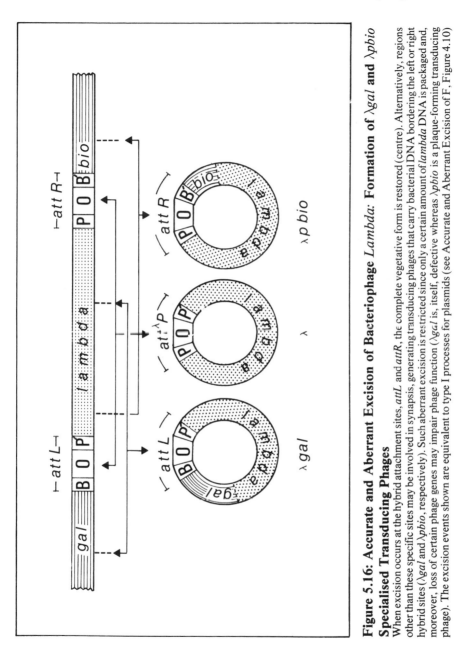

Figure 5.16: Accurate and Aberrant Excision of Bacteriophage *Lambda*: Formation of *λgal* and *λpbio* Specialised Transducing Phages

When excision occurs at the hybrid attachment sites, *attL* and *attR*, the complete vegetative form is restored (centre). Alternatively, regions other than these specific sites may be involved in synapsis, generating transducing phages that carry bacterial DNA bordering the left or right hybrid sites (*λgal* and *λpbio*, respectively). Such aberrant excision is restricted since only a certain amount of *lambda* DNA is packaged and, moreover, loss of certain phage genes may impair phage function (*λgal* is, itself, defective whereas *λpbio* is a plaque-forming transducing phage). The excision events shown are equivalent to type I processes for plasmids (see Accurate and Aberrant Excision of F, Figure 4.10) essential genes are lost. (Note that the bacterial gene(s) carried by a *p*laque-forming tranducing phage is prefixed with a '*p*'. Early notation required the prefix '*d*' for a *d*efective transducing phage, but this has since been dropped.) This imprecise excision has been exploited for the isolation of specialised transducing phages; either the *gal* or *bio* regions (but not both due to the restrictions in amount of *lambda* DNA accommodated and the fact that phages carrying *aHL* and *attR* are 'self-excisers') are covalently joined to phage DNA. That is, in analogy with F-prime formation (Section 4.1.3 (b)), type I but not type II events occur. Note that both λ*gal* and λ*pbio* lack *att*λ*P* and carry instead a hybrid attachment site (either *attL* or *attR*), which unlike the normal site combines poorly with *att*λ*B*.

In practice, the lysate from induction of a *lambda* lysogen (Section 5.1.3 (b)) is a heterogeneous population of viral particles.[9] While the majority of phage genomes are intact, a small proportion, say 10^{-5}, are transducing phages for either the *gal* or *bio* region. It is, therefore, a *low-frequency transducing (LFT) lysate* since the bacterial genes are present in only a few of the phages and are transferred at low frequency to recipient cells. Infection of, say, a sensitive, phenotypically Gal⁻ recipient with this LFT lysate will give a small number of Gal⁺ *transductants* (about 1 per 10^6 active phages). The survivors, able now to utilise the sugar galactose, are, typically, diploid for *gal*[10]; the transductant carries both the original chromosomal *gal* mutation (the lesion responsible for the Gal⁻ phenotype) and a defective λ*gal* prophage. Transductants are, therefore, generally immune to superinfection by homologous phages.

The *gal⁻/gal⁺* merodiploid can have arisen by Rec-mediated integrative recombination between homologous *gal* regions on λ*gal* and the bacterial genome (a *gal⁻*/λ*gal⁺* lysogen, Figure 5.17). More commonly, as the LFT lysate consists of a large fraction of non-transducing wild-type phage, the transductant will be a double lysogen (*gal⁻*/λ/λ*gal⁺*, Figure 5.18). This is likely to have arisen by a two stage phage-mediated process: insertion of the infective phage DNA at the normal attachment site, *att*λ*B*, followed by Rec- (or Int-) mediated reciprocal recombination between DNA of the resident prophage and λ*gal⁺*. The wild-type helper, therefore, supplies homologous DNA for recombination. The resultant lysogen carries the two phages in tandem, formed by λ*gal* insertion within the resident prophage.

When this double lysogen is now induced, the phage lysate consists of approximately equal numbers of *lambda* wild-type helper and transducing phages. Moreover, this phage population will be homogeneous for the specialised transducing particle present in the double lysogen (a number of different

---

9. Clearly, formation of specialised transducing phages from induced lysates requires prior isolation of the lysogen (rather than just infection as in propagation of P1 *vir*).
10. A small number of survivors will be haploid, apparently formed by Rec-mediated substitution of the chromosomal *gal⁻* region with *gal⁺* DNA carried on the transducing phage. These transductants are, thus, not lysogens and lack immunity to superinfection (unless they have been independently lysogenised by the normal phage).

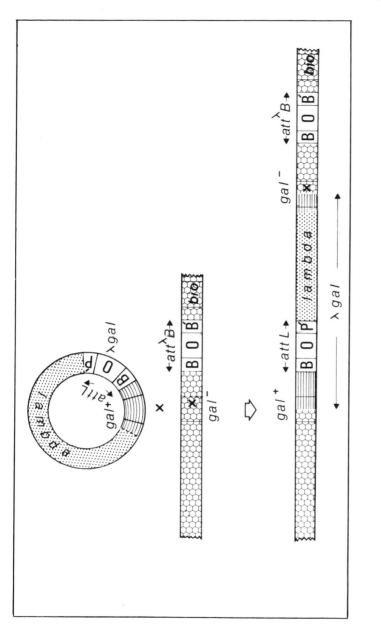

Figure 5.17: Formation of a λ*gal* **Lysogen**

Integrative recombination between the bacterial and phage-carried *gal* DNA generates a λ*gal* lysogen. In the case of a *gal*⁻ ::λ*gal*⁺ cointegrate (the tandem colons represent additive recombination), a Gal⁺ lysogenic transductant is formed. Genetic exchange rather than integrative recombination would result in a Gal⁺, non-lysogenic transductant (not shown). See also Formation of a λ/λ*gal* Dilysogen, Figure 5.18

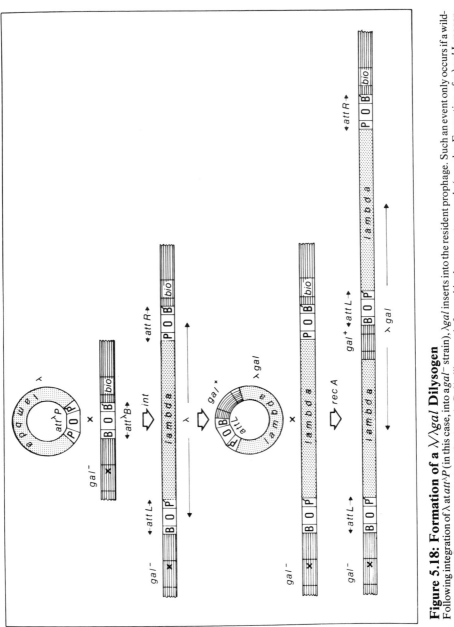

Figure 5.18: Formation of a λλ*gal* Dilysogen

Following integration of λ at *att*λ*P* (in this case, into a *gal*⁻ strain), λ*gal* inserts into the resident prophage. Such an event only occurs if a wild-type helper phage is present during infection. A Gal⁺ dilysogen is formed in the present example (see also Formation of a λ*gal* Lysogen, Figure 5.17) transducing particles are likely to be present in an LFT lysate, formed by independent aberrant excision events in the lysogenic population upon phage induction). In addition, the lysate will transduce Gal⁻ strains to Gal⁺ at high frequency (as high as 100 per cent) — it is a *high-frequency transducing (HFT) lysate* (summarised in Figure 5.19).

What are the crucial differences between this HFT *lambda* lysate and a P1 stock? The vector in specialised transduction is *lambda* DNA, itself; viral and host chromosomal DNA are covalently associated in the *lambda* transducing fragment. Both phage and bacterial genes are transferred intact. This preservation ensures genetic continuity of the phage and bacterial material. Transduction by an HFT *lambda* lysate is usually accomplished by integration of the specialised transducing phage into the bacterial chromosome. The transductant is, therefore, a lysogen (normally double,[11] λ/λ*gal* type, see above and Figure 5.18), immune to superinfection by homologous phages and prone to spontaneous curing at low frequency, segregating Gal⁻-type colonies. Only a small proportion of phage particles in a P1 lysate (about 0.3 per cent), on the other hand, carry foreign DNA and this is a random population of fragments of bacterial DNA. The P1 protein capsid, not the viral genome, acts as the vector. Moreover, gene transfer is accomplished by exchange with, rather than integration into, the recipient's chromosome. Concomitantly, loss of both (protein) vector and non-recombined bacterial DNA occurs. In conclusion, an HFT *lambda* lysate will generally give rise to a lysogenic *cointegrate* (producing a homogeneous population of transducing phages upon induction), whereas gene transfer by generalised transduction results in a non-lysogenic recombinant.

Specialised transducing phages, therefore, allow the transfer of specific bacterial DNA from one strain (from which the LFT lysate was originally obtained) to another (forming a partial diploid recipient). DNA transfer is not limited to the *gal-bio* region. There are a limited number of specific *secondary attachment sites* showing partial homology with *att*λ*B* at which phage *lambda* integrates (in the absence of the primary attachment site, *att*λ*B*). These sites are distributed at random around the bacterial chromosome (see Section 6.2.2 (ai)). Thus, the range of (potential) *lambda* transducing particles can be extended with phages carrying DNA bordering these secondary sites (for example, λ*precA*, λ*para*, λ*mal*, λ*ptrp*). Moreover, bacterial genes can be transposed from their normal locus to different regions of the bacterial chromosome (making use of the secondary-site integration phenomenon), thereby allowing the isolation of deletion and fusion strains — spontaneous loss of DNA sequences separating the transposed bacterial genes from chromosomal cistrons may be selected for, giving rise to deletions of this region or even fusion of prophage and bacterial genes.

---

11.   In the absence of helper phage (i.e. at low m.o.i. when the probability of co-infection is slight), transducing phages lysogenise at very low frequency due to the nature of the hybrid attachment sites. (In the Rec-mediated mode, lysogenisation without helper is possible using bacterial homology.)

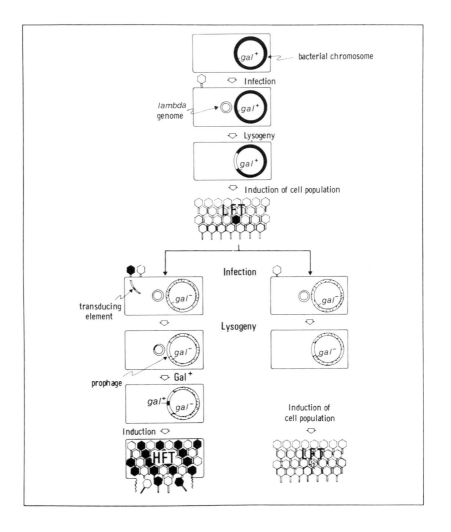

Figure 5.19: Stages in Specialised Transduction

A small proportion (10^{-5}) of phages in a lysate from induction of a *lambda* lysogenic cell population carry *gal* DNA due to aberrant excision (Figure 5.16) If this low frequency transducing lysate (LFT) is used to infect a *gal⁻* strain, only injection of λ*gal⁺* DNA will give rise to Gal⁺ transductants (in this case a λ/λ*gal* dilysogen, see Figure 5.18). Entrance of viral DNA, on the other hand, results in a Gal⁻ lysogen and/or cellular lysis (right), depending upon whether the phage enters the lysogenic or lytic phase (see Figure 5.6). In the high-frequency transducing lysate (HFT) produced after induction of the λ/λ*gal* dilysogen, roughly 50 per cent of phages are λ*gal*

Specialised transducing phages offer the potential for amplication of expression of specific bacterial genes. Increased gene dosage (100-fold or more) is a consequence of replication of *lambda* transducing phages within a bacterium. A further increase may be obtained by employing *lambda* derivatives carrying mutations (e.g. *Sam7*) that prevent lysis from within. In the absence of the phage-coded endolysin system, phage multiplication continues to the extent that the phage-encoded bacterial proteins represent a significant proportion of the total cellular protein. (Upon infection by λ*ptrp* derivatives, for example, synthesis of *trp* enzymes is boosted to comprise about 50 per cent of the total soluble protein of the host cell.) Manipulation of the phage's genetic apparatus can improve this amplification and, thus, overcome inherently low-level expression of certain bacterial genes.

Most importantly, transducing phage offer a source of concentrated bacterial DNA (an average-sized gene of 1200 base pairs, for instance, coding for a polypeptide of 40 000 daltons, is approximately 0.02 per cent of the size of the bacterial chromosome but 2.5 per cent of the *lambda* genome, an increase of about 100-fold). Pure DNA may be obtained in large quantities from phage lysates (Section 7.3.1). It may be used directly for protein-nucleic acid studies (*in vitro* transcription and translation analysis, regulatory protein binding, etc.) or as a source of bacterial DNA for cloning (see Section 7.3.3).

Bibliography

5.1 Bacteriophages and their Life-Cycles

Bukhari, A.I. (1976) 'Bacteriophage Mu as a transposition element', *Ann. Rev. Genet., 10,* 389-412.

Cairns, J., Stent, G.S. and Watson, J.D. (eds.) (1966) *Phage and the Origins of Molecular Biology* (Cold Spring Harbor Laboratory, New York).

Denhardt, D.T., Dressler, D. and Ray, D.S. (eds.) (1978) *'The Single-Stranded DNA Phages* (Cold Spring Harbor Laboratory, New York).

Echols, H. (1979) 'Bacteriophage and bacteria: Friend and foe', in J.R. Sokatch and L.N. Ornston (eds.) *The Bacteria VII: Mechanisms of Adaptation* (Academic Press, New York), pp. 487-516.

Fiers, W. (1979) 'Structure and function of RNA bacteriophages', in H. Fraenkel-Conrat and R.R. Wagner (eds.), *Comprehensive Virology, 13: Structure and Assembly* (Plenum Press, New York), pp. 69-204.

Fraenkel-Conrat, H. (1974) 'Descriptive catalogue of viruses' in H. Fraenkel-Conrat and R.R. Wagner (eds.), *Comprehensive Virology, I* (Plenum Press, New York).

Hershey, A.D. (ed.) (1971) *The Bacteriophage Lambda* (Cold Spring Harbor Laboratory, New York).

Luria, S.E., Darnell, J.E., Baltimore, D. and Campbell, A. (1978) *General Virology*, 3rd edn (Wiley, New York).

Zinder, N.D. (ed.) (1975) *RNA Phages* (Cold Spring Harbor Laboratory, New York).

5.1.1 Anatomical Considerations

Air, G.M. (1979) 'DNA sequencing of viral genomes', in H. Fraenkel-Conrat and R.R. Wagner (eds.), *Comprehensive Virology, 13: Structure and Assembly* (Plenum Press, New York), pp. 205-292.

Bukhari, A.I., Shapiro, J.A. and Adhya, A. (eds.) (1977) *DNA Insertion Elements, Plasmids and Episomes* (Cold Spring Harbor Laboratory, New York).

Caspar, D.L.D. and Klug, A. (1962) 'Physical principles in the construction of regular viruses',

Cold Spring Harbor Symp. Quant. Biol., 27, 1-24.

Crowther, R.A. and Klug, A. (1975) 'Structural analysis of macromolecular assemblies by image reconstruction from electron microscopes, *Ann. Rev. Biochem., 44,* 161-182.

Eiserling, F.A. (1979) 'Bacteriophage structure', in H. Fraenkel-Conrat and R.R. Wagner (eds.), *Comprehensive Virology, 13: Structure and Assembly* (Plenum Press, New York), pp. 543-580.

Thomas, C.A. (1967) 'The rule of the ring', *J. Cell. Physiol., 70,* suppl. 1, 13-34.

Williams, R.C. and Fisher, H.W. (1974) *An Electron Micrographic Atlas of Viruses* (Thomas, Springfield).

Wood, W.B. and King, J. (1979) 'Genetic control of complex bacteriophage assembly', in H. Fraenkel-Conrat and R.R. Wagner (eds.), *Comprehensive Virology, 13: Structure and Assembly* (Plenum Press, New York), pp. 581-633.

5.1.2 Phage Infection

Barksdale, L. and Arden, S.B. (1974) 'Persisting bacteriophage infections, lysogeny and phage conversions', *Ann. Rev. Microbiol., 28,* 265-299.

Bayer, M.E. (1979) 'The fusion sites between outer membrane and cytoplasmic membrane of bacteria: Their role in membrane assembly and virus infection', in M. Inouye (ed), *Bacterial Outer Membranes: Biogenesis and Function* (John Wiley, New York), pp. 167-202.

Duckworth, D.H., Glenn, J. and McCorquodale, D.J. (1981) 'Inhibition of bacteriophage replication by extrachromosomal genetic elements', *Microbiol. Rev., 45,* 52-71.

Hohn, T. and Katsura, I. (1977) 'Structure and assembly of bacteriophage *Lambda*', *Curr. Topics Microbiol. Immunol., 78,* 69-110.

Konisky, J. (1979) 'Specific transport systems and receptors for colicins and phages', in M. Inouye (ed.), *Bacterial Outer Membranes: Biogenesis and Function* (John Wiley, New York), pp. 319-359.

Murialdo, H. and Becker, A. (1978) 'Head morphogenesis of complex double-stranded deoxyribonucleic acid bacteriophages', *Microbiol. Rev., 42,* 529-576,

Randall, L.L. and Philipson, L. (eds.) (1980) *Receptors and Recognition, B7: Virus Receptors (Part I, Bacterial Viruses)* (Chapman and Hall, London).

Rabussay, D. and Geiduschek, E.P. (1977) 'Regulation of gene action in the development of lytic bacteriophages', in H. Fraenkel-Conrat and R.R. Wagner (eds.), *Comprehensive Virology, 8: Regulation and Genetics* (Plenum Press, New York), pp. 1-196.

Streisinger, G., Emrich, J. and Shahl, M.M. (1967) 'Chromosome structure in phage T4, III: Terminal redundancy and length determination', *Proc. Nat. Acad. Sci. USA, 57,* 292-295.

Weisberg, R.A., Gottesman, S. and Gottesman, M.E. (1977) 'Bacteriophage λ: The lysogenic pathway', in H. Fraenkel-Conrat and R.R. Wagner (eds.), *Comprehensive Virology, 8: Regulation and Genetics* (Plenum Press, New York), pp. 197-258.

5.1.3 Phage Methodology

Adams, M. (1959) *Bacteriophages* (Interscience Publishers, New York).

Miller, J.H. (1972) *Experiments in Molecular Genetics* (Cold Spring Harbor Laboratory, New York).

5.2 Bacteriophage-Mediated Gene Transfer

Hayes, W. (1968) *The Genetics of Bacteria and their Viruses,* 2nd edn (Blackwell, Oxford).

Low, K.B. and Porter, D.D. (1978) 'Modes of gene transfer and recombination in bacteria', *Ann. Rev. Genet., 12,* 249-287.

Morse, M.L., Lederberg, E.M. and Lederberg, J. (1956) 'Transduction in *Escherichia coli* K12', *Genetics, 41,* 142-156.

Zinder, N.D. and Lederberg, J. (1952) 'Genetic exchange in *Salmonella', J. Bacteriol., 64,* 679-699.

5.2.1 Generalised Transduction

Ikeda, H. and Tomizawa, J.-I. (1965) 'Transducing fragments in generalised transduction by phage P1: Molecular origin of the fragments', *J. Mol. Biol., 14,* 85-109.

Novick, A. (1966) 'Phenotypic mixing', in J. Cairns, G.S. Stent and J.M. Watson (eds.), *Phage and the Origins of Molecular Biology* (Cold Spring Harbor Laboratory, New York), pp. 133-141.

Wu, T.T. (1966) 'A model for three-point analysis of random general transduction', *Genetics, 54,* 405-410.

5.2.2 Specialised Transduction

Campbell, A.M. (1977) 'Defective bacteriophages and incomplete prophages', in H. Fraenkel-Conrat and R.R. Wagner (eds.), *Comprehensive Virology, 8: Regulation and Genetics* (Plenum Press, New York), pp. 259-328.

Nash, H.A. (1977) 'Integration and excision of bacteriophage λ', *Curr. Topics Microbiol. Immunol., 78,* 174-199.

6 Reactions of DNA

The bacterial genome encodes about one to two thousand different protein species, the majority of which are likely to be enzymes rather than structural components. Of the numerous cytoplasmic processes carried out by these biological catalysts, those that involve nucleic acid, in particular DNA, are crucial for cellular survival and, in the long term, for preservation of genetic information. Thus, although perturbation of a biosynthetic reaction at the transcriptional or translational level may result in auxotrophy, such an effect is phenotypic (rather than genotypic) since it is compensated by *de novo* RNA and protein production once the block is removed. When a change occurs at the level of DNA, itself, it is heritable and will be passed on to all progeny cells (unless otherwise repaired). Reactions of DNA are, therefore, limited in the present context to those that directly affect the genetic material — the breaking and making of internucleotide bonds or the modification of DNA nucleotides — and not the passive use of DNA as a template, as in transcription (Chapter 2). These processes, DNA replication, recombination and repair (mutation has been discussed in Chapter 3), have many gene functions in common.

Genetic DNA, whether chromosomal or extrachromosomal in nature, is not a static entity. The mere fact of its continued presence in succeeding generations of cells indicates that there is a bacterial machinery for copying DNA templates. This replicative apparatus produces accurate copies despite the complexity of the process since the genotype of an organism is almost invariably maintained. DNA synthesis, in the case of a conjugative plasmid (or an Hfr strain, one carrying an integrated genetic element, Chapter 4), is necessary for both production of faithful replicas *(vegetative replication)* and their transfer to a permissive host *(transfer replication)*. Moreover, bacteriophage production also requires replicative DNA synthesis.[1] Thus, DNA replication is vital for genetic perpetuation, for the maintenance of *replicons* (Section 6.1).

Reactions of DNA require the breaking and making of internucleotide bonds. Nuclease, DNA polymerase and ligase functions are, therefore, required. Just as these activities are essential for DNA replication, they also mediate in the exchange of DNA between duplexes. The nature of the product of *genetic recombination* depends upon the DNA substrate (for instance, whether linear or

---

1.   The genome of some bacterial viruses consist of single-stranded RNA rather than of DNA (Section 5.1.1). Replication involves an RNA replicase and not a DNA polymerase function as in the case of the DNA viruses (see Blumenthal, T. and Carmichael, G.G. (1979), *Ann. Rev. Biochem., 45,* 525-548).

covalently-closed circular duplexes are involved), the recombination system and the number of crossover points. Synapsis between homologous regions of DNA is responsible for most recombination events *(general recombination)*, although it may apparently occur at nucleotide sequences dissimilar in nature *(non-homologous recombination)*, giving rise to deletions, duplications and DNA insertions. It is genetic recombination that is responsible for the 'jumbling' of genetic material associated with chromosome evolution (Section 6.2).

Damage to DNA may stem from mutagens or from an error in the replication process itself. Discontinuities, in addition to single base changes, are generated in DNA duplexes by the reactions of DNA. A *gap* in a double-stranded DNA molecule may be defined as the loss of both a phosphodiester bond and nucleotide, or multiples thereof, whereas a *nick* refers to the loss of just a phosphodiester linkage (note that only the latter is directly repairable by a single enzyme species, DNA ligase). Not all mutations are neutral: some are lethal or lead to auxotrophy. Though changes in genetic DNA lead to a heritable change and, hence, a mutation, such alterations may be repaired before their deleterious effect is realised. Replication and recombination functions are involved in the repair of DNA damage. The intrinsic exonuclease activities of the replicative apparatus are important both for the excision of mismatched base pairs during proof reading and for DNA repair (Section 6.3).

These reactions of DNA — replication, recombination and repair — all involve nucleic acid-protein interactions. It should be remembered that DNA duplexes present a fairly uniform surface to possible effector molecules since the invariant sugar-phosphate backbone resides on the outside of the double helix (Section 1.1.1 (b)). Any specificity in interaction, prior to melting of double-stranded DNA, relies upon groups in the major or minor grooves. Alternatively, direct access to the base pairs, themselves, may proceed in some DNA reactions through 'freezing' of a randomly melted region — unwinding brought about spontaneously (thermally) by specific DNA unwinding proteins or by other DNA transactions which involve single-stranded chains.

6.1 DNA Replication

Replicative DNA synthesis is one of the most complex reactions involving protein-nucleic acid interactions. Difficulty in the analysis of this process stems from the number of proteins required and the fact that the components of the replication apparatus appear to dissociate upon purification. Its study has been greatly facilitated by the use of cell extracts which support DNA synthesis. Unfortunately, *in vitro* replication of the bacterial chromosome is still restricted to ill-defined systems using crude cell lysates. Replication in soluble, membrane-free extracts is only observed with more simple replicons, such as certain plasmid and phage DNAs. In the case of single-stranded viral genomes, particular stages in the replication processes — initiation and elongation — can be studied using reconstituted systems from defined, purified proteins. It is, therefore, now possible to assign a number of different components of the replicative machinery to specific bacterial- (and phage-) coded gene products.

The *modus operandi* for investigation of replication has been to isolate mutant strains defective in DNA synthesis and to determine the respective gene products affected. Since lesions in essential genes, the replicative cistrons, result in cellular death, such studies rely on conditional-lethal, commonly temperature-sensitive, mutants. The component lacking (or non-functional) in a mutant organism is defined by employment of an *in vitro* complementation assay. Briefly, this involves the isolation (and purification) of a protein fraction from a wild-type strain that allows continued DNA synthesis to occur under non-permissive conditions in cell preparations of bacteria temperature-sensitive for DNA synthesis (genetic and *in vitro* complementation techniques are discussed in Section 7.1.1 (a)). An alternative approach for the identification of components of the replicative apparatus has been to search for, and purify, proteins with activities logically necessary for DNA production, for instance DNA binding proteins, enzymes that ligate adjacent nucleotide chains and, of course, DNA polymerases.

The basic reaction, the polymerisation of deoxyribonucleotides, is catalysed by the enzyme *DNA-dependent DNA polymerase*. At least three of these polymerases are present in the bacterium *E. coli*. Certain bacterial viruses, for instance, phage T4, encode their own replicative systems: others rely upon host enzymes (the single-stranded bacteriophages M13, ϕX174 and G4 are cases in point). All DNA polymerases are unable to initiate new chains and require a primer with a 3′-hydroxyl group (supplied by separate enzymes). They have in common the restriction that polymerisation of deoxyribonucleoside triphosphates only proceeds in the 5′ → 3′ direction. (The strand which can be synthesised continuously in this direction is referred to as the *leading* strand: the complementary chain is designated the *lagging* strand, Figure 6.1.) Novel mechanisms need to be invoked for copying both strands of DNA duplexes because of their antiparallel nature. The lagging strand is, in fact, synthesised discontinuously in short stretches, termed *Okazaki fragments*, in a direction opposite ('backwards') to movement of the *replication fork*. Whether both strands are synthesised discontinuously, or just the lagging one, is still contentious.

In addition to the 5′ → 3′ polymerisation function, DNA polymerases have exonuclease activity, either in the 5′ → 3′ or in the 3′ → 5′ direction (some enzymes have both functions). The 3′ → 5′ exonuclease activity is important in 'proof reading', the correction of intrinsic errors in replication occurring, for instance, through mismatched base pairs. (Exonuclease activity of the replicative machinery also seems to play an important role in the repair of damage to DNA caused by external influences such as UV irradiation.) Clearly, fidelity of replication is of utmost importance for the preservation of genetic information.

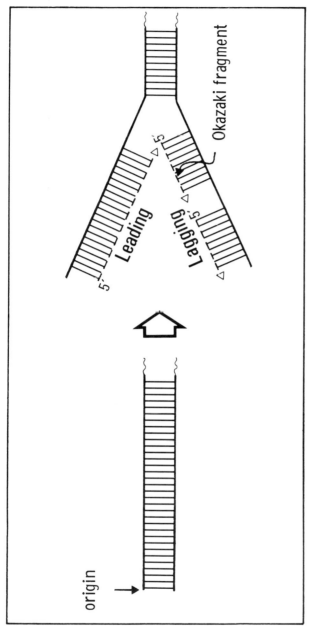

Figure 6.1: The Replication Fork

Replication from the origin site, initially proceeds along one strand (the leading DNA strand), since DNA polymerases can only polymerise dNMPs in the 5' → 3' direction. As the other strand (the lagging strand) is exposed, it becomes available as a template for discontinuous synthesis. These 'forward' and 'backward' processes generate a replication fork (and 'bubble' or 'eye' when replication is initiated within a DNA duplex rather than at one terminus, see Figure 6.8)

6.1.1 The Elements of Replication

DNA synthesis upon DNA templates requires a *DNA polymerase* function for catalysis of polymerisation of the four common 5'-deoxyribonucleoside triphosphates (dATP, dGTP, dTTP and dCTP). DNA replication appears, in general, to be initiated at a specific site(s) on each replicon, the *origin of replication* (phage ϕX174 single-strand DNA has no single *ori* sequence, Section 6.1.3 (a)). The oligonucleotide primer that enables the DNA polymerase to initiate DNA synthesis may be supplied by the cellular *DNA-dependent RNA polymerase* or by a specific bacterial-coded *DNA primase*. Movement of the replication fork is mediated by DNA binding proteins that allow *helix destabilisation* to occur and the enzyme *DNA gyrase*, responsible for changes in the superhelicity of duplex DNA. Finally, sealing of polydeoxyribonucleotide chains is catalysed by *DNA ligase*. The complexity of template-dependent DNA production is emphasised by the number of components required for specific replicative synthesis, particularly at the initiation stage, and the multifunctional nature of the DNA polymerases, themselves.

The majority of replication elements are diffusible proteins: they are *trans*-active. Replication of bacteriophage genomes and autonomous plasmids may, thus, proceed through the complete or partial use of the host replicative system, though their control is commonly mediated by extrachromosome-encoded elements. Since the origin of replication is a structural feature, a defined non-coding DNA sequence, it is *cis*-specific. Some replicons also seem to encode a specific, cis-active *terminator* of replication.

(a) DNA-Dependent DNA Polymerases. Three different bacterial enzymes, DNA polymerase I, II and III (encoded by *polA, polB* and *dnaE (polC)*, respectively) have been identified in *E. coli*. That these polymerases are disparate is clearly indicated by their molecular size, subunit composition and enzymatic properties, in addition to the distinct nature of the three genetic loci (Table 6.1). DNA polymerases I and II are single polypeptide chains (molecular weight 109 000 and 120 000, respectively), whereas DNA polymerase III is active as an oligomeric complex. (At the time of writing, the actual identity of the *polC (dnaE)* gene product is not known since DNA polymerase III preparations contain other proteins; it is thought to be the 140 000 dalton subunit.) With respect to their enzyme function (see Figure 6.2), the three bacterial enzymes have a $5' \rightarrow 3'$ polymerase activity dependent upon a 3'-OH primer and require Mg^{2+} for correct action (the latter requirement is in keeping with other nucleic acid enzymes, Section 2.1.1 (a)). Though both polymerase I and III have $5' \rightarrow 3'$ exonuclease activity, the polymerase III catalysed reaction is single-strand specific (all three enzymes are able to excise nucleotides in the $3' \rightarrow 5'$ direction, a property necessary for removal of incorrectly inserted bases). Thus, only the *polA* gene product is capable of removing distortions in the DNA template — such as UV-induced pyrimidine dimers — that would interfere with base pairing during replication. Finally, DNA polymerase II and III appear to be selective in

Table 6.1 Properties of the Three *E. coli* DNA Polymerases

Enzyme	Size	Structural Gene	Map Location (min)	Molecules/ Cell	Exonuclease Activity			Template Preference	Active Species	Physiological Role
					3'→5'	5'→3'				
DNA polymerase I	109,000	polA	86	300	+ [a]	+ [a]		flexible	single polypeptide	DNA repair [b]
DNA polymerase II	120,000	polB	2	~10	+			gapped DNA	single polypeptide	?
DNA polymerase III	140,000 [c]	polC (dnaE)	4	~10	+	+ [d]		gapped DNA	oligomeric complex	DNA replication

a   The 3' → 5' and 5' → 3' exonuclease activities of DNA polymerase I were initially termed 'exonuclease II' and 'exonuclease VI', respectively. This has since been abandoned.

b   The *polA* product has also been implicated in primer removal (see text).

c   Although the actual identity of the '*polC* product' is not known (see text), it is likely to be the largest (140,000 dalton) subunit of DNA polymerase III holoenzyme: six polypeptides are associated with this complex in addition to the α subunit (β, γ, δ, ε, θ and τ of molecular weight 40 000, 52 000, 32 000, 25 000, 10 000 and 83 000).

d   The 5' → 3' exonuclease activity of DNA polymerase III differs from that of DNA polymerase I in that the reaction requires a single strand in the first instance.

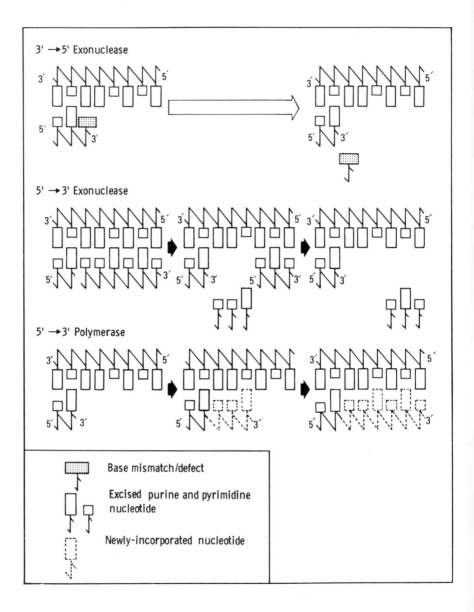

Figure 6.2: Activities of Bacterial DNA Polymerases
In addition to the 5′ → 3′ polymerase function, *E. coli* polymerases can excise deoxyribonucleotides in the 3′ → 5′ and 5′ → 3′ direction (polymerase I and III but not II possess both exonuclease activities, albeit with different specificities, Table 6.1). These other properties are important in proof reading and DNA repair (Section 6.3). Strand extension from a nick concomitant with 5′ → 3′ exonuclease degradation leads to displacement of the nick along the DNA template (see Nick Translation, Figure 6.3)

their template requirements, in addition to the common necessity of all DNA polymerases for a 3'-OH primer, in that replication proceeds most rapidly on gapped, duplex DNA in the absence of additional proteins. Little is known about the structure of these important cellular constituents. The most intensely studied enzyme, DNA polymerase I, was, not surprisingly, the first to be isolated (by Kornberg and his group).

DNA polymerase I, like many nucleic-acid enzymes (cf. RNA polymerase, Section 2.1.1 (a)) has a bound zinc ion. Within this single molecule, the 5' → 3' exonuclease activity is associated with the N-terminal region while the larger C-terminal portion (representing 70 per cent of the complete molecule) carries both polymerase and 3' → 5' exonuclease(-proof-reading)-activities. Furthermore, the presence of two such functional domains on a single polypeptide chain is demonstrated by the isolation of a group of conditional-lethal *polA⁻* mutants, termed *polAex⁻*, defective only in the 5' → 3' exonuclease activity. Kornberg has proposed a model for DNA polymerase I function which involves at least six distinct sites on the enzyme molecule: a site for template binding, for the primer and its 3'-OH terminus, for the 3' → 5' proof-reading function, for the substrate (one of the four dNTPs) and, lastly, a site for the 5' → 3' exonuclease activity.

It is now considered that DNA polymerase I and II have no direct role in replicative DNA synthesis. Their slow elongation rates (about 17 and 3 nucleotides per second, respectively) and, most importantly, the viability of *polA⁻ polB⁻* double mutants, indicate that these two polymerase activities are not intimately connected with replication of the bacterial chromosome *per se*. DNA polymerase I is, however, required for replication of certain small plasmids, for example, ColE1 (Section 4.2). The physiological defects of various *polA* mutant derivatives suggest that this gene product has essentially a DNA repair, rather than replication, role.[2] Certainly, it would be hard to reconcile the number of molecules of polymerase I with a simply synthetic function (Table 6.1). The Kornberg enzyme can use its 5' → 3' exonuclease activity to sequentially remove nucleotides — either ribo- or deoxyribonucleotides — from a 5'-terminus, while simultaneously filling in the gap with deoxyribonucleotides (using the 5' → 3' polymerase function, Figure 6.3). This property, termed *nick translation*, may be important in gap filling as well as in RNA primer replacement following discontinuous replication (discussed in Sections 6.3.2 (ai) and 6.1.2 (a)). The presence of DNA polymerase II remains a puzzle.

DNA polymerase III is, however, necessary for continued DNA synthesis. The cellular species active in elongation, *DNA polymerase III holoenzyme*, appears to be a complex of at least four polypeptide chains, termed α, β, γ and δ (molecular weight 140 000, 40 000, 52 000 and 32 000 respectively, Table 6.1).

---

2.   In addition to the sensitivity of *polA* alleles to mutagens such as UV or MMS, *polA⁻ recA⁻* (or *recBC⁻*) double mutants are inviable. As a repair enzyme, DNA polymerase I perhaps functions in conjunction with other proteins (Section 6.3.1). It has, for example, been recovered in association with exonuclease V (the *recBC* gene product).

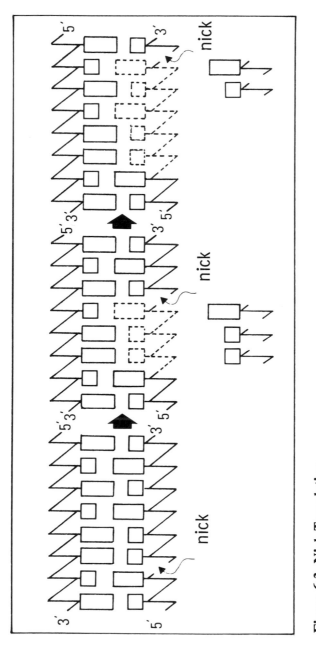

Figure 6.3: Nick Translation

The 5′ → 3′ polymerase function of DNA polymerase I adds nucleotides at the 3′-hydroxy side of a nick (*dotted lines*) and, at the same time, existing nucleotides are removed from the other side by the integral 5′ → 3′ exonuclease activity. The consequence of these two concerted reactions is to 'translate' the nick along the DNA duplex in the 5′ → 3′ direction. The sequence produced is complementary in composition to the intact strand (it will be radiolabelled if the reaction is carried out in the presence of dNTP isotopes *in vitro*). Nick translation has been implicated in primer removal and DNA repair (Section 6.3)

The role of the holoenzyme components other than α in the propagation stage has led to the term *elongation factor*. Thus, β is *elongation factor I*, and γ (or δ) is equivalent to *factor III* (see Table 6.2).

A comparison of the two major classes of nucleic acid enzymes, RNA polymerase and DNA polymerase, is pertinent at this stage. RNA synthesis in *E. coli* requires a large, multisubunit enzyme, the DNA-dependent RNA polymerase. The holoenzyme form is sufficient for initiation at the majority of promoter sites (catabolite sensitive operons require, in addition, cAMP and CAP protein, see Sections 2.1.1 (b) and 9.1). Moreover, once started, RNA synthesis proceeds in the absence of even the accessory factor, the σ subunit. Interestingly, bacteriophage T3 and T7 RNA polymerases are able to fulfil all the basic functions of their larger counterparts although only about one fifth the molecular size, and single polypeptide chains. The three bacterial DNA-dependent DNA polymerases are similar in molecular weight to these viral RNA polymerases but are multifunctional in nature (they have polymerase, $3' \rightarrow 5'$ and, in the case of DNA polymerase I and III, $5' \rightarrow 3'$ exonuclease activities). However, even DNA polymerase III holoenzyme, a complex of at least four different polypeptides, is unable to initiate in the absence of additional proteins (emphasising, perhaps, that strict control over replicative DNA synthesis is crucial for cellular growth). Presumably, the economy in the viral RNA polymerase reflects the stringency in promoter selection on these phage DNAs (in contrast to the bacterial genome), with a mere fraction of the bacterial multisubunit complex being required for the actual RNA polymerase reaction (as in the case of DNA synthesis). In summary, the size of an enzyme complex and the number of accessory proteins is, clearly, important for both specificity and fidelity.

(b) Protein Components of the Replication Fork. A number of bacterial gene products, in addition to DNA polymerase III holoenzyme, are required for replication. Enzymes that play an important part in the actual catalysis of DNA production include *RNA polymerase* and *DNA primase*, which act at the initiation stage (the latter is also responsible for priming Okazaki fragment synthesis) and *DNA ligase*, necessary for sealing of discontinuities. Other proteins have a more topological role, being involved in helix unwinding and single-strand stabilisation. Specific plasmid- and phage-coded replication factors are mentioned in Section 6.1.3 (see also Chapter 10).

The bacterial *DNA-dependent RNA polymerase*, unlike the replication enzymes, is able to initiate *de novo* nucleic acid synthesis on duplex DNA (Section 2.1.1 (ai)). Its role in the initiation (but not elongation) of chromosome replication is indicated by the sensitivity of only the former process to the antibiotic rifampicin (discussed in Section 6.1.2 (d)). A specific *DNA primase* (originally described as a rifampicin-resistant RNA polymerase), the *dna G* gene product (molecular weight 60 000), is also involved: primase can polymerise either ribo- or deoxyribonucleotides on single-strand templates *in vitro*. It seems likely, at least with respect to replication of the bacterial chromosome, that RNA polymerase — perhaps, in concert with DNA primase — functions at the initiation stage for the production of origin RNA, whereas DNA primase acts at the level of Okazaki fragment synthesis. Note that certain phages make use of primase and not RNA polymerase for initiating overall DNA synthesis (for example, ϕX174 and G4, whose replication is discussed in Section 6.1.3 (b)).

Discontinuous DNA synthesis generates a DNA duplex carrying a number of nicks. *DNA ligase* promotes the formation of a phosphodiester bond between these unlinked, adjacent nucleotides in a DNA duplex: it requires a free 3'-OH and 5'-phosphate (Figure 6.4). The reaction proceeds via an AMP-ligase intermediate, formed in a NAD^+-dependent step (the bacteriophage T4-encoded DNA ligase, also a well-characterised enzyme, utilises ATP rather than NAD^+ as cofactor; it has a molecular weight of 60 000). In *E. coli*, the enzyme (a single polypeptide of molecular weight 75 000, the *lig* gene product) appears to be in vast excess: the 300 molecules per cell can cope with a potential 7500 single-strand breaks per minute per cell, whereas only an estimated 100 sealing events per minute are necessary for discontinuous replication on the lagging strand (see legend to Figure 6.4). The reason for this apparent abundance is not known. Certainly, DNA ligase is an essential component of the replicative apparatus, as indicated by the isolation of conditional-lethal mutations at the *lig* locus (a build-up of Okazaki fragments occurs at the non-permissive temperature before the ensuing cellular death).

The process of DNA replication requires extensive melting of the DNA duplex, exposing single-strand DNA for template function. Such 'bared' regions must be stable for some time to allow both continuous and discontinuous replication to occur. Moreover, as single strands, they are particularly susceptible to nuclease attack. *E. coli* encodes a gene product that stimulates replication by DNA polymerase III holoenzyme through binding to single-stranded DNA (the active species is an 80 000 dalton tetramer of identical 20 000 molecular weight subunits). Originally isolated as a *single-strand DNA binding protein*, it has since been termed a *helix destabilising* (or unwinding) *protein* (the T4 gene 32 product of molecular weight 33 000, isolated by Alberts and co-workers, is the prototype for this type of DNA binding protein). The locus for the *E. coli* gene product, *ssb* (*s*ingle-*s*trand *b*inding), has recently been mapped at 91 min on the bacterial map (Figure 6.6).

The bacterial *rep* gene product (about 67 000 daltons) is required for phage ϕX174 replication though it is not apparently crucial for *E. coli* DNA synthesis. This enzyme seems to aid duplex unwinding by movement of a long single-stranded DNA into the fork, in an ATP-dependent reaction.

Topological constraints are imposed on the replication of covalently closed circular molecules of double-stranded DNA by their helical nature due to the lack of freedom of rotation of one strand about the other. As the replication fork proceeds, duplex unwinding generates positive supercoils (Figure 6.5). Movement of the growing point, thus, becomes increasingly difficult in the absence of a

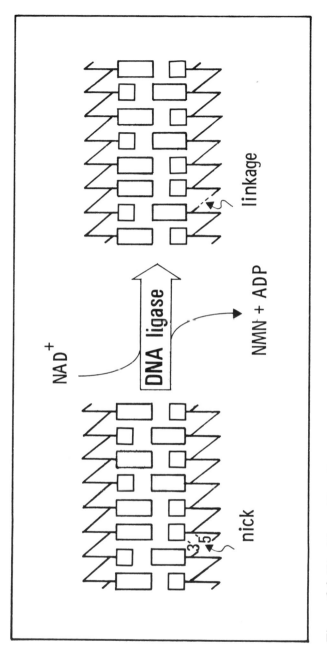

Figure 6.4: The Ligase Reaction

If one strand of the bacterial chromosome (3.8×10^6 bp, 2.5×10^9 daltons) is replicated continuously, about 3.8×10^6 nucleotides will be polymerised in the discontinuous mode. Since Okazaki fragments are, on the average, about 1000 nucleotides in length, there must be about 3.8×10^3 such 10S fragments per chromosome, equivalent to a synthetic rate of about 100 fragments per minute (the entire chromosome is replicated in approximately 40 minutes). Though each fragment supplies two ends for joining, each sealing event requires two termini. About 100 sealing events per minute are, therefore, necessary. Note that *E. coli* DNA ligase will not accept 5'-RNA termini as substrate

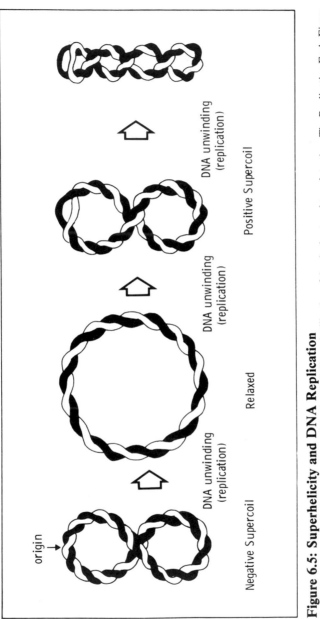

Figure 6.5: Superhelicity and DNA Replication

As replication proceeds through melting of the double helix and utilisation of the single-strand templates (see The Replication Fork, Figure 6.1), the strain generated is distributed throughout the covalently closed circular molecule of DNA by supertwisting in the direction of the helix, thereby removing negative coils and producing positive ones. The enzyme DNA gyrase is responsible for the return to the negative supercoiled state (see also legend to Figure 6.7) concomitant nicking-closing reaction — the introduction of a break in at least one DNA strand, rotation and subsequent resealing — to relax this positive superhelicity (a temporary disruption permits an alteration in the topological linkage of the two strands of the double helix).

DNA gyrase (variously called *Eco DNA topoisomerase II*, swivelase or untwistase) is responsible for this decrease in positive superhelicity of covalently-closed circular DNA molecules.[3] The *E. coli* enzyme is a tetramer, consisting of two copies of two non-identical polypeptides, the *gyrA* and *gyrB* (previously *nalA* and *cou*) gene products (molecular weights 105 000 and 95 000, respectively); these two subunits are targets for the antibiotics nalidixic acid and coumermycin (inhibitors of DNA replication are discussed in Section 6.1.2 (d)). Gyrase, in particular, the *gyrA* component, relaxes negative supercoils by a breakage-and-reunion action. Alternatively, in the presence of ATP, the oligomeric enzyme introduces negative superturns into closed circular duplex DNA, which is initially relaxed or positively supercoiled, a reaction also requiring nicking-closing.[4] It has recently been shown that gyrase cleaves at specific sites to make a staggered double-stranded break of four base pairs: the consensus sequence is 5'YRT↓*GNYN*NY (where the arrow indicates the position of the gyrase cut in the strand shown, the four bases between the staggered cuts are italicised and T and A are commonly found at the pyrimidine (Y) and purine (R) positions).

The main *E. coli* loci implicated in chromosomal replication are shown in Table 6.2/Figure 6.6.

6.1.2 The Mechanism of Bacterial Replication

The double-helix model of Watson and Crick readily suggests how essential information for genetic propagation is contained in DNA, itself: the duplex unwinds, thereby exposing two single-stranded templates upon which the complementary chains may be synthesised. Although the Watson-Crick model allows an instant comprehension of the self-generation of genetic material, the actual mechanism involved in replication of the bacterial chromosome has, to a large extent, evaded complete understanding. In fact, the majority of information on replicative systems has been gleaned from small replicons such as plasmids and bacteriophage genomes (discussed in Section 6.1.3).

3.   A topoisomerase affects the topology of a DNA duplex without altering the chemical structure. Clearly, enzymes that influence supercoiling come under this category. Note that helix-destabilising proteins also act without affecting DNA chemical structure. Another enzyme that catalyses the interconversion of topological isomers of DNA, ω protein or Eco DNA topoisomerase I, of molecular weight 110 000 has no known role in replication. It catalyses the energetically favourable, ATP-independent relaxation of negative superhelical DNA (one of the reactions of DNA gyrase).
4.   Expression from some promoters, for instance the P_L promoter of phage *lambda*, is sensitive to gyrase inhibitors, suggesting a role for DNA gyrase in transcriptional initiation. Supercoiling may aid local melting of double-stranded DNA at these sites and, hence, facilitate RNA polymerase binding. Certainly, supercoiling has been implicated in general promoter usage (Section 2.1.1 (b)). This topoisomerase is also involved in integrative recombination of phage *lambda* (see Section 6.2.2 (ai)).

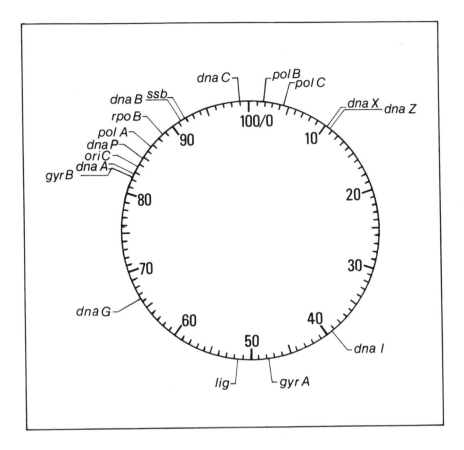

Figure 6.6: *E. coli* **Genetic Map Showing Major Genes Involved in Replication of the Bacterial Chromosome**
The properties of these host-coded elements are listed in Tables 6.1 and 6.2. Map locations from Bachmann and Low (Bachmann, B.J. and Low, K.B. (1980) *Microbiol. Rev., 44*, 1-56)

The mere size of the bacterial chromosome is largely responsible for this apparent complexity. There is, clearly, a handling problem: the susceptibility of this enormous molecule of nucleic acid to artifactual damage during its isolation severely restricts its potential use for priming *in vitro* replicative systems (DNA isolation is discussed in Section 7.3.1). To some extent, this problem can be circumvented by making use of 'holey' cells — bacteria made permeable to small exogenous substances by treatment with organic solvents such as toluene that act on the lipid bilayer, or by plasmolysis with sucrose (Section 7.3.1). Destruction of the membrane permeability barrier uncouples phosphorylation from electron flow and results in a dependence upon exogenous ATP for an energy supply (as well as for a substrate pool). Large substances, however, are

Table 6.2ᵃ Chromosomal Genes Involved in Replication of the Bacterial Genome

Gene	Map Location (min)	Name/Role	Size	Function in Replication	
				Initiation	Elongation
dnaA	82	replication-specific transcriptional regulator	54,000	+	
dnaI	39	nk	nk	+	
dnaP	85	nk	nk	+	
polC (dnaE)	4	DNA polymerase III, α subunit b	140,000 b	+	+
dnaZ	10	DNA polymerase III, γ subunit	52,000	(+) c	+
dnaX	10	DNA polymerase III, δ subunit	32,000	(+) c	+
dnaB	91	nk	55,000	+	+
dnaC	99	nk	25,000	+	+
dnaG	66	DNA primase	60,000	+	+
rpoB	90	RNA polymerase, β subunit d	150,000	+	–
gyrA (nalA)	48	DNA gyrase, α subunit	105,000	(+) c	+
gyrB (cou)	82	DNA gyrase, β subunit	95,000	(+) c	+
ssb	92	single strand DNA binding protein	20,000	+	+
polA	86	DNA polymerase I	109,000	(+) c	+
lig	52	DNA ligase	75,000	+	+

a Abbreviation: nk, not known.
b Identity not proven.
c May also be involved in initiation at *oriC*.
d See Section 2.1.1 (a).

not readily accepted. Alternatively, gentle lysis on cellophane discs allows the addition of macromolecules such as protein fractions.[5] These preparations of holey or lysed cells appear to mimic 'natural' replication but neither accept exogenous DNA nor initiate new rounds of replication. It would seem that the DNA synthesis observed in even the best of these systems reflects propagation of replication forks initiated prior to treatment.

Each round of replication starts at a specific site, *oriC*, on the bacterial chromosome. Replication proceeds bidirectionally around the circular genetic element, terminating at a region opposite to the initiation point (Figure 6.7). The topological constraints of helix unwinding (of covalently closed circular duplex molecules) are satisfied by a DNA gyrase that maintains negative superhelicity in the path of the replicative apparatus. Further constraints are imposed by the nature of the enzyme complex involved, DNA polymerase III holoenzyme (and, indeed, of all known DNA polymerases). Its requirement for a free 3'-OH for initiation excludes single-stranded DNA molecules as templates *per se*. Synthesis at the origin is 'primed' by a short, evidently removable, polyribonucleotide chain — itself produced by other cellular components, the bacterial RNA polymerase (and, possibly, DNA primase, see below). Primer excision involves a nuclease activity (perhaps, RNase H and DNA polymerase I).

DNA replication is discontinuous on at least one strand, the lagging strand, since DNA polymerase-mediated extension of polydeoxyribonucleotide chains is limited to the 5' → 3' direction (all known DNA polymerases share this polarity of synthesis). Okazaki fragment synthesis on this lagging strand is apparently also initiated with RNA, though in this case it is synthesised exclusively by DNA primase, an enzyme disparate from RNA polymerase. (Note the distinction between initiation of rounds of replication and initiation of DNA stand synthesis during elongation.) DNA ligase is responsible for sealing the fragments following primer removal.

At termination, the products of this reaction — *semi-conservative* DNA synthesis — are two covalently closed circular molecules of DNA, each duplex consisting of one parental, and one newly synthesised strand. This entire process (as well as segregation of daughter molecules) takes place in a dense milieu, packed with components of the transcription-translation apparatus (see Chapter 2) and the enormous mass of the bacterial chromosome. Moreover, although the genome is, itself, in a highly condensed state, it has to accommodate rapid movement of the replication forks (and local melting for transcription).

Considering the speed of the two replication forks (about 800 nucleotides per second, see legend to Figure 6.7), the overall process is remarkably accurate: the error frequency is estimated to be about 1 in 10^{10} base pairs replicated (from the

---

5.   The importance of this method is that cell lysates can be kept concentrated and be fed with small substances through the disc.

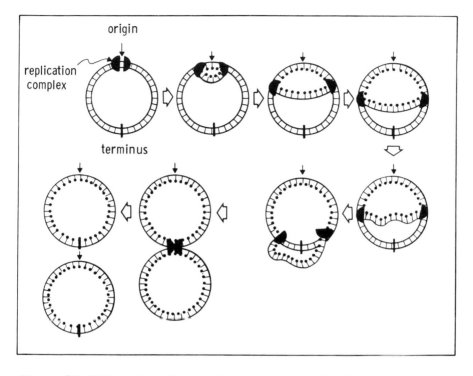

Figure 6.7: Bidirectional Replication of the Bacterial Chromosome

Replication is initiated at *oriC* (83 min, Figure 6.6) and terminated opposite to this point on the circular genetic element. The bacterial chromosome (3.8×10^6 bp) is replicated in a constant time (40 minutes), independent of growth rate. Since chromosomal DNA synthesis proceeds bidirectionally, the two replication forks each copy 1.9×10^6 bases in 40 minutes, a rate for one strand, and hence for one replication complex, of about 800 nucleotides per second (equivalent to 80 turns of the double helix). This is at least 15 times faster than the transcription process (Section 2.1.2). It seems unlikely that gene expression takes place in conjunction with DNA replication without a mechanism that physically confines the disturbance (5000 rpm) caused by replication to a narrow region of DNA (perhaps, through DNA gyrase acting close to the replication fork)

spontaneous mutation rate, see Section 3.2.1 (a)).[6] This precision emphasises the importance of replicative DNA synthesis in cellular survival (a higher error rate can be tolerated in transcription or translation since errors incurred are phenotypic — they are not generally passed on — rather than at the level of the genetic material, itself). Though accurate base pairing is important, in the first instance, for maintaining the fidelity of DNA replication, the $3' \rightarrow 5'$ editing function of DNA polymerase III holoenzyme reduces the chance of perpetuating

6.   It has been noted before, however, that not all changes at the level of the genetic material are necessarily deleterious (Section 3.1). Since cryptic mutations by their very nature lack a detectable phenotype and will pass unnoticed, the actual error rate for DNA replication *in vivo* — calculated from the spontaneous mutation frequency — may, in fact, be higher (the observed mutation rate will also depend upon the efficiency of DNA repair, see Section 6.3).

a mismatch (polymerase I and II also have this activity, Section 6.1.1 (a)). The mutator activity associated with certain *dnaE* alleles stresses the involvement of DNA polymerase III in base selection; the defects may lie in the proof-reading activity in these cases.

(a) Bidirectional Replication of the E. coli Chromosome. Genetic and biochemical studies have shown that replication, initiated at a unique region, proceeds bidirectionally around the bacterial chromosome. Replicative intermediates of the *E. coli* genome were first visualised by Cairns using gentle lysis techniques to obtain intact DNA molecules for autoradiographic analysis. The topology of these intermediates (see Figure 6.7) has led to their name, θ *(theta)-structures*, and to θ — or *Cairns* — *type replication* for this mode of DNA production (transfer replication of Hfr derivatives and conjugative plasmids, the rolling-circle mode, is discussed in Section 6.1.3 (a)). Replicative DNA synthesis is a processive reaction and can be described in terms of three major stages, initiation, elongation and termination.

(i) *Initiation.* Each round of replication of the bacterial chromosome is initiated at a specific region, *oriC*. Genetic studies have placed the origin of replication between *uncB* and *asn* at 83 min, near to *ilv* (see Figure 6.6). It has recently been narrowed down to a region of about 420 base pairs using *in vitro* recombinant DNA methods (see Restriction Technology, Section 7.3.3).[7] The precise limits of *oriC* have as yet to be established. The cloned sequence, of lower G+C content than overall bacterial DNA, has extended A+T-rich regions (this is true, in fact, for a number of different replication origins and may provide for ready strand separation at initiation). There are numerous repeat sequences, potential binding sites for protein components of the replicative apparatus; secondary structure may also be important.

A number of putative RNA polymerase 'binding' sites (Pribnow consensus sequences TATAATR, Section 2.1.2 (ai)) are also present, one of which has the associated recognition sequence, TTGACA, found at, or near to, position −35 in several promoters (see Section 2.1.2 (ai) and Part IV). The *oriC* region has little potential for encoding a protein product since translational chain-termination codons are frequent in all phases. Any functional RNA polymerase binding sites are, thus, likely to be involved in primer formation (rather than in the production of a translatable mRNA species).

The cloned bacterial origin shows close similarity with sequences on the

7.   Briefly, DNA fragments of *E. coli* are inserted into a cloning vector that lacks a replication origin but carries a readily selected marker (say, antibiotic resistance). They are transferred to competent cells by transformation. Recombinant molecules — hybrids of vector and *E. coli* DNA — that are maintained (thereby imparting drug resistance) must, therefore, carry a functional replication origin derived from the bacterial replicon. Recombinant DNA methods are discussed in Section 7.3.3. This technique has identified a second replication origin, *oriJ*, distinct from *oriC*. It appears to be the origin of the cryptic Rac prophage integrated at about 30 min (see *recE* and *sbcA*, Section 6.2.1 (aii)).

complementary strand of phage G4, close to the start site for primer synthesis (and the replication origin of *S. typhimurium* and bacteriophage *lambda*). The involvement of the *dnaG* gene product, DNA primase, in initiation of phage G4 replication might imply a similar role in the case of the bacterial chromosome, itself. However, it should be pointed out that whereas replication of phage *lambda* proceeds bidirectionally, the G4 growing point moves in only one direction from the origin. The physiological significance of these sequence similarities is, therefore, not known.

The exact role of RNA polymerase and DNA primase in the initiation of rounds of replication is contentious. One plausible model, taking into account the promoter-like sequences in *oriC*, is that initiation *per se* is primed by an RNA product of RNA polymerase. DNA primase might then be relegated to initiation of Okazaki fragment synthesis on the lagging strand (see Figure 6.8). An alternative model is suggested by the homology between the bacterial replication origin and the *dnaG*-dependent initiation sequence of phage G4. In this, RNA polymerase acts simply as a sequence-specific DNA-melting protein (again at one of the promoter-like regions), with DNA primase actually responsible for primer synthesis (*transcriptional activation* may be necessary solely to supply a single-stranded DNA region for primase action). Either model could explain the observed inhibition of initiation by rifampicin (at a time when inhibitors of protein synthesis have no effect, Section 6.1.2 (d)).

A number of other gene products, for instance, *dnaA, dnaB* and *dnaC* (as well as *dnaI* and *dnaP*), have been implicated in the initiation of replication of the bacterial chromosome. A role for the *dnaA* product in initiation is demonstrated by temperature shift experiments with conditional-lethal mutations: a switch to the non-permissive temperature is followed by completion of rounds of replication before DNA synthesis ceases (for those products involved in the elongation reaction, the block is imposed immediately upon temperature shift). The apparent suppression of *dnaA* lesions by certain RNA polymerase mutants carrying an altered β subunit suggests an interaction between these two cellular constituents at the primer stage. The *dnaA* function can be replaced in *cis* (but not *trans*) by insertion of F (and other) plasmids into the bacterial chromosome (*integrative suppression*, see Section 4.1.3 (ai)). The F plasmid presumably supplies both an origin and a *dnaA*-like function since replication proceeds from the site of F integration in Hfr strains under conditions of *dnaA* inactivation, the F-coded factor(s) acting on its own origin sequence. The accumulated evidence would support a model in which *dnaA* acts before RNA polymerase, together generating origin RNA required for initiation of a cycle of replication. Recent work has, in fact, identified the *dnaA* product (subunit molecular weight of 54 000) as a replication-specific transcriptional regulator. The *dnaB* and *dnaC* proteins (molecular weight 50 000 and 25 000, respectively) act in concert at initiation, probably subsequent to RNA polymerase: they also function in elongation. Note that, in addition to the necessity of RNA production, continued protein synthesis is essential for initiation of chromosomal replication (but not

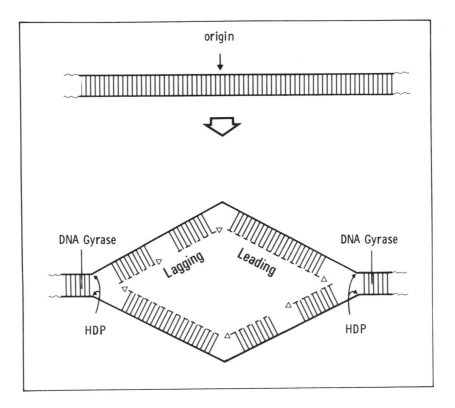

Figure 6.8: Bidirectional Replication from *oriC*
There are two replication forks, each with its leading and lagging strand. Movement is aided by DNA gyrase, which maintains negative superhelicity in the path of each form and helix destabilising proteins (HDP), such as the *ssb* gene product, responsible for duplex melting

for relaxed plasmids, discussed in Section 6.1.3 (a)).

(ii) *Elongation.* Successful initiation results in partial separation of parental strands, thereby permitting Okazaki fragment synthesis. It is still contentious whether both strands, or just the lagging strand, are subject to this discontinuous mode of replication. The apparent discontinuity of synthesis on the leading strand may stem from uracil-containing DNA — formed by the misincorporation of dUMP in the place of dTMP[8] — acting as a target for uracil-DNA glycosylase. The repair enzyme removes the incorrect base, leaving an apyrimidinic site susceptible to further correction by specific incision enzymes (see DNA Repair, Section 6.3). The presumptive Okazaki fragments of the leading strand are, therefore, likely to be repair, rather than replicative, intermediates.

8.   Some dUMP incorporation is known to be inevitable, for dUTP is an essential precursor of dTTP and its pool size cannot be many orders of magnitude less.

Strand separation relies upon helix-destabilising proteins as well as topoiso-merases (Figure 6.8). The enzyme DNA gyrase presumably functions at, or beyond, the replication fork, maintaining negative superhelicity. The elongation reaction is catalysed by the enzyme complex DNA polymerase III holoenzyme, and requires, in addition, the *dnaB, C* and *G* gene products.

DNA synthesis on the lagging strand is, clearly, very much dependent upon movement of the replication fork due to the $5' \rightarrow 3'$ direction of DNA synthesis and to the chains being of opposed polarity (whereas extension on the leading chain may be 'driven' by the polymerase reaction). It is thought that fork migration 'uncovers' naked DNA upon which a short RNA transcript is synthesised by the *dnaG* gene product (Figure 6.9). Discontinuous synthesis (and chain elongation, in general) is, therefore, not inhibited by the antibiotic rifampicin. The presence of a free 3'-OH terminus permits extension of the RNA primer by DNA polymerase III holoenzyme, covalently joining the two dissimilar polynucleotide chains (Okazaki fragments are characteristically about 1000 nucleotides (10S) in length). It has been suggested, following studies on the role of the *dnaB* gene product as a 'mobile replication promoter' for initiation of phage ϕX174 replication (Section 6.1.3 (b)), that this protein has a similar function in activating DNA primase to prime Okazaki fragment synthesis.

As one complementary strand is synthesised discontinuously, the primer of the adjacent polynucleotide chain, upstream of the replication form, must be removed prior to ligation (otherwise a hetero-oligomer consisting of both dNMPs and rNMPs would be produced, see Figure 6.9; actually, the *E. coli* ligase cannot join RNA to DNA). Ribonuclease H (an endonuclease of molecular weight 40 000) is capable of partial primer excision since it specifically degrades RNA present in RNA-DNA hybrids. However, it cannot break a 5'-rNMP-3'dNMP linkage. The accumulation of RNA primers in conditional-lethal *polA ex* mutants at the non-permissive temperature suggests that the task is completed by DNA polymerase I. Nick translation may, thus, extend *de novo* Okazaki fragment synthesis at the 3'-end making use of the $5' \rightarrow 3'$ polymerase function for gap filling and the $5' \rightarrow 3'$ exonuclease activity to remove RNA primers from the 5'-end of Okazaki fragments. The enzyme DNA ligase then ligates adjacent polynucleotide chains (Figure 6.9).

(iii) *Termination.* In *E. coli* the replication fork moves bidirectionally (Figure 6.8) to complete synthesis within about 40 minutes. The two forks meet opposite to *oriC* at a point about 30 min on the bacterial genetic map (near to *trp*, see Figure 6.6); they, presumably, arrive simultaneously at this terminus. Since initiation of a round of replication occurs independently of its termination, a number of replication origins may be present on one replicating chromosome (Figure 6.10).

This ability of partially replicated molecules to serve as a substrate for initiation prior to their completion tends to further increase the concentration of

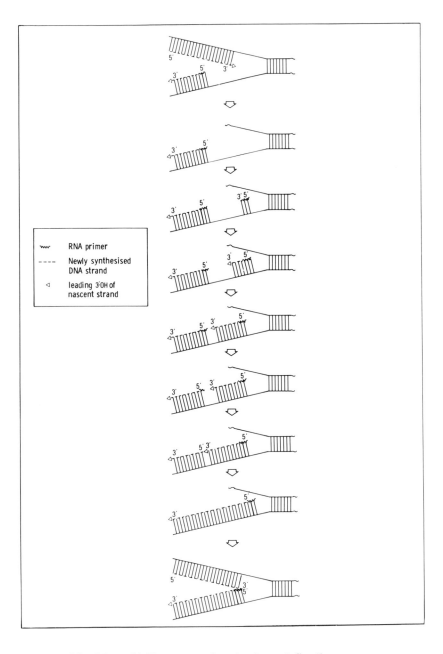

Figure 6.9: Okazaki Fragment Synthesis and Sealing

Only discontinuous DNA synthesis on the lagging strand is shown. Since RNase H is unable to remove the final ribonucleotide, it is proposed that DNA polymerase I erases this residue ($5' \rightarrow 3'$ function) while extending the Okazaki fragment ($5' \rightarrow 3'$ polymerase function). Finally, fragments are sealed by DNA ligase

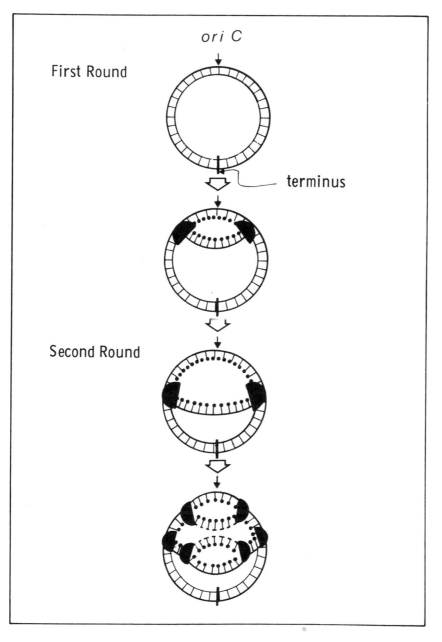

ori C

First Round

terminus

Second Round

Figure 6.10: Multiple Initiation Events on a Single Replicon
Initiation of replication need not await completion of previous rounds. The present example shows a second initiation event at the bacterial origin of replication, *oriC*, half-way through the previous round of replication. Since *oriC* is, itself, duplicated upon replicative initiation, the number of growing forks increases from two to six in the second round (see also Figure 6.11)

genes near to the origin relative to those at the replication terminus (the number of origins increases with growth rate, generating a proportional increase in origin-proximal genes). Clustering of essential genes involved in replication and gene expression around *oriC* would, thus, appear to be significant (see Figures 2.3, 2.20 and 6.6). Moreover, the lack of known genetic markers in the terminator region seems pertinent: genes in this part of the bacterial chromosome are the last to be copied. That the terminus exists as a defined DNA sequence is suggested by its function even when replication is initiated at a novel site distinct from *oriC*.

Successful termination requires both complete replication of the bacterial chromosome and covalent linkage of the synthesised molecules — with the formation of supercoiled, monomeric chromosomal DNA (rather than a catenane). As a circle, continued DNA synthesis brings the 5'- and 3'-terminus together for ligation (perhaps, an explanation for the common occurrence of circular genetic elements). Though ring closure is not a necessary prerequisite for initiation of further rounds of chromosomal replication (witness the multiple initiation events described above), the bacterial genome is known to be a covalently closed circular duplex. Gene functions, if any, involved in this final step await characterisation.

(b) The Role of the Cell Envelope in Replication. Functional replication of the bacterial chromosome requires the production of two covalently closed circular molecules of DNA, faithful copies of the parental molecule, that are segregated correctly at cell division. Completion of one round of replication is, therefore, not sufficient for the creation of two viable, identical daughter cells. Correct partition of the chromosomal genetic elements (and, in fact, of extrachromosomal elements as well, see Section 6.1.3 (a)) is thought to involve the inner bacterial membrane. Certainly, electron micrographs show the bacterial DNA attached to the cell envelope. It is pertinent at this stage to briefly review the structure of the cytoplasmic membrane (Section 1.2.1).

On the fluid mosaic model, the inner membrane consists of a phospholipid bilayer carrying different proteins (embedded either totally or partially). It is, therefore, likely that any DNA-membrane association involves both polar interactions with specific membrane proteins and non-polar interactions with hydrophobic groups present on the lipid components. In fact, a *dnaP* mutant resistant to phenylethyl alcohol has an altered cell envelope and is thermosensitive with respect to DNA synthesis. That such an association — between DNA and membrane — exists is substantiated by analysis of the bacterial chromosome.

Within the bacterial cytoplasm, the chromosome is present in a supercoiled form (about 100 loops of supercoiled DNA per chromosome), condensed into a tight mass: the complete molecule is many hundreds of times the length of the bacterial cell. This folded structure, termed the *nucleoid*, is associated with both protein and nucleic acid. A large proportion of the protein component is represented by the enzyme DNA-dependent RNA polymerase (core form). It is, therefore, not surprising that the nucleoid also contains nascent RNA, about half of which is stable RNA. When isolated from gently lysed spheroplasts at low temperature (see Section 7.3.1), the intact nucleoid structure sediments as a massive 6000S complex (compare, for example, with the sedimentation coefficient of 70S for the 3 Mdal bacterial ribosome). Under these conditions, the nucleoid is separated from the soluble cellular components, but is bound to the inner membrane.

The current favoured model for chromosome segregation is pleasingly simple: separation is achieved by growth of the cell envelope between the attachment points of the two chromosomes. The 'driving' of partition by cell envelope synthesis is difficult to support if membrane outgrowth is polar, occurring from only one 'end' of the bacterium (rather than a central growth zone), unless it is at this site that one of the genetic elements is bound (this would leave the other replicon in contact with the membrane growth point from the previous division cycle). Which part of the chromosome is membrane-attached? It now seems likely that there are multiple attachment points as indicated by the isolation of DNA-membrane complexes in which the nucleic acid is either from the replication origin or from growing points. (A 31 000 dalton outer membrane protein has been implicated in the binding of the initiation region to the cell envelope; furthermore, a 65 000 molecular-weight cell-envelope component interacts specifically with the cloned *oriC* region). The existence of such complexes supports the early model of Jacob, Brenner and Cuzin for the control of DNA replication (but see below).

(c) Control of DNA Replication. E. coli exhibits the phenomenon of growth-dependent gene dosage: cells from exponential cultures in minimal medium contain about two copies of the chromosome but at faster growth rates (say, in nutrient medium containing glucose) the amount of chromosomal DNA increases (relative plasmid copy number decreases with growth rate). What is responsible for this control over chromosome content? The time taken to complete one round of replication appears to be constant (about 40 minutes at 37°), independent of growth conditions. This is the case even in cells with a doubling time of 20 minutes, in which the entire cell cycle represents only one half of the replication period. Moreover, cell division commences at a fixed time, about 22 minutes after completion of replication and, thus, approximately 60 minutes after initiation of that particular round.

This constancy led to the suggestion by Cooper and Helmstetter that replication starts at a fixed time (about 60 minutes) before cell division, with the initiation event not necessarily occurring in the same cell cycle as its associated division sequence. On this model, those cells that are rapidly growing — with a doubling time less than the constant value of 60 minutes — carry a completed chromosome at division by virtue of initiation of the round of replication in a previous generation. Thus, for cells with a doubling time of 40 minutes, chromosomal material destined to finish replication in one round must have initiated 20 (60 – 40) minutes before — in fact, half-way through (20/40) — the previous division cycle (Figure 6.11). This implies that each division cycle encompasses two replication events (under these growth conditions), completion of one round and initiation and partial replication of another. A cornerstone of the model is that initiation occurs prior to completion of a preceding round of replication (though each initiation event is associated with a specific division event, occurring 60 minutes thereafter): replication is *dichotomous* (or multi-forked). Rapidly dividing cells, therefore, exhibit multiple growing points (in the example in Figure 6.10, a maximum of six), with origin-proximal genes present in considerably higher concentrations than those situated close to the terminator. Thus, it is the number of initiation events per cell cycle that varies with growth rate. Moreover, it is at the initiation stage that overall DNA synthesis is controlled.

Control mechanisms concerned with replicative initiation have been discussed with respect to plasmid production and incompatibility of, in particular, the F extrachromosomal genetic element (Section 4.1.2 (b)). The bacterial chromosome may be viewed simply as a large replicon with its own specific origin of replication. On the positive control model, the effector is perhaps supplied by the cytoplasmic membrane: each round of DNA replication is initiated upon inter-action with a specific membrane site (specific according to the replicon concerned). Control is, therefore, mediated through membrane outgrowth, the duplication of the attachment site. Certainly, the isolation of a DNA-membrane complex supports a role for the cell envelope in chromosomal replication (but not necessarily its control, see below). Incompatibility on this basis would stem from competition between genetic elements of a single Inc group (see Section 4.1.2 (b)) for the same membrane attachment site. That is, regulation of rounds of replicative initiation and incompatibility, are mediated through a single control element. These two properties should not, therefore, be separable.

Consider, however, a composite genetic element consisting of two replicons in tandem, the chromosome and the F plasmid. In such an Hfr strain, replication is apparently initiated from the chromosomal origin, *oriC*, in the presence of a functional *dnaA* gene product (and only at the origin supplied by the integrated plasmid when the initiator protein is inactive). Irrespective of this apparent lack of expression of F replication functions, F incompatibility is still exhibited since F-prime (and F) plasmids are unable to exist stably in an Hfr strain. Incompati-bility may, thus, apparently be divorced from replicative control. It is, further-more, generally hard to reconcile the positive effector model with incompatibility of multicopy plasmids (see also Section 4.2). It seems more likely now that the cell envelope plays a structural role.

Of the two major hypotheses for replicative control systems, regulation by membrane attachment or through repressor control, the latter is generally favoured (see also Section 4.1.2 (b)). In this model, a pulse of inhibitor is synthesised upon the start of a new round of DNA synthesis due to replication of

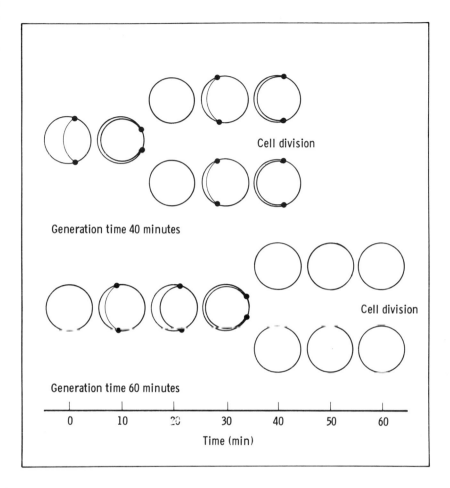

Figure 6.11: The Number of Replication Points as a Function of Growth Conditions

Time course for initiation events for bacterial cells with a doubling time of 40 and 60 minutes (black dots indicate replication points). Since replication and division (or rather the time between replicative initiation and cell division) takes a fixed period of about 60 min, independent of culture conditions (40 min for replication and a further 20 min for division, see bottom scheme), cells with a generation time less than this constant time must carry partially replicated chromosomes at division. Thus, for cells with a doubling time of 40 min, the two chromosome copies segregated at the end of each cell cycle are half replicated (top scheme). After Cooper and Helmstetter (Cooper, S. and Helmstetter, C.E. (1968) *J. Mol. Biol., 31,* 519-540) the repressor gene(s) (presumably close to the origin). Further initiation events are inhibited until the concentration of this control protein, reduced by cell growth or inherent instability, attains a critical level. Dilution of this negative control element upon bacterial growth readily explains synchrony between replicative initiation and cell division: a new round of replication is only initiated when the repressor reaches the critical concentration. The attainment of this level early in the division cycle through reduced activity of the replication inhibitor would explain the phenomenon of high-copy number plasmids (see Sections 4.2 and 6.1.3 (a) for stringent and relaxed replication). Moreover, the negative control model permits a ready understanding of incompatibility — a process mediated through a diffusible and, therefore, *trans*-active protein molecule(s) (discussed with respect to F and other plasmids in Chapter 4). Finally, it allows a plausible explanation for the apparent dissociation of replicative control and incompatibility noted above for Hfr strains.

In the autonomous state of F, the bacterial chromosome is, in fact, of higher copy number than the plasmid. The negative control model, therefore, predicts that the critical dilution of the chromosomal-encoded repressor is attained prior to that for the extrachromosomal genetic element, irrespective of whether these two replicons exist separately in the cytoplasm or are covalently joined. However, when integrated, more frequent initiations at the chromosomal origin, *oriC*, increase the number of F repressor gene copies and, thus, the concentration of this F control element, thereby inhibiting replication from *oriV*, the plasmid origin. Although the F plasmid remains 'silent' as a replicon in Hfr strains (being passively replicated by the chromosomal system), expression of the F-specific repressor — the key element in plasmid incompatibility — leads to the IncFI phenotype. Moreover, the F replicon may be activated, promoting replication of the entire Hfr chromosome from its site of integration: when the chromosomal *dnaA* function is inoperative, DNA synthesis no longer starts from *oriC* to increase the F repressor above the critical level that blocks F-initiated replication.

What is the molecular basis for initiation control? Available data tend to support an active system involving a negative element. That is, a diffusible substance (or substances) capable of maintaining copy number, whether of a small extrachromosomal genetic element or of the bacterial chromosome, by regulating the frequency of initiation events in the cell cycle. Various cell-free systems have been devised for the study of DNA replication, both of small plasmids and simple bacteriophage genomes. Although these soluble extracts have been crucial in defining essential components of the replication complex (discussed in Section 6.1.1), the actual regulatory elements have not as yet been identified (there is no indication that the replicative proteins are involved in control). Clearly, the isolation and sequencing of replication origins, the site at which control is thought to be mediated, is a major step towards an understanding of the mechanism(s) concerned. Moreover, according to the proponents of the negative control model, the structural gene for the repressor should be situated close to the origin of replication (to allow a burst of synthesis upon replicative initiation, see Section 10.1.1 (b)).

(d) Antibiotic Inhibitors of Replication. Numerous antibiotics interfere with the replication process. DNA synthesis, crucial for cellular growth and survival, may be inhibited at a number of different levels. Drugs may act directly, by binding to the DNA template or by inactivation of one of the (enzyme) components of the replicative apparatus; or, indirectly, by preventing the synthesis of these components.

The size of the bacterial chromosome makes it a prime target for antibiotics that have some specificity for DNA. A number of such inhibitors of template function have been discussed previously: the antibiotic actinomycin D, in relation to inhibition of transcription (Section 2.1.1 (c)), acridines for their ability to produce frame-shift mutations and for 'curing' of certain plasmids (Sections 3.1.1 (d) and 4.1.1) and the use of ethidium bromide for the separation of supercoils from relaxed molecules (introduced in Section 1.1.1 (c), the method is discussed further in Section 7.3.1). These DNA binding substances have in common the ability to intercalate between DNA bases, thereby sterically inhibiting movement of the replication fork. Whereas the above-mentioned compounds bind non-covalently, reduced mitomycin C (Figure 6.12) forms actual cross-links to the DNA template.

Inhibitors of RNA polymerase function, such as rifampicin, streptovaricin or streptolydigin (Section 2.1.2 (c)), are also potent on DNA replication when the RNA primer is supplied by this enzyme, rather than by DNA primase. Upon binding to RNA polymerase, these antibiotics prevent initiation of rounds of replication without interfering directly with movement of the growing point: discontinuous synthesis is not inhibited by rifampicin. The 'curing' of F by low concentrations of rifampicin presumably reflects preferential inhibition of plasmid DNA initiation (cf. acridine orange 'curing', Section 4.1.1). Neverthe-less, initiation of replication at the bacterial origin is also sensitive to this antibiotic. Other inhibitors of enzyme function include nalidixic acid and coumermycin (Figure 6.12), which act on the two subunits — the *gyrA* and *gyrB* gene products — of DNA gyrase, an essential component of the replicative apparatus, necessary for the elimination of positive supercoils arising from DNA synthesis.

Cordycepin triphosphate (3'-dATP, Section 2.1.2 (c)) also interferes with enzyme function, in this case by the prevention of RNA polymerase- or primase-mediated RNA chain elongation (whereas rifampicin and streptovaricin inhibit initiation). This analogue of ATP lacks the 3'-OH group and upon incorporation in RNA (in this case, primer RNA), cannot be extended. It leads to premature termination of primer synthesis. 2', 3'-dideoxy dNTPs inhibit DNA polymeri-sation *in vitro* in the same way.

DNA synthesis, because of its dependence on a number of cellular components (Section 6.1.1), is especially susceptible to indirect perturbation. Clearly,

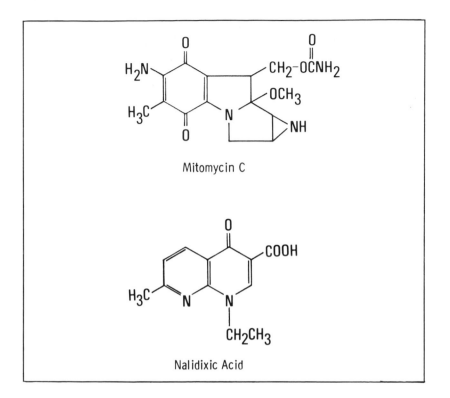

Mitomycin C

Nalidixic Acid

Figure 6.12: Structure of Some Inhibitors of Replication

maintenance of the substrate pool is important for chain elongation (the relative concentrations of deoxyribonucleotides also appear to be involved in maintaining a low-level error rate). The involvement of the cell envelope, in particular, the cytoplasmic membrane, in chromosome replication seems to make replicative DNA synthesis sensitive to external influences (for instance, ethanol derivatives).

6.1.3 Replication of Small Genetic Elements

As intimated in the preceding section, DNA replication is still something of a mystery, at least with respect to large replicons such as the chromosome of *E. coli*. The size of the bacterial genome and a requirement of an intact cell envelope for *de novo* initiation of DNA synthesis (rather than just continued propagation of replication) is responsible for this incomplete understanding. Replication of small circular DNA molecules, for instance plasmids or simple viral genomes, are most amenable for study *in vitro* (particularly the latter): their DNA is readily isolated intact. Note that these studies involve purified or partially purified components. Indeed, the initiation and elongation complexes, both multicomponent systems, have been reconstituted from their individual constituents in the case of single-strand phage replication. Such systems are, therefore, truly cell free since they allow DNA synthesis to occur in the test-tube in the absence of even cell-envelope fragments (indicating that replication of these simple genomes, in so far as the process takes place under these conditions, does not involve membrane-DNA attachment, whereas replication of bacterial DNA generally requires the cell envelope). Replication of small genetic elements is estimated to be an order of magnitude slower than chromosomal DNA synthesis (Section 6.1.2). The latter requires continued protein synthesis for initiation unlike the relaxed group of plasmids. The apparent overall similarity between replication of these small circular elements and the bacterial chromosome (as indicated by their dependence almost entirely on the cellular machinery for replication enzymes) makes such studies particularly relevant. Thus, these systems have proved most fruitful in defining components of the replicative apparatus and in the investigation of certain stages in replication. Moreover, they have also given *in vitro* information on another type of replication mode.

Replication of the bacterial chromosome (and vegetative, though not transfer replication, of Hfr derivatives and some conjugative plasmids) proceeds by bidirectional movement of the growing point, initiated from a unique origin; it generates θ-structures (Cairns replicative mode, Section 6.1.2 (a)). Small double-stranded genetic elements may also replicate by the rolling-circle mechanism (see Figure 6.13). This mode is thought to be involved in transfer replication of plasmid DNA. These two types of replicative DNA synthesis — Cairns and rolling-circle — differ essentially at the initiation and termination stages: at each stage, a nucleolytic reaction is necessary in the rolling-circle mechanism. Initiation in the rolling-circle mode occurs by nicking (unnecessary in the case of bidirectional replication from *oriC*). A further nucleolytic event is required to generate a DNA chain of one genome length since the newly synthesised strand is covalently attached to the parental molecule. Finally, it should be noted that θ-intermediates can have two replication forks.[9]

It is worthwhile, at this stage, reconsidering the replicon concept (Section 1.1.2 (b)). It applies to a molecule of DNA that is replicated and conserved during cell division (either through the use of cellular components, in part or *in toto*, or by encoded gene functions). A replicon is an autonomous unit of replication. What, then, are the minimum requirements for continued survival of such a genetic element? Although a replicon need not encode protein product(s) such as a (specific) DNA polymerase or primase, it must be capable of allowing both successful initiation of replication and the control of this process. A defined origin sequence may apparently not be necessary (witness the ability of the *dnaB* + *dnaG* gene products to catalyse multiple initiations of phage ϕX174 SS → RF

---

9.   The existence of two structural forks in the Cairns mechanism is independent of whether DNA synthesis is bidirectional or unidirectional. However, only in the former case do both forks move relative to the origin of replication.

replication, see below). A replicon-encoded repressor gene, situated close to the origin, is implicit in the negative control model.

(a) Plasmid Production. Replication of small genetic elements may now be studied in membrane-free extracts in which both the cell envelope and endogenous DNA are removed by centrifugation (after gentle lysis, see Section 7.3.1). Soluble systems offer a further refinement over those available for investigation of chromosomal replication.

There is no clear-cut generalisation about replicative DNA synthesis on covalently closed circular duplex templates. This is clearly exemplified by the apparent difference in the movement of the growing point: bidirectional in the case of the prototypic plasmid, F, and unidirectional for the small extrachromosomal genetic element, ColE1 (see also Section 4.2.2). Some discrepancies may arise from natural fusion of two different replicons (in addition to cointegrates of the bacterial chromosome and F, the plasmid R6K, molecular weight 25 Mdal, is a case in point since it seems to have two or three different origins of replication). The majority of information on plasmid production to date stems from the study of just a few genetic elements and their derivatives, particularly of small, multicopy plasmids.

Small plasmids appear to differ in their dependence upon the chromosomal *dnaA* functon for initiation, some absolutely requiring this gene product, others apparently able to continue replication for some time without it. Large plasmids, on the other hand, for instance, F or R100, generally replicate independently of the bacterial *dnaA* product, as exemplified by their property of integrative suppression (in which the defect in initiation of chromosomal replication due to a *dnaA* lesion is suppressed upon plasmid integration, Sections 4.1.3 (ai) and 6.1.2 (ai)). As far as it is known, the role of the *dnaB, dnaC* and *dnaG* proteins as well as the bacterial RNA polymerase is the same in plasmid and chromosomal replication.

The overall similarity between replication of chromosomal and extrachromosomal genetic elements (and, hence, the validity of the use of these smaller replicons as model systems for the study of bacterial replicative mechanisms) is supported by the dependence of plasmid replication on the DNA propagation functions of the bacterium, for instance, *dnaB, dnaC, dnaE (polC), dnaZ* and DNA gyrase.[10] This suggests that the mechanism of discontinuous replicative DNA synthesis is the same (or much alike) in these genetic elements. Any differences presumably lie in their mode of initiation (and its control) rather than

10.   Surprisingly, a number of different plasmids require the bacterial DNA polymerase I for their replication. Of the three activities associated with the *polA* gene product (see Section 6.1.1 (a)), the polymerase and $3' \rightarrow 5'$ exonuclease function seem to be the most likely candidate: *polAex* mutants that retain these two activities but are unable to excise nucleotides in the $5' \rightarrow 3'$ direction remain suitable hosts for such plasmids. Note that chromosomal replication requires the *polAex* product for joining Okazaki fragments. Thus, a catalogue of the differences between plasmid and chromosomal replication should include the role of the *dnaA* and *polA* gene products. In addition, certain plasmids exhibit relaxed control of replication (see below).

at the elongation stage. Although membrane attachment is unlikely to play a role in replication (witness the ability of cell-free preparations to support plasmid DNA synthesis) or its control, some sort of association with the bacterial envelope appears to exist, possibly necesary for accurate plasmid partition.

It is generally accepted that vegetative plasmid replication proceeds via some sort of Cairns mode (with either unidirectional or bidirectional movement of the growing points), while transfer replication, in which DNA passage is effected between two bacterial cells, involves the rolling-circle mechanism (Figure 6.13). Transfer replication has been described in Section 4.1.2, mainly with respect to the conjugative plasmid F. Some signal, perhaps cell-cell contact, triggers replacement of the θ-mechanism by a rolling-circle mode. In this mode, the transfer origin, *oriT*, rather than the vegetative site *(oriV)* is active; these two regions are distinct on F (see Section 10.1.1). Note that upon nicking of one strand, and with transfer of its 5'-end, the 3'-OH terminus can serve as primer for extension through continuous DNA production. Synthesis of a complementary chain upon the transferred strand must be discontinuous (primed, usually, by the bacterial *dnaG* gene product). Neither replicative process is apparently necesary for single-strand transfer of the F plasmid (Section 4.1.2 (a)).

A 15-20 Mdal contiguous DNA sequence encodes F transfer functions (see *tra* Operon, Section 10.1.1 (a)). It is, therefore, not surprising that small plasmids less than this minimum size lack the ability to promote transfer — they, clearly, lack the genetic capacity to encode these functions.[11] Such non-conjugative plasmids may, however, be mobilised by transfer-proficient extra-chromosomal genetic elements, for instance, by F, itself. Mobilisation has been shown not to be dependent upon direct covalent interaction between the two replicons. It is presumed that the same (or similar) mechanism — the rolling-circle mode — is involved irrespective of whether a second plasmid is present, with the exception that conjugative plasmids supply some function(s) *in trans* necesary for mobilisation (protein(s) not encoded by the non-conjugative, mobilised genetic element). Some transfer-defective plasmids also seem to supply essential elements. Such a model for transfer replication of small conjugation-deficient plasmids would suggest that they possess both origins of replication, *oriV* and *oriT*, and are unable, perhaps, to supply the necessary triggering or initiation functions (let alone the genes for pilus production). Some light is shed on this by analysis of the phenomenon of repressed and derepressed plasmids (Section 4.2), discussed in Section 10.1.2.

Certain plasmids, termed *relaxed*, are readily distinguishable from their

---

11.   Regions of DNA involved in vegetative replication have been defined by *in vitro* manipulation (see Section 7.3.3, Restriction Technology). The 0.80 Mdal (1.225 kb) plasmid designated pBR345, derived from the small, multi-copy genetic element ColE1, is capable of autonomous replication. This 'mini' ColE1 plasmid is the smallest covalently closed circular duplex so far constructed (a minimum size of 225 bp has been estimated on steric considerations for such duplex molecules of DNA). Other replication origins have been isolated by cloning (Section 6.1.2 (ai)). The ColE1 region required for its replication and maintenance, located on a 580 bp fragment of DNA, is different from these origin sequences.

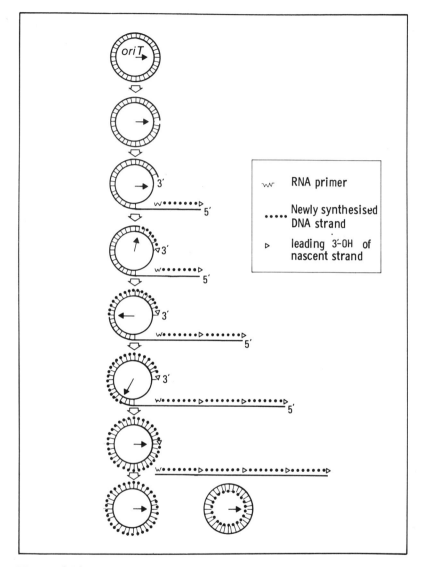

Figure 6.13: The Rolling-Circle Mechanism of Replication

The rolling-circle or transfer mode of replication is initiated with a nick (at *oriT*, in the case of the F plasmid, see Chapter 10). The leading strand carries the 5′-OH and is, therefore, replicated discontinuously (arrowheads, representing the leading 3′-OH, have been left in to emphasise this fact, see also Okazaki Fragment Synthesis and Sealing, Figure 6.9). The 3′-OH, on the other hand, serves as a primer for continuous replication on the intact, circular strand. Rolling-circle replication produces concatamers which require strand scission to produce unit chain lengths, and subsequent circulation *stringent* counterparts since replication continues in the absence of ongoing protein biosynthesis. Vegetative DNA synthesis can apparently proceed in the absence of any continuing production of plasmid- or chromosomal-coded proteins though it is dependent upon endogenous bacterial functions present before imposition of the block. In the presence of the translational inhibitor, chloramphenicol (Section 2.2.2 (d)), the number of copies of the relaxed plasmid ColE1 increases to about 3000 per cell. The presence of RNA sequences in these plasmid circles under conditions of inhibition of protein synthesis is indicated by the sensitivity of these genetic elements to alkali or RNase treatment. Whether this RNA results from failure in primer excision is, as yet, unsure. It is possible that these particular RNA stretches are a product of the non-permissive conditions and unconnected to the natural initiation process. Note that the terms 'relaxed' and 'stringent' strictly refer to the replication mode of these plasmids under particular conditions. That is, a 'relaxed' genetic element demonstrates relaxed replication in the presence of a translational inhibitor but not in its absence. Amplification of plasmid DNA molecules during relaxed growth has proved useful for their isolation and purification (considered in Section 7.3.3).

Another form of plasmid amplification has been noted in *Proteus mirabilis*, a member of the Enterobacteriaceae. The two components of certain R plasmids, encoding the transfer functions and drug resistance (the *RTF* and *r-determinants* components, respectively), are able to replicate autonomously. Moreover, polymerised versions of the small r-determinants component, consisting of tandem copies of the DNA region, exist both as separate entities and in conjunction with a single RTF component (a covalent association in either case). The molecular basis for such R plasmid amplification is not understood. It would seem that R plasmids exhibiting this property evolved by the fusion of two separate replicons, one of which, the r-determinants component, is no longer able to replicate as a separate entity in *E. coli*. On this basis, *P. mirabilis* supplies (or lacks) some function that regulates replication of this component in *E. coli*. Interestingly, the r-determinants component is bounded by two copies of a DNA sequence known to promote fusion events (see Insertion Sequences and Transposons, Sections 6.2.2 (aii), 7.1.2 (cii) and 10.1.1 (c)).

(b) Viral DNA Synthesis. The majority of information on the components of the replication complex has come, in fact, from *in vitro* studies on bacteriophage replication, particularly those bacterial viruses whose genomes consist of covalently closed circular molecules of single-stranded DNA. The filamentous phage M13 and the two closely related isometric viruses ϕX174 and G4 have been most studied, largely because single-stranded replication requires no 'unwinding' system, hence, greatly simplifying the job of *in vitro* reconstruction and because of size of their genetic material (5000-6000 nucleotides, about three orders of magnitude smaller than the *E. coli* chromosome). Though morphologically distinct (Section 5.1.1 (a)), these two groups of single-stranded

DNA viruses have similar replication mechanisms. Moreover, unlike the larger bacteriophages (for example, *lambda* or T7), replication is dependent upon many chromosomal-coded factors and is, therefore, likely to reflect mechanisms involved in replicative DNA synthesis of the bacerial chromosome, itself (replication of the double-stranded DNA phage *lambda* is discussed in 10.2.1 (c)).

The approach for these studies has been to isolate from *E. coli* extracts the minimum components necessary for phage DNA production. Such investigations have been greatly aided by the preparation of membrane-free soluble systems in which replication is dependent upon exogenous DNA. Initial studies employed crude isolates but painstaking work has led to the purification of a number of replication proteins and the reconstitution of active replication complexes. It has, for example, identified the components of the elongation reaction, DNA polymerase III holoenzyme, a multisubunit aggregate consisting of at least four non-identical polypeptides (Section 6.1.1 (a)). Once obtained in homogeneity, replication proteins may be analysed for their precise role in DNA replication. Thus, replication of single-stranded phage DNA — at least the initiation and elongation reactions — can now be studied using both defined templates and purified proteins.

Replication of these two morphological groups of single-stranded DNA phages, M13, ϕX174 and G4, proceeds in three main stages (Figure 6.14). The viral genome, '+' strand (the SS form), is first used as a template for synthesis of the complementary '–' molecule (apparently redundant at least with respect to coding potential). The initial product of this reaction, a replicative intermediate (termed RF II) that contains a nick, is subsequently converted to a covalently closed circular molecule of duplex DNA (the RF I form). The second stage involves multiplication of the RF I form; whether by the Cairns or rolling-circle mode is not clear, although the latter is favoured. Finally, 'productive' replication — viral '+' single-strand synthesis — proceeds via a rolling-circle-like mechanism. Three different replicative processes are, therefore, amenable for study: SS → RF, RF multiplication (RF → RF) and RF → SS, the first of which is dependent exclusively on host proteins.

A primer is required for initiation of replication on the three single-stranded DNA bacteriophages (as it is for initiation of chromosomal and extrachromosomal replication, Sections 6.1.2 (ai) and 6.1.3 (a), above). Whereas this origin RNA, in the case of the filamentous phage M13, is synthesised by the rifampicin-sensitive bacterial RNA polymerase, both phage ϕX174 and G4 require DNA primase for initiation (and, hence, are not susceptible to specific inhibitors of RNA polymerase, see Section 6.1.2 (d)). The two isometric viruses are distinguishable with respect to replicative initiation in that phage ϕX174 primer formation occurs at multiple, apparently random, sites and requires *dnaB, dnaC* and other proteins, whereas G4 initiation is at a single specific region on the viral genome and needs only primase (this G4 sequence has been discussed in relation to *oriC* in Section 6.1.2 (ai)). The DNA primase product on

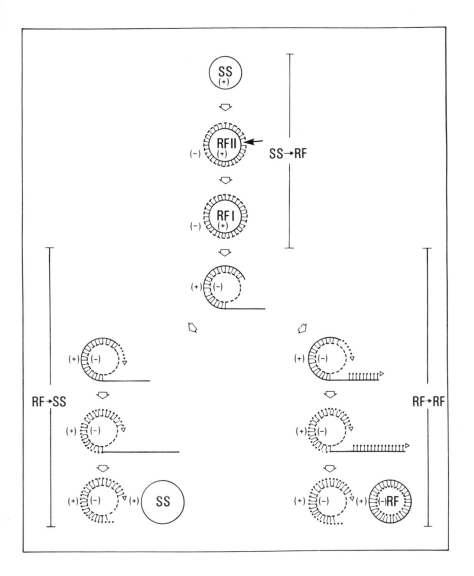

Figure 6.14: Stages in Replication of Single-Stranded Circular Phage DNA
A much simplified scheme for SS → RF, RF → RF and RF → SS modes of replication
(abbreviations: SS, single-stranded or '+' form; RF II, I, replicative intermediate with or without a
nick, arrowed). The double-stranded form produced from replication on the '+' strand (SS → RF)
supplies the template for SS and RF production via the rolling-circle mode (RF → SS and RF →
RF). Although the two icosahedral single-stranded phages G4 and φX174 show great similarity in
their primary DNA sequence, there are clear differences in replication mechanisms. Thus, φX174
requirements include a number of proteins over and above those necessary for G4 replication.
Moreover, whereas G4 replication is initiated at a fixed position on the viral DNA, φX174 appears
make use of a 'mobile replication promoter' (but see footnote 12 on page 268). (Note that the G4
origin, the sequence for which is absent, not surprisingly, from the φX174 chromosome, has been
taken as the prototype DNA primase-promoted initiation site)

ϕX174 SS DNA *in vitro* is a series of short oligonucleotides (15-50 bases in length), containing ribonucleotides or a mixture of ribo- and deoxyribonucleotides. It is still uncertain whether multiple primers serve as initiation sites for DNA synthesis *in vivo* or whether in a coupled system where both primer and elongation components are present, only a single origin RNA is extended.[12] Studies on the involvement of the *dnaB* gene product in ϕX174 priming has led to the suggestion that this protein serves as a 'mobile replication promoter', allowing DNA polymerase III holoenzyme to initiate *ad hoc* DNA synthesis in the presence of primase. On this model, the *dnaB* protein binds to, and migrates along, the single-strand template, occasionally promoting primase action.

The 20 000 dalton DNA binding protein (the *ssb* gene product) is required for both initiation and elongation, apparently covering most of the DNA template. Ligation completes the conversion of single-stranded DNA to the covalently closed circular duplex form, RF I (see legend to Figure 6.14 for a comparison of replication mechanisms employed by the two isometric phages). Multiplication of this double-stranded molecule and synthesis of viral '+' strands on the complementary '−' chain is a particularly complex process, involving both chromosomal- and viral-encoded proteins.

As can be seen, a major difference between the replication mechanisms of these single-stranded viruses lies at the initiation stage (supporting the contention that since this event is most susceptible to variation it is a likely candidate for control, Section 6.1.2 (c)). In addition to primer formation — whether synthesised by RNA polymerase or DNA primase at a specific site or at random — there are clear differences in requirements in that both the *dnaB* and *dnaC* gene products are essential for initiation of ϕX174 replication (other proteins, as yet uncharacterised in detail, seem also to be necessary). The involvement of at least seven different proteins in ϕX174 primer synthesis in the presence of the *ssb* gene product (only one of these are involved in G4 initiation), perhaps, reflects the extra components required for *ad hoc* initiation as compared with starting at a defined sequence (but see footnote 12, below). Unfortunately, there is not scope in this text to do justice to the elegant work carried out on these phage systems: the reader is referred to the excellent reviews listed at the end of the chapter.

6.2 Genetic Recombination

Although segregation frequently occurs soon after completion of a round of DNA replication, the bacterium may carry more than one copy of a particular

12.   Although the multicomponent priming system is able to move rapidly around the circular DNA molecule, resulting in random primer and DNA synthesis, recent studies have shown that a 55 nucleotide sequence of ϕX174 DNA, located in the same intergenic region as the G4 origin, is responsible for initiation of complementary strand synthesis. Thus, in the presence of the *ssb* product, ϕX174 DNA (like that of M13 and G4) appears to have a unique origin of replication.

DNA segment — if not the entire bacterial chromosome — at any one time in the division cycle. Even after partition, diploid or partial diploid status can be created upon gene transfer mediated by bacteriophages or through cell-to-cell contact (transduction and conjugation has been discussed in Sections 5.2 and 4.1.3, respectively). In addition, infection of a permissive host by high phage numbers (m.o.i. < 1) generates a similar polyploidy situation for viral genomes. In short, despite the fact that *E. coli* is an haploid organism, more than one copy of a particular sequence of nucleotides may be present in the cytosol. These sequences, moreover, may show extensive homology.

In the discussion on DNA synthesis, emphasis has been placed on the capacity of DNA to act as its own template for reproduction through the specificity of base pairing (Section 6.1.2). The fact that one strand encodes the information for a complementary molecule and that the resultant double-stranded helix is maintained by exact hydrogen-bonding implies that two complementary sequences have the capacity to base pair irrespective of their origins: there is nothing unique about a DNA template and its associated single-strand product that distinguishes it from a second complementary polynucleotide chain. Complementary regions may be present as components of a single genetic element — the two antiparallel strands — or may arise from partial duplications on separate genetic elements (or fragments of DNA). When two similar nucleotide sequences exist, interaction between DNA duplexes, *synapsis*, may occur. It is this pairing of chromosomes that leads to the exchange of DNA, a process called *genetic recombination*. Recombination, in fact, occurs between both separate DNA molecules (*interchromosomal* or *interstrand*) and sites on a single genetic element (*intrachromosomal* or *intrastrand*, Figure 6.15). Nevertheless, irrespective of the nature of the DNA 'substrate', recombination involves the breaking and making of internucleotide bonds. It is a reaction of DNA.

What defines a region of homology? Clearly, two true copies of a genetic element are identical in their entirety: they will exhibit total homology. Any region on two such elements is, therefore, available for synapsis in an interstrand event. As the two sequences diverge through spontaneous (or induced) mutations, their similarity is reduced. Note, however, that if every gene of the *E. coli* chromosome carried a single base change (the organism would, of course, not survive!), this would represent a difference of only about 2000 base pairs out of a total of 3.8×10^6, a 0.05 per cent divergence from wild-type. Thus, even after extensive changes through single-site mutations, two genetic elements will still exhibit extensive homology. Multisite lesions — deletions, inversions or duplications — have a gross effect on chromosomal material in so far as reducing the potential for synapsis. Consider, for example, recombination between an F*lac* plasmid and a bacterial chromosome carrying a *lac* deletion: there is little, if no, homology between the two genetic elements with respect to bacterial DNA. The range in sequence homology — from 100 per cent to just a few base pairs — defines, in fact, the substrates for various recombination systems.

Figure 6.15: Viable Crossover Events

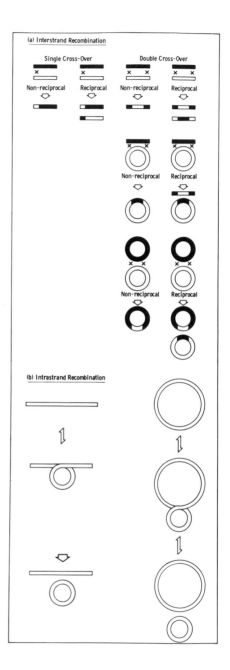

When extensive homology exists, exchange between DNA is catalysed by the *general* (or homologous) recombination system, a key component of which is the ubiquitous *recA* gene product (ubiquitous because of its common involvement in many reactions of DNA). This system is responsible for synapsis between any (two) regions that exhibit extensive sequence complementarity: hence, the term 'general'. Although the *recA* protein plays a key role in general recombination, genetic exchange still occurs in its absence, albeit reduced by as much as six orders of magnitude. These are *non-homologous* events since little sequence similarity is necessary. *recA*-independent modes of recombination may be divided into two types according to the specificity of the reaction: *site-specific recombination*, so-called because exchange occurs only at specific regions present on one or both DNA molecules and *illegitimate recombination*, covering other *recA*-independent events. It should be emphasised that both types of *non-homologous recombination* require only very limited sequence homology, as little as five or six base pairs.

6.2.1 General Recombination

By far the majority of exchange reactions of bacterial DNA in a standard genetic cross are mediated by the general recombination system, as exemplified by the massive reduction in recombinants when this system is non functional.[13] This is not surprising as regions of homology are present in 'high' concentration upon conjugation or transduction (an F' plasmid or phage P1 fragment carrying 2 per cent of the bacterial chromosome will have DNA that is virtually identical — bar a few single-site changes — to one fiftieth, about 80 kb, of the chromosome, and as much as 50 min or greater may be transferred in an Hfr cross). It should be emphasised that the general recombination system catalyses recombination between regions of extensive homology. This system will not promote, for instance, integration of wild-type phage *lambda* into the chromosome since the viral genome is virtually distinct from that of the bacterium (any similarity is at the level of just a few base pairs):[14] conversely, integration of *lambda* occurs in the absence of a functional general recombination system; an example, in fact, of site-specific recombination.

Genetic exchange in general (and non-homologous) recombination is not limited by the actual location of the regions of homology. It may occur within a single strand when gene duplications are present (the 7 rRNA gene clusters, for instance, serve as a substrate for general recombination). Duplicate copies *in*

13.   This reduction in recombination frequency in a *recA*⁻ background is a useful property since it allows the stable presence of two homologous regions without loss by genetic exchange. An F-prime plasmid, for example, exists solely as an autonomous genetic element — despite the partial diploid status — when the general recombination system is non-functional (see Complementation Analysis, Sections 7.1.1 (a) and 7.1.2 (aii)). Note that certain phage encode their own recombination system(s); the *lambda red* genes (*redX* and *redB*, Section 10.2.1 (c)) are a case in point.

14.   The probability of a sequence of n bases occurring by chance, based upon a random distribution of nucleotides, is proportional to $1/(4^n)$. Thus, a particular pentanucleotide stretch may be present about 3700 ($3.8 \times 10^6/1024$) times on the bacterial chromosome (about once per gene), whereas the number of heptanucleotide sequences is only likely to be about 230 ($3.8 \times 10^6/16300$).

trans, either on another replicon or transferred through conjugation or trans-duction, also readily recombine. Whereas intrachromosomal recombination results in deletions, inversions or duplications, interchromosomal exchange produces one or two hybrid molecules, *recombinants*, derived in part from the two parental DNA duplexes.

A single-gene product, the *recA* protein, plays a key role in general recombination. Lesions at this site abolish all exchange events catalysed by the homologous sytem without affecting non-homologous recombination. More-over, such mutants show a markedly reduced mutation frequency when subjected to ultraviolet light, indicating the importance of genetic recombination, or at least the *recA* protein, in DNA repair (see Section 6.3). In fact, *recA*⁻ mutations are pleiotropic.

The *recA* gene and its encoded protein have now been isolated. Other recent developments in this field include the construction of *in vitro* recombination systems that allow the rapid screening of cell extracts for protein fractions possessing DNA fusion activity and the visualisation of intermediates. Viral DNAs are particularly useful for the study of recombination, both *in vivo* and *in vitro*, since the products of the reaction may be readily analysed (see DNA Isolation, Section 7.3.1).

(a) The Elements of General Recombination. By screening mutagenised cultures for derivatives unable to support recombination in genetic crosses, Clark isolated mutants carrying lesions in *recB* and *recC*, as well as in *recA*. Further systems have been constructed that allow an indirect selection of recombinant deficient strains (see also Recombination Pathways, below). Loci identified in this manner are generally prefixed by the acronym '*rec*' to indicate their involvement in genetic exchange (see Table 6.4). It should be emphasised that the inability of certain mutants to produce recombinants in genetic crosses reflects a defect in genetic exchange rather than disruption of the DNA transfer process. Many *rec* mutations show pleiotropic effects and frequently have increased sensitivity to mutagens such as UV or γ-rays. (Note that lesions at certain loci cause augmented sensitivity to UV but are not *rec*-like since they do not show any effect on recombination; these *uvr* genes are discussed in Section 6.3.) Mutants with lesions in *recA*, *recB* or *recC* also grow more slowly and have reduced viability. The *rec* gene products *per se* are not, however, essential.

That numerous other proteins, in addition to the *recA* product, participate in genetic recombination is suggested by the nature of the process itself. Strand transfer requires firstly pairing of complementary chains from non-sister chromosomes. This single-strand invasion of a DNA duplex appears to be promoted by *helix destabilising proteins* and the *topoisomerase*, DNA gyrase. Secondly, since hydrolysis of internucleotide bonds is required for DNA exchange, it is not surprising that the components of the general recombination apparatus include a number of specific *nucleases*. Finally, enzyme constituents of the *replicative machinery*, as well as the components described above, also seem to be involved in genetic recombination.

(i) *The recA Gene Product.* A protein initially termed *protein X* (the active species is a tetramer of 38 000 dalton subunits) and later shown to be the *recA* product is synthesised in greater amounts in physiologically or genetically derepressed strains (see below). Identification of the gene product was made possible by the isolation of a specialised *lambda* transducing phage, *λprecA*, carrying wild-type and mutant *recA* alleles (making use of *secondary site insertion*, Section 6.2.2 (ai), in the *srl* locus, the genes responsible for the use of sorbitol as a carbon source). The purification of the *recA* protein to homogeneity has allowed the *in vitro* investigation of its role in recombination and SOS induction. The *recA* protein is a DNA-dependent ATPase (one of the many potentially involved in genetic exchange, see (ii) below) that unwinds DNA in the presence of ATP. The protein appears to bind homologous single strands of DNA, thereby forming a DNA duplex, while unwinding the parental molecule. Thus it catalyses the formation of two putative early stages in recombination: *displacement loops (D-loops)* — melted duplex regions carrying a single-strand invader — and *figure eight* structures — interstrand recombinants of two covalently closed circular DNA molecules (Figure 6.16.).

Mutations at the *recA* locus are pleiotropic, giving rise, in addition to a recombination-deficient phenotype, to: extreme sensitivity to mutagens such as UV, MMS or mitomycin C; reduced susceptibility to mutation by UV; a decrease in UV-mediated phage *lambda* induction and in expression of colicinogenic plasmids; aberrant regulation of cell division; and rapid degradation of damaged DNA (as well as spontaneous breakdown of native DNA). It was, in fact, the latter phenotype that led to the punning term '*rec*kless' for this class of bacterial lesion. It should be emphasised that although other mutations may give rise to one or two of these responses, only *recA⁻* mutations show all these properties (and see *lexA*, below). These pleiotropic effects can be ascribed in part to aberrant regulation of *recA* expression. Studies have shown, for example, that the level of *recA* protein is mediated by the state of the bacterial cell and is apparently inducible by agents that damage DNA structure or interfere with its synthesis, for instance, UV irradiation, thymine starvation or nalidixic acid treatment (it may also be induced genetically, see below). It would seem that these DNA perturbations lead to activation of a series of functions, termed as a group the *SOS functions*, which consist of error-prone repair, phage *lambda* and colicin induction, and uncoupling of cell division from DNA replication (the role of *recA* in DNA repair is discussed in Section 6.3.2 (b)). The common element in these apparently unrelated but inducible processes is the *recA* gene product, as indicated by their abolition (or lack of control) in strains carrying lesions at this locus.

Gudas and Pardee (1975) have presented a model for the control of *recA* expression in which the gene product, itself, is required to remove a repressor, the *lexA* protein, which blocks transcription (and hence translation) of *recA*

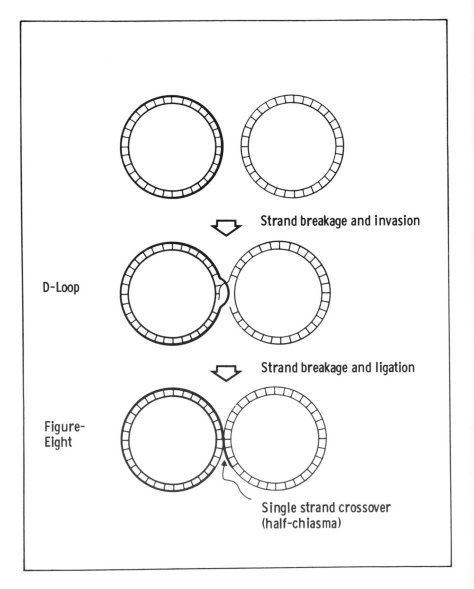

Strand breakage and invasion

D-Loop

Strand breakage and ligation

Figure-
Eight

**Single strand crossover
(half-chiasma)**

Figure 6.16: Intermediates in General Recombination

Invasion of a duplex by a single strand generates a D-loop, a double-stranded region with complementary single-stranded loop. Following strand breakage and reunion, a figure eight structure may be created from two D-loops formed between circular elements. Both interstrand events, putative early stages in general recombination (see Models for General Recombination, Figure 6.26), are catalysed by the *recA* protein in the presence of ATP

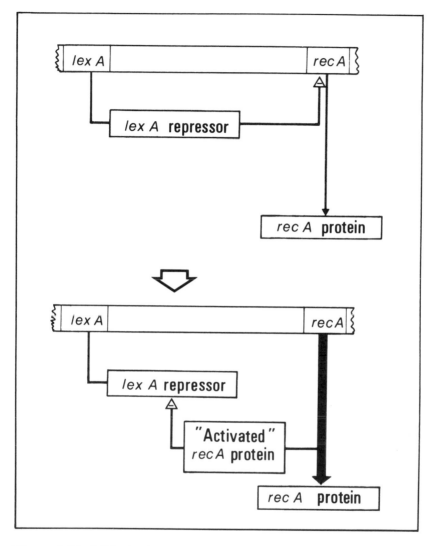

Figure 6.17: A Model for *recA* Control

It is postulated that the *lexA* gene product is a protein repressor that inhibits expression of *recA* by binding to an adjacent regulatory site (see Operon Control, Chapter 8). The basal level (thin arrow) of *recA* gene product, previously termed protein X, is sufficient for normal recombination processes (and for protection of any single-stranded DNA against *recBC*, Exo V, action). During conditions of DNA damage (or interference to DNA synthesis), some effector molecule(s) produced by DNA breakdown activates the *recA* protein as an anti-repressor (it has recently been shown that *recA* protein cleaves *lexA* product in the presence of ATP and polynucleotide *in vitro*). Once the *lexA* repressor is no longer functional, *recA* synthesis becomes induced (thick arrow). After Gudas and Pardee (Gudas, L.G. and Pardee, A.B. (1975) *Proc. Nat. Acad. Sci., USA, 72,* 2330-2334)

(Figure 6.17). That is, *recA* protein regulates its own expression, induction (whether physiological or genetic) requiring a functional *recA* allele. In this autoregulatory scheme, a low level of the *recA* protein is produced under normal physiological conditions (this constitutive expression would maintain the normal level of general recombination) but DNA damage promotes, in a manner as yet not understood, *recA*-dependent inactivation of the *lexA* repressor, thereby inducing protein X synthesis.

How may induction of SOS functions be mediated? It is known that *recA*-dependent induction of phage *lambda* from its prophage state is the result of a specific proteolytic reaction. The *lambda* repressor, *c*I (26 000 daltons in size), necessary for maintaining lysogeny, is split into two defined fragments (about 12 000 and 14 000 daltons). This cleavage reaction can be mimicked *in vitro* using defined components and is catalysed by purified *recA* protein: it requires ATP and polynucleotide. It is tempting to speculate that the common role for the *recA* gene product in the inducible SOS functions is one of a protease. On this model, *recA*-mediated regulation of gene expression would be indirect, limited to the endopeptidic cleavage of specific transcriptional control elements. The common requirements of ATP and polynucleotide for both *recA*-directed *c*I cleavage and unwinding of duplex DNA *in vitro* suggests rather that the *recA* protein is activated both as a specific endopeptidase and for its role in genetic exchange upon binding to single-strand DNA. That is, *recA* function requires direct interaction (alone or in association with other proteins) with DNA.

Different lesions within a single gene may lead to markedly different phenotypes. This is particularly true for *recA* and *lexA* (Table 6.3). At least three

Table 6.3 Alleles of *recA* and *lexA*

Gene	Allele	Phenotype
recA	tif⁻	thermal induction of SOS functions.
	zab⁻	intragenic suppression of Tif phenotype.
	lexB⁻	reduced UV sensitivity ; some recombination proficiency.
lexA	lexA⁻	non-inducibility of SOS functions.
	tsl⁻	thermal induction of some SOS functions.
	spr⁻ ᵃ	constitutive expression of SOS functions.

a   The *spr⁻* allele requires the *sfi⁻* (since renamed *sul⁻*, see DNA Repair, Section 6.3.1) background, otherwise the cell filaments to death. That is, light microscopic examination reveals long strings of cells that arise from lack of septation due to uncoupling of cell division from DNA replication (penicillin-induced filaments can be seen in Figure 1.17 (c)).

different classes of mutations map in *recA*, the structural gene for protein X, and are designated, *tif, zab* and *lexB*. The *tif-1* lesion makes synthesis of *recA* protein thermoinducible (the constitutive expression of SOS functions is also particularly evident at high temperature). *zab* mutations, on the other hand, suppress this Tif phenotype and are, therefore, an example of intragenic suppression. *lexB⁻* mutations are similar to *recA⁻* but differ in that they retain some proficiency for recombination, are less sensitive to UV and their DNA is not spontaneously degraded. Similarly, *lexA⁻* or *tsl⁻* strains, which carry base changes in the same gene, exhibit disparate properties. Whereas *lexA⁻* mutants appear to encode a 'super repressor' that renders *recA* expression totally non-inducible (without affecting recombination proficiency), *tsl⁻* lesions seem to result in a thermo-sensitive repressor, allowing induction (and, thus, SOS functions) upon switching to the non-permissive temperature. Finally, *spr⁻* mutants exhibit constitutive expression of SOS functions (equivalent to *tsl⁻* at high temperature, Table 6.3). It was, in fact, synthesis of protein X (*recA* gene product) in these allelic derivatives that led to the Gudas-Pardee model for *recA* regulation (Figure 6.17). (Gene regulation is discussed in Part IV.)

(ii) *The Involvement of Other Gene Products in Recombination.* A number of *rec* genes have been identified;[15] other gene products are involved in regulation of their expression (see Table 6.4). Mutations, for example, that appear to define a control element (molecular weight 25 000) for the *recA* gene map at the *lexA* locus. The *sbc* locus is so called since lesions at this site are necessary for intergenic suppression: *sbc* mutations suppress *recB⁻* or *recC⁻* alleles. The recombinational activity, at a low level in *recB⁻* or *recC⁻* strains (about 1 per cent the wild-type rate), is restored to near wild-type by mutations at *sbc*; these suppressors also alleviate the reduced viability associated with *recB⁻* or *recC⁻*. Strangely, whereas *sbcA⁻* results in increased levels of an exonuclease, Exo VIII (the *recE* gene product), the *sbcB⁻* suppressor mutation inactivates an exonuclease, Exo I (the *sbcB* gene product).

Nucleases appear to play an important role in genetic exchange, both for single-strand production and for DNA removal. It should, therefore, come as no surprise that of those characterised loci involved in general recombination, the majority encode either a specific nuclease or their controlling element (note that the central recombination component, the *recA* protein, appears to lack such an activity). Thus, *sbcB* is the structural gene for exonuclease I. Exonuclease V is coded for by *recB* and *recC* (often abbreviated to *recBC*). This enzyme binds to duplex DNA, unwinding it in an ATP-dependent reaction, cleaving large oligonucleotide fragments in the process (Exo V, in addition to its exonuclease activity on double-stranded DNA, cleaves single-stranded molecules, both exo- and endonucleolytically). Exonuclease VIII is encoded by the *recE* locus,

15.   In some cases, divergent terminology has developed for one and the same locus and only recently has their co-identity been ascertained. Thus, *recA⁻* or *recH⁻* mutations define the same genetic locus. Similarly, *recF⁻* and *uvrF⁻* lesions map at the same site (see Table 6.4).

expression of which is controlled by the *sbcA* product. Finally, it should not be forgotten that the bacterial polymerases possess exonuclease activities (both 3′ → 5′ and 5′ → 3′, Section 6.1.1 (a)). It now seems likely that both the nucleolytic and polymerase functions of these enzymes (or possibly of only the *polA* product) are used in strand exchange.

Other DNA-dependent ATPases,[16] in addition to the *recBC* nuclease, may be involved in strand displacement during recombination. As a class, they all consume energy in the form of ATP. An 180 000 dalton species, the largest monomeric polypeptide present in the bacterium, denatures duplex DNA adjacent to a gap. DNA gyrase (composed of the *gyrA* and *gyrB* gene products) and the *rep* protein should also be included in this list (discussed in Section 6.1.1 (b)). Certainly, DNA gyrase may play a vital role in D-loop formation (and branch migration, see below).

Strand displacement is likely to be mediated by specific proteins that bind to single-strand DNA. These *helix destabilising proteins* promote unwinding by stabilising melted regions (see Section 6.1.1 (b) for their role in replication). They, presumably, have a second function, that of protecting these single-strand regions from endonucleolytic attack.

In summary, genetic recombination appears to require numerous gene products (Table 6.4/Figure 6.18) in addition to components of the replicative apparatus and transcription elements (there is, for example, an *rpo* recombination pathway, see (bii) below). They have in common activities for DNA melting, stabilisation of the product and excision and synthesis of DNA, all functions necessary for the single-strand invasion of duplexes. Further gene products whose role is restricted to repair of DNA damage is discussed in Section 6.3.1.

(b) Recombination Pathways. Although inactivation of the *recA* product totally abolishes all general recombination (to a level 10^{-3}-10^{-6} of that of the wild-type), lesions at other loci implicated in homologous genetic exchange are less dramatic in their effect. This observation has led to the search for alternative pathways of recombination.

(i) *recA-mediated Pathways.* Indirect suppression of *recB⁻* or *recC⁻* mutations in the form of additional lesions in the *sbcA* or *sbcB* genes restores the recombination frequency to near wild-type (an increase from about 1 to 50 per cent of the normal rate). However, starting with the double mutant, *recBC⁻sbc⁻*, it is possible to isolate derivatives carrying lesions that map at regions distinct from *recA* (or the above loci) again showing a low-level recombination frequency.

16.   The term DNA-dependent ATPase refers to those enzymes that hydrolyse ATP in a DNA-mediated reaction. Included in this class, therefore, are ATPases that have nucleolytic or other activities. Thus, the *recBC* nuclease, Exo V, is a DNA-dependent ATPase in the first instance and, more specifically, an ATP-dependent DNase. DNA gyrase, on the other hand, also in this class, is, in particular, a topoisomerase.

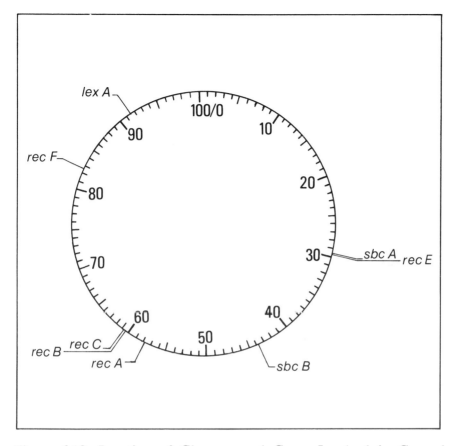

Figure 6.18: Location of Chromosomal Genes Involved in General Recombination
Properties of these genes and their allelic derivatives are given in Tables 6.3 and 6.4. Map locations from Bachmann and Low (Bachmann, B.J. and Low, K.B. (1980) *Microbiol. Rev., 44,* 1-56)

This method of searching for other genes in recombination — the isolation of recombinant defective strains of indirectly suppressed *recA⁺ recBC⁻* derivatives — has identified the *recF* (and *recK* and *recL*) loci. Mutations in any one of these *rec* genes brings down the yield of viable recombinants to about 0.05% of the wild-type level. Thus we have the sequence

recA⁺ ⟶ recBC⁻ (1%) → sbc⁻ (50%) ⟶ recF⁻/K⁻/L⁻ (0.05%)

where lesions at distinct sites either restore (solid arrows) or reduce (broken arrows) recombination ability. This might, at first sight, seem to imply that the *recF* product is necessary in these *sbc* strains. However, though *recF⁻* mutations dramatically block recombination with a *recBC⁻ sbc⁻* strain, they do not affect recombination in a *recBC⁺ sbcB⁺* (or *sbcB⁻*) background. Clark has proposed

Table 6.4[a] Chromosomal Genes Involved in Genetic Recombination[b]

Gene	Map Location (min)	Gene Product Name	Gene Product Size	Function and/or Mutant Phenotype
recA (recH)	58	protein X	38,000	DNA-dependent ATPase; specific endopeptidase.
recB[c]	60 ⎫	ExoV	270,000(N)	ATP-dependent 3'⎱→5' and 5'→3' ss and ds.
recC[c]	60 ⎬			exonuclease (DNA-dependent ATPase); ss-endonuclease.
recE[d]	30	ExoVIII	nk	recE-specific recombination pathway.
recF (uvrF)	82	nd	nk	recombinant-deficient in recBC⁻ sbcB⁻ background.
recG	(82)[e]	nd	nk	recombinant-deficient in recBC⁻ sbcB⁻ background.
lexA	91	nd	25,000	controlling gene for recA expression.
sbcA[d]	30	nd	nk	controlling gene for recE expression.
sbcB	44	ExoI	nk	suppressor of recBC⁻.

a   Abbreviations: nd, not designated; nk, not known; N, native molecular weight; ss and ds, single strand and double strand, respectively.

b   Adapted from Hanawalt *et al.* (1979).

c   Mutations affecting ExoV that map at 60 min fall into two complementation groups (see *Cis-Trans* Complementation Section 7.1.1 (a)) suggesting two gene products. A further component of the nuclease is apparently necessary for function since lesions at, as yet, unmapped loci effect the enzyme.

d   *recE* and *sbcA* are encoded by the Rac prophage (Section 5.1.3 (b)).

e   Approximate map location.

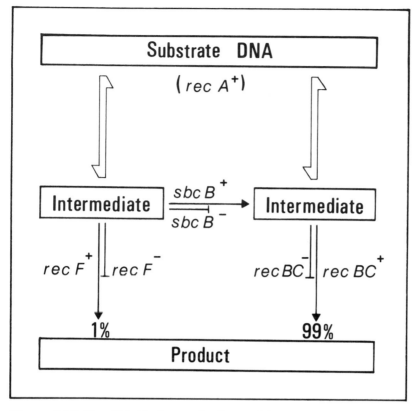

Figure 6.19: Possible Pathways in General Recombination
A scheme for the *recF* (left) and *recBC* (right) recombination pathways is show (solid arrows, effective pathways; incomplete arrows, ineffective pathways). With *sbcB* functional, recombination is shunted through the efficient *recBC* pathway. A *sbcB+ recBC−* strain shows, therefore, only a low level recombination proficiency (about 1 per cent the wild-type rate), representing flux through the *recF* pathway. When *sbcB* is non-functional, the reaction can proceed via the *recF* or *recBC* pathway. *recA* is central to both pathways and is likely to function early in these processes. After Horii and Clark (Horii, Z. and Clark, A.J. (1973) *J. Mol. Biol., 80,* 327-34)

that there are (at least) two distinct recombination pathways, the *recBC* pathway and the *recF* pathway (defined by lesions at *recBC* or *recF/K/L*, respectively, Figure 6.19).[17] On this model, *sbc* is indirectly rather than directly involved in the *recBC* pathway. That this is the major pathway of recombination in bacterial recombination is exemplified by the 100-fold reduction in recombinants in *recBC−* strains. The low level *recF* pathway, on the other hand, is only operational when Exo I is absent (for instance, through *sbcB−* lesions).

---

17. A further pathway, the *recE* pathway, is non-functional in a wild-type background since exonuclease VIII is blocked by *sbcA+*. It does not require the *recA+* product in the case of phage *lambda* recombination. *recE* and *sbcA* are encoded by the lambdoid prophage Rac (Table 6.4).

Moreover, it is detectable only when the major *recBC* pathway is inactive since it is so inefficient.

The actual position of *recA* in this scheme is still somewhat contentious: it is presumably common to both pathways since *recA*⁻ lesions block all general recombination (but see footnote 17 on page 281). Through analysis of recombinants produced during conjugation between two genetically marked strains carrying known mutations in *rec* genes, it has been established that the *recA* protein must act before the *recBC* nuclease.

(ii) *The rpo Pathway.* The importance of single-stranded regions in genetic recombination would indicate a role for proteins possessing helix destabilising properties. The fact that the DNA-dependent RNA polymerase is a sequence-specific melting protein (Section 2.1.2 (ai)) might suggest involvement of the transcription complex in recombination. Early studies on transcriptionally induced and uninduced cultures failed to show a difference in recombination frequency in *E. coli*. However, it has recently been shown that the formation of recombinant *lambda* molecules is dependent upon a rifampicin-sensitive step when the other recombination pathways are not available (see (i), above). The localisation of crossovers in regions on the viral genome known to be actively transcribed is in keeping with a scheme in which RNA polymerase mediates exchange through helix destabilisation. The physiological significance of this recombination pathway, dubbed the *rpo pathway*, requires clarification.

(c) The Mechanism of General Recombination. Any recombination event requires the breaking of phosphodiester bonds and their subsequent rejoining to form a hybrid product. In this *heteroduplex*, one stretch of single-stranded DNA would be covalently linked to a second sequence, both derived from distinct DNA regions (these two regions may be intra- or inter-chromosomal). That is, it is hybrid at the level of the individual polynucleotide chains. Certainly, there is genetic and biochemical evidence for the formation of such a heteroduplex joint. It is not known what actually initiates DNA exchange nor is it understood how the strands are assimilated to form the joint. Nevertheless, the process of general recombination can be broken down into only a few events, strand breakage, strand pairing, strand invasion/assimilation, chiasma formation (crossover), breakage and reunion, and, finally, mismatch repair (processing) of the resultant joint. These steps are described below in relation to four models proposed for interchromosomal recombination, those of Holliday, Whitehouse, Hotchkiss and Meselson-Radding (see Figure 6.26).

(i) *Strand Breakage.* An initial nucleolytic event (and breaking of hydrogen bonds) is necessary to produce a single-strand molecule. It is this stretch of DNA carrying either a 3'- or 5'- terminal hydroxyl that is 'active' in recombination (note that only the former is a potential primer for DNA extension, Section 6.1.1 (a)). Exchange of DNA between duplexes requires effectively four 'cuts' and, thus, four joins (see Figure 6.24: each breakage and reunion event is represented by a heteroduplex joint), two of which are generally postulated to occur near, or at the end, of the recombination process (see (iii), (iv) and (v), below). Of the initial incision events, it is not known whether two cuts occur simultaneously (and if so, why) or whether it is the production of one that later induces a second event (Figures 6.20 and 6.21).

When there are two initial cuts, they may occur on like strands — say, the 5' → 3' chains — of the two chromosomes participating in recombination (Holliday model, Figure 6.20 (ai)) or on strands of opposite polarity (Whitehouse model, Figure 6.20 (aii)). It is, in fact, hard to imagine why two breaks should occur simultaneously in non-sister strands, whether co-parallel or not.[18] Rather, a hypothesis in which the product of a random nucleolytic event 'invades' a homologous region of a second chromosome to cause the second cut, appears more reasonable (see Hotchkiss and Meselson-Radding models, Figure 6.21 (b)). Note also, an initial cut may not always be necessary: a gap or nick present on a DNA duplex (presumably present from incomplete ligation of Okazaki fragments) may be directly active in strand invasion or serve as a primer for DNA extension and, thus, allow strand displacement.

At what site does recombination occur? It has been proposed that *inverted repeats* or *palindromes* are targets for cleavage and lead to initiation of strand exchange. Certain DNA sequences, termed *chi*, naturally present in *E. coli* (and occurring on the viral genome of phage *lambda* after mutagenesis) actually enhance recombination, both at regions neighbouring the sequence and at a distance. This recombinogenic sequence appears to be an octomer (5'GCTGGTGG 3'), present in wild-type *E. coli* every 5-10 kb — a role as a universal recombinator has been proposed. Chi recombination is a specific process as it proceeds only via the *recBC* pathway (and, thus, is solely dependent on both *recA* and *recBC* function, Section 6.2 (bi)). It would seem, in general, that the breakage event which produces a donor strand may be random but any subsequent cleavage reaction is dictated by pairing (see D-loop Formation, below).

(ii) *Strand Pairing.* Formation of a heteroduplex joint requires pairing of non-sister strands. The actual order of events, whether strand breakage precedes pairing or *vice versa*, is still unclear. Although it is possible to build three-dimensional physical models of 4-stranded helices, the stimulation of recombination by gaps and single-strand tails — single strands with free (unpaired) ends — suggests an initial nucleolytic event (or a single-stranded replicative intermediate active in recombination). Helix destabilising proteins are reported to promote pairing at high concentration of counterions (and melting at low ionic

18. *Non-sister* strands refers to strands of separate duplexes: they may be of like polarity or antiparallel in nature. The term *daughter* is used to refer to the newly synthesised chain.

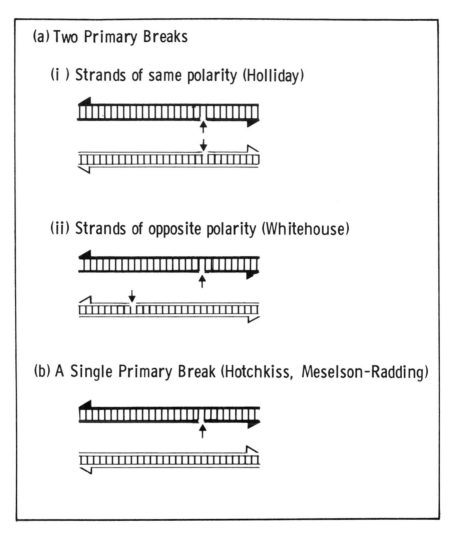

(a) Two Primary Breaks

(i) Strands of same polarity (Holliday)

(ii) Strands of opposite polarity (Whitehouse)

(b) A Single Primary Break (Hotchkiss, Meselson-Radding)

Figure 6.20: Stages in General Recombination; Strand Breakage
Exchange of genetic material is likely to be initiated by a single-stranded tail(s), generated either by two (a) or one (b) primary break(s). In the former case, the two breaks may be on strands of like polarity (Holliday model) or be anti-parallel in nature (Whitehouse scheme). The Hotchkiss and Meselson-Radding models propose an initial single-strand break, leading to a second co-parallel event through D-loop formation (Figure 6.21)

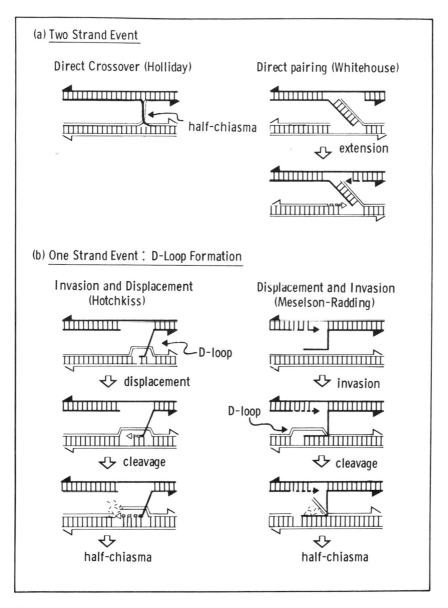

Figure 6.21: Stages in General Recombination; Strand Pairing and Invasion
When two strands are 'cut' at the onset of recombination (a), they may either cross over to pair with non-sister chains (breaks co-parallel, Holliday model) or form a complementary bridge (breaks anti-parallel, Whitehouse scheme). In the case of a one-strand event (b), a single-stranded tail invades a second duplex molecule, thereby generating a second, co-parallel break. Whereas the Hotchkiss model postulates strand cleavage by virtue of extension of the invading chain, rupture in the Meselson-Radding hypothesis is the result of D-loop formation (see Figure 6.16), itself, and DNA synthesis is responsible for 'driving' strand displacement

strength), possibly by removing intrastrand base pairs.

Following two (simultaneous) incision events, and strand unravelling, the Holliday model proposes that the two single strands pair with homologous regions on intact non-sister molecules and are ligated (Figure 6.21 (a)). This pairing creates a *half-chiasma*, a crossover of two rather than four strands (discussed in (iv) below). In this way, Holliday explained reciprocal single-chain exchange (upon breakage and reunion, see (v)). In the Whitehouse model, since cleavage is at anti-parallel strands, their non-sister tails may pair together (rather than pairing with non-sister intact strands) when the cuts are at non-homologous sites (Figure 6.21 (a)). The two most recent recombination models considered in this text, those of Hotchkiss and Meselson-Radding, make use of pairing and strand invasion of a non-sister duplex (that is, *D-loop* formation) for the production of the second nick (Figure 6.21 (b)).

(iii) *Strand Invasion.* What type of molecular event is likely to break a single-strand molecule of DNA? Clearly, distortion of a DNA duplex may create sufficient strain to cleave internucleotide bonds. When an invading strand pairs with a non-sister complementary region, the second chain of the DNA duplex is displaced. Structures termed *displacement loops (D-loops)*, putative inter-mediates in recombination, have been visualised in the electron microscope; they appear to consist of a *triple-strand joint* caused by invasion of a duplex by a single strand (see Figure 6.21 (b)). D-loops seem to be a precursor to the heteroduplex joint since cleavage of the unpaired chain, followed by ligation, would create this hybrid structure.

It is postulated that either direct strand invasion triggers a second nucleolytic event, perhaps the unpaired region becomes a target for endonuclease attack (Meselson-Radding model), or that a second break occurs only following DNA synthesis (Hotchkiss model, see Figures 6.21 (b) and 6.26). For synthesis to be along the heterologous template, the latter hypothesis requires, of course, that the invading strand carries a 3'-OH. Although the Meselson-Radding model does not require DNA synthesis for strand breakage, it proposes that DNA extension is responsible for strand displacement to produce the D-loop intermediate (Figure 6.21 (b)). Thus, both recombination models include an important (albeit different) role for DNA replication in genetic exchange. The fact that certain *dna* lesions cause a decrease in recombination frequencies supports the involvement of the replicative apparatus. Moreover, there is evidence for physical association of the *recBC* nuclease with DNA polymerase I (the inviability of *polA⁻recBC⁻* double mutants has also been mentioned above, Section 6.1.1 (a)), although this may reflect an enzyme complex active in essential DNA repair (see Section 6.3.1).

The formation of D-loops, themselves, may be aided by helix-destabilising elements since these proteins promote strand displacement in replication. Certainly, helix-destabilising proteins protect single-strand regions from endo-nucleolytic attack, and D-loops are known to be susceptible to the *recBC* nuclease, Exo V. Uptake of single-strand fragments by homologous duplex DNA occurs *in vitro* when the recipient molecule is superhelical, suggesting a role for DNA gyrase — this enzyme introduces negative superhelical turns in the presence of ATP (the bacterial topoisomerase was initially isolated for its ability to stimulate site-specific recombination *in vitro*, see Phage *Lambda* Integrative Recombination, Section 6.2.2 (ai)). The reaction with superhelical DNA is catalysed by purified *recA* protein. The single-strand binding protein (and phage T4 gene 32 product) stimulates this strand assimilation reaction suggesting a role for helix destabilising elements in general recombination *in vivo*; genetic evidence supports this contention. DNA-dependent ATPases other than DNA gyrase and the *recBC* nuclease may also be involved in strand displacement. Uptake of the single strand into the recipient duplex — termed *strand invasion* (or *assimilation*) — may, thus, proceed through coupling of DNA chain extension with excision or it may be promoted by DNA gyrase-mediated superhelicity, perhaps in the absence of replication.

(iv) *Chiasma Formation.* With the occurrence of two single-strand breaks, a further two incision events (or their equivalent) are required for recombination. The most obvious mechanism to allow DNA exchange would be the crossover of genetic material (followed by breakage and rejoining, see (v), below). This type of recombination intermediate, a *chiasma*, has been observed in light micro-scopic examination of dividing eukaryotic cells: non-sister chromatids form cross-shaped structures.

When the two breaks are simultaneous, and on parallel strands (as postulated in the Holliday model), partial melting and rejoining — albeit with chains from non-sister chromosomes — will result in a crossover that consists of just two single strands (Figure 6.22). This two-strand crossover is termed a *half-chiasma* in that a (full) chiasma consists of both strands of the two DNA duplexes. Molecular models show that, in the limit, a crossover may be a single phosphodiester bond. The half-chiasma of Holliday is generally accepted as an intermediate in prokaryotic recombination (see Hotchkiss and Meselson-Radding models, Figure 6.26), particularly because the predicted *figure eight* molecules, formed in a single-strand crossover between two covalently closed circular DNA duplexes, has been found in electromicrographs (see Figure 6.16). Moreover, a functional *recA* gene is required for their formation, the action of the protein being possibly, in this instance, to promote D-loop formation. This early role for the *recA* function in general recombination is in keeping with genetic studies of conjugal crosses between pairs of *recA⁻* and *recBC⁻* strains (see Recombination Pathways, above).

As a structure, the cross-bridged intermediate is important since helix swivel may occur about the bridge (Figure 6.22). This *isomerisation* generates structures in which the two intact rather than broken strands are involved in a crossover. Breakage and reunion, thus, allows recombination of flanking regions (discussed in (v), below). As an intermediate in genetic exchange, the crossover

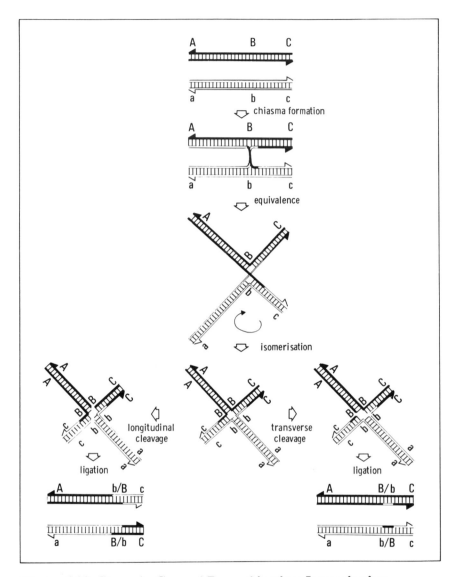

Figure 6.22: Stages in General Recombination; Isomerisation

Strand breakage, pairing and invasion on the Holliday-like models of Hotchkiss and Meselson-Radding generate a half-chiasma, a crossover of two rather than four strands. Rotation about this two strand crossover converts it into a four-fold junction. Studies with molecular models indicate that hydrogen bonds need not be broken in this isomerisation step. In the planar structure, where the four termini are 'pulled out', breakage and reunion may be regarded as occurring in either one axis or the other (North-South or East-West) to produce two different recombinants (equivalent to further rotation of the planar structure to produce a second Holliday intermediate — the two intact strands, rather than the heteroduplex molecules, involved in the crossover — followed by cleavage and ligation, see also Figure 6.24). Flanking markers *(A/a, C/c)* are exchanged following longitudinal but not transverse cleavage is also important because it is free to migrate in both directions, the position of the breakage and reunion event dictating the type of recombinant formed (Figure 6.23). *Branch migration* extends the region of heteroduplex DNA symmetrically on both chromosomes: only on the Meselson-Radding model is there asymmetric strand transfer in the first instance (Figures 6.21 (b) and 6.26). Although migration requires, at first sight, little if any energy, as hydrogen bonds separated in the process may be resealed in the aftermath of the half chiasma, it now seems likely that migration is enzymatically driven (either by DNA replication or by a topoisomerase).

Holliday was the first to propose the half-chiasma as an intermediate in recombination. The Hotchkiss and Meselson-Radding models also include a half-chiasma as a prelude to genetic exchange (Figure 6.21); they rely upon DNA replication and removal of parental DNA. In the first, both the invading strand and subsequent cross-bridged species are used as primers for DNA synthesis. In the second recombination hypothesis, DNA extension from a single polynucleotide chain is responsible for strand displacement to allow D-loop formation and, thus, a crossover. Although the Whitehouse model proposes a crossover of double rather than single-stranded DNA, no actual breakage and reunion occurs at this complementary bridge (see Figure 6.26).

(v) *Breakage and Reunion.* Breakage of internucleotide bonds and subsequent ligation of non-sister strands would appear to complete the process of exchange of genetic material (see Figure 6.26). The two Holliday-like models, those of Hotchkiss and Meselson-Radding, both require the same final step (although their products may be different, see below) since, in each case, a half-chiasma is postulated as an intermediate in recombination. On the Whitehouse model, only an incipient chiasma is present at the terminal stages in recombination.

The segment of DNA transferred depends both on the initial cut(s) and on the final breakage-reunion reaction. Assuming that the initial event is random (but see Chi, above), it seems likely that the location of the half-chiasma is crucial in dictating the extent of genetic exchange and the types of products formed. When, for example, the crossover event occurs close to the initial breakage point (Figure 6.23), only a small region of DNA is exchanged (DNA exchanged according to the Whitehouse scheme is either bordered by the position of the asymmetric cuts or is random, dictated by the degree of DNA loss, see Figure 6.26). In addition, the type of recombinant depends upon whether breakage involves the strands originally cut or the other two (equivalent to exchange between either one or two pairs of duplexes, respectively; Figures 6.22 and 6.24). That is, when only a single exchange takes place (in the absence of an isomerisation event), there is no recombination of flanking markers: separation only occurs for exchange between two duplexes.

(vi) *Mismatch Repair.* The initial product of exchange and ligation is a heteroduplex molecule in which single-strand sections are derived from non-

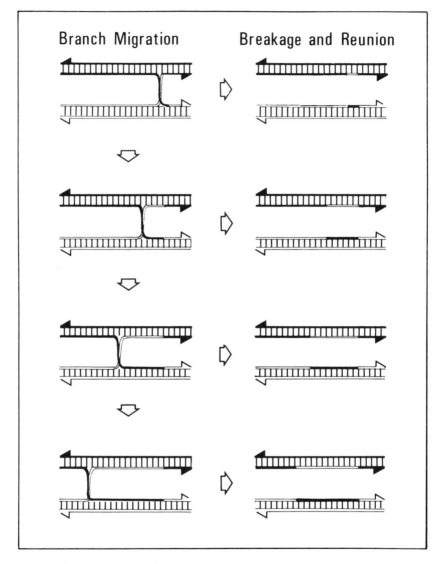

Figure 6.23: Stages in General Recombination; Branch Migration
Chiasma movement dictates the extent of DNA exchanged (note that the cross-bridged structure may involve either the heteroduplex molecules or the intact strands, depending upon isomerisation, see Figures 6.22 and 6.24)

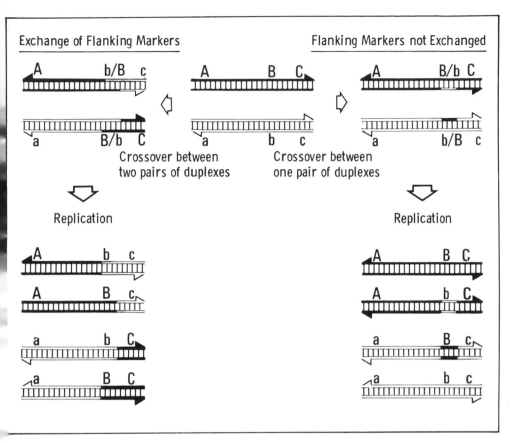

Figure 6.24: Heteroduplex Intermediates in General Recombination
Crossover between two pairs of duplexes takes place when, following isomerisation (Figure 6.22), cleavage of the intact rather than hybrid strands occurs (in the latter case, there is only crossover between one pair of duplex). Replication of the recombinant products yields homoduplexes in which the flanking markers *(A/a, C/c)* are separated (molecules that carry *A, c* and *a, C* rather than *A, C* and *a, c*)

sister chains. Although the chromosomes involved in general recombination will show extensive homology, there may be unpaired regions in the heteroduplex products that arise from slight genetic differences (Figure 6.25). These non-complementary sequences are, presumably, unstable and prone to mismatch repair. The process seems to consist of excision of the mismatched bases and subsequent repair through DNA extension; one strand, frequently the recipient (note that only the donor strand has a free end), appears to act as template for DNA synthesis. Correction of mismatches, in which tracts of up to 3000 nucleotides in length are removed, requires a functional *uvrD* gene product suggesting a mechanism similar to *excision repair* (DNA repair is discussed in Section 6.3). It is thought that the formation and repair of heteroduplex joints are responsible for the non-reciprocal recombination of genetic markers (Figure 6.25).

It is worthwhile at this stage making a comparison of the four recombination models discussed above (Table 6.5, Figure 6.26). Both the Hotchkiss and Meselson-Radding hypotheses are similar to the archetypal theory of Holliday. They propose, in fact, a mechanism for the second incision event based upon D-loop formation. In both these models, the second cut is co-parallel with the first as initially proposed by Holliday. Moreover, both postulate that DNA synthesis (and removal) is an intrinsic part of genetic recombination. They differ mechanistically in four main aspects. Firstly, whereas the invading strand carries the 3'-OH in the Hotchkiss model and may, thus, serve as a primer for DNA extension, Meselson-Radding propose 5'-OH invasion, with DNA synthesis from the residual 3'-OH terminus displacing the invading strand. Secondly, DNA synthesis occurs on both duplexes according to Hotchkiss but only on one, that supplying the invading strand, in the hypothesis of Meselson and Radding. Certainly, replication and recombination may be very closely allied, particularly in phage-phage crosses. Thirdly, if a single enzyme complex is responsible for concomitant DNA extension and nucleolytic degradation — the bacterial DNA polymerases are well fitted for this task (Section 6.1.1 (a)) — it must, presumably, 'straddle' both duplexes in the Meselson-Radding model. Finally, there is definite asymmetric strand transfer in the scheme of Meselson and Radding, which helps to explain non-reciprocal recombination events. At present there is little to choose between these two models.

The Whitehouse model is quite distinct from those of Holliday, Hotchkiss and Meselson-Radding since it involves double nucleolytic attack of anti-parallel strands (Table 6.5/Figure 6.26). A full but incipient chiasma is formed. Note that DNA synthesis and nucleolytic degradation cannot occur simultaneously in this model as it is template DNA that is removed. There is no further breakage and reunion (other than the initial cuts and the ligation step to seal loose ends). Moreover, a number of more novel enzyme activities are required.

6.2.2 Non-Homologous Recombination

Previous discussion on genetic recombination has centred on reactions between

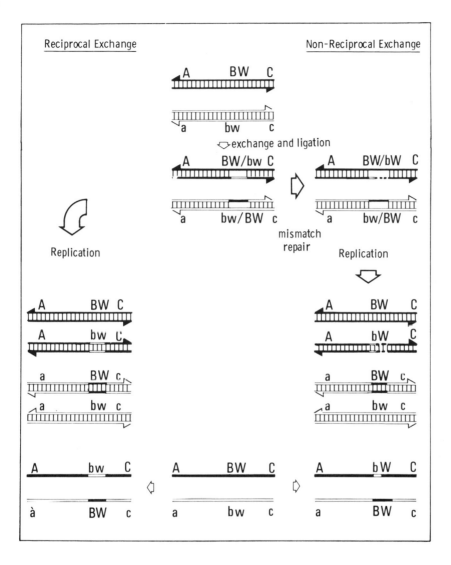

Figure 6.25: Resolution of a DNA Heteroduplex; Reciprocal and Non-Reciprocal Exchange

A possible route for mismatch repair in which part of the donor strand is excised (top). Repair synthesis on the recipient chain (dotted line) results in only one of the two markers (*BW/bW* rather than *BW/bw*) being exchanged. In summary (bottom), whether non-reciprocal or reciprocal genetic exchange takes place is dictated by mismatched repair

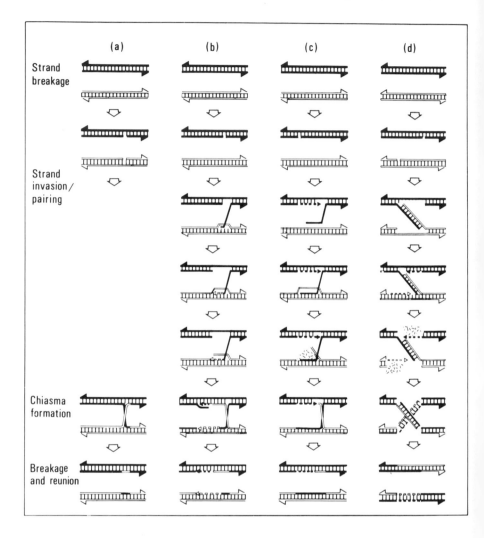

Figure 6.26: Models for General Recombination

The models of Holliday (a), Hotchkiss (b), Meselson-Radding (c) and Whitehouse (d) are compared (see also Table 6.5). The Holliday-type models involving a half-chiasma, those of Hotchkiss and Meselson-Radding, invoke D-loop formation for the second co-parallel breakage event. According to Hotchkiss, extension of the invading tail is responsible for strand scission. On the Meselson-Radding model, DNA synthesis on the donor duplex is necessary for strand displacement and, thus for strand invasion, and once formed, the D-loop becomes susceptible to, say, nucleolytic attack. The single-stranded tail generated by the second breakage event can then participate in a cross-bridged intermediate (and can be extended on the Hotchkiss scheme). Whether flanking markers are exchanged or not depends upon an isomerisation step prior to the final breakage and reunion (see Figure 6.22). The two anti-parallel cut strands of the Whitehouse scheme pair together to form a complementary bridge. A second bridge is created following replication of the remaining single-stranded regions (and destruction of these templates after use), to give a full but incipient chiasma. Genetic recombination is, thus, completed in all instances by a ligation step

Table 6.5 A Comparison of Four Models for General Recombination[a]

Stage in Recombination	Model Characteristics			
	Holliday	Hotchkiss	Meselson–Radding	Whitehouse
Strand breakage				
number of strands in initial event	2	1	1	2
polarity of strands in initial event	co-parallel	–	–	anti-parallel
causes of secondary event	–	DNA extension	endonuclease attack	–
polarity of strands in secondary event	–	co-parallel	co-parallel	–
Strand invasion/pairing				
requirement for DNA synthesis	–	invasion	displacement	chiasma
extent of DNA synthesis	–	2 strands	1 strand	2 strands
initiation of strand transfer	symmetric	symmetric	asymmetric	symmetric
Chiasma formation				
type	half	half	half	incipient
potential for isomerisation	✓	✓	✓	–
Breakage and reunion				
resolution at crossover	✓	✓	✓	–

a   See Figure 6.26.

DNA regions of extensive homology. These are *recA*-promoted events and are, consequently, not apparent in the absence of this essential general recombination component. Other substrates, however, are available for genetic exchange that neither show a requirement for marked sequence similarity nor for the *recA* protein. *Non-homologous* recombination is, therefore, the sole type of exchange process occurring in *E. coli* in the absence of a functional *recA* allele.

Although *recA*-independent events characteristically involve DNA regions of only little sequence homology, it should not be thought that the difference between homologous and non-homologous recombination is just one of degree (the extent of DNA available for pairing). A mechanistic similarity between these two processes is hard to reconcile with the lack of dependence on the *recA* protein since this gene product has been shown to function early in genetic exchange. Non-homologous recombination requires few cellular components — proteins, moreover, that are not necessary for general events. In short, non-homologous recombination is a unique recombinational phenomenon distinct from general recombination: it is not a 'variation on a theme'.

Non-homologous events may be classified according to whether recombination takes place at specific sites or at random. Specificity in *site-specifc recombination* is apparently maintained through interaction of a particular enzyme system with a defined DNA sequence. In contrast, the chance occurrence of any two identical runs of nucleotides on separate genetic elements, be it just a few base pairs, may serve as a substrate for *illegitimate recombination.*

(a) Site-Specific Recombination. This category includes any recombination event that takes place in the absence of the *recA* protein but is limited to particular DNA sequences present on one or both of the genetic elements involved. Site-specific recombination refers, in general, to the insertion of foreign DNA into chromosomal material. The four main types of *DNA insertion elements*, bacteriophages, plasmids, insertion sequences (IS) and transposons (Tn), have in common the ability to 'jump' from one site to another, independent of the presence of a functional *recA* product; the site of insertion may be random in some cases. It is now apparent that certain of these *transposable genetic elements* share a recombination mechanism that differs markedly from the phage *lambda* integrative/excisive system: whereas insertion of phage *lambda* requires two DNA sequences, a functional bacterial and phage attachment site ($att^\lambda B$ and $att^\lambda P$, respectively), insertion of phage Mu, IS and Tn elements seems to involve only one specific site, that on the target DNA concerned. Moreover, transposition appears to occur without loss of the original element. It should be emphasised, however, that in all these insertion events, the entire genetic content of the two elements undergoing recombination is retained (DNA is exchanged in general recombination and may be lost in the absence of a reciprocal process). Such events may, thus, be called *integrative* (or additive) *recombination.*

(i) *Bacteriophages.* Phages that lysogenise by integrative recombination normally have specific attachment site(s) on the bacterial chromosome (bacteriophage Mu is a notable exception in that insertion is random and not site-specific; see (ii) below). Coliphage *lambda*, for example, integrates at high efficiency (about 90 per cent) into $att^\lambda B$ (situated between the *gal* and *bio* operons, Section 5.2.2), flanked by a direct repeat of the common core sequence. However, when this region is deleted (a situation made readily feasible by the absence of essential bacterial genes close to $att^\lambda B$), *lambda* integrates at a low rate, two to three orders of magnitude reduced, into other sequences. Note that these *secondary attachment sites* exist naturally on the bacterial chromosome and may, thus, frequently fall within a coding region: the presence of foreign DNA has the potential of both causing gene inactivation and exerting a polar effect on expression of genes of the same operon downstream from the site of insertion (see Mutagenesis by DNA Insertion elements, Section 7.1.2 (c)).

Since integration is still site-specific in $att^\lambda B$ deletion strains — the viral genome is inserted only at particular regions in the absence of the primary attachment site — it has been proposed that these secondary sites show some, albeit reduced, homology with the primary $att^\lambda B$ sequence. The components of the secondary site have been designated $\Delta O \Delta'$ (in analogy with BOB'), where Δ and Δ' represent bacterial sequences to the left and right of the common core, O. The secondary *attL* and *attR* hybrid attachment sites formed upon integration consist, therefore, of $\Delta OP'$ and $PO\Delta'$, respectively (equivalent to BOP' and POB'). Recent evidence suggests that only the components of the phage attachment site, P, O and P', dictate the specificity of *int*-dependent site-specific recombination, supporting the above description (where only the core or a partial sequence, is common).

Independent of whether phage *lambda* inserts at the primary site, $att^\lambda B$, or a secondary site, integrative recombination occurs in the absence of the *recA* protein. It has, however, an absolute requirement for the phage-coded *int* gene product, *integrase*, as exemplified by the abolition of site-specific insertion in λint^- mutants (other, as yet unidentified, host genes are required). It is, presumably, through the common core region, O, that site-specific recombination occurs. Note that this 15 bp sequence is highly A + T-rich (about 80 per cent, see Section 5.2.2), which may reduce helix stability to facilitate local denaturation. This region will, therefore, tend to exist as single strands when present in a supercoiled molecule. In fact, site-specific recombination *in vitro* requires superhelical *lambda* DNA — at least $att^\lambda P$ must be located on a negatively supercoiled circle of DNA (it was this dependence that led to the isolation of Topo II, DNA gyrase); supertwisting is also required *in vivo*. It has now been shown that the *int* gene product has nicking-closing activity, in other words, that it is a topoisomerase.

Additive recombination of *lambda* is reversible. Excision of the phage from the bacterial chromosome requires a second phage protein, the *xis* gene product ('excisase'), in addition to the *int* protein. The DNA substrate, however, need not be in a supercoiled state for excisive recombination. It is possible that excision requires less DNA unwinding either through the action of the *xis* gene product or because of the nature of the hybrid *att* sites, *attL* and *attR*.

It is interesting to note that the essential elements of phage *lambda* site-specific recombination, *att$^\lambda$P*, *int* and *xis*, are all clustered on a small region of the viral genome (about 1500 bp), a cluster similar in size to the IS elements (discussed in (ii), below). Moreover, the OP' region of the *lambda* attachment site shows a 10 bp perfect match with an arm of one such element (IS1). These similarities may be fortuitous, although a phylogenetic relationship cannot be discounted, since the recombination mechanisms of *lambda* and transposable genetic elements appear distinct.

That there is genetic conservation in site-specific recombination is exemplified, in the case of *lambda*, by the reverse *(int + xis)*-mediated reaction in which *lambda* excision restores the exact attachment sequence (and the original circular phage genome). Though both the integrative and excisive processes are highly accurate, errors do occasionally occur (at a frequency of about 10^{-5}) resulting in the formation of specialised transducing phages carrying bacterial sequences bordering the attachment site (this *recA*-independent process, an example of *illegitimate recombination*, is discussed below). The fidelity of excision is reduced for secondary attachment sites and many bacterial survivors carry deletions (the extent of DNA loss being dictated by the location of essential genes relative to the point of insertion). This may be ascribed, presumably, to a lack of homology between the phage 15 bp core and that in the secondary site, a difference that results in the formation of two dissimilar sequences flanking the integrated phage, rather than a direct repeat of the common core (Figure 6.27).

Secondary-site insertion may, thus, be employed both for directed mutagenesis and for the isolation of specialised transducing phages carrying bacterial genes other than those adjacent to the primary attachment site. This aspect of bacteriophage technology is discussed in Section 7.1.2 (c) (see Mutagenesis by DNA Insertion Elements).

(ii) *Insertion Sequences and Transposons.* These genetic elements exhibit *recA*-independent translocation between DNA sequences of no apparent homology. (Note that *translocation* and *transposition* are used synonymously.) They have defined nucleotide sequences (Table 6.6). *Insertion sequences* are short pieces of DNA (about 700-1500 bp) which may have terminal inverse repeats (Figure 6.28). They may carry transcriptional terminators or promoters (see below) but no gene product has been unambiguously identified (the genetic evidence for an IS1-encoded 'recombinase' enzyme is supported by sequence data for this insertion element — it potentially codes for a protein as large as 96 amino acids — and a protein of 315 residues could be encoded in the IS2 sequence in orientation II). Thus, IS elements have no known function (other than that of translocation).

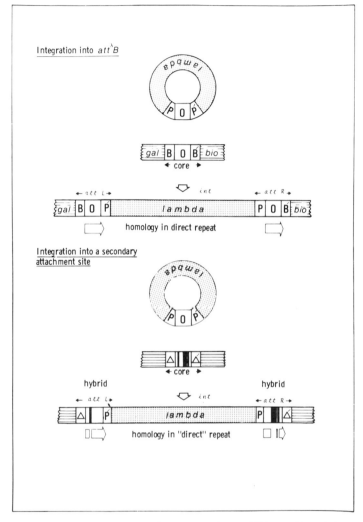

Figure 6.27: Secondary-Site Insertion of Phage *Lambda*
The same phage structural elements, POP', are employed whether *lambda*
integrates into the primary attachment site or a secondary one (consisting of BOB'
and ΔOΔ', respectively). Moreover, in both cases, the reaction is catalysed by the
phage-coded *int* gene product (and, as yet, unspecified host proteins). Sequencing of
the hybrid attachment sites, *attL* and *attR*, formed upon secondary-site insertion
has shown that the common core region of ΔOΔ' is, in fact, different from that in
att$^\lambda$*P* or *att*$^\lambda$*B*. Although the core-like region of secondary *att* sites are not
completely homologous with the 15 bp *att*$^\lambda$*B/P* O sequence, they retain both the
A·T-rich characteristics and an interrupted homology of a minimum of 8 bp (these
differences are represented by *three black lines of varying thickness* in the bottom
diagram). This sequence variation in secondary-site O regions generates only a
'partial' direct repeat upon secondary-site integration, which may explain the lack of
accurate excision from such sites

Table 6.6[a] Properties of Some Transposable Genetic Elements[b]

Genetic Element	Size (bp)	Terminal Repetition		Polarity [d]	Drug Resistance [e]
		Element	Target [c]		
Insertion sequences					
IS1	768	30 bp inverted	9 bp	both orientations	-
IS2	1,327	32 bp inverted	5 bp	orientation I	-
IS3	1,400	32 bp inverted	3 or 4 bp	one orientation	-
Transposons [f]					
Tn1 (Tn2, Tn3)	4,957	38 bp inverted	5 bp	one orientation	ampicillin
Tn5	5,400	1,450 bp inverted	9 bp	both orientations	kanamycin
Tn9	2,638	IS1 direct	9 bp	both orientations	chloramphenicol
Tn10	9,300	1,400 inverted	9 bp	both orientations	tetracycline
Bacteriophages					
Mu	38,000	11 bp inverted	5 bp	both orientations	-
λ	48,600	-	15 bp	both orientations	-

a   Abbreviation: bp, base pairs.

b   After Calos and Miller (1980).

c   Refers to direct repetitions flanking inserted element. The 15 bp terminal repeat associated with prophage *lambda* represents the common core O present in both $att^\lambda B$ and $att^\lambda P$.

d   Insertion of foreign DNA into a coding region generally inhibits expression of distal genes of the same operon. This is only true for IS2, IS3 and Tn1 in one orientation. Moreover, whereas IS2 is polar in one orientation (termed orientation I), there is a functional promoter in orientation II.

e   Refers to antibiotic resistance encoded by transposon.

f   Tn1, Tn2 and Tn3 are very similar (the complete DNA sequence of Tn3 is known).

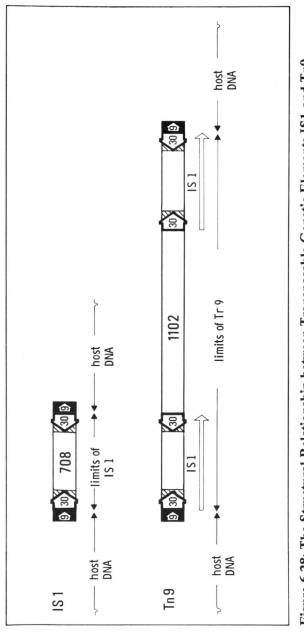

Figure 6.28: The Structural Relationship between Transposable Genetic Elements IS1 and Tn9

The structure of the terminal regions of IS1 and Tn9 are shown (not to scale). Arrows represent inverse repeats (hatched) and flanking direct repeats (black). The 30 bp invert repeat is constant and is considered part of the transposable genetic element (this repeat is not, in fact, perfect). IS1 (and Tn9) is found bordered by a direct repeat of 9 bp, the sequence of which depends upon the site of insertion. The transposon Tn9 consists of two direct repeats of IS1 flanking a region of 1102 bp (at least 855 bp are involved in the expression of choramphenicol acetyl transferase, responsible for imparting resistance against the translation-specific antibiotic). It, therefore, also carries a 9 bp repeat at either end of the inserted element. See Structural Features of Phage Mu, Figure 5.29

Transposons, on the other hand, are considerably larger in size (say, between 4 and 21 kb, Table 6.6) and consist of coding regions that confer resistance to one or more antibiotic — tetracycline in the case of Tn10 — flanked, invariably, by insertion sequences (note, therefore, the structural relationship between these two classes of transposable elements). The flanking elements, either in direct repeat or inverted (Figure 6.28), are required for Tn transposition (see below). A characteristic 'hairpin' (or 'stem-and-loop', Figure 1.12) is formed in hetero-duplex analysis when the terminal repeats are in an inverted orientation (Section 7.3.2). Although transposons were initially identified as encoding drug resis-tance, recent studies on Tn3 have identified a 'transposase', essential for transposition (about 110 000 daltons), and a repressor protein (about 20 000 daltons) that regulates its expression (and is, itself, autogenously controlled), in addition to the penicillin-hydrolysing enzyme, β-lactamase. Transposons are, in truth, 'jumping' genes.

Insertion sequences and transposons are ubiquitous in prokaryotic systems. Insertion sequences are normal constituents both of the bacterial chromosome and plasmids (the number of estimated copies per *E. coli* K12 chromosome is eight for IS1 and five for IS2 and IS3; the F plasmid carries two IS3 elements in the same orientation, and one IS2, Section 10.1.1) and many R plasmids appear to owe the presence of their drug-resistant markers, at least in part, to transposons or transposon-like structures. The presence of IS elements on both chromosome and plasmid may, therefore, provide homology for *recA*-promoted recombination (see also Section 7.1.2 (cii)). There is direct support for this contention from the analysis of the structure of some F' derivatives. Moreover, one rationale for *recA*-independent Hfr formation, a rare event, is that these IS sequences serve as sites for integrative recombination (see Section 10.1.1 (c)). IS1 or phage Mu, for example, when present as a single copy on one replicon, have been shown to fuse their vector to a second replicon: there is duplication of the transposable element which lies on either side of the junction of the two replicons (see Figure 6.32, Section 6.2.3).

IS transposition is quite rare (about 10^{-7} per copy), comparable with the spontaneous mutation rate. Indeed, since the bacterium harbours a number of copies of these elements (Table 6.6), IS-mediated mutagenesis may represent a significant proportion of spontaneous events. The frequency of transposon transposition varies between element (for instance, 10^{-3}, 10^{-5} and 10^{-6} per transposon for Tn5, Tn9 and Tn10, respectively) but is generally higher than that for IS sequences. Transposition by either insertion sequences or transposons is relatively nonspecific but not entirely at random. This apparently conflicting statement stems from the observation that though there are many sites at which these elements insert — certainly, they may be found in any gene given the right selection — there are preferred sites, even clustering of sites, at which integration takes place when a cell population is 'mutagenised'. This apparent non-randomness depends upon the insertion element concerned. The two extremes, specificity and randomness, are well exemplified by secondary site insertion of phage *lambda* and Mu lysogeny. The reason for this variation is not understood.

The polarity of IS and Tn elements may depend upon the orientation of insertion (Table 6.6). Integration of IS2 in one orientation allows transcription of distal genes, presumably because of a functional promoter site on the IS element (certainly, sequences similar to the canonical Pribnow box, Section 2.1.2(ai), are present on this element). When present in the opposite orientation, no readthrough is observed due to the presence of a rho-dependent transcriptional termination signal. IS2 may, thus, be regarded as the prototype genetic 'switch'. The polar effects associated with these transposable elements have proved useful in genetic mapping (see Section 7.1.2 (c), Mutagenesis by DNA Insertion Elements).

Although both IS and Tn elements promote insertional inactivation, precise excision reactivates the target gene. The fidelity of the excision event is indicated by the fact that insertion mutants always revert (the reversion frequency, between 10^{-6} and 10^{-10} per cell, is apparently dictated by the site of insertion, suggesting an important role for the junction of the transposable element and the adjoining chromosome; see below). Thus, in a manner analogous to phage *lambda* integrative recombination (but via a different mechanism, see below), transposition neither irreversibly damages the site of insertion nor the insertion element, itself. This is not to say, however, that excision is always perfect: just that faithful events may occur. In fact, these transposable elements promote the formation of site-specific deletions, extending continuously from one of the direct repeat sequences found at the junction between element and target DNA into the adjacent chromosomal material (the frequency of deletion formation varies between element but is generally higher than the translocation rate). The majority of elements appear to remain intact and, thus, retain the ability to produce further chromosomal rearrangement. Note that these deletions, like the insertion and excision events, occur independently of the host *recA* function.

Insertion sequences and transposons, as portable genetic elements, are distinct from temperate phages (or plasmids that exhibit integrative recombination), in that they have no vegetative mode of existence. They replicate solely in covalent association with their host genome. In short, these transposable elements are not replicons. Phage Mu also, in fact, has no true autonomous existence since its lytic cycle involves random integration into the host chromosome; no free replicating molecules of Mu DNA have been observed (even the encapsidated viral form may be regarded as the integrated state as it is bordered by random stretches of host material, Figure 6.29). The fact that transposition, i.e. insertion, normally occurs at a markedly higher rate than reversion — precise excision — suggests that these two processes are different. Movement, in the cases that have been examined, appears to occur without the transposable element actually physically leaving the original insertion site. Donor DNA, in other words, is retained in the translocation process.

How can transposition be effected in the absence of an autonomous state? Recent models suggest that transposition is intimately connected with the replication process, in conjunction with DNA nicking and ligation steps (note that *lambda* integration occurs independently of replication). Staggered cuts to produce (5 or 9 bp) cohesive ends are important features of these proposals. Transposition has been observed in the absence of vector replication though nothing is known about replication of target DNA or any replication that may be initiated within the transposable genetic element, itself. It is, perhaps, pertinent, in this context, that a short, duplicated sequence brackets the insertion element. Although the size of the sequence is element-dependent (five base pairs in the case of IS2, Tn3 and Mu, and nine base pairs for IS1, Tn9 and Tn10, Table 6.5 and Figures 6.28 and 6.29), the exact sequence may vary in composition. (Recent studies on Tn9 insertion show a strong preference for G·C base pairs at each end of the 9 bp sequence.) Thus, transposition would seem to lead to duplication of a nucleotide sequence which pre-exists at the site of insertion (but not on the transposable element). That is, only the target DNA supplies the genetic information for the flanking direct repeats. This is in direct contrast to a Campbell-type integrative recombination event in which both target DNA and the incoming genetic element provide an attachment sequence (*att*$^\Lambda$*B* and *att*$^\Lambda$*P*, in the case of phage *lambda*). Moreover, transposition does not require supercoiling of donor DNA.

The natural division of transposable genetic elements into two groups according to the size of the flanking repeats is not understood (Tn 9 presumably falls into the same group of IS1 by virtue of the presence of this flanking insertion sequence on the transposon). It has been suggested that this may reflect the substrate specificity of two different classes of enzymes involved in transposition, enzymes responsible for the production of cohesive ends. Elucidation of the molecular processes leading to this sequence duplication should be greatly aided by the identification of a transposon-encoded transposase.

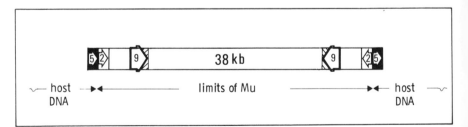

Figure 6.29: Structural Features of Phage Mu
The structure of the terminal region of Mu is shown (not to scale). Inverse and direct repeats are represented by arrows (hatched and open, and black, respectively). The left and right junction points have been termed att^{Mu}L and att^{Mu}R in analogy with phage *lambda* (Figure 6.27), since these neighbouring sequences seem to be of a defined nature, namely two inverted (imperfect) repeats. That is, Mu would appear to have a specific attachment site. There is some similarity in the host DNA adjacent to the Mu ends (a direct repeat of 5 bp). Whether this represents splitting of a specific host sequence upon phage Mu integration (Campbell-type event, see Figure 5.16) or an actual duplication process (as in the case of IS elements, Figure 6.28) is not known

(b) Illegitimate Recombination. The translocation of Is and Tn elements, initially classified as an illegitimate event, has since been placed under the category of site-specific recombination (see Insertion Sequences and Transposons, above). This 'legitimisation' stems, of course, from an increased understanding of the processes involved. A number of recombination events still, however, appear to be illegitimate in that neither the *recA* product nor specific DNA sequence homology are required. These include spontaneous DNA arrangements that occur in the absence of a transposable genetic element such as deletions and duplications, as well as the formation of specialised transducing phages. Little, in fact, is known about their generation.

Deletions were shown, in one study, to occur between non-tandem direct repeat sequences (5 or 8 bp). These may, therefore, arise in a manner analogous to *lambda* prophage excision (Figue 6.30 (b)). This scheme is, however, representative of only some intramolecular events (see also Section 3.2.1 (a)). The general consensus is that deletion termini are distributed at random along the bacterial chromosome although a hotspot for a deletion endpoint may occur infrequently (transposable genetic elements, in contrast, promote site-specific deletion formation, the endpoints defined by the internal junction of the element, Section 6.2.2 (aii)).

Duplications occur at a higher frequency than deletions (see also Section 3.2.1 (a)). This does not necessarily imply, however, separate mechanisms owing to the lethal potential of chromosome loss. Duplications are characteristically unstable (presumably because they supply a substrate for the general recombination system): a *recA*⁻ background is required to maintain the partial diploid state. These chromosomal rearrangements may, themselves, arise by legitimate recombination when direct repeat sequences are naturally present on the bacterial chromosome (Figure 6.30 (a)). Thus, the multiple non-contiguous rRNA loci (coding for 16S, 23S and 5S rRNA) at minutes 80-100 on the bacterial genetic map (the four gene clusters are present in the same orientation, see Figure 2.20) seem to be responsible for tandem duplication of the *glyT* region (89 min). Homology in such *recA*-dependent duplication events may also, presumably, be supplied by IS elements.

Though the majority of specialised *lambda* transducing phages are formed after (spontaneous) induction, abnormal excision from the attachment site is independent of the phage *int* and *xis* gene products (the constraints on survival of the transducing phage and host cell have been discussed in Section 5.2.2 and in (i), above). Endpoints are, in many cases, randomly distributed in the prophage and in the adjacent bacterial chromosome.

In conclusion, it is clear that a single model cannot explain all illegitimate events. Mispairing through short regions of homology, possibly present on two arms of a replication fork, may be responsible for some of these processes. Specificity in endpoints may reflect synapse stabilisation through the interaction of DNA binding proteins or, perhaps, staggered cuts created by endonucleases.

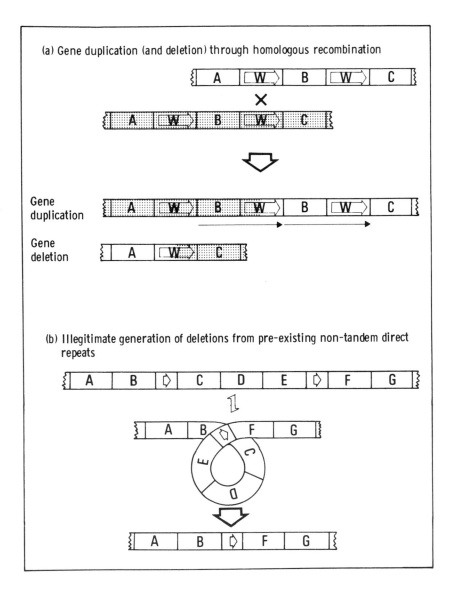

Figure 6.30: Models for Chromosomal Rearrangement
The presence of non-tandem repeats supplies a substrate for the general recombination system (a) and may be responsible for *recA*-independent events (b). The sequences may arise from multiple gene copies pre-existing on the bacterial chromosome (the numerous tRNA operons have, for instance, been implicated in gene duplication events). Alternatively, IS duplications may be active in exchange reactions. Whereas *recA*-mediated recombination between homologous sequences, say on the two arms of a replication fork, generates both duplications and deletions, illegitimate excision of short repeats in a Campbell-type event (see Figure 5.16) creates deletion mutants

In this context, it is important to determine what bacterial gene(s), if any, is involved in illegitimate recombination.

6.2.3 Phylogenetic Implications of Recombination

In the absence of any functional recombination system, genetic variation would rely solely upon mutational events. By itself, single site changes in the genetic material will only slowly cause divergence of species (although such changes may extensively alter the function of individual gene products): the majority of deletions will, of course, be deleterious when only one gene copy is present. Thus, the evolutionary potential for genetic variation stems from the interplay of mutagenesis and DNA exchange in conjuction with selection pressures.

General recombination allows some 'jumbling' of genetic material to occur through the exchange of alleles. More significantly, the homologous crossing-over between identical circular chromosomes can lead to *chromosome multi-merisation* (Figure 6.31). It has been suggested, in fact, that the present-day *E. coli* chromosome is, itself, the product of two sequential duplication events (of an ancestral chromosome one quarter the size). This proposition comes from the observation that metabolically related genes, particularly those involved in glucose catabolism, tend to lie either 90° or 180° (25 or 50 min) apart on the circular genetic map (the four gene clusters concerned with glucose breakdown are located at approximately 17, 42, 62 and 89 min). Similarly, a 180° relationship has been noted for the linkage maps of *Streptomyces coelicolor* and *Nocardia mediterrani* (Gram-positive bacteria), suggesting a single duplication event. That multimerisation is not without precedent in the prokaryotic kingdom is exemplified by the amplification of certain drug-resistant plasmids in *P. mirabilis* (see Section 6.1.3 (a)). Moreover, duplications as large as one fifth of the chromosome have been found in *E. coli* laboratory strains.

Multimerisation can benefit an evolving organism through the increase in gene dosage (an effect immediately realised) or because of decreased susceptibility to deleterious mutations. Subsequent divergence of the multiple copies to create further useful, even essential genes, serves to ensure maintenance of duplicated regions. Note that in the absence of such a selection, general recombination would remove tandem duplications (only a very small fraction of non-laboratory strains are likely to carry the $recA^-$ allele!). Although multimerisation readily explains the symmetrical distribution of certain *E. coli* genes, it is possible that this lack of randomness represents constraints on gene expression. The location of a coding region on the genetic map may have some importance, perhaps, in regulatory terms. The clustering of essential genes involved in replication and the expressive machinery about *oriC* are cases in point (Figures 2.3, 2.20 and 6.6). Certainly, it would now appear that transcription and translation are 'coupled' at the genetic level through the placement of genes involved in these processes under direct coordinate control (discussed in Section 9.3).

Despite this reservation about multimerisation of the entire bacterial chromosome, there are clear instances of gene duplication in *E. coli* K12. Notable

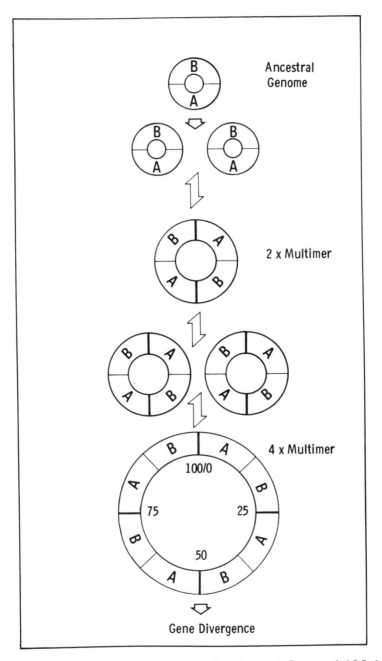

Figure 6.31: Chromosome Evolution through Sequential Multimerisation
A four-fold increase in genome size (and all the evolutionary benefits of gene redundancy) is achieved through two sequential multimerisation steps. The organisation of the *E. coli* chromosome supports such a scheme, suggesting an ancestral element of 1×10^6 bp in size

examples are the two ornithine transcarbamylase loci, *argI* and *argF* (96 and 6 min), the two gene copies for the translational factor EF-Tu, *tufA* and *tufB* (73 and 89 min), the tandem duplications of the tRNA^{Tyr} gene, *tyrT, V* (27 min), and the multiple rRNA gene clusters (56, 72, 74, 84, 86, 89 and 90 min). Such duplications are likely to have arisen through illegitimate events or by transposable genetic elements (a number of IS elements are present on the bacterial genome, Section 6.2.2 (aii)). The *rrn* duplications are of particular interest since the presence of different tRNA spacer genes (between the genes for 16S and 23S rRNA, Section 9.3.1 (bi)) indicates sequence divergence after the duplication event. Moreover, their asymmetric position on the *E. coli* genetic map indicates that if the chromosome evolved by multimerisation, this process occurred prior to rRNA duplication. The generation of tandem duplications by such non-contiguous multiple loci has been discussed above (see Illegitimate Recombination, Section 6.2.2 (b)).

Direct fusion of two replicons by the general recombination system requires extensive homology and is, therefore, likely to produce multimers of the same (or similar) genetic elements. Novel DNA may be introduced through transposition (mediated by intra- and intergenic plasmid transfer), either directly in the case of a transposon encoding, say, drug resistance or indirectly when the transposable element acts as a portable region of homology (Figure 6.32). Moreover, as regulatory 'switches' (witness the presence of promoters and transcriptional terminators on IS2), IS and Tn elements allow the activation and shut-off of specific gene expression. These elements have, clearly, a marked evolutionary potential with respect to the wide variety of replicons present in prokaryotes. The tetracycline-resistant transposon, Tn10, for example, has been passed in *S. typhimurium* from the drug-resistant plasmid R 100 to phage P22, from the phage to the *leu* operon, from this operon to coliphage *lambda* and from *lambda* to the *E. coli trp* operon. Moreover, the presence of flanking IS elements on R plasmids suggests that these genetic elements acquired their resistance determinants through transposition. Replicons can, therefore, no longer be viewed as static entities in the light of transposable genetic elements. A comparison of bacterial genetic maps serves to illustrate this point.

The linkage maps of the members of the family Enterobacteriacea are, in general, fairly similar. Those of *E. coli* K12 and *S. typhimurium*, for which the most is known, show extensive similarities (Figure 6.33). One large inversion of the *trp-cysB* region (about one tenth of the chromosome) appears to have taken place since divergence of these Gram-negative bacteria. The insertions and/or deletions that mark the differences between *E. coli* and *S. typhimurium* may have resulted from movement of transposable genetic elements. The fact that the *lac* operon, which is apparently absent from *S. typhimurium*, is flanked by two copies of IS1 (in an inverted orientation) suggests a transposition event for its presence on the *E. coli* chromosome. This contention finds support in the recent finding of a *lac* transposon, Tn951 (16.6 kb long), which carries at least *lacI, Z* and *Y* (the *lac* operon is considered in Section 9.1). Over and above these

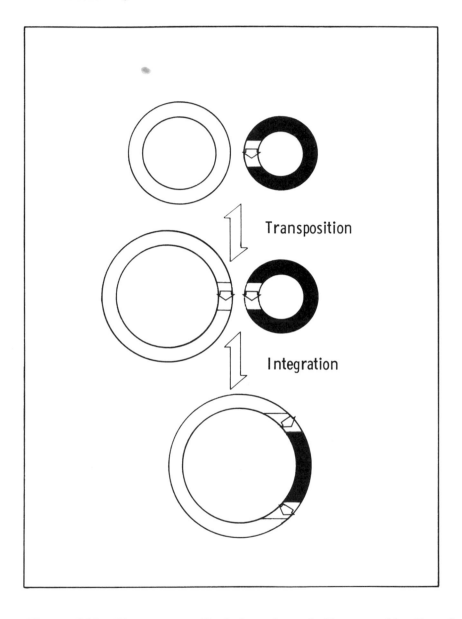

Figure 6.32: Chromosome Evolution through Transposable Genetic Elements

Since transposable elements such as IS and Tn sequences appear to translocate without physically moving — transposition is, in other words, a duplication process — entrance of a replicon carrying such an element into the bacterial cytoplasm is sufficient to create the conditions for *recA*-mediated insertion. Transposition of the IS element to the bacterial chromosome supplies a region of homology that serves as a substrate for the general recombination system (cointegrates may also be formed through illegitimate recombination)

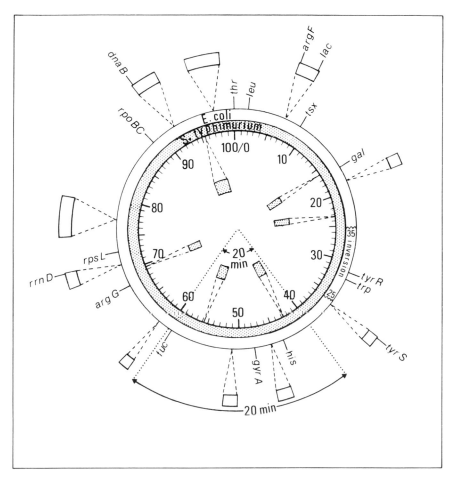

Figure 6.33: A Comparison of the Genetic Maps of *E. coli* **and** *S. typhimurium*

Homologous regions of the two circular genetic maps are aligned (making use of the 1976 maps), with putative insertions/deletions external to the scheme (only regions of non-homology of 1 min or greater are shown; Riley and Anilionis (1978) report 12 loops on the *E. coli* map with respect to *S. typhimurium* and 13 loops on the *S. typhimurium* map). Interestingly, the second ornithine transcarbamylase gene, *argF*, absent from *E. coli* strains B and W, and the *lac* operon, missing from *S. typhimurium*, fall on insertion/deletion loops. There is also an inversion between 25.6 and 35.4 min. About 86 per cent of the two maps can be aligned in this way (note that genetically homologous regions are likely to show sequence differences since hybridisation studies give only a 46 per cent relatedness). There appears, therefore, to be a strong selective pressure on these bacterial genomes. After Riley and Anilionis (Riley, M. and Anilionis, A. (1978) *Ann. Rev. Microbiol., 32,* 519-560) apparent insertions/deletions (the largest represents about 4 per cent of the chromosome), there appears to be a high degree of relatedness between *E. coli* and *S. typhimurium* (approximately 86 per cent of the two maps can be aligned and a figure of 46 per cent relatedness is obtained from analysis of the degree of association of their DNAs). Whether this reflects constraints on the genetic information encoded and/or the location of this coding material on the genetic map is not known.

6.3 DNA Repair

In discussing DNA replication, strong emphasis was placed on fidelity. Clearly, inheritance of a true copy of a genetic element, whether it be the massive bacterial chromosome or a plasmid one hundredth its size, demands accurate DNA synthesis upon the parental template. This, of course, applies to both productive (and transfer) replication and gap-filling DNA synthesis. Despite the remarkable accuracy of the replication process, errors do occasionally occur, generating lesions at a site of mispairing. Spontaneous mutations may arise, in addition, at the post-replicative stage, to change the chemical structure of DNA: spontaneous deamination of 5-methylcytosine to thymine is a case in point. Finally, genetic lesions can be induced externally by a variety of mutagens (Section 3.2.1 (b)). The end product, be it a single-site or multisite change, is a potential hazard to continued cellular growth. The fact that genetic elements do not become overrun with lesions can be attributed to efficient repair systems.

DNA synthesis, unlike any other enzymatic process, is buffered extensively against errors. The first line of defence, against mistakes incurred by the replicative complex, itself, is the $3' \rightarrow 5'$ proof-reading function associated with all prokaryotic DNA polymerases. Moreover, *E. coli* DNA polymerase I possesses an additional $5' \rightarrow 3'$ exonuclease activity that allows excision of damaged DNA in the path of the replication fork. Also encoded by the bacterium, but distinct from the replication apparatus, are enzymes that remove incorrect bases, sever phosphodiester bonds and excise runs of nucleotides. That these repair functions afford effective protection against mutagen-induced DNA damage is demonstrated by the marked increase in mutation rate (in response to, say, UV light) in their absence. Interestingly, this increase can, itself, sometimes be ascribed to repair system(s) — correction pathways that make use of the so-called SOS functions appear to exist to the detriment of the cell in that they are highly error prone, the process of repair actually augmenting the mutation frequency. Although the below discussion is limited to the repair system of *E. coli* other than the DNA polymerase copy-editing function (Sections 3.2.1 (a), 6.1.1 (a) and 6.1.2), it should be noted that repair systems are ubiquitous in nature, both in cellular organisms and some of their viruses. Repair mechanisms operating in genetic recombination have been considered in Section 6.2.1 (cvi) (see Mismatch Repair).

6.3.1 The Elements of DNA Repair

The importance of accurate replication and the maintenance of integrity of the completed product is exemplified by the numerous genetic loci that affect the cell's response to perturbations in genetic structure (Table 6.7/Figure 6.34) — perturbations, moreover, that may stem from a variety of causes. Bacterial repair mechanisms are economic in that they make use of pre-existing functions. Thus, the components of the replication and recombination apparatus participate in DNA repair. These represent polymerase, nuclease and ligase functions, clearly functions that could logically play a vital part in the cutting and patching of DNA (see Figure 6.35). In addition, the cell codes for enzymes whose role is essentially limited to the repair of DNA damage.

(a) Specific Repair Enzymes. Two main types of activity, other than the replication and recombination functions, have been implicated in DNA repair: *endonuclease* activity, responsible for chain cleaving, and *glycosylase* activity, which removes nitrogenous bases *in situ* (portrayed in Figure 6.35). It is the specificity of action of these components that dictate the repair pathway taken (see Figure 6.36).

A number of bacterial loci have been identified thanks to the UV-sensitive phenotype imparted by their mutant alleles (Table 6.7 and Figure 6.34). These may be distinguished from the *rec* genes in that only mutations at the latter sites affect the recombination frequency in conjugational and transductional crosses (Section 6.2.1 (a)). To date, four distinct genes, *uvrA, uvrB, uvrC* and *uvrD* (*uvr* for *UV* resistance), have been implicated in the dark repair of pyrimidine dimers (Figure 3.6, Section 3.2.1 (bi), shows the structure of a cyclobutylthymine dimer). The *uvrA* gene product (molecular weight 100 000) exhibits ATP- and GTP-dependent binding to superhelical DNA; *uvrBC* seems to exist as a protein aggregate (molecular weight 70 000). The individual *uvr* products have no endonuclease activity since an ATP-dependent incision function, breaking at the 5'-side of a pyrimidine dimer, requires a complex of at least these three proteins. That the UV protection afforded by the *uvrABC* system comes, in fact, from the action of an endonuclease, which is not specific for dimers, is indicated by the requirement for functional *uvr* gene products in excision repair of gross DNA damage (for example, mitomycin C crosslinks). This would suggest that the repair system recognises covalent distortion in the DNA structure, a specificity that should prevent natural melting of the DNA duplex becoming subject to the action of the *uvrABC* endonuclease. The gene *v* (or *denV*) product of bacteriophage T4 (molecular weight 18 000) is, on the other hand, an endonuclease truly specific for UV damage.

A separate enzyme pathway exists for the repair of pyrimidine dimers. The *phr* gene product (35 200 daltons) is a *photoreactivating enzyme* that requires light energy for action on the intrastrand lesion. Unlike the *uvr* system, which encodes an endonuclease function, the *phr* pathway cleaves the dibutyl ring, to generate two free, unlinked pyrimidine bases. *Monomerisation* is apparently

Table 6.7 [a] Chromosomal Genes Implicated in DNA Repair[b, c]

Gene	Map Location (min)	Gene Product Name	Size	Function and/or Mutant Phenotype
Mutagen Sensitivity				
phr	16	photoreactivation enzyme	35,200	light-dependent pyrimidine dimer monomerisation.
uvrA	92		100,000	excision repair
uvrB	17	non-specific ATP-dependent endonuclease	70,000(N)	
uvrC	42			
uvrD (mutU, recL, uvrE)	85	nk	nk	excision repair
alk	43	nk	nk	sensitivity to MMS (not UV or X-rays)
ras	(9) [d]	nk	nk	sensitivity to UV and X-rays
Other Repair Enzymes				
dam	74	adenine methylase	nk	DNA modification
dcm	43	cytosine methylase	nk	DNA modification
dut	81	dUTPase	16,000	reduction of accessible dUTP pool
tag	47	3-methyladenine glycosylase	20,000(N)	excision repair
ung	56	uracil-DNA glycosylase	24,500	excision repair
xseA	53	Exo VII [e]	nk	excision repair
xthA	38	Exo III (Endo II) [f]	28,000	excision repair
Filamentation Functions [g]				
lon (capR, deg)	10	protease ?	94,000	filamentation, UV and X-ray sensitivity, aberrant proteolysis
sulA (sfiA)	2	nk	nk	suppressor of filamentation
sulB (sfiB)	2	nk	nk	suppressor of filamentation

a   Abbreviations: nk, not known; N, native molecular weight; MMS, methylmethane sulphonate.

b   Adapted from Hanawalt *et al.* (1979).

c   For replication (*polA,B,C, dnaB, lig*) and recombination (*rec* loci, *lexA, sbcA,B*) functions implicated in DNA repair, see Tables 6.1 to 6.3.

d   Approximate map location.

e   Exo VII is a 5′ → 3′ and 3′ → 5′ exonuclease.

f   Exo III has both 3′ → 5′ exonuclease and AP (apurinic or apyrimidinic) endonuclease activity (as well as RNaseH and DNA 3′-phosphatase function); the AP Endo II activity has also been referred to as Endo VI.

g   These loci have been implicated in cellular survival after DNA damage but are not thought to govern DNA repair.

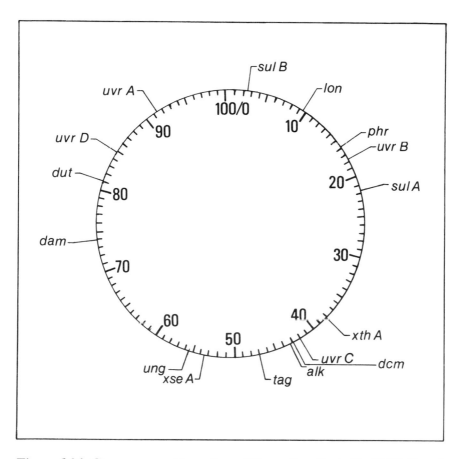

Figure 6.34: Chromosomal Location of Genes Implicated in DNA Repair
Properties of these repair elements are listed in Table 6.7 (for recombination functions, see Table 6.4/Figure 6.18). Map locations from Bachmann and Low (Bachmann, B.J. and Low, K.B. (1980), *Microbiol. Rev., 44,* 1-56)

dependent upon a nucleic acid cofactor since the enzyme, itself, lacks a chromophore. Other loci where mutations impart increased radiation (or mutagen sensitivity), *alk* and *ras*, have not as yet been characterised.

A novel class of enzymes, *glycosylases*, are active in certain repair processes. Glycosylases remove *in toto* a specific nitrogenous base from a polynucleotide chain, leaving the glycosyl bond between the base and the deoxyribose ring. The action of these enzymes is to create an apurinic or apyrimidinic (AP) site, depending upon their specificity (Figure 6.35). Thus, *uracil-DNA gycosylase* (a tetramer of 24 500 daltons subunits, the *ung* gene product), releases uracil from uracil-containing single-stranded or double-stranded DNA (but not from dUMP). Its cellular role is not to prevent formation of uracil-containing DNA through misincorporation of uridine (A·U in the place of A·T) as the presence of

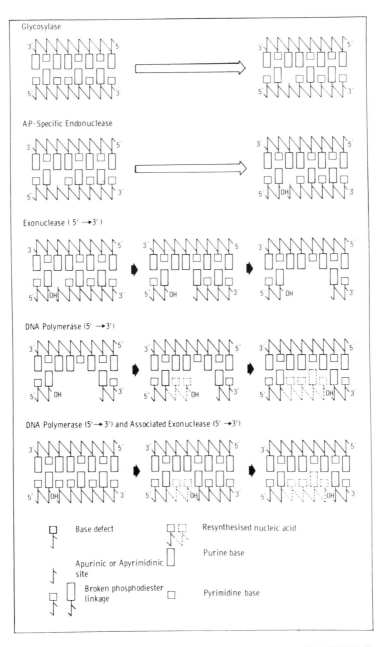

Figure 6.35: Some Enzyme Activities Implicated in DNA Repair

These functions are presented in the sequence in which they are thought to act (see Schemes for Pre-Replication DNA Repair, Figure 6.36). Resynthesis (DNA polymerase 5′ → 3′ activity) may or may not take place in conjunction with nucleotide excision (associated 5′ → 3′ exonuclease activity of DNA polymerase I). DNA ligase (Figure 6.4) would, of course, be required to complete repair uridine is not, in itself, mutagenic (*dut⁻* mutants deficient in *dUTPase* are viable, despite the increased incorporation of uridine into DNA). Rather, uracil-DNA glycosylase appears to be an anti-mutator for G·C → A·T transitions, the product of spontaneous, *in situ* deamination of cytosine residues to uracil (G·C → G·U) and rounds of replication (G·U → A·U → A·T). Such a model is in keeping with the hypothesis of Miller and co-workers for mutational hotspots: since thymine is a natural component of DNA and will not be the target of a specific glycosylase, thymine will not be removed when formed in spontaneous deamination of 5-methylcytosine (Section 3.2.1 (c)). Unfortunately, it is not known whether uracil-DNA glycosylase is essential as all *ung⁻* mutants to date are leaky, containing low levels of activity.

The *tag* locus encodes a *3-methyladenine glycosylase* (native molecular weight 20 000) that releases both 3-methyl- and 3-ethyladenine derivatives. Although this imparts protection against agents that alkylate primarily at the N-3 position, these alkyl derivatives do not appear to be important in alkylation mutagenesis (Section 3.2.1 (b ii)). The physiological role of the *tag* product is, therefore, not clear (particularly as the presence of a 3-methyl group actually labilises the adenine-glycosyl bond). A *hypoxanthine-DNA glycosylase* activity has also been identified in *E. coli* extracts (about 30 000 daltons in size). It is presumably reponsible for the removal of hypoxanthine bases formed in the non-enzymatic deamination of adenine (thereby protecting against a transition mutation, see Section 3.1.1).

The product of N-glycosylic bond hydrolysis, an apurinic or apyrimidinic site, may be acted upon by a specific *AP endonuclease* that cleaves at the 5'-side of the AP site (Figure 6.35). Exo III, the *xthA* gene product (molecular weight 28 000) has an associated AP endonucleolytic activity, termed *Endo II* (or VI), incising at the 5'-side of the lesion; it is a non-essential enzyme. The only other specific AP endonuclease characterised to date is *Endo IV* (molecular weight 33 000). *Endo III* (27 000 daltons),[19] *Endo V* (20 000 daltons) and the associated single-strand incision activity of the *recBC* nuclease appear to act non-specifically at many lesions (apart from the latter, Section 3.3.2 (a), all endonucleases attack sites of base loss on double-stranded DNA).

A repair mechanism which acts directly by removing damaged material, following a 5' cut, requires a 5' → 3' function. Suitable candidates are the *polA ex* activity (the 5' → 3' function of DNA polymerase I, Section 6.1.1 (a)), the single-strand specific nuclease, *Exo VII* (*xseA* gene product), and the *recBC* nuclease, *Exo V*; these proteins have, in fact, both 5' → 3' and 3' → 5' activity. (The essential replication enzyme, DNA polymerase III, has, of course, a single-strand specific 5' → 3' function (Section 6.1.1 (a)) but whether this is required for *in vivo* repair is not known.) Although these 5' → 3' exonucleases

19. A DNA glycosylase activity specific for 5, 6-hydrated thymine co-purifies with Endo III (and one for pyrimidine dimers in the case of the T4 UV endonuclease); it is not known whether these two functions reside on a single polypeptide chain. DNA repair involving such enzyme (complexes) consists of base removal, followed by strand scission.

remove pyrimidine dimers *in vitro*, lesions at any one genetic locus do not appear to severely affect physiological dimer excision. Double-mutant studies suggest a possible role for *polAex* function, as well as possibly the *xseA* and *recBC* exonucleases.

The large number of DNA polymerase I molecules per cell, the apparent association of this enzyme with *ExoV* and the inviability of *polA⁻recA⁻* (or *recBC⁻*) double mutants has already been mentioned as implicating the *polA* product in DNA repair (Section 6.1.1 (a)). Further support for this contention comes from the UV-sensitive phenotype of *polA⁻* mutants (Figure 3.7, Section 3.2.1 (b i)) and the ability of the enzyme, unlike DNA polymerase II and III, to bind to nicks generated by a dimer-specific endonuclease (for example, the gene *υ* product of phage T4). Finally, DNA polymerase I is well suited for repair synthesis by virtue of its ability to promote *nick translation* — strand synthesis concomitant with strand excision (Figure 6.3, Section 6.1.1 (b)).

6.3.2 The Mechanism of DNA Repair

Just as damage to DNA may occur during or after replication, so too can DNA repair commence directly on the altered site or on the lesion formed on the daughter strand. Thus, DNA correction processes can be conveniently divided into *pre-replication* and *post-replication* events (it should be emphasised that replication in this instance refers strictly to daughter-strand production and not repair synthesis). Correction of DNA damage prior to replication is, clearly, the most efficient form of repair in that the lesion is present for a fraction of the cell cycle. A second-generation repair process will be tolerated only in the case of a lethal mutation if there is a delay in the realisation of the mutant phenotype.

It is important to distinguish between two kinds of DNA damage: that in which the damaged strand is identifiable (e.g. a pyrimidine dimer) and that in which it is not (e.g. a base change). Repair mechanisms can reliably correct only the former (hence, the importance of copy-editing systems, Section 6.1.1 (a)).

In the main, lesions induced by UV irradiation have served as model systems for the study of repair processes *in vitro* and *in vivo*. This stems from the very nature of the primary UV photoproduct, dimers of pyrimidine bases, whose production and removal can be easily quantified. A suitable substrate, for example, is radiolabelled bacteriophage DNA (the virus having been grown in the presence of radioactive thymidine and/or cytosine) that is subjected to UV irradiation. It can be used directly as a target for repair enzymes *in vitro* or introduced into different mutant hosts by transfection to determine dimer removal under a variety of physiological conditions. The dimer content before and after 'repair' is readily ascertained by hydrolysis, chromatography and counting for radioactivity. Alternatively, the size distribution on DNA agarose gels (Section 7.3.1), after treating with a dimer-specific endonuclease such as phage T4 gene *υ* protein, is an indication of dimer content, as is the infectivity of the DNA, itself.

(a) Pre-Replication Repair. Correction mechanisms that have the potential of functioning upstream of the replication fork may still require *de novo* DNA synthesis, *repair synthesis* (or resynthesis). *Excision repair*, in which a stretch of nucleotides is removed, is of this type (Figure 6.36). Briefly, it involves the breakage of a phosphodiester linkage at the 5'-side of the site of damage, excision of the lesion and neighbouring sequence (on the same strand), and repair synthesis of the deleted portion using the intact single strand as template and 3'-OH of the severed chain as primer. Covalent linkage of these *repair patches* to parental DNA is performed by the enzyme, DNA ligase. Alternatively, a lesion may be repaired directly in the absence of any resynthesis; examples include *photoreactivation*, the light-dependent scission of pyrimidine dimers, and direct *base-replacement*, a process as yet unknown in prokaryotes (Figure 6.36). It should be emphasised that replication *per se* is not necessary for these proposed pre-replication repair mechanisms.

(i) *Excision Repair.* A model involving nucleotide excision encompasses the correction of the three main types of DNA damage, missing base, base defect (including mismatch) and a structural alteration that spans more than one nucleotide. It is generally proposed that excision repair starts with an *incision* event, the endonucleolytic scission of a phosphodiester bond at, or close, to the primary DNA lesion (Figure 6.36). The endonuclease concerned may be specific for an apurinic or apyrimidinic site, an AP endonuclease, or may be one that acts at structural defects such as the *uvrABC* endonuclease, which appears to recognise distortions created by cyclobutyl rings and other covalent modifications (Figure 6.36).

The second stage in excision repair would seem to involve an exonuclease which excises pyrimidine dimers *in toto* as part of a small oligonucleotide. That repair excision is not restricted just to the mutant site is indicated by the size of the repair patches, about 25 or several hundred nucleotides in length (see below). Three exonucleases, *polAex*, *xseA* and *recBC* gene products, have been implicated in pyrimidine dimer excision (rather than their monomerisation, see (ii), below). Exonucleolytic removal of damaged DNA (and adjacent material) is likely, at least in the case of *polAex*-mediated repair, to be concomitant with DNA synthesis, making use of the associated *polA* 5' → 3' polymerase function. Such coupling may serve to 'drive' excision and would protect against the formation of exonuclease-susceptible single-strand regions. Certainly, the multifunctional DNA polymerase I is eminently suitable for repair synthesis.

It should be noted that since DNA synthesis is required for the production of repair patches, excision repair mechanisms cannot function on single-stranded DNA. The replicative form of single-stranded phages such as ϕX174 and M13 would, therefore, serve as substrate for excision repair (as well as supplying the template for productive replication, Section 6.1.3 (b)).

The combination of DNA excision and patch synthesis, whether mediated by a single multifunctional enzyme or by a protein complex, results in the

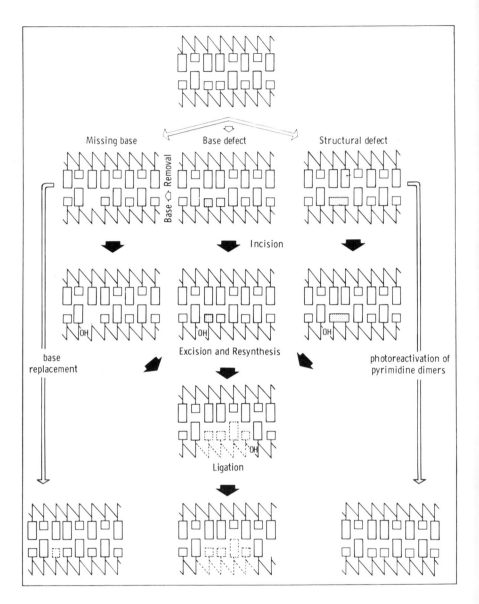

Figure 6.36: Schemes for Pre-Replication DNA Repair

In pre-replicative excision repair, correction is initiated with an incision event 5' to the damaged region (stippled) — by an AP-specific endonuclease in the case of a missing base or by an endonuclease, such as the *uvrABC* complex, that recognises distortions in the DNA duplex — and completed by repair synthesis (dotted lines) and ligation. UV-induced diadducts may be specifically repaired by an enzyme system that cleaves the pyrimidine photodimers *in situ* (termed photoreactivation since light energy is necessary for function). A further pre-replication repair system has been identified in eukaryotes which inserts bases at vacant sites. After Hanawalt*et al.* (Hanawalt, P.C., Cooper, P.K., Ganesan, A.K. and Smith, C.A. (1979) *Ann. Rev. Biochem., 48,* 783-836) juxtaposition of a 3'-hydroxyl and 5'-phosphoryl group. Ligation by DNA ligase completes the repair process. Since the nick generated at the end of excision repair is chemically identical with the initial 'repair' nick created by the repair endonuclease, it would seem there is a cellular mechanism that avoids the indiscriminate and premature ligation of incisions. Perhaps, repair nicks are blocked by removal of the terminal phosphate (by a DNA phosphatase) or through association with a specific DNA-binding protein.

There are two qualitatively different pathways of excision repair, *short-patch repair* and *long-patch repair*, the distinction based upon the length of nucleic acid replaced (about 25, and several hundred nucleotides, respectively). That these two processes are, in fact, mechanistically distinct is shown by the primary involvement of the *polA* product in short-patch repair. Furthermore, the strict requirement of a *recA+ lexA+* genotype for the long-patch process and its inducible nature places long-patch repair as one of the SOS functions. It is, apparently, an *error-prone repair* process, in keeping with this contention (discussed in Post-Replication Repair, below).

(ii) *Photoreactivation.* The *E. coli phr* gene product binds specifically to pyrimidine dimers in the dark. Visible light (300-600 nm) activates monomerisation, hence the term *photoreactivation*. The enzyme is released after rupture of the cyclobutyl ring. Since the photoreactivating enzyme, itself, lacks absorption above 300 nm, the significance of a light-dependent step in monomerisation was initially unclear. It has, however, recently been shown that the absorption responsible for photoreactivation is developed only when the enzyme is bound to UV-damaged DNA. The intimate role of polynucleotide in light activation is indicated by the disappearance of this new absorption spectrum upon dissociation of the binary *phr* protein-DNA complex.

Other pre-replication, non-excision mechanisms exist for the repair of UV damage. Photolysis of pyrimidine dimers, for example, occurs by direct absorption of UV light in the range 200-300 nm, the wavelength responsible for their production in the first instance. This reaction can also become *photosensitised* in that the light-mediated process depends upon light absorbed by a *sensitiser*, a molecule such as an oligopeptide containing aromatic residues.

(iii) *Base Replacement.* A plausible scheme for the repair of a defect resulting from a missing (or aberrant) base would be direct transfer and insertion of the appropriate moiety. As yet, no such *insertase* has been found in *E. coli* although there is a eukaryotic enzyme, active specifically at apurinic sites.

(b) Post-Replication Repair. When dimer-containing DNA is used as a template for DNA synthesis, processing DNA polymerase stalls at the diadduct. Under these conditions, there appears to be an 'idling' reaction in which dNTPs are incorporated and removed (in the form of NMPs), the $3' \rightarrow 5'$ copy-editing function presumably unable to allow permanent incorporation because of

distortion in base pairing. Rather than skip a few nucleotides before resuming synthesis (an unlikely event since no DNA polymerase can start DNA synthesis *per se*, Section 6.1.1 (a)), it would seem that an Okazaki fragment is initiated at the next available primer to leave an unpaired region of 1000 bases or less (if, in fact, replication is continuous on the leading strand, this region could be considerably larger). Replication of UV-irradiated DNA in the absence of excision repair (or other pre-replication process) creates, therefore, a *secondary lesion*, a defect on the daughter strand opposite the site of the primary lesion. The magnitude of this secondary lesion depends, of course, both on the nature of the primary mutation and the exact mechanism responsible for daughter-strand gap production. Two main mechanisms have been proposed for post-replication repair, *daughter-strand gap repair*, involving correction by genetic exchange, and *error-prone repair*, in which suppression of the idling reaction creates mismatches.

(i) *Daughter-strand Gap Repair.* Recombination of a gapped region with an intact sister chain, followed by resynthesis to replace the exchanged segment of DNA, would allow the repair of lost genetic information in the daughter strand (Figure 6.37). Repair of secondary lesions would require, in other words, both a functional general recombination system and repair synthesis components. The involvement of the bacterial genes *recA, lexA, recBC, recF, polA* and the *uvr* system (as well as, possibly, *dnaG*) in *daughter-strand gap repair* is, therefore, consistent with this model (the requirement for a *recA*$^+$ allele has also led to the term post-replication recombination repair, though note that *lexA*$^+$ and *recF*$^+$ are not necessary for general recombination).

(ii) *Error-prone Induced Repair.* The functions associated with error-prone repair have already been discussed in relation to their inducibility and a common dependence upon a functional *recA* allele (Section 6.2.1 (a i)); they include phage *lambda* and colicin induction and uncoupling of cell division from DNA replication. The current model for *recA* control over these *SOS functions* is that degradation of newly synthesised DNA, perhaps at, or near to, a stalled replication complex or at an incision site, activates *recA*-dependent inactivation of the *lexA* product (a repressor of *recA* protein production, Section 6.2.1 (ai)). Thus, certain *lexA*$^-$ lesions result in a lack of inducibility (a super repressor?), whereas other mutations at the same locus, *tsl*$^-$, allow thermal induction of SOS functions (see Section 6.2.1 (a ii)).

It was the dramatic reduction in the frequency of UV-induced mutations in a *recA*$^-$ background that initially suggested a role for post-replication repair in UV mutagenesis (Section 3.2.1 (b i)). There is now good evidence that a proportion of mutagenic events stem from error-prone replication past pyrimidine dimers. Since error-prone repair is an inducible phenomenon requiring protein synthesis, it has been proposed that some polypeptide species produced in response to UV damage permits reduced specificity in base pairing (Figure 6.38). It is this

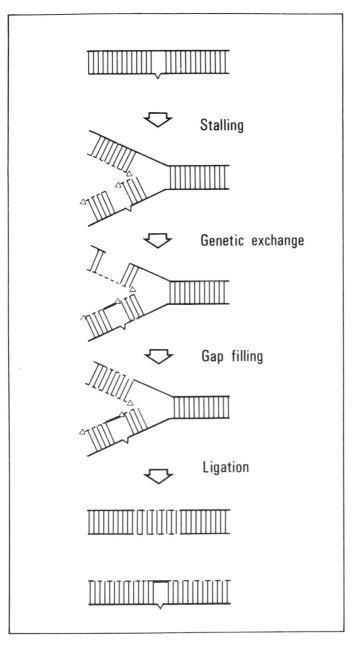

Figure 6.37: Daughter-Strand Gap Repair
DNA polymerase is left stalled at the dimer (but see Error-Prone Induced Repair, Figure 6.38) and replication is continued from the next available primer. The gap created in the daughter strand following recombination with this unreplicated region can be repaired by resynthesis and ligation

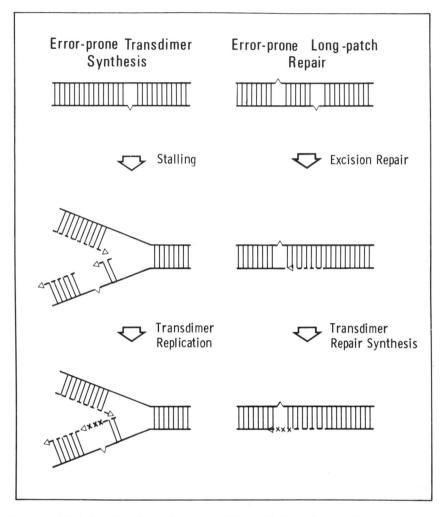

Figure 6.38: Mechanisms for Error-Prone Induced Repair
Inducton of the SOS functions allows DNA replication past pyrimidine dimers (left, see also Daughter-Strand Gap Repair, Figure 6.37), an error-prone process (crosses represent mismatches). Transdimer synthesis could also play a role in replication past a diadduct during excision repair (right). Note the distinction between these two processes; transdimer replication in post-replication error-prone repair and transdimer synthesis in pre-replication error-prone long-patch repair

reduction in specificity, allowing extension past a dimer (rather than the normal idling reaction), that is responsible for aberrant base insertion. If the site of action of this hypothetical SOS protein is the $3' \rightarrow 5'$ copy-editing function of DNA polymerase III holoenzyme, the modified polymerase may possibly show an overall reduced fidelity of replication (rather than just opposite the region of DNA distortion), leading to the generation of mutations at sites other than the primary lesion. That the target is DNA polymerase III, itself, is supported by the finding that certain *dnaE* mutants lack the inducible error-prone pathway. *Error-prone transdimer synthesis* has been put forward as a model for the increased survival of UV-damaged phage is irradiated, as compared with unirradiated, bacteria *(Weigle reactivation)*.

Another low-fidelity repair process may operate at the pre-replication stage. An error-prone mode of excision repair is thought to act where two closely spaced pyrimidine dimers are situated on separate strands (Figure 6.38). Since lesion excision and repair synthesis can only occur concurrently up to the site of the second dimer, transdimer synthesis, albeit error prone, is required to complete patch repair (in the absence of other repair processes). The inducibility of *error-prone long-patch repair*, and its dependence on the *recA* protein, suggests that this mode of repair is part of the SOS error-prone system.

In conclusion, at least two aberrant repair mechanisms exist. One, error-prone transdimer synthesis, operates at the post-replication stage, whereas error-prone long-patch repair functions on the primary lesion, itself. Both models, in fact, incorporate error-prone DNA synthesis past a dimer: transdimer replication and transdimer repair synthesis (for the post-replication and pre-replication events, respectively).

Bibliography

6.1 DNA Replication

Kornberg, A. (1980) *DNA Replication* (Freeman, San Francisco).

6.1.1 The Elements of Replication

Champoux, J.J. (1978) 'Proteins that affect DNA conformation', *Ann. Rev. Biochem., 47,* 449-479.
Cozzarelli, N.R. (1980) 'DNA gyrase and supercoiling', *Science, 207,* 953-960.
Geider, K. (1976) 'Molecular aspects of DNA replication in *Escherichia coli* systems', *Curr. Topics Microbiol. Immunol., 74,* 55-112.
Gross, J.D. (1972) 'DNA replication in bacteria', *Curr. Topics Microbiol. Immunol., 57,* 39-74.
Lehman, I.R. (1974) 'DNA ligase: Structure, mechanism and function', *Science, 186,* 790-797.
Lehman, I.R. and Uyemura, D.G. (1976) 'DNA polymerase I: Essential replication enzyme', *Science, 193,* 963-969.
Wechsler, J.A. (1978) 'The genetics of *E. coli* DNA replication', in I. Molineux and M. Kohiyama (eds.), *DNA Synthesis: Present and Future* (Plenum Press, New York), pp. 49-70.
Wickner, S.H. (1978) 'DNA replication proteins of *Escherichia coli', Ann. Rev. Biochem., 47,* 1163-1191.

6.1.2 The Mechanism of Bacterial Replication

(a) *Bidirectional replication of the E. coli chromosome*

Alberts, B.A. and Sternglanz, R. (1977) 'Recent excitement in the DNA replication problem', *Nature, 269,* 655-661.
Cairns, J. (1963) 'The chromosome of *Escherichia coli', Cold Spring Harbor Symp. Quant. Biol., 28,* 43-46.
Kolter, R. and Helinski, D.R. (1979) 'Regulation of initiation of DNA replication', *Ann. Rev. Genet., 13,* 355-391.

Ogawa, T. and Okazaki, T. (1980) 'Discontinuous replication', *Ann. Rev. Biochem., 49,* 421-457.
Tomizawa, J. and Selzer, G. (1979) 'Initiation of DNA synthesis in *Escherichia coli', Ann. Rev. Biochem., 49,* 999-1034.
Zechel, K. (1978) 'Initiation of DNA synthesis by RNA', *Curr. Topics Microbiol. Immunol., 82,* 71-112.

(b) The role of the cell envelope in replication

Pettijohn, D.E. (1976) 'Prokaryotic DNA in nucleoid structure', *CRC Crit. Rev. Biochem., 4,* 175-202.

(c) Control of DNA replication

Cooper, S. and Helmstetter, C.E. (1968) 'Chromosome replication and the division cycle of *Escherichia coli* B/r', *J. Mol. Biol., 31,* 519-540.
Donachie, W.D., Jones, N.C. and Teather, R. (1973) 'The bacterial cell cycle', *Symp. Soc. Gen. Microbiol., 23,* 9-44.
Donachie, W.D. (1979) 'The cell cycle of *Escherichia coli'*, in J.H. Parish (ed.), *Developmental Biology of Prokayotes* (Blackwell, Oxford), pp. 11-35.
Helmstetter, C.E., Pierucci, O., Weinberger, M., Holmes, M. and Tang, M.-S. (1979) 'Control of cell division', in J.R. Sokatch and L.N. Ornston (eds.), *The Bacteria, VII: Mechanisms of Adaptation* (Academic Press, New York), pp. 517-579.
Jacob, F., Brenner, S. and Cuzin, F. (1963) 'On the regulation of DNA replication in bacteria', *Cold Spring Harbor Symp. Quant. Biol., 28,* 329-347.
Koch, A.L. (1977) 'Does the initiation of chromosome replication regulate cell division?', *Adv. Microb. Physiol., 16,* 49-98.
Lark, K.G. (1979) 'Some aspects of the regulation of DNA replication in *Escherichia coli'*, in R.F. Goldberger (ed.), *Biological Regulation and Development, I: Gene Expression* (Plenum Press, New York), pp. 201-217.
Pritchard, R.H., Barth, P.T. and Collins, J. (1969) 'Control of DNA synthesis in bacteria', *Symp. Soc. Gen. Microbiol., 19,* 263-298.
Pritchard, R.H. (1978) 'Control of DNA replication in bacteria', in I. Molineux and M. Kihiyama (eds.), *DNA Synthesis: Present and Future* (Plenum Press, New York), pp. 1-26.

(d) Antibiotic inhibitors of replication

Cozzarelli, N.R. (1977) 'The mechanism of action of inhibitors of DNA synthesis', *Ann. Rev. Biochem., 46,* 641-668.
Sarin, P.S. and Gallo, R.C. (eds.) (1980) *Inhibitors of DNA and RNA Polymerases* (Pergamon Press, New York).

6.1.3 Replication of Small Genetic Elements

(a) Plasmid production

Rowbury, R.J. (1977) 'Bacterial plasmids with particular reference to their replication and transfer properties', *Prog. Biophys. Molec. Biol., 31,* 271-317.
Rownd, R.H. (1978) 'Plasmid replication', in I. Molineux and M. Kohiyama (eds.), *DNA Synthesis: Present and Future* (Plenum Press, New York), pp. 751-772.
Staudenbauer, W.L. (1978) 'Structure and replication of the colicin E1 plasmid', *Curr. Topics. Microbiol. Immunol., 83,* 93-156.

(b) Viral DNA synthesis

Denhardt, D.T. and Hours, C. (1978) 'The present status of ϕX174 replication *in vivo*', in I. Molineux and M. Kohiyama (eds.), *DNA Synthesis: Present and Future* (Plenum Press, New York), pp. 693-704.
Hurwitz, J. (1979) 'Analysis of *in vitro* replication of different DNAs', *CRC Crit. Rev. Biochem., 7,* 45-74.
Kornberg, A. (1979) 'The enzymatic replication of DNA', *CRC Crit. Rev. Biochem., 7,* 23-43.

6.2 Genetic Recombination

6.2.1 General Recombination

(a) The elements of general recombination

Clark, A.J. (1971) 'Toward a metabolic interpretation of genetic recombination of *E. coli* and its phages', *Ann. Rev. Microbiol., 25,* 437-464.

Eisenstark, A. (1977) 'Genetic recombination in bacteria', *Ann. Rev. Genet., 11,* 369-396.

Gudas, L.J. and Pardee, A.B. (1975) 'Model for regulation of *Escherichia coli* DNA repair functions', *Proc. Nat. Acad. Sci. USA, 72,* 2330-2334.

(b) Recombination pathways

Clark, A.J. (1973) 'Recombinant deficient mutants of *E. coli* and other bacteria', *Ann. Rev. Genet., 7,* 67-86.

(c) The mechanism of general recombination

Chipchase, M. (1976) 'Recombination by strand assimilation and strand crossover', *J. Theor. Biol., 57,* 249-279.

Fox, M.S. (1978) 'Some features of genetic recombination in prokaryotes', *Ann. Rev. Genet., 12,* 47-68.

Hotchkiss, R.D. (1974) 'Models of genetic recombination', *Ann. Rev. Microbiol., 28,* 445-468.

Low, K.B. and Porter, D.D. (1978) 'Modes of gene transfer and recombination in bacteria', *Ann. Rev. Genet., 12,* 249-287.

Radding, C.M. (1978) 'Genetic recombination: Strand transfer and mismatch repair', *Ann. Rev. Biochem., 47,* 847-880.

Stadler, D.R. (1973) 'The mechanism of intragenic recombination', *Ann. Rev. Genet., 7,* 113-127.

Stahl, F.W. (1979) 'Special sites in generalised recombination', *Ann. Rev. Genet., 13,* 7-24.

Stahl, F.W. (1979) *Genetic Recombination: Thinking About it in Phage and Fungi* (Freeman, San Francisco).

6.2.2 Non-homologous Recombination

(a) Site-specific recombination

Bukhari, A.I. (1976) 'Bacteriophage Mu as a transposition element', *Ann. Rev. Genet., 10,* 389-412.

Bukhari, A.I., Shapiro, J.A. and Adhya, S.L. (eds.) (1977) *DNA Insertion Elements, Plasmids and Episomes* (Cold Spring Harbor Laboratory, New York).

Calos, M.P. and Miller, J.H. (1980) 'Transposable elements', *Cell, 20,* 579-595.

Grindley, N.D.F. and Sherratt, D.J. (1978) 'Sequence analysis at IS1 insertion sites: Models for transposition', *Cold Spring Harbor Symp. Quant. Biol., 43,* 1257-1261.

Kleckner, N. (1977) 'Translocatable elements in prokaryotes', *Cell, 11,* 11-23.

Schwesinger, M. (1977) 'Additive recombination in bacteria', *Bacteriol. Rev., 41,* 872-902.

Shapiro, J.A. (1979) 'Molecular models for the transposition and replication of bacteriophage Mu and other transposable elements', *Proc. Nat. Acad. Sci. USA, 76,* 1933-1937.

Starlinger, P. and Saedler, H. (1976) 'IS-Elements in microorganisms', *Curr. Topics Microbiol. Immunol., 75,* 112-152.

Starlinger, P. (1977) 'DNA rearrangements in prokaryotes', *Ann. Rev. Genet., 11,* 103-126.

Starlinger, P. (1980) 'Transposition', *Plasmid, 3,* 241-259.

Weisberg, R.A. and Adhya, S. (1977) 'Illegitimate recombination in bacteria and bacteriophage', *Ann. Rev. Genet., 11,* 451-473.

(b) Illegitimate recombination

Anderson, R.P. and Roth, J.R. (1977) 'Tandem genetic duplications in phage and bacteria', *Ann. Rev. Microbiol., 31,* 473-505.

Franklin, N.C. (1971) 'Illegitimate recombination', in A.D. Hershey (ed.), *The Bacteriophage Lambda* (Cold Spring Harbor Laboratory, New York), pp. 175-194.

6.2.3 Phylogenetic Implications of Recombination

Cohen, S.N. (1976) 'Transposable genetic elements and plasmid evolution', *Nature, 263,* 731-738.

Nevers, P. and Saedler, H. (1977) 'Transposable genetic elements as agents of gene instability and chromosomal rearrangements', *Nature, 268,* 109-115.

Reanney, D.C. (1977) 'Genetic engineering as an adaptive strategy', *Brookhaven Symp., 29,* 248-271.

Reanney, D.C. (1978) 'Coupled evolution: Adaptive interactions among the genomes of plasmids, viruses and cells', *Int. Rev. Cytol.*, suppl. *8,* 1-68.

Riley, M. and Anilionis, A. (1978) 'Evolution of the bacterial genome', *Ann. Rev. Microbiol., 32,* 519-560.

Sanderson, K.E. (1976) 'Genetic relatedness in the family Enterobacteriaceae', *Ann. Rev. Microbiol., 30,* 327-349.

Sparrow, A.H. and Nauman, A.F. (1976) 'Evolution of genome size by DNA doublings', *Science, 192,* 524-529.

6.3 DNA Repair

6.3.1 The Elements of DNA Repair

Friedberg, E.C. (1975) 'Dark repair in bacteriophage systems', in P.C. Hanawalt and R.B. Setlow (eds.), *Molecular Mechanisms for DNA Repair* (Plenum Press, New York), pp. 125-133.

Lindahl, T. (1979) 'DNA glycosylases, endonucleases for apurinic/apyrimidinic sites, and base excision-repair', *Prog. Nucl. Acid Res. Molec. Biol., 22,* 135-192.

6.3.2 The Mechanism of DNA Repair

Bernstein, C. (1981) 'Deoxyribonucleic acid repair in bacteriophage', *Microbiol. Rev., 45,* 72-98.

Hanawalt, P.C., Cooper, P.K., Ganesan, A.K. and Smith, C.A. (1979) 'DNA repair in bacteria and mammalian cells', *Ann. Rev. Biochem., 48,* 783-836.

Grossman, L., Braun, A., Feldberg, R. and Mahler, I. (1975) 'Enzymatic repair of DNA', *Ann. Rev. Biochem., 44,* 19-43.

Lehman, A.R. and Bridges, B.A. (1977) 'DNA repair', in P.N. Campbell and G.D. Greville (eds.), *Essays in Biochemistry, 13* (Academic Press, London), pp. 71-119.

Moseley, B.E.B. and Williams, E. (1977) 'Repair of damaged DNA in bacteria', *Adv. Microbiol. Physiol., 16,* 99-156.

Sutherland, B.M. (1978) 'Photoreactivation in mammalian cells', *Int. Rev. Cytol.*, suppl. *8,* 301-334.

Witkin, E.M. (1976) 'Ultraviolet mutagenesis and inducible DNA repair in *Escherichia coli'*, *Bacteriol. Rev., 40,* 869-907.

7 Investigation of Gene Structure and Function

An understanding of the principles underlying cellular growth, of the interplay of regulatory circuits with physiological factors, and of the myriad processes that involve gene expression, relies ultimately on a knowledge of chromosome structure. There is a wide variation in the complexity of prokaryotic genomes. Nevertheless, each genetic element consists, in the simplest terms, of a series of genes (with their associated control regions), perhaps interspersed with non-coding — 'silent' — stretches: and, at the chemical level, of a macromolecule of nucleic acid.

The region responsible for a particular property may be defined making use of natural processes (Section 7.1). These *in vivo* prokaryotic techniques lead to the ordering of genes on a chromosome and, thus, to the construction of a *genetic map*. The expression of specific genes, both at the transcriptional and translational level, can also be studied *in vivo* (discussed in Section 7.2). Moreover, methods such as gene transposition and gene fusion, in conjunction with specialised transduction and recombinant DNA technology (Section 7.3), allow *in vitro* fine structure analysis, as well as the investigation of gene control with respect to the regulatory sites concerned. In short, present-day technology permits study of structure-function relationships at the level of the individual base pairs.

7.1 Genetic Analysis of Bacteria and their Viruses

It is now well accepted that one phase of a single stretch of nucleotides encodes one product (see Chapter 2). [With the discovery of overlapping genes (and with gene splicing in eukaryotes), such concepts as one gene-one protein (and even colinearity) require more definition. The translation reading-frame is, clearly, important in dictating the expression of genetic material.] Lesions which affect expression, resulting in gene disfunction, demonstrate the existence of such a coding region. The types of mutations responsible distinguish between the involvement of a structural gene coding for a protein product or a non-translatable locus (Sections 3.1.2 and 3.4). Lesions at a particular site are genetic markers for that locus; their position defines the region responsible for a particular property. The chromosome may, thus, be considered as a linear array of functional units — genes.

The location of such functional genetic loci and their control regions by mapping techniques, and the determination of the number of different genes associated with a certain property, is the essence of genetic analysis. DNA transfer between mutant strains is necesary for such studies. Prokaryotic genomes are not always readily manipulable in their entirety (witness, for example, the random shearing of single-stranded donor DNA during an Hfr cross, Section 4.1.3 (a ii)). It is the distinct phenotype associated with each individual gene (or group of genes) that allows a specific 'handle' on portions of the chromosome. Genes are ordered relative to known markers making use of this property. The genetic map is, therefore, a functional representation of the hereditary material. It rests with *in vitro* techniques to construct a comparable physical description (discussed in Section 7.3).

7.1.1 The Elements of Genetic Analysis

Consider the situation in which a strain of *E. coli* is mutagenised (Section 3.2) and independent mutants all deficient in the same function, say lactose utilisation, are obtained (the number should be large to ensure that all possible loci involved in the function have been 'pinpointed'). Are these lesions in the same gene or in different genes? *Complementation analysis* answers just this by defining the gene as a unit of function. Moreover, the *recombination frequency* between pairs of different mutations is, by and large, an indication of their distance apart. In the first approach, a functional gene copy is supplied in *trans*, whereas the second involves the reconstitution of an active gene. Clearly, then, complementation acts at the level of gene expression while recombination requires direct DNA-DNA interaction. Both are universal, independent of the mutant species (but see below). In each case, it is the isolation of mutant derivatives that is the starting point for genetic analysis.

(a) The cis-trans Complementation Test: The Gene as a Unit of Function. Complementation analysis is based upon the premise that two different mutations affecting the same function lie in the same gene if their presence on duplicate gene copies is unable to restore the wild phenotype (Figure 7.1). When two different lesions do complement in *trans*, they are considered to reside in different genes: each mutation, therefore, defines one particular gene. There are two main conditions for the test to be applicable. Firstly, recombination beween the two gene copies must be negligible (reduced by using, say, a *recA*⁻ strain, or conditions under which recombination is rare). And, secondly, complementation is expected to occur only if the gene product is diffusible. The latter is, clearly, a characteristic property of protein or RNA molecules. Structural genes and those encoding RNA species are, thus, defined by this test. On the other hand, mutations in regulatory sequences are *cis*-specific — they affect expression of

adjacent genes, those contained within the same polycistronic operon — and are not subject to complementation (unless, of course, it is a wild-type copy of the entire operon that is present in *trans*).

An alternative name for the gene, the cistron, derives from the *cis-trans* complementation test (Figure 7.1).[1] Mutations within the same gene complement in *cis* but not in *trans* (an active gene product is absent in this latter configuration but see Intracistronic Complementation, below). Complementation occurs when the two lesions reside in different genes, irrespective of whether they are in *cis* or *trans*. Thus, alleles — different mutant derivatives of a single gene — are defined by the dependence of their phenotype on configuration (Figure 7.1).

The *cis-trans* test relies on the construction of strains carrying different pairs of mutations on duplicate chromosomal copies (or partial copies). This powerful

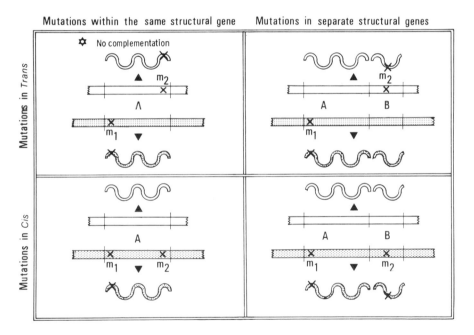

Figure 7.1: The *Cis-Trans* Complementation Test
Schematic representation of a partial diploid situation (for either bacterial or phage genes, see text). When (different) mutations, m_1 and m_2, are present in the same gene (left), there is complementation in *cis* but not in *trans* since no active gene product is produced in the latter case. Complementation occurs when lesions reside in different genes, irrespective of whether they are in *cis* or *trans* (right). Complementation analysis places mutations into groups, representing one gene, one cistron (but see Intracistronic Complementation, Figure 7.2)

1.   'Gene' and 'cistron', initially units of heredity and function, respectively, are now synonymous in their meaning. There is still strong preference, however, for the classical term. Use of 'cistron' tends to be limited to such terms as 'polycistronic' and 'intracistronic'.

genetic tool may be applied to the study of any organism that exhibits some form of diploidy (even the transient diploid state created by phage mixed infection is sufficient, Section 7.1.3). It is for this reason that the discovery of bacterial conjugation was so important. The production of stable partial diploid strains of *E. coli* using F-prime plasmids allows complementation analysis of this haploid organism. With the minimum of manipulation (Section 7.1.2), independent mutations can be readily arranged into complementation groups, each group equivalent to one gene (see also *in vitro* complementation, below).

Two main complications are encountered in this type of genetic analysis; pleiotropic mutations and intracistronic complementation. Point mutations that affect expression of two or more genes behave as if multisite — deletions or insertions rather than single-site changes — on the complementation test. This apparent pleiotrophy may stem from polar nonsense (or frameshift) mutations in a structural gene (Sections 3.1.1 (c) and 8.1.2 (b)) or from regulatory lesions. In each case, the genes affected are part of a polycistronic operon — they are genetically linked and have a common control region.

Structural genes that encode the subunits of a multimeric complex occasionally give spurious results in so far as the presence of two lesions in *trans* may allow wild-type or pseudo-wild function although they reside in the same gene. The quaternary structure of the active species is, in this case, a hetero–oligomer consisting of the two mutant polypeptides, in which the mutant site on one polypeptide chain is complemented *in situ* (rather than in *trans*) by the presence of the correct polypeptide sequences on the other strand (Figure 7.2). Interaction between two inactive polypeptide species carrying different amino acid changes, thus, appears to allow some function. It seems likely that this *in situ* complementation arises from constraints in protein structure imposed by the presence of the correct amino acid sequence. *Intracistronic* (or *intragenic*) *complementation* — compensation at the level of individual polypeptide chains — is restricted essentially to missense mutations (nonsense lesions, for example, give rise to incomplete proteins). Its occurrence may be taken as a sign of a protomeric gene product.

The discussion so far presupposes that a non-functional gene product does not interfere with the active species in *trans*. Not all lesions are recessive. Dominant mutations occur that reduce or inactivate the wild-type function. *Negative complementation*, an example of this, is another form of intracistronic complementation: the hybrid oligomer is rendered inactive by the presence of certain missense polypeptides (see Section 9.1, *lac* Repressor).

It is worthwhile at this stage considering the *in vitro* counterpart to the *cis-trans* test. The *in vitro complementation assay* is akin to the genetic complementation test with the exception that the gene product rather than the gene is supplemented in *trans*. Entrance of protein species requires, of course, breakdown of the natural permeability barrier (through cell lysis or partial solubilisation of the cell envelope).

The *in vitro* complementation assay has been mentioned with regard to the

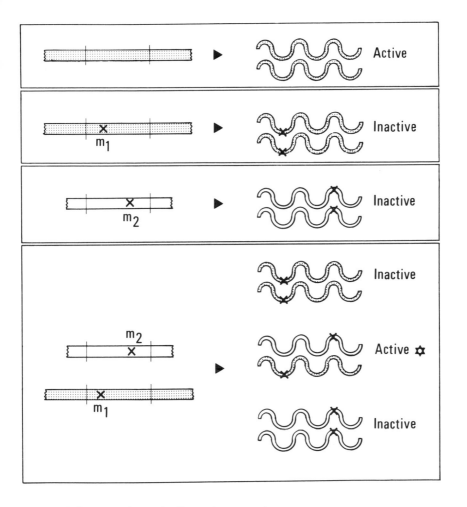

Figure 7.2: Intracistronic Complementation
When a gene product is active as an oligomer (in this instance, a homodimer), a hybrid complex consisting of two different mutant subunits may be active (bottom) despite the fact that the homo-oligomers, themselves, are non-functional. Such *in situ* complementation would suggest that the lesions, m_1 and m_2, are in separate genes despite their allelic nature

isolation of components of the replicative apparatus (Section 6.1.2). Consider a cell lysate obtained from a *dna⁻* (Ts) mutant. Although it will be defective in replicative DNA synthesis at high temperature, it may be supplemented with crude fractions from a wild-type culture to restore *in vitro* DNA production. The essential cellular constituent responsible for this *in vitro* complementation is narrowed down, first to partially purified fractions and, subsequently, to a homogenous isolate. When the same purification procedure is applied to extracts from the mutant strain, the resultant protein isolate should exhibit thermolability for complementation activity (it may also show different electro-phoretic mobility due to charge changes, Section 3.4). In conclusion, the *in vitro* complementation assay allows correlation between a structural gene (as defined by certain genetic lesions), physiological defects associated with lesions in this gene and the actual protein product encoded.

(b) Recombination: An Indication of Genetic Distance. Though genetic recombination must be low or absent for the *cis-trans* test, this molecular process is of great importance. It has been used classically for gene mapping and fine structure analysis — the study of the gene at the level of the individual nucleotide pairs. Such analysis was exploited, for example, in demonstrating the colinearity of the gene and its product (the advent of restriction technology, Section 7.3.3, largely supersedes this classical approach). In fact, recombination was originally employed for precise genetic studies in the absence of any detailed understanding of the molecular processes involved (discussed in Section 6.2.1).

Whereas complementation does not normally occur between alleles (cases of intracistronic complementation have been noted above), recombination allows the formation of a wild-type gene, with the single *proviso* that the two lesions do not reside in exactly the same nucleotide. Since there is no necessity for stability, the partial diploid or diploid status required for recombinational analysis can be created by employing conjugational or transductional crosses in the case of bacterial studies, or by mixed infection with two genetically marked bacterio-phages.

The distance separating two genetic lesions (affecting the same function in the case of fine structure analysis, or in different cistrons for gene ordering by P1 cotransduction, see below) is determined from the frequency with which the pair of lesions gives functional recombinants.[2] The direct correlation between these two parameters, genetic distance and recombination frequency, stems from the increase in the probability of a crossover between two lesions on different DNA duplexes (to generate a wild-type recombinant) the further they are apart (Figure 7.3). When the distance is small, there is only a short region of DNA within which recombination must occur to separate the two mutations. The extreme case, with the two genetic defects within the same nucleotide pair (albeit on different DNA molecules), gives no functional recombinants at all. Recombination between adjacent nucleotide pairs, that is within a single codon, has been demonstrated (at a frequency of 10^{-5} in the case of *trpA*, the structural gene for the α subunit of tryptophan synthetase).

The accuracy of mapping by recombinational analysis is limited by the rate at which point mutations revert to wild (or pseudo-wild) phenotype. When the

2.   Such studies are feasible in so far as there is a means for the selection of recombinants. Two *lac⁻* lesions, for example, may give recombinants able to utilise the sugar lactose (growth of the parents is maintained only through the use of another carbon source). It is more difficult to select for recombination between lesions not affecting essential or auxotrophic markers, such as derivatives lacking a particular phage receptor.

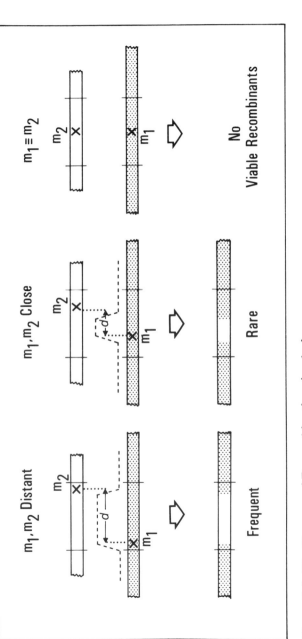

Figure 7.3: The Essence of Recombination Analysis

The presence of mutations, m_1 and m_2, in both gene copies precludes complementation with an active gene product (Figure 7.1). Recombination allows the formation of a functional gene (and, hence, products) so long as the lesions are non-identical (that is, only if $m_1 \neq m_2$). The greater the intragenic distance, d (intergenic in the case of non-allelic recombination), the more likely a crossover (dashed lines) can occur between m_1 and m_2 (white, donor fragment; stippled, recipient chromosome). The frequency of viable recombinants, therefore, increases as d increases. Note that the position of the leftward crossover is not important (so long as it is to the left of m_1).

reversion rate is high in relation to the recombination frequency, a significant proportion of apparent recombinants may be revertants. Genetic distances calculated from spuriously high recombination frequencies would, thus, be overestimated. Other limitations arise from the complexity of the recombination process (Section 6.2.1). Since synapsis requires homology between the two gene copies, the probability of recombination is very much dictated by the DNA sequences bordering the mutant sites. Frequency analysis of recombinants that arise, say, between a point lesion and a deletion are beset by difficulties (but see Deletion Mapping, below) because of the lack of homology. (Interestingly, inserted DNA present in the form of a transposon or IS element, Section 6.2.2 (a ii), acts as a point mutation in recombinational studies). Moreover, since the recombination frequency appears to depend on the mutational change, point mutations may be ordered incorrectly if only two factor crosses are considered (Figure 7.4; see also Figure 7.9). Thus, though recombination frequencies between different pairs of lesions are generally additive (except over very short distances), allowing their linear arrangement on the chromosome (see Figure 7.10), an alternative method that is both fast and accurate is favoured.

Recombination analysis, in which one gene copy carries a deletion rather than a point mutation, allows rapid screening of single-site changes since it avoids the necessity of pair-wise crosses.[3] It requires, instead, knowledge of the deletion endpoints. A single-site change is mapped between two such endpoints simply in terms of viable recombinants being obtained with one deletion derivative but not the other. The sensitivity of this method depends upon the size of the deletion intervals — the minimum distance separating any two deletion endpoints (Figure 7.5). Large intervals are used for rapid grouping of unknown mutations; subsequent analysis, in which the intervals are small, allows their accurate placement. Fine structure studies become possible when the region of interest is divided into a large number of deletion intervals such that the minimum distance between each endpoint is only a few base pairs (deletion mapping is discussed in greater detail with reference to the use of F-primes, Section 7.1.2 (a ii)).

Thus, recombinational analysis allows the determination of genetic distances at the molecular level with an accuracy bordering upon individual nucleotides. It has been used to order regulatory mutations, in addition to lesions in coding regions. There can be gross differences between the actual physical map and the comparable genetic description, a complication particularly prevalent for bacteriophage genomes (See Section 7.1.3).

7.1.2 Mapping Bacterial Genes

Numerous techniques for the determination of genetic linkage in bacterial

---

3.   With N different mutants, the number of *combinations* and, hence, of pair-wise crosses is $N!/(N-2)\,!2! = \tfrac{1}{2}N(N-1)$ $(=\tfrac{1}{2}N^2$ when N is large). Clearly, there is a geometric increase in the number of crosses with each new mutant strain available for study. A modest collection of 100 different mutants would require at least 4950 quantitative crosses, whereas, with a deletion collection of, say, nine that covers the region concerned, 900 qualitative crosses are necessary.

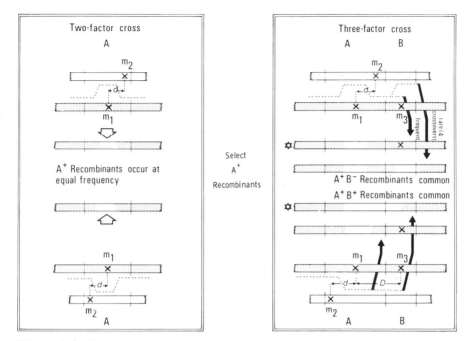

Figure 7.4: Ordering Mutations by Three-Factor Crosses

A two-factor cross gives no indication of the orientation of the mutations, m_1 and m_2 in gene A, since viable A^+ recombinants occur at the same frequency, irrespective of whether m_2 lies to the left or right of m_1 (note that the intragenic distance, d, is the same in both cases). When a third mutation (m_3) is introduced (in gene B), the frequency of A^+B^- recombinants is different from that for A^+B^+ and depends upon the placement of m_2 relative to m_1. If the order is m_1-m_2-m_3, four crossovers (dashed lines) are required to generate A^+B^+ and, thus, A^+B^- is more common: for an order m_2-m_1-m_3, crossovers are more likely to occur to the right of m_3 (to generate A^+B^+) than in the intergenic region D (assuming that D is small in relation to the size of donor fragment, white)

systems (chromosomal and plasmid genes) are available, based in the main upon either complementation or recombination (see also Section 7.3). Both types of analysis require the transfer of genetic material, effected in *E. coli* by conjugation and transduction. An unmapped lesion can be rapidly located using Hfr interrupted mating; alternatively, a collection of F-primes, representing the entire bacterial genome, is employed. (This requires, of course, knowledge of the direction of Hfr transfer, or the extent of bacterial sequences carried on each substituted F plasmid.)

Mutations affecting a particular character are placed into complementation groups. The pairs of mutations required for the *cis-trans* test are created by transfer of chromosomal lesions to F-prime plasmids (or *vice versa*) making use of recombination between the homologous regions on the two genetic elements. (Frequent recombination between the diploid regions converts the initial heterogenote carrying two different alleles into a homogenote, in which both gene copies are the same — hence, the term *homogenotisation* for this process.)

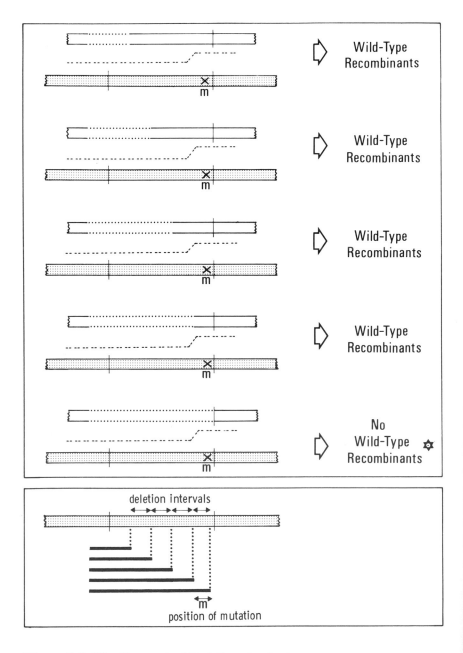

Figure 7.5: The Concept of Deletion Analysis
Crossover (dashed lines) between donor fragment (white) and recipient chromosome (stippled) gives rise to viable recombinants except when the former carries a deletion that extends into the mutant region (top). Knowledge of the deletion endpoints allows the construction of a genetic map placing the mutant site (bottom)

Lesions isolated initially on F-primes can be transferred to the chromosome in this manner or by, say, P1 transduction.

Once a genetic defect has been positioned on the bacterial map, its site can be pinpointed more accurately using F′ plasmids that carry deletions of the bacterial region concerned. Transduction has also proved very important for fine genetic analysis.

(a) Conjugation. The F plasmid is a convenient vector for bacterial sequences. Hfr mapping (and P1 transduction, see below) was the classical approach to genetic analysis of *E. coli* strains. Use of the present-day large collection of substituted plasmids has, to a large extent, superseded this.

(i) *Hfr Mating.* The bacterial chromosome of an Hfr strain is transferred at a uniform rate, the origin of transfer depending upon the position of F plasmid integration (Section 4.1.3 (aii)). In this polar process, a large part, if not the entire chromosome, can transfer to a suitable recipient. The time of entry of bacterial markers allows the determination of genetic distance in terms of units of time, assuming that the recombination frequency is independent of the region of DNA involved (but see Section 4.1.3 (a iii)). The position of an unknown marker is, therefore, determined from the minimal mating time required to generate viable recombinants (Figure 7.6).

Accuracy in the measurement of transfer time is, clearly, important. Moreover, since there is a natural gradient of transfer due, apparently, to random breakage of the single-stranded donor DNA, markers proximal to the origin (say, within 30 min) are most accurately located by interrupted mating (use of a group of Hfrs with different origins of transfer circumvents this problem). Markers close to regions of non-homology, for example, adjacent to the leading F sequence or near to a deletion, may give spurious results. Hfr mapping normally involves relative genetic distances — the difference in entry time between the unknown marker and a known locus — as this avoids the necessity of taking into account the period required for the formation of stable mating pairs (presumed constant for both markers).

Mapping by interrupted mating is disadvantageous in that the analysis is neither fine nor particularly accurate, problems that stem from the mere difficulty of assessing actual entry time (as discussed above, and Section 4.1.3 (aiii)). An alternative approach, though certainly no more accurate, is to employ the natural gradient of transmission. In this case, the relative proportion of different markers is determined rather than the time of entry. Cultures are mated for a fixed time and a proximal marker selected. The decline in the proportion of unselected distal markers is dependent upon their genetic distance from the selected one. An unknown marker can, thus, be placed in order relative to known loci.

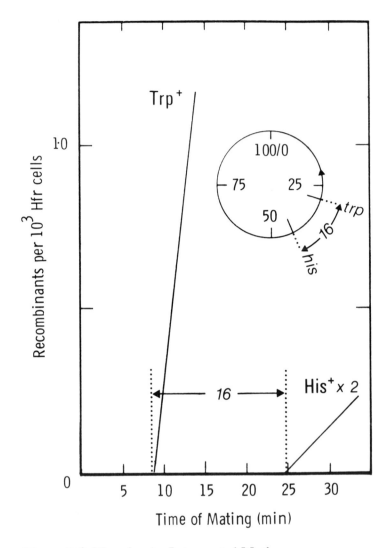

Figure 7.6: Mapping by Interrupted Mating

The *his* locus can be located relative to *trp* in a Hfr cross (the Hfr, KL99, transfers from about 22 min, with *trp* close to the 5'-leading end, see also Section 4.1.3 (a)). Determination of the relative time of entry allows for the period necessary for mating-pair formation. The inset shows the relative position of *his* and *trp* on the bacterial map ('His⁺ × 2' refers to the fact that the actual slope for the His⁺ recombinants is half that shown). After Bachmann *et al.* (Bachmann, B.J., Low, K.B. and Taylor, A.L. (1976) *Bacteriol. Rev., 40,* 116-167)

(ii) *F-Prime Plasmids.* A large number of substituted plasmids have been isolated from Hfr strains in which F is present at different sites in the bacterial chromosome (Figure 4.8, Section 4.1.3 (a)). Indeed, at present, the bacterial geneticist has in his 'arsenal' a collection of F-primes of various sizes and points of origin covering the entire bacterial map; and, most importantly, the extent of bacterial sequences carried on these hybrid plasmids is known (see Figure 4.13, Section 4.1.3 (b)). They offer a powerful tool for genetic analysis of bacterial systems, for rapid location of unknown markers as well as fine structure studies.

F-prime plasmids are conveniently employed to approximately position the site of a genetic lesion (this is analogous to Hfr mapping but lacks the associated enumeration). Different F' donors are crossed with the mutant strain, selecting transconjugants (or recombinants) in which the mutant phenotype is no longer expressed (Figure 7.7). Viable derivatives are not obtained with substituted plasmids that carry regions of the bacterial chromosome distant from the genetic

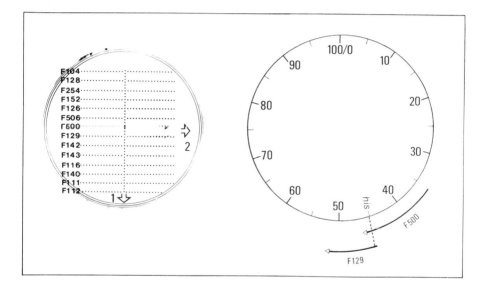

Figure 7.7: The Use of a Family of F-Prime Plasmids in Mapping
In a series of crosses between a *his⁻* recipient and 14 different F-prime donors (see Figure 4.13), only F500 and F129 donor strains give His⁺ transconjugants. (In the simulated cross, left, '1' and '2' refer to the order in which cultures are applied to the agar plate; i.e. the *his⁻* recipient is streaked onto the plate and then cross-streaked with the donor strains.) This places *his* in the region common to the two F-primes (right). F-prime mapping, clearly, requires both that the mutant strain is able to act as a recipient — the absolute frequency of transfer is not important — and that there is a selection against the F-prime donor (see F-Prime Transfer: Counter-Selection against Donor and Recipient, Figure 4.15). The presence of a plasmid in the recipient strain does not interfere with this test since compatible genetic elements coexist within the same cytoplasm while the unselected element is lost in the case of plasmids of the same Inc group (Section 4.1.2 (b))

lesion (see Counter-Selection, Figure 4.15). Only that plasmid with the wild-type bacterial sequence allows survival. The location of the genetic lesion is, thus, defined by the extent of bacterial DNA carried on this plasmid — the larger the bacterial region, the less precisely is the lesion positioned. F' mapping does not hinge upon whether the viable derivative is a stable transconjugant (complementation in the absence of recombination) or a recombinant (replacement of the original genetic defect by the wild-type sequence) since the mere fact of either indicates that the essential bacterial sequence is present.

The second application of F-primes in mapping bacterial genes requires a functional recombination system (Rec$^+$). It involves derivatives of a single-substituted plasmid that carry deletions of varying extent in the region under study (see Figure 7.5), rather than a collection of different F-primes. The deletions are ordered with respect to each other by determination, in each case, of the amount of DNA lost and, hence, their endpoints — this normally involves mapping against known point mutations or the employment of physical techniques such as heteroduplex mapping or restriction technology (Section 7.3). In practice, a set of isogenic strains carrying these F-prime deletion derivatives is crossed with each mutant: some allow viable recombinants, while others do not (the actual number of recombinants and the transfer frequency need not be considered in deletion mapping). The minimum bacterial region required for recombination, as defined by the largest deletion able to give survivors and the consecutive one that does not, pinpoints the stretch of DNA in which the lesion lies (see Figure 7.5). This allows ordering of unknown mutations, and their location relative to known markers. The technique is both easy to apply and powerful. Its sensitivity depends upon the number of deletion intervals covering a particular region — the greater the number, the shorter the distance between endpoints. Recent studies on the *lac* repressor gene (Section 9.1) have made use of a deletion collection with an average endpoint separation of less than one nucleotide triplet.

(b) Transduction. In the preceding discussion on mapping techniques, transfer of bacterial genetic material was dependent upon cell-cell contact. *E. coli* DNA can also be conveniently transferred by certain temperate bacteriophages (Section 5.2). The two main methods of phage-mediated transduction involve bacterial viruses that either transfer any region of the chromosome — *generalised transduction* — or those whose nucleic acid is covalently joined to particular bacterial sequences — *specialised transduction*.

(i) *P1 Generalised Transduction*. Bacteriophage P1 packages random fragments of host DNA upon infection (Section 5.2.1). The limitation in the size of foreign DNA contained in the phage capsid (about 91.5 kb) may restrict the use of P1 in strain construction. However, it proves to be advantageous for genetic mapping since only loci separated by a distance not greater than about 2 minutes of genetic map are transferred together in a single transducing particle (Figure 7.8). Moreover, the frequency at which both loci are transduced together, when only one marker is selected, increases with decreasing distance from each other. There is, therefore, an inverse correlation between *cotransduction frequency* and genetic distance (note that whereas the chance of recombination between

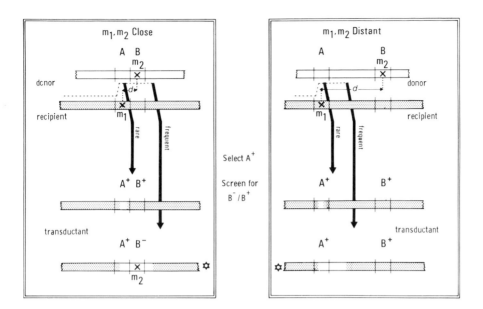

Figure 7.8: The Correlation between Cotransduction Frequency and Genetic Distance

A^+ transductants are selected and tested for the presence of the unselected marker B^-/B^+, that is, for cotransduction of the donor B^- allele (on the transducing fragment): for recombination to separate the two lesions, m_1 and m_2, the second crossover (black arrows) must occur within the intergenic distance, d (the first crossover, dashed lines, may occur anywhere on the DNA fragment, to the left of m_1). When m_1 and m_2 are close (i.e. d is small, left), most crossovers occur after m_2 and, hence, A^+B^- transconjugants are common. When the distance separating m_1 and m_2 is large (right), the second crossover event is likely to take place between these two lesions to generate A^+B^+ transconjugants. Thus, the probability that two markers are transferred on one fragment together, their frequency of cotransduction, decreases with their distance apart. Wu (Wu, T.T. (1966) *Genetics, 54*, 405-410) has proposed that the frequency of cotransduction $= (1-d/L)^3$, where d is the distance between markers and L is the length of the transducing particle (both in the same units of genetic distance, min or kb). Though it now appears that the extent of DNA carried is greater than the 2.0 min initially proposed (91.5 kb is equivalent to 2.4 min of genetic map), this value is still applied in the above mapping function because of end effects (Section 5.2.1). Genetic distances are convertible to actual units of DNA length, base pairs, since 2 min is equivalent to about 76 kb. Thus, from Wu's equation, markers separated by 0.26 min (10 kb), 0.78 min (30 kb) and 1.58 min (60 kb) are cotransduced at a frequency of 65, 22 and 1 per cent, respectively. Note that since an average gene is about 1 kb, markers within a single gene are cotransduced at high frequency (>96%). The original assumption upon which the mapping function was derived, namely that DNA fragments are cut at random from the bacterial genome, is only approximately true since some regions appear to be preferentially packaged

two markers increases with distance apart, Section 7.1.1 (b), their cotransduction frequency decreases with distance). A mathematical formula has been constructed to equate these two variables; from it, cotransduction frequency can be converted to separation distances in minutes of genetic map, or kilobase pairs (see legend to Figure 7.8).

Cells infected by normal phage particles — those carrying viral DNA — are killed (Section 5.2.1). Whereas encapsidation of bacterial DNA avoids this lethal event, cellular survival of a mutant recipient occurs only with a particular transducing particle. Moreover, isolation of a stable transductant (rather than an abortive one) necessitates integration of the transducing fragment into the host chromosome (use of a low m.o.i. at the transduction stage avoids lysis by additional non-defective phages and reduces the chance of lysogeny). Thus, though the propagation step can proceed in a *recA⁻* strain, transduction requires a recipient with a functional recombination system. Mapping by P1 cotransduction is applicable to any bacterial (or plasmid) gene with the single *proviso* that markers are not further apart than 91.5 kb. It allows rapid ordering of closely clustered genes (Figure 7.9). P1 cotransduction has shown, for example, that the five genes responsible for tryptophan synthesis (Section 9.2) are linked.

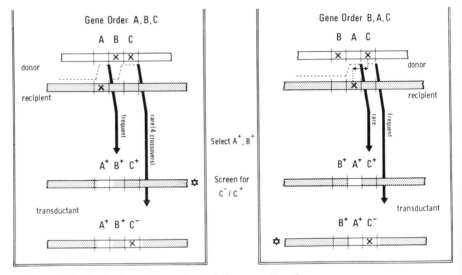

Figure 7.9: Ordering Genes by P1 Cotransduction

A⁺B⁺ transductants are selected and screened for C⁻/C⁺ (previous experiments, Figure 7.8, will have shown that *A* and *B* are close). When the gene order is *A–B–C* (left), a further crossover is required to generate A⁺B⁺C⁻ and, hence, such transductants are rare (first and second crossovers are represented by dashed lines and black arrows, respectively). For a gene order *B–A–C* (right), A⁺B⁺C⁻ transductants are more common than A⁺B⁺C⁺, provided that *C* is not far from *A*

Bacteriophage P1 is extremely important for the accurate ordering of lesions within a single gene (Figure 7.10). In this case, the phage capsid is used simply as a vector for the transfer of different alleles. The proportion of wild-type transductants formed upon intra-allelic recombination — recombination between lesions that lie within a single gene — is a measure of their distance apart (Section 7.1.1 (b)). A fine-structure map showing the position of various mutational changes may, thus, be constructed. Yanofsky and his group demonstrated colinearity between the *trpA* gene and its protein product, tryptophan synthetase

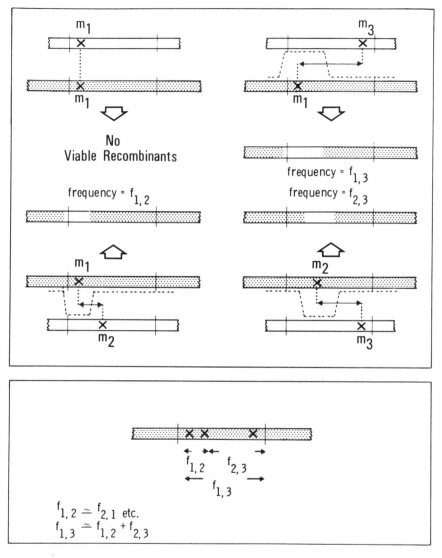

Figure 7.10: Fine-Structure Mapping with Bacteriophage P1

Phages grown on one mutant strain carrying a particular allele, say, the m_3 derivative (donor, white) is used to infect another strain bearing a different allele, m_1, for example (recipient, stippled). The number of wild-type transductants is an estimation of the distance separating the two lesions (m_1, m_3). The process is repeated for different mutant pairs (m_1, m_2 and m_2, m_3). The frequency of recombination between lesions (crossovers, dashed lines) is proportional to their distance apart (Figure 7.3) and, hence, the frequency of wild-type transductants increases in the order $f_{1,2}, f_{2,3}, f_{1,3}$. Mapping by P1 transduction is useful in that the values obtained are essentially additive, allowing the construction of genetic maps (that is, $f_{1,3} \simeq f_{1,2} + f_{2,3}$)

α subunit, by comparison of the relative order of mutationally altered sites in the *trpA* gene — obtained from deletion mapping and recombination frequencies (two- and three-point genetic tests using P1) — with the order of the resultant amino acid substitutions. In fine-structure analysis, where transduction frequencies are extremely low (a few per cent and less) causing difficulty in relating absolute frequencies with accurate genetic distance, it is necessary to allow for variation in the efficiency of the transduction process itself (because, say, of a reduction in the proportion of phage in a P1 lysate carrying bacterial sequences). Transductional data are, therefore, expressed relative to some outside (unlinked) marker (the internal control in the case of *trpA* was *his*).

(ii) *Specialised Transduction.* Bacteriophages, whose genome (rather than the protein head) contain bacterial sequences, are vectors for these sequences alone. The isolation of such specialised transducing phages has been described with reference to *lambda* (Section 5.2.2). Foreign DNA, covalently joined to the viral genome, arises during faulty excision from the site of integration. The range of bacterial sequences present on these transducing particles has been extended, making use of *secondary site insertions* (Section 6.2.2 (a i)). Bacterial DNA transferred by such vectors may exist stably in the recipient in one of three main ways: a haploid recombinant (Rec-mediated substitution); or, a lysogen formed by Rec-mediated integration either into the region of bacterial DNA homology or into a resident prophage (Section 5.2.2). When both sets of bacterial sequences — chromosomal and viral — carry different mutations affecting the same gene, a wild-type recombinant is created in only the first two instances (Figures 5.17 and 5.18 in Section 5.2.2 illustrate *gal⁻/λgal⁺* and *gal⁻/λ/λgal⁺* lysogens). This intra-allelic event, clearly, requires a functional recombination system.

It is their specialised nature that makes such phages so useful in genetic analysis. Each transducing phage serves as a mobile source of a defined region of bacterial DNA (analogous to substituted F plasmids, Section 5.1.3 (b)). Derivatives that carry different lesions — point mutations, deletions, or insertions — are readily constructed by homogenotisation or by induction of a lysogen carrying the phage integrated close to the mutant site. Above all, specialised transducing phage offer a source of pure bacterial DNA for *in vitro* studies (discussed in Section 7.3).

(c) Mutagenesis by DNA Insertion Elements. Previous discussion on gene inactivation centred on chromosomal defects that occur spontaneously or are brought about with mutagens; chemical agents and radiation are powerful inducers of mutational change (Section 3.2.1 (b)). In addition, DNA, itself, can also act directly as a mutagen. DNA-mediated mutagenesis makes use of portable genetic elements — bacteriophages, insertion sequences and transposons (and plasmids, Section 4.1.3 (a i)) — which integrate in the absence of a functional recombination system (Section 6.2.2 (a)).

The presence of a foreign piece of DNA in a coding region is, in general, strongly *polar* on distal genes: transcription (and translation) often terminates within the transposable genetic element resulting in the inhibition of expression of genes within the same operon, downstream from the site of insertion. Such an effect, unlike polarity resulting from nonsense mutations (Section 3.1.1 (c)), is not suppressed by nonsense suppressors (but see Polarity and Its Suppression, Section 8.1.2 (a)). Mutagenesis mediated by insertion elements reveals, therefore, those genes contained within a single polycistronic operon. Moreover, these elements can be used to generate deletion mutants since they can promote DNA loss adjacent to the site of integration (Section 6.2.2 (a ii)). Finally, as portable regions of DNA homology, insertion elements facilitate site-specific integration.

(i) *Bacteriophages as Transposable Genetic Elements.* In the absence of a functional chromosomal attachment site *(att^λB)*, phage *lambda* integrates at secondary sequences (Sections 5.2.2 and 6.2.2 (ai)). Unlike the genetic material of higher organisms, the majority of prokaryotic DNA is codogenic — hence, a high proportion of lysogens that carry the phage integrated into these naturally occurring variants of *att^λB* are deficient in a particular function (about 3 per cent of the lysogenic survivors of *E. coli* deleted for *att^λB* are auxotrophs). Secondary site insertion is, therefore, a very powerful form of directed mutagenesis, in so far as the target carries an *att^λB*-like sequence. (Secondary site mutagenesis is equally applicable to genes carried on autonomous plasmids.)

In practice, secondary lysogens are isolated from a Δ*(gal-att^λB-bio)* strain of *E. coli*, selecting homoimmune survivors (immune to the lethal action of *lambda* derivatives carrying the same immunity region). They are subsequently screened for insertion into the site of interest, say, Trp⁻; alternatively, a single procedure can be applied if there is a positive selection for gene inactivation. Secondary *lambda* lysogens allow the isolation of specialised transducing phage that carry bacterial sequences (or episomal material, in the case of a *plasmid-lambda cointegrate*) adjacent to the secondary site. The range of DNA carried is, thus, vastly extended.

As a tool for probing operon structure, phage *lambda* is limited because of the apparent non-randomness of secondary insertion sites. Bacteriophage Mu is far more powerful in this respect. It has been used successfully in mapping ribosomal structural genes and the components of the F plasmid transfer system (although its ability to promote deletions in adjacent operons has proved, in this case, to be a nuisance).

(ii) *Insertion Sequences and Transposons.* These amazing transposable elements (Section 6.2.2 (a ii)) provide a means for genetic engineering *in vivo*. Here, with no apparent limitations, is DNA transfer par excellence! Though their mechanism of transposition is not understood, the ability of IS and Tn elements to integrate at random, in two opposing orientations, provides a means for the generation of non-leaky polar mutations at any site (both bacterial and viral genomes).[4] Moreover, once inserted, reversion is at low-frequency (10^{-6}-10^{-10}), a process also independent of *recA* function.

Since the large transposable elements carry drug resistance markers, there is a direct selection for their presence. The regions carrying transposons can, therefore, be readily transferred from strain to strain (using the antibiotic-resistance determinant as a genetic 'handle'). Like most transposable genetic elements, IS and Tn sequences can generate deletions when 'leaving' their sites of insertion, of great use in mapping (the insertions, themselves, behave as point mutations). The sequences also serve as substrates for Rec-promoted recombination: in the absence of any other regions of homology, cointegrates are generated at these sites alone. These portable regions of homology, thus, serve to direct integration of one replicon (be it a phage or plasmid) into another, at locations specified by the transposable genetic elements themselves (the same in each case, see Section 6.2.3). This is useful for the construction of specialised transducing phages or novel Hfr strains (alternatively, when a single replicon carries two such elements, deletions, inversions or duplications can be generated).

(d) Pathway Analysis. Biological systems are of an ordered nature. A series of defined chemical reactions are responsible for a particular metabolic process. The elucidation of the enzymes used and the reaction order is made readily possible by isolation of mutants defective in the process.

Consider, for instance, a biosynthetic reaction in which the essential nutrient σ is produced from the natural substrate α via two intermediates, β and γ. In this scheme (Figure 7.11), each step is catalysed by a particular (but as yet unknown) enzyme: αase converts the substrate α to β, βase forms γ from β and γase acts upon γ to produce the final product, δ. Each enzyme is presented as consisting of one polypeptide chain (or multiples thereof),[5] encoded, therefore, by a single cistron (gene A, gene B and gene C for αase, βase and γase, respectively). Mutants that carry lesions in any one of these three genes, no longer synthesise σ — they are auxotrophs, requiring supplementation of the culture medium with this nutrient for growth. In fact, analysis of the pathway (Figure 7.11) suggests that the supplement need not be σ, provided that the bacterium has a transport system for (or is permeable to) the two intermediates.

Gene A mutants, for example, are able to grow with both β or γ (in addition, to growth on δ) since this class still encode functional enzymes (βase and γase) for

---

4.   The detection of low-level constitutive expression of distal *trp* genes upon integration of *lambda* into this operon initially suggested the presence of a low-level constitutive promoter in the left arm of the phage. This low-level expression from *lambda* P_1 indicates that the orientation of phage integration, in relation to the bacterial promoter, dictates whether the insertion is completely polar (rather than leaky) on distal gene expression. IS and Tn elements, on the other hand, tend to be polar in either orientation (except for IS2, see Section 6.2.2 (a ii)).

5.   The one gene-one enzyme hypothesis is not generally applicable (tryptophan synthetase, for example, consists of two different subunits, encoded by *trpA* and *trpB*). It is used in the present discussion for clarity.

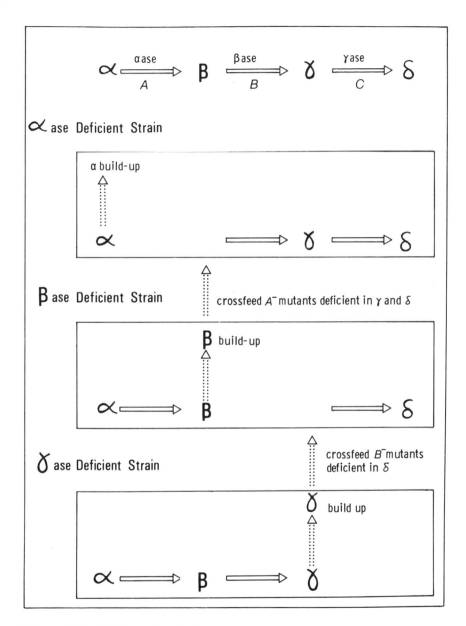

Figure 7.11: Pathway Analysis

The order of events in a biochemical process can be ascertained from the products accumulated in mutant strains. In the present example, B^- derivatives can crossfeed A^- strains and build-up of γ in C^- derivatives can support growth of both A^- and B^- mutants, as would be expected for a pathway order of $\alpha \rightarrow \beta \rightarrow \gamma \rightarrow \delta$ (see also The *trp* Operon, Section 9.2.1) intermediate conversion — αase (the gene *A* product) is required only for the first step of the pathway. Moreover, strains that carry defects in gene *B* cannot convert β to γ and are, thus, potential β sources (assuming that there is no other pathway for β breakdown, that the presence of β is not severely inhibitory on gene expression, and that it is secreted efficiently into the growth medium). Such auxotrophs allow *crossfeeding* of gene A^- mutants (Figure 7.11). Similarly, build-up of γ in gene C^- strains crossfeeds both αase- and βase-deficient derivatives.

In practice, the steps of an unknown pathway are determined by the isolation of independent mutants unable to grow in the absence of a certain nutrient (Trp^- auxotrophs, for example) — a large number is required to ensure that all structural genes responsible for that trait are defined. An understanding of the nature of the chemical processes concerned allows an 'intelligent' guess at the intermediates. Lesions may be arranged into complementation groups, with each group representing defects in one gene. Alternatively, they are classed according to their synthetic block, ascertained from their nutrient requirements (gene B^- mutants, for instance, utilise γ when supplemented in the growth medium) and crossfeeding characteristics. This defines both the steps of the synthetic pathway ($\alpha \rightarrow \beta \rightarrow \gamma \rightarrow \delta$) and the genes responsible *(A,B,C)*. Such studies, in combination with genetic mapping, have demonstrated that though functionally related genes are frequently clustered in prokaryotes (witness the *trp* system, Section 9.2: the *arg* genes, on the other hand, are widely dispersed), their order of expression is not necessarily the same as the sequence of the biochemical pathway. Crossfeeding by *trp* mutants is considered in Section 9.2.1.

7.1.3 Mapping Phage Genes

Complementation and recombination studies are applicable for the construction of bacteriophage maps.[6] Phage genetics is simplified, both in terms of genetic complexity and in that no additional vector is required to create, albeit transitorily, a diploid status (whereas conjugational or transductional crosses are necessary for mapping bacterial genes, Section 7.1.2). In addition, plaque morphology (Section 5.1.3 (a)) is a useful phenotypic trait. Finally, those phages that exhibit both the lysogenic and lytic response may be studied in either of these states.

Complementation analysis is carried out by mixed infection with two defective phages under non-permissive conditions (Figure 7.12). Recombination is negligible since their lesions render the individual phages unable to propagate on the particular host strain (a productive step on a permissive host would be

---

6.   Of course, those phages that do not undergo recombination (such as the small RNA bacterial viruses) are not amenable to this type of study. Only complementation analysis is, therefore, possible. This, itself, is difficult since one F pilus only seems to allow one phage to inject its nucleic acid. It is fortuitous that the genomes of these RNA phages are small enough to be readily analysed by biochemical techniques rather than genetic methods.

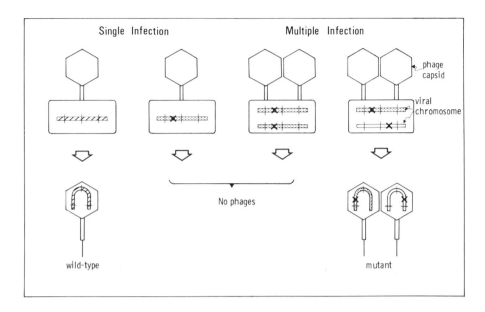

Figure 7.12: Phage Complementation

Coinfection by mutant phage derivatives carrying lesions in separate genes — complementation in *trans* (Figure 7.1) — is necessary for phage production. Note that progeny particles carry the parental mutations and are defective *per se* (see also Phage Recombination, Figure 7.13)

necessary prior to plating, see below). Each plaque must, consequently, be the result of successful lysis of a single cell by a phage pair. Under these conditions, only the existence of two non-allelic mutants within a single bacterium leads to phage production (a mutant pair with lesions in the same cistron does not complement *in trans*, Section 7.1.1 (a)). In this way, phage mutants are quickly placed into functional groups, each representing a single cistron. It should be emphasised that complementation studies on phage derivatives require the presence of both viral genomes in the cytoplasm. Lesions that block phage adsorption or DNA entrance cannot be studied by mixed infection. Rather, transfection or phenotypic mixing is made use of (Section 5.2.1).

Phage crosses are also readily performed. The initial mixed infection of a permissive host allows several rounds of propagation (Figure 7.13). The resulting lysate consists of both parental types as well as phage recombinants formed during this productive period. It is screened on a suitable indicator strain, able to differentiate between these types (one that permits growth only of wild-type recombinant phages, say, but neither of the defective input particles).[7]

7.   This is analogous to the stage in the *cis-trans* complementation test of mixed infection under non-permissive conditions. A high m.o.i. should, therefore, be avoided when intergenic crosses are screened.

Thus, plaques formed under these conditions represent recombinants in which the mutant site has been substituted by the wild-type region — their formation requires an intact gene rather than just a functional gene product. Their frequency is a measure of phage genetic distance.

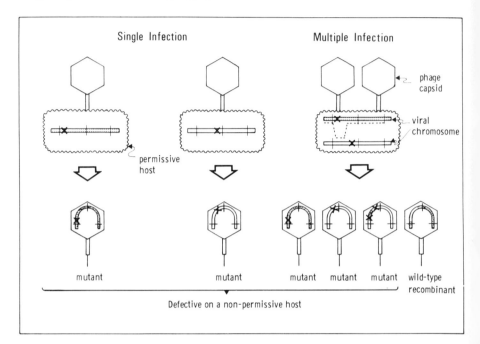

Figure 7.13: Phage Recombination
Infection by two genetically marked phages of a permissive host — one that supports growth of these bacterial viruses despite their mutant nature — allows viral production and genetic exchange. Only those recombinants (crossover, dashed lines) that have lost both lesions are, however, competent on a defective host. The frequency of recombination is a measure of the distance separating the mutant markers (Figure 7.3)

Bacteriophage maps constructed by recombination analysis can be quite distinct from the physical entity. The genetic map of the T-even phage T4, for example, is circular because the mature linear form is a random population of circular permutations (Section 5.1.1 (b)). Bacteriophage *lambda*, on the other hand, has both a linear physical and genetic map (excluding the vegetative form, which is circular): adjacent markers are separated by cutting of phage concatamers at the *cos* site upon packaging (the prophage map is a circular permutation of this due to site-specific integration at a site distinct from *cos*, Sections 5.2.2 and 10.2).

The classical work of Benzer aptly demonstrates the finesse of phage mapping. A large collection of T4 derivatives was isolated as displaying aberrant plaque morphology. The mutants, termed *r*, (for *r*apid lysis), produce large clear plaques, whereas those of wild-type phages have a clear centre surrounded by a turbid halo. The *rII* group are further distinguished by their inability to plate on *E. coli* K12 lysogenic for bacteriophage λ^+ (expression of the *lambda rex* gene is responsible for this inhibition of growth of T4 rII mutants, Section 10.2.1 (a)). Benzer showed by complementation analysis that this rII phenotype is the result of lesions in two distinct cistrons, *rIIA* and *rIIB*. Deletion mapping then ordered individual *rIIA* mutations — use of large intervals, and then successively smaller ones, rapidly positioned them, thereby avoiding a multiple of pair-wise crosses. Here, indeed, was fine-structure analysis supreme.

7.2 Monitoring Gene Function

Discussion of gene function has previously been in terms of an all-or-nothing response — cellular growth or death. In some cases, however, a particular gene may be expressed but at a level too low to sustain growth. Alternatively, mutant strains may survive with markedly different concentrations of the same gene product. An understanding of physiological processes at the level of gene regulation is, therefore, unlikely to be obtained solely from analysis of extreme effects. It requires study of rates of expression, both instantaneous and steady state (the efficiency in the inactivation and/or removal of a gene product can be as important as its synthesis, witness phage *lambda N* protein, Section 10.2.3 (b)).

Structural-gene expression is assayed either at the level of the protein product or its mRNA template. Measurement of these latter, unstable species must distinguish between functional (in terms of template capacity) and inactive messages, formed during rapid turnover. Nevertheless, such measurements allow a direct estimation of the efficiency of gene expression at the transcriptional level that is independent of other controlling factors (for instance, the ribosome loading frequency). Comparison of RNA and protein synthesis rates for a particular gene gives the proportion of functional mRNA molecules present. The change in their respective levels when subject to various physiological or genetic perturbations has demonstrated, in fact, the major role of the transcription process in the modulation of gene expression.

Some mention has been made of a major analytical tool, namely polyacrylamide gel electrophoresis (Section 3.4). This, and enzyme assays, are the two main methods for the determination of the degree of expression of a particular genetic locus.

7.2.1 Protein Biosynthesis. Of the two main categories of polypeptide species, structural and non-structural, only the latter are readily assayed for function (structural proteins have an intrinsically passive role — they maintain, but do not disturb the *status quo*, frequently in conjunction with numerous other components). Measurement of enzyme activity is dependent solely upon the availability of a suitable substrate: the product of the *lacZ* gene, for instance, converts lactose to glucose and galactose (lactose analogues with chromophoric side groups are used in practice). Non-enzymatic proteins of the second class that are not directly responsible for chemical change require more circuitous methods for their analysis. (For example, certain transcriptional regulatory elements can be assayed making use of their ability to alter the retention of DNA on nitrocellulose filters, see *lac/trp* Repressor, Sections 9.1 and 9.2.)

The membrane permeability barrier must first be overcome when dealing with cellular (rather than extracellular) enzymes, to allow passage of both substrate and product molecules. Alternatively, cells are lysed under mild conditions (say, EDTA/lysozyme, Section 7.3.1) that favour enzyme stability. These *whole cell extracts* also have their problems since they contain specific inhibitors and interference factors. Endogenous ribonucleases, for example, destroy RNA — the product of the RNA polymerase reaction. (Moreover, RNA synthesis may proceed in the absence of exogenous template because of the presence of residual bacterial DNA.) Partial purification of cell-free extracts may, therefore, be necessary.

A major drawback to any approach dependent on protein activity lies in the monitoring of those polypeptides with no direct function: apart from structural proteins (as indicated above), precursor molecules, polypeptide fragments, missense proteins as well as individual subunits of oligomeric species are inactive *per se*.[8] How are these species assayed? In practice, synthesis of non-functional proteins, as well as active species, is monitored by electrophoretic separation of radiolabelled whole-cell extracts on thin vertical slabs of poly-acrylamide.[9] Identification of the protein band(s) of interest (two-dimensional electrophoresis resolves about 1100 different bacterial species though there are fewer phage-coded proteins) may be achieved from comparison with purified markers, by precipitation with specific antiserum, and through use of character-ised mutant derivatives (Section 3.4).

Radiolabelling is carried out in the following manner. A short *pulse* with tritium (^{3}H)- or ^{14}C-labelled leucine (the most common amino acid and, therefore, present in the highest molar amounts) results in the isotope's incorporation into all growing polypeptide chains.[10] Since these nascent chains are at different stages of extension — some are just started, others nearly

---

8.   The *E. coli* DNA-dependent RNA polymerase is a case in point. Polymerisation of acid-soluble ribonucleotides into an acid-insoluble form has a minimum requirement for core enzyme, an oligomeric complex of three different polypeptides (Section 2.1.1 (a)). The individual subunits have no discernible function.

9.   Gene products encoded by transducing phages are routinely studied by infection of UV-irradiated cells. This inactivates the bacterial chromosome (causing premature transcriptional termination) and, hence, proteins radiolabelled at this stage are phage-coded rather than bacterial-coded. They are separated by electrophoresis. Moreover, in infection of a UV-irradiated *lambda* lysogen by a *lambda* transducing phage, phage promoters are repressed (Section 5.2.2) and only bacterial genes accompanied by bacterial promoters are expressed.

10. ^{35}S-Methione is obtainable at high specific radioactivity and is, hence, extremely useful (despite the low occurrence of this amino acid in proteins).

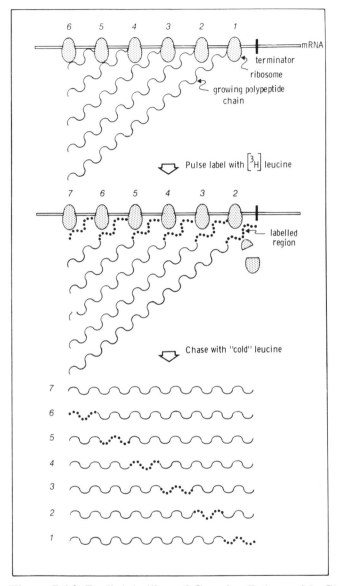

6 5 4 3 2 1

mRNA

terminator

ribosome

growing polypeptide
chain

Pulse label with $\left[^{3}\text{H}\right]$ leucine

7 6 5 4 3 2

labelled
region

Chase with "cold" leucine

7

6

5

4

3

2

1

Figure 7.14: Radiolabelling of Growing Polypeptide Chains
A short pulse with [³H]-leucine results in the radiolabelling of polypeptides at various stages of
extension. Chasing with 'cold' leucine permits the completion of every chain receiving the isotope
(further incorporation is stopped at the end of each chase with sodium azide, an electron transport
inhibitor, and by lowering the temperature). Unstable species are studied using long chase periods. A
discrete population of cellular proteins are labelled during the short pulse. If any of these break down
in the presence of cold leucine, further *de novo* synthesis cannot serve to replace this fraction: only
unlabelled species are produced in the chase period. Thus, degradation reduces the amount of
radiolabel associated with a particular gene product, whereas the counts retrieved in a stable species
are unaffected by the length of chase

complete — termination of the reaction at this stage would produce a spectrum of radiolabelled, foreshortened species (Figure 7.14). In order to ensure that these labelled chains are completed, 'cold' (unlabelled) leucine is subsequently added in excess and the incubation continued for a further few minutes *(chase)*. After washing off any unincorporated label, cells are lysed with sodium dodecyl-sulphate (SDS) to liberate their cellular contents (as well as solubilising the membrane proteins). It is this whole-cell extract that is applied to a polyacryl-amide gel (in the presence of the same ionic detergent, and a reducing agent such as 2-mercaptoethanol to cleave disulphide bridges).

Since the amount of bound SDS is proportional to the size of the polypeptide (about 1.4 ng/ng of protein), these SDS-proteins have relatively similar charge to mass ratios (all are highly negatively charged due to the bound detergent, thus minimising any intrinsic charge difference). They, therefore, migrate towards the anode at a rate dependent upon their size (rather than both size and charge), the smaller species moving most rapidly. SDS gel electrophoresis allows the estimation of protein size (using nanogram quantities) since there is generally a good inverse logarithmic relationship between molecular weight and migration distance. The pore size of the polyacrylamide gel is adjusted to maximise resolution in the molecular weight range of interest (say, 5% acrylamide for $2\text{-}35 \times 10^4$ daltons; 10% for $1\text{-}10 \times 10^4$ daltons). This simple technique has, in many ways, superseded conventional centrifugal methods for the determination of protein size. It should be noted, however, that not all proteins obey the inverse relationship (glycosylated species, for example, migrate anomalously for their size). Moreover, since SDS denatures proteins, this technique can be applied only to individual subunits.

Protein bands, normally visualised by staining (using protein-specific reagents such as Coomassie blue), are sliced up, solubilised and counted for radio-activity in a liquid scintillation spectrometer.[11] Alternatively, a fluor can be incorporated into the gel slab, which, after drying, is placed upon a sheet of sensitised X-ray film (for periods of 24h or more). *Fluorography* results in dark, exposed bands, quantifiable by densitometric scanning. (In *autoradiography* of ^{14}C- and ^{35}S-containing gels, the fluor step is omitted.) It should be emphasised that these bands (whether identified by staining, scintillation counting or fluorography/autoradiography) need not represent active species. Thus, electro-phoretic analysis reveals the synthesis of mutant and precursor derivatives, as well as wild-type proteins.

7.2.2 RNA Production. RNA labelled with, say, ^{3}H-uridine or ^{32}P-phosphate in a short pulse is isolated from SDS-lysed cells by phenol extraction. The nucleic acid partitions into the aqueous phase and can be concentrated by ethanol

---

11.   The isotopes ^{3}H and ^{14}C are low-energy emitters (the energy of electrons produced by ^{14}C decay is at least an order of magnitude greater). Liquid scintillation counting as well as fluorography both employ fluors that give off light upon collision with these β particles. It is this light that is detected by the scintillation spectrometer or film.

precipitation. It, too, can be fractionated by gel electrophoresis. However, because of the large size of most cellular RNAs (Section 1.1.1 (a)), large-pore gels are employed. Species of approximately 1.2 Mdal or less are routinely separated on 2-10 per cent gradient gels, whereas larger molecules require a low concentration of acrylamide stabilised with agarose (say, 2 per cent acrylamide with 0.5 per cent agarose): some small RNAs, such as the tRNAs, are separable on 10 per cent gels. Migration distance is dependent upon RNA size in these electrophoretic systems since the charge properties of the four common ribonucleotides are similar. There is again an inverse logarithmic relationship between size and distance migrated. Specific stains are available for detection after electrophoresis (for instance, methylene blue or pyronine), though the strong absorption at 260 nm allows direct visualisation. The bands are quantified by scintillation-counting methods or by autoradiography; a fluor is not necessary with the high-energy isotope ^{32}P (films are exposed directly for short intervals, say 1h).

RNA species analysed in this manner are separated solely according to size. (This type of fractionation is often of little use for mRNA samples obtained from intact cells due to their continuous degradation by endogenous enzymes.) There is, of course, no simple relationship between size and function (though stable RNAs have characteristic molecular weights and non functional messengers tend to be smaller than their active counterparts). Their nature may be investigated in two main ways: *in vitro* translation and RNA-DNA hybridisation. In the first, RNA extracted from the gel matrix, a process, itself, fraught with difficulties because of the presence of gel impurities that interfere with its template capacity, is used to prime an *in vitro* cell-free translation system (a sonicated cell extract, fractionated by centrifugation that is entirely dependent upon exogenous RNA for protein production). The protein products are subsequently analysed by SDS polyacrylamide gel electrophoresis (Section 7.2.1).

The alternative method makes use of complementarity between an RNA molecule and its coding region. *RNA-DNA hybridisation* involves the *in vitro* formation of an RNA-DNA duplex, termed a *hybrid*, upon incubation of RNA with a purified DNA *probe*. The latter is frequently a single strand of a transducing phage (see Section 7.3.2): it works equally well with restriction fragments of DNA, separated by electrophoresis (Section 7.3.3). The enzyme ribonuclease is used to enrich for RNA-DNA hybrids. Since free, but not DNA-bound, RNA is susceptible to nuclease attack, only hybridised RNA is present in the RNase-resistant fraction (these operations are greatly aided by trapping RNA to DNA immobilised on nitrocellulose filters). Hybridisation against DNA deletions of known length or characterised restriction fragments (Section 7.3.3) both assign RNAs to particular segments of the genome as well as allowing their quantitation. Thus, *transcriptional mapping* localises the region of DNA from which a transcript arose.

7.3 Physical Analysis of Gene Structure

Techniques that involve strain manipulation (Section 7.1.2) rely upon natural processes for the transfer of genetic material, its recombination and expression. In no case is the actual DNA responsible for a particular function isolated. Clearly, then, methods based upon purified DNA adds a further dimension to structure-function studies. The very homology (or its lack) between related sequences allows both direct visualisation of gene structure, as well as quantitation of gene expression (RNA-DNA hybridisation has been discussed in Section 7.2.2). Moreover, the advent of specific enzymes that cleave double-stranded DNA at particular sites gives rise to the technology for genetic dissection and resection *in vitro*. The starting point for such studies is the isolation of DNA.

7.3.1 DNA Isolation

The large size of the bacterial chromosome makes isolation and handling of the intact element a formidable task. Particularly gentle lysis techniques are required to avoid spontaneous shearing of the 1250 nm-long piece of covalently-closed circular DNA: careful lysis of EDTA/lysozyme-produced *spheroplasts* with non-ionic detergents yields a DNA-membrane complex.[12] To date, studies on the complete chromosome are restricted essentially to its overall organisation: the *nucleoid* contains nascent RNA and some protein, a high proportion of which is represented by the enzyme, DNA-dependent RNA polymerase (Section 6.1.2 (b)). Transfer of specific sequences onto smaller replicons — a purification step in itself (mention has been made of the effective concentration of bacterial genes by the construction of specialised transducing phage, Section 5.2.2) — is by far the most convenient starting point for DNA isolation.

Isolation of phage DNA is a remarkably straightforward process. An initial low-speed clearing spin removes the membrane fraction and any residual, intact cells. Phage particles are banded by *equilibrium sedimentation centrifugation* in caesium chloride (CsCl). In this *isopycnic centrifugation*, sedimenting species move through the *buoyant density gradient* until a position is attained at which their density and that of the gradient is equivalent. This technique, consequently, separates particles according to density and not size. It is particularly applicable to nucleic acids and phages because species not separable by size may band at different densities due to a slight change in base composition. Thus, lysates of a defective transducing phage may be purified of the wild-type helper (Section 5.2.2) owing to disparate densities. Components

---

12. Though the peptidoglycan layer protects against osmotic shock, it can be removed by lysozyme (cleaving at β1,4-linkages, see Figure 1.17). EDTA treatment is, however, first necessary to allow the enzyme to penetrate the outer membrane. The spheroplasts formed in the presence of EDTA-lysozyme may be stabilised by isotonic sucrose solutions. Alternatively, in addition to mild lysis with non-ionic detergents, they can be disrupted by suspension in hypotonic medium (under hypertonic conditions *plasmolysis*, or cytosol shrinkage, occurs).

separated in this manner — phage bands appear blue — are withdrawn through a needle inserted into the centrifuge tube or recovered by drop collection. The DNA is stripped of its associated protein by a series of phenol extractions (Section 7.2.2), any residual organic solvent being removed by dialysis.

Plasmid DNA is obtained in a similar manner, except that the cytoplasm is released by artificial lysis (SDS or chloroform). The difficulty in plasmid purification lies in their separation from the bacterial DNA present (the 'coating' of viral or viral-associated sequences permits the use of DNase to remove other nucleic acids). Moreover, in addition to covalently-closed circular molecules, the open circular (nicked) and linear plasmid forms are often present. Small plasmid DNA is routinely separated by isopycnic centrifugation in the presence of the intercalative dye, ethidium bromide. Since the two strands of intact plasmid DNA are physically inseparable and topologically constrained, these covalently-closed circular molecules are subject to restricted intercalation compared, weight-for-weight, with similarly sized nicked or linear molecules. The reduction in density associated with ethidium bromide binding is, therefore, smaller with covalently-closed circular DNA, which bands at a considerably higher density than the mass of chromosome material (and any broken plasmid DNA) in ethidium bromide -CsCl gradient centrifugation. (The DNA may be visualised directly with a UV light source.) Rapid methods for plasmid isolation that avoid density centrifugation include high salt precipitation and acid-phenol partition.

Plasmid DNA obtained by centrifugation (or by these latter methods) may be further fractionated by electrophoresis in large-pore, low-concentration agarose slabs (phage and plasmid size may be determined from either agarose electrophoresis or sedimentation rate analysis). Though agarose gel electrophoresis is, in general, applied to supercoiled DNA of 100 Mdal or less (F, for example, has a molecular weight of 62 Mdal), linear molecules of up to 500 Mdal are separated with low agarose concentrations (supercoiled DNA migrates faster than linear DNA, whereas nicked molecules show a reduced mobility). The bands are visualised by ultraviolet light illumination after ethidium bromide staining.

A further problem that should be noted is concerned with the absolute plasmid DNA concentration. Unlike phages, which are produced in, perhaps 100-200 copies per cell, stringent plasmids have a low copy number of about 1-2. Relaxed plasmids, on the other hand, are both present in greater quantities (copy numbers of 20 or more) and may be amplified by chloramphenicol treatment to about 3000 copies per cell (Section 6.1.3 (a)).

7.3.2 Heteroduplex Analysis

The degree of similarity between DNA molecules of related origins can be assessed without knowledge of their sequences. Formation of interstrand hybrids — making use of the antiparallel and duplex nature of DNA — is a clear indication of related sequences. Such physical mapping may be achieved by heat denaturation of an equal mixture of the two DNAs, followed by reannealing. The resultant duplexes are visualised in the electron microscope (using a magnification of about 10 000): single-stranded regions, prevented from collapse by formamide, appear thinner than paired segments. Plasmid DNA is first nicked (by aging, or cleavage with a restriction enzyme, Section 7.3.3, or low level of DNase) to generate a linear molecule.

Consider, for example, a deletion mutant of bacteriophage *lambda*. Such a derivative has an identical sequence to the wild-type parent apart from the region of lost DNA. Complementary strands from the different phages will, therefore, be able to pair. (Separation of the two strands of phage *lambda* DNA, termed the *l* and *r* strand, prior to mixing avoids the presence of parental pairs — *homoduplexes* — formed by reannealing of the two original sets of antiparallel strands.)[13] The *heteroduplex* formed carries a single-stranded 'bubble' arising from the presence of the complete strand at the site of the missing DNA; this type of structure is termed a *deletion* (or *insertion*) loop since it is generated by a deletion in one strand or an insertion in the other (Figure 7.15). Conversely, replacement of DNA results in a *substitution loop* in which both strands are unpaired.

The visualisation of structural modifications to DNA by heteroduplex analysis allows direct measurement of the genetic material, accurate to within about 50 base pairs. DNA molecules of known size are included on the same grid as internal markers (lengths are measured on the electron micrograph with a map measurer or electronic graphic calculator). Heteroduplex analysis is, thus, an accurate technique for physical mapping. It has proved of great use in the analysis of viral chromosome structure and of bacterial sequences carried on specialised transducing phages: the physical map of *lambda*, for example, and many of its early derivatives was constructed in this way. It has also allowed comparative plasmid studies (exemplified by the classic work of Davidson and his group on the origin of certain substituted F plasmids). The novel structures generated by transposons and insertion sequences in intrastrand reannealing experiments, in which a single DNA species is subject to denaturation, were a key factor in the elucidation of the basic structure of these transposable genetic elements. Moreover, the type of IS element present in a gene (as well as its orientation, see Section 6.2.2 (a ii)) has been identified by heteroduplex formation with *lambda* derivatives carrying known IS elements (sequence homologies of as few as 12 base pairs have been reported).

7.3.3 Restriction Technology

It now appears that different bacterial species have a well-defined defence system against foreign DNA that 'breaches' the cell membrane and enters the

13.   The different G + C content of the l and r strands allow their fraction by CsCl centrifugation in the presence of poly (U,G), a method dependent upon the difference in binding of the synthetic polyribonucleotide to the two strands. However, heteroduplexing is often carried out without such refinement.

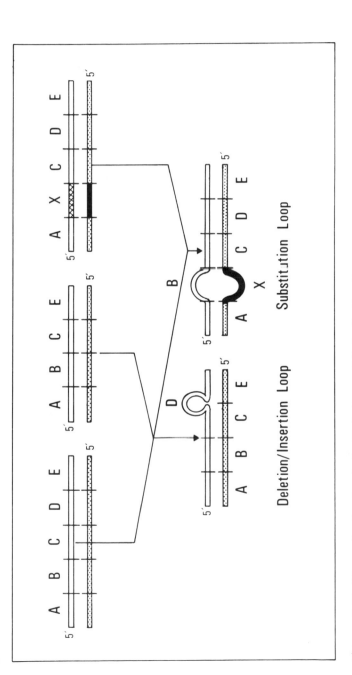

Figure 7.15: Heteroduplex Analysis: Deletion and Substitution Loops

Duplexes are denatured and allowed to reanneal in mixed solution (or, better, separated strands are renatured to avoid formation of homoduplexes). When one duplex lacks a region of DNA or contains an additional sequence, but is similar in other aspects to a second duplex, base pairing of antiparallel chains gives rise to deletion/insertion loops (left). A substitution loop (right) is formed if one of otherwise two identical DNA molecules contains different genetic material (X) rather than an insertion/deletion cytoplasm. Specific nucleases encoded by the threatened cell (or by extra-chromosomal genetic elements contained therein) cleave this DNA intruder, thus rendering it susceptible to degradation by other non-specific enzymes. These specific endodeoxyribonucleases have been termed, generically, *restriction enzymes* since they restrict, but do not totally inhibit, the survival of exogenous DNA (Figure 7.16).

Restriction enzymes may be divided into two main groups according to their cleavage properties.[14] *Type I* enzymes interact with DNA at particular sites (that presumably supply some sort of 'entry' signal) but make a double-strand scission at random, elsewhere. *Type II* enzymes, on the other hand, cleave directly at specific nulceotide sequences in double-stranded DNA (this useful property is discussed further, below). In general, cleavage-site specific restriction enzymes are small in size and only require Mg^{2+} as cofactor (type I enzymes require, in addition, ATP and S-adenosylmethionine).

How does the genome of the host organism survive the presence of restriction enzymes? Strains producing these restriction nucleases encode specific *modification enzymes* that alter the susceptible molecule. That is, restriction and modification enzymes have a common DNA recognition site. Methylases that protect without altering DNA function are responsible; their site of action is the target sequence, itself, in the case of type II enzymes (see Table 7.1). This overall system is known as *host controlled restriction-modification (R-M)*. It should be noted that formally speaking, a site-specific endonuclease may be classed as a restriction enzyme only if a modifying enzyme of the same specificity is shown to coexist in the same organism.

(a) Restriction Cleavage Maps. Sites susceptible to type II restriction enzymes consist of only a few base pairs (four to six, Table 7.1). These sequences generally possess two-fold rotational symmetry and are, consequently, palindromic: the base sequence is the same when read from, say, the 5'-end of each strand (Figure 7.17)). The cut is frequently displaced in either direction from the dyad axis of symmetry, creating staggered breaks, with the two resultant DNA molecules having 'sticky' (or 'cohesive') single-stranded termini (Figure 7.17). When the cut is at the axis, 'flush' ends are created. In either case, 3'-hydroxyl and 5'-phosphoryl ends are produced.

The sequence-specific nature of the type II restriction enzymes (particularly those that produce sticky ends, see Cloning, below) has led to *in vitro* genetic dissection and resection. The application of these enzymes to DNA studies is analogous to peptide analysis using proteases such as chymotrypsin and trypsin (the former acts mainly at aromatic residues, whereas trypsin catalyses hydrolysis of both arginyl and lysyl peptide bonds). Isolated DNA (Section 7.3.1) is cut into a series of defined fragments, readily analysed by gel

---

14.　A third class has recently been identified. These type III restriction enzymes cleave DNA a specific number of residues 3' from the recognition sequence.

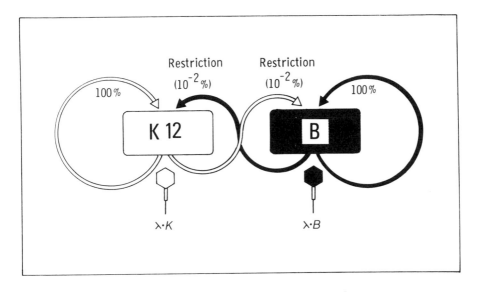

Figure 7.16: Host-Controlled Restriction and Modification of Bacteriophage *Lambda* **in Two Closely-Related Bacterial Strains,** *E. coli* **K12 and B**

Restriction is not an all-or-nothing phenomenon. Phage *lambda*, for instance, grown on *E. coli* K12 (i.e. K-modified, λ.K) will exhibit reduced plating efficiency if titred on the closely-related B strain. The few plaques formed result from viral DNA that has escaped the B restriction system. Survival normally requires concomitant modification by the host restriction-modification system (in this case, *E. coli* B) and, thus, further propagation of λ.B is no longer restricted when grown on this particular strain. These λ.B phages are now, however, susceptible to attack by K-specific restriction enzymes since they lack K-modification. (To be precise, those of the first cycle progeny phages that have inherited a single strand of parental DNA through semi-conservative replication are still able to plate on K12 strains since modification of one strand is sufficient protection. In the case of an extrachromosomally coded RM system, such as that of phage P1, *lambda* particles able to grow on K12(P1) — *E. coli* strain K12 lysogenic for P1 — plate with equal efficiency on K12, itself, since the P1 system is additional rather than an alternative as B is for K12.) Moreover, upon regrowth in the original host, *E. coli* K12, their ability to plate on the B strain is lost. DNA modification is, clearly, a phenotypic phenomenon in that continued growth on one particular host is required to sustain protection against the restriction system of that strain (unless, of course, the target sites are lost by mutation). The symbol *hsd* (*h*ost *s*pecifity for *D*NA) is used for the genetic components of the restriction-modification system. The phenotype is designated by the letters 'r' and 'm', with subscripts indicating the host/genetic element. An *E. coli* K12 strain that lacks, say, its restriction but not its modification system is written $r_K^- \, m_K^+$

electrophoresis. The fine resolution of this latter technique has proved of great importance: it allows both preparative and analytical studies, resolving species over a wide molecular weight range (from a few base pairs to as much as one million).

There is now an extensive battery of these sequence-specific restriction endonucleases, many of which have different target sites (Table 7.1). Thus, fragments produced from the reaction may be further restricted with enzymes of

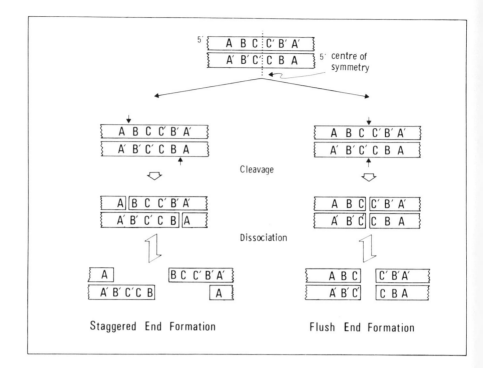

Figure 7.17: Types of Termini Created in Double-Stranded DNA by Site-Specific Endonucleases
The targets of site-specific, type II, restriction enzymes are generally palindromes — DNA sequences that are the same when 'read' in either direction — and, thus, have two-fold symmetry. When endonuclease attack is at the centre of symmetry, flush end fragments are generated (right), while cleavage at sites off the centre produces restriction fragments with 'sticky' ends (left)

different specificities. The ordering of fragments in this way generates a *restriction cleavage map* — a representation of the DNA region in terms of its restriction targets (Figure 7.18). It is this knowledge that is applied to the determination of the presence, and location, of DNA changes such as deletions or insertions. The restriction targets serve as reference points in such studies since the presence of new sequences may supply further cleavage sites, whereas a deletion removes them. Physical mapping in this manner allows the *in vitro* analysis of plasmid and phage genetic structure (and bacterial sequences carried thereon) as well as that of eukaryotic genomes.

(b) DNA Cloning. The process is reversible. DNA 'clipped' from one source may be inserted into another replicon, such as a small plasmid or bacteriophage, making use of complementarity between single-stranded termini (Figure 7.19). Base pairing requires that a single type II restriction enzyme — one that cuts away from the dyad axis — is used both for DNA excision and vector opening.

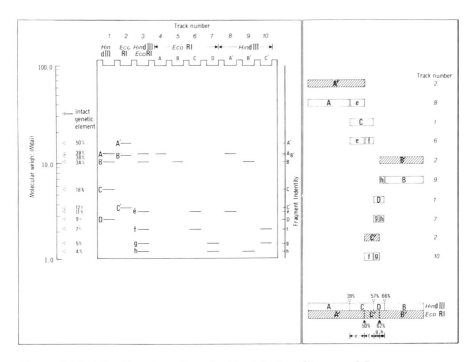

Figure 7.18: The Construction of a Restriction Cleavage Map

*Hin*dIII- and *Eco*R1-treatment of a hypothetical 31 Mdal viral DNA generates a number of fragments readily separated by agarose electrophoresis (left). There are 4 *Hin*dIII fragments (A, B, C, D, track 1), 3 *Eco*R1 bands (A', B', C', track 2) and 4 additional ones (e, f, g, h, track 3) in a double digest. Each band can be excised and treated with a second enzyme, a procedure that allows band identification. Thus, *Eco*R1 band A' consists of A and e (track 8) while *Hin*dIII fragment C is composed of e and f (track 6). In this manner, it is possible to construct a cleavage map showing the various restriction targets (right). Such a map is of paramount importance for cloning particular regions. (If the small h fragment, for instance, contains a gene of interest, the locus will be present in at least 10-fold higher concentrations on this fragment than on B). Multisite lesions, moreover, can be identified by restriction analysis since the removal (or addition) of target sites markedly changes the digestion products

Ligation of the loose ends *in vitro* with bacteriophage T4 or *E. coli* DNA ligase produces an intact *recombinant DNA molecule* (direct transformation in the absence of the ligation step relies on the host's endogenous ligase, see below). Flush-ended fragments can also be joined by ligation *in vitro*.

How does this *chimaera* enter the bacterium? Chemical-induced competence through calcium/magnesium chloride treatment is necessary prior to *transformation* (the process of transfer of naked viral nucleic acid is referred to as *transfection*). An r_K^- m_K^+ host (see Figure 7.16) is normally used to avoid destruction of heterologous DNA. Knowledge of the *E. coli* genetic system — host-controlled restriction-modification, recombination, and plasmid maintenance among others — and suitable phage and plasmid vectors makes it a particularly useful host for *cloning*.

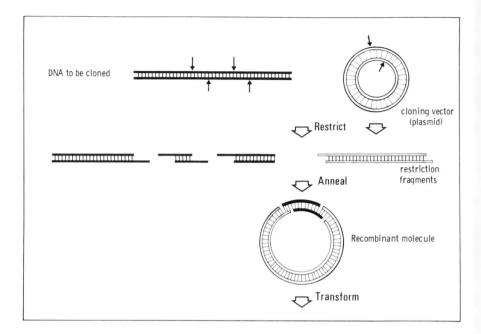

Figure 7.19: Cloning a Fragment of DNA in a Plasmid Vector
DNA to be cloned (black) and cloning vector (white) are treated with the same type II restriction
enzyme to generate molecules with complementary, 'sticky' sequences (the cloning vector in the
present example has only a single restriction target, a situation that may only be achieved by genetic
or *in vitro* manipulation). A recombinant molecule is formed by *in vitro* annealing (DNA ligase is
necessary where base pairing is not extensive). It is isolated after transformation of a suitable host
(generally r⁻m⁺). Clones carrying recombinants may be selected when the DNA inserted
complements a mutant defect (cf. cloning of *trp* genes) or through insertional inactivation when the
presence of foreign DNA inactivates a readily screened marker (colicin production, for example). *In
situ* DNA-DNA hybridisation (Section 7.2.2) using labelled DNA 'probes' allows the detection of
clones carrying specific foreign sequences

A wide variety of plasmid and phage vectors are now available for DNA
cloning. They have in common the ability to propagate in *E. coli* with (or
without) the DNA insertion. The use of small plasmids stems from the ease of
handling these stably inherited genetic elements. They carry convenient genetic
markers, such as resistance determinants (Section 4.2), that allow direct
selection for transformants. Moreover, relaxed plasmids are readily amplified
by chloramphenicol treatment. Their small size, indicating that few proteins are
encoded, makes these plasmids useful for studying the expression of cloned
sequences. With phage vectors, on the other hand, there is no requirement for
host survival as it is the progeny particles, themselves, that contain the foreign
DNA. Phage DNA is readily obtained (Section 7.3.1). There is, however, a
limit (as well as a minimum) to the size of the DNA fragment inserted since only
about 10 per cent more than the normal *lambda* genome — a linear molecule —

Table 7.1[a] Properties of Some Type II Site-Specific Restriction Endonucleases[b]

Microorganism[c]	Enzyme[d]	Restriction Characteristics	
		Sequence[e]	$\underline{\lambda}$ Cleavage sites[f]
Escherichia coli (R)	Eco RI	G A*A T T C C T T A A*G	5
Escherichia coli (R)	Eco RII	→C C A/T G G G G T/A C C← *	>35
Haemophilus aegyptius	Hae III	G G*C C C C G G *	>50
Haemophilus influenzae	Hind II	G T Y R*A C C A R Y T G *	34
Haemophilus influenzae	Hind III	* A A G C T T T T C G A A *	6

a   Abbreviations: R, purine base; Y, pyrimidine base.

b   Data from Roberts (1981).

c   The presence of a plasmid (or phage) is shown in parenthesis.

d   A note on nomenclature of restriction-modification systems: An italicized three-letter symbol derived from the genus and species name identifies the host organism, thus *Eco* and *Hin* (for *E. coli* and *H. influenzae*). A single, non-italicized letter following designates the strain (for instance, *Eco*k, *Hin*d) or plasmid (e.g. *Eco* RI, *Eco* PI) responsible for the host-mediated effect. In the cases where a single host (or plasmid) codes for a number of different systems they are designated by Roman numerals (for example, *Eco* RI, *Eco* RII, *Hin*d II, *Hin*d III). Restriction enzymes are formally prefixed endonuclease (or endo) R as in endo R·*Eco* RI. Clearly, there is not an unlimited number of cleavage sequences. Enzymes from different bacterial species may, therefore, share common specificities (e.g. *Bsu* RI from *B. subtilis* has the same recognition sequence as *Hae* III). Such enzymes have been termed *isoschizomers*. Note that despite sharing the same sequence, they may cleave at different positions.

e   The sequences are written 5' → 3'/3' → 5'. The sites of cleavage are given with an arrow. An asterisk indicates which bases are modified by corresponding methylases: C, 5-methylcytosine; A, N⁶-methyladenine.

f   The number of cleavage sites in bacteriophage *lambda* DNA is given. Thus, there are 5 *Eco* RI sites generating 6 fragments. Note how the number decreases as the cleavage sequence increases in size/specificity.

is packaged; the use of deletion derivatives circumvents this to a certain extent (Section 5.2.2). Conversely, in the case of plasmid vectors, a minimum size of 225 base pairs in length is required for circularisation.

DNA cloning is analogous to the *in vivo* construction of F-prime derivatives or specialised transducing phages (Sections 4.1.3 (b) and 5.2.2). The *in vitro* technique extends the range of foreign sequences carried, thus avoiding the limitations imposed by phage primary or secondary attachment sites (but see DNA insertion elements, Section 6.2.2 (a i)). The cloned fragment may, moreover, be specifically excised by cleaving with the same restriction endonuclease employed for vector opening and fragment production.

Recombinant technology is applied for the isolation of specific regions such as promoter sites or replication origins. In the 'shot-gun' approach, the whole DNA of an organism, be it a bacterium or eukaryote, is cut into defined fragments using a specific type II restriction enzyme to create a clone 'bank' containing the entire genomic material (akin to a generalised transducing lysate, Section 5.2.1). Thus, DNA from one source may be investigated in a markedly different background.

There is some controversy over the dangers inherent in the use of recombinant DNA techniques. It is feared that pathogenic DNA (be it for some known toxin such as cholera or a cryptic sequence), amplified by cloning, may invade the human system via the innocuous bacterium, *E. coli* (or one of its viruses). This particularly applies to 'shot-gun' experiments where there is no control over the nature of the DNA inserted into the cloning vector (apart from its source). For this reason, stringent physical and biological containment conditions are applied, the severity depending upon the nature of the DNA under study (cloning of *E. coli* DNA is considered safe!). This includes genetically disabled *E. coli* strains that are unable to grow under the normal physiological conditions found in the intestinal tract.

It has been estimated that a minimum of 1440 plasmids in the case of *E. coli* and, perhaps, 184 000 for *Drosophila* are required to include the total genome content (respective size of 3.8 and 180×10^6 base pairs). This clearly demonstrates the need to ensure that only recombinants (rather than just the parental molecule) are present in the clone bank. Complementation has been used to select for a particular gene (or group of genes): when the recipient has a deletion of the region concerned, only transformants (or transfectants) carrying the required sequences may survive. Alternatively, composite vectors are recognised by insertional inactivation, in which the insertion of foreign DNA inactivates a gene product encoded by the cloning vector (integration into the colicin E1 structural gene, for example, produces chimaeric plasmids that fail to synthesise the bacterial toxin). *In situ* DNA-DNA hybridisation (Section 7.2.2) using highly labelled DNA probes (prepared by nick translation *in vitro*, Section 6.1.1 (a)) allows the detection of clones (or plaques) carrying particular sequences.

Restriction technology, coupled with DNA sequencing methods, is an enormously powerful approach for the study of gene structure at the molecular level. It has provided a wealth of information on the structure of large eukaryotic genomes, and on prokaryotic sequences involved in coding and the regulation of gene expression. Indeed, many of the sequences discussed in Part IV were obtained in this manner.

Bibliography

7.1 Genetic Analysis of Bacteria and their Viruses

7.1.1 The Elements of Genetic Analysis

Hayes, W. (1968) *The Genetics of Bacteria and their Viruses*, 2nd edn (Blackwell, Oxford).

7.1.2 Mapping Bacterial Genes

Bachmann, B.J. and Low, K.B. (1980) 'Linkage map of *Escherichia coli* K-12, edition 6', *Microbiol. Rev., 44*, 1-56.
Beckwith, J. (1978) '*lac*: the genetic system', in J.H. Miller and W.S. Reznikoff (eds.), *The Operon* (Cold Spring Harbor Laboratory, New York), pp. 11-30.
Beckwith, J. and Rossow, P. (1974) 'Analysis of genetic regulatory mechanisms', *Ann. Rev. Genet., 8*, 1-13.
Clowes, R.C. (1960) 'Fine genetic structure as revealed by transduction', *Symp. Soc. Gen. Microbiol., 10*, 92-114.
Franklin, N.C. (1978) 'Genetic fusions for operon analysis', *Ann. Rev. Genet., 12*, 193-221.
Kleckner, N., Roth, J. and Botstein, D. (1977) 'Genetic engineering *in vivo* using translocatable drug-resistance elements: New methods in bacterial genetics', *J. Mol. Biol., 116*, 125-159.
Lederberg, E.M. (1960) 'Genetic and functional aspects of galactose metabolism in *Escherichia coli* K12', *Symp. Soc. Gen. Microbiol., 10*, 115-131.
Miller, J.H. (1972) *Experiments in Molecular Genetics* (Cold Spring Harbor Laboratory, New York).
Ullman, A. and Perrin, D. (1970) 'Complementation in β-galactosidase', in J.R. Beckwith and D. Zipser (eds.), *The Lactose Operon* (Cold Spring Harbor Laboratory, New York), pp. 143-172.
Yanofsky, C. (1967) 'Gene structure and protein structure', *Sci. Am., 216*, 80-94.
Yanofsky, C. (1971) 'Tryptophan biosynthesis in *E. coli*: Genetic determination of proteins involved', *J. Am. Med. Assoc., 218*, 1026-1035.

7.1.3 Mapping Phage Genes

Benzer, S. (1962) 'The fine structure of the gene', *Sci. Am., 206*, 70-84.
Campbell, A. (1971) 'Genetic structure', in A.D. Hershey (ed.), *The Bacteriophage Lambda* (Cold Spring Harbor Laboratory, New York), pp. 13-44.
Edgar, R.S., and Epstein, R.H. (1965) 'The genetics of a bacterial virus', *Sci. Am., 212*, 70-78.

7.2 Monitoring Gene Function

Coakley, W.T., Bater, A.J. and Lloyd, D. (1977) 'Disruption of microorganisms', *Adv. Microb. Physiol., 16*, 279-341.
Cooper, T.G. (1977) *The Tools of Biochemistry* (Wiley, New York)
Gaál, Ö., Medgyesi, G.A. and Vereczkey, L. (1980) *Electrophoresis in the Separation of Biological Macromolecules* (Wiley, Chichester).
Kennell, D.E. (1971) 'Principles and practices of nucleic acid hybridisation', *Prog. Nucl. Acid Res. Molec. Biol., 11*, 259-301.

7.3 Physical Analysis of Gene Structure

7.3.1 DNA Isolation

Brown, D. and Stern, R. (1974) 'Methods of gene isolation', *Ann. Rev. Biochem., 43*, 667-693.

7.3.2 Heteroduplex Analysis

Davidson, N., Deonier, R.C., Hu, S. and Ohtsubo, E. (1975) 'Electron microscope heteroduplex studies of sequence relations among plasmids of *Escherichia coli*, X: Deoxyribonucleic acid sequence organisation of F and F-primes, and their sequences involved in Hfr formation', in D. Schlessinger (ed.), *Microbiology 1974* (American Society for Microbiology, Washington), pp. 56-65.

Fiandt, M., Hradecna, Z., Lozeron, H.A. and Szybalski, W. (1971) 'Electron micrographic mapping of deletions, insertion and homologies in the DNAs of coliphage lambda and phi 80', in A.D. Hershey (ed.), *The Bacteriophage Lambda* (Cold Spring Harbor Laboratory, New York), pp. 329-354.

Fisher, H.W. and Williams, R.C. (1979) 'Electron microscopic visualisation of nucleic acids and of their complexes with proteins', *Ann. Rev. Biochem., 48*, 649-679.

7.3.3 Restriction Technology

Air, G.M. (1979) 'DNA sequencing of viral genomes', in H. Fraenkel-Conrat and R.R. Wagner (eds.), *Comprehensive Virology, 13: Structure and Assembly* (Plenum Press, New York), pp. 205-292.

Arber, W. (1974) 'DNA modification and restriction', *Prog. Nucl. Acid Res. Molec. Biol., 14*, 1-37.

Arber, W. (1979) 'Promotion and limitation of genetic exchange', *Science, 205*, 361-365.

Benzinger, R. (1978) 'Transfection of *Enterobacteriaceae* and its applications', *Microbiol., Rev., 42*, 194-236.

Brammar, W.J. (1977) 'The construction *in vitro* and exploitation of transducing derivatives of bacteriophage λ', *Biochem. Soc. Trans., 5*, 1633-1652.

Collins, J. (1978) 'Gene cloning with small plasmids', *Curr. Topics Microbiol. Immunol., 78*, 122-173.

Curtiss, R. (1976) 'Genetic manipulation of microorganisms: Potential benefits and biohazards', *Ann. Rev. Microbiol., 30*, 507-533.

Khorana, H.G. (1979) 'Total synthesis of a gene', *Science, 203*, 614-25.

Lacks, S.A. (1977) 'Binding and entry of DNA in bacterial transformation', in J.L. Reissig (ed.), *Receptors and Recognition, B3: Microbial Interactions* (Chapman and Hall, London), pp. 177-232.

Low, K.B. and Porter, D.D. (1978) 'Modes of gene transfer and recombination in bacteria', *Ann. Rev. Genet., 12*, 249-287.

Old, R.W. and Primrose, S.B. (1980) *Principles of Gene Manipulation* (Blackwell, Oxford).

Roberts, R.J. (1976) 'Restriction endonucleases', *CRC Crit. Rev. Biochem., 4*, 123-164.

Roberts, R.J. (1981) 'Restriction and modification enzymes and their recognition sequences', *Nucl. Acids Res., 9*, r75-r96. (This list is updated annually.)

Smith, H.O. (1979) 'Nucleotide sequence specificity of restriction endonucleases', *Science, 205*, 455-462.

Wu, R. (1978) 'DNA sequence analysis', *Ann. Rev. Biochem., 47*, 607-634.

Wu, R., Bahl, C.P. and Narang, S.A.C. (1978) 'Synthetic oligodeoxynucleotides for analysis of DNA structure and function', *Prog. Nucl. Acid Res. Mol. Biol., 21*, 101-141.

Part IV
Gene Regulation

8 Operon Control

Gene expression consists of two major stages, DNA transcription and RNA translation (Chapter 2). Although cellular modulation of either stage is sufficient to regulate protein production, control at the post-transcriptional level is energetically wasteful. Moreover, the transcription process supplies a ready means for the *coordinate regulation* of both linked and unlinked genes. Prokaryotic gene regulation acts predominantly, in fact, at the transcriptional level.

Transcriptional control is mediated through variation in the RNA chain initiation and termination rates. It involves, therefore, specific sequences of DNA, *regulatory regions*, at which certain *regulatory elements* interact to modulate the production of RNA molecules. A single regulatory region supplies the necessary information for either monocistronic or polycistronic transcripts. Moreover, it is the diffusible nature of the regulatory elements concerned, be they proteins or small organic molecules, which enables them to exert control over either linked or dispersed genes.

Other forms of control, those acting on the transcript or its translation, would seem to be in the minority in prokaryotic systems. Nevertheless, post-transcriptional control appears to be important to allow fine adjustment. This is particularly the case in the expression of simple genetic elements such as bacteriophages.

Whether control acts at the transcriptional or post-transcriptional level, genetically encoded sequences are involved. On the one hand, modulation in the RNA chain initiation or termination rates is effected through interactions with DNA. On the other, post-transcriptional control involves a specific RNA target. In either case, control over one or more (linked) genes results from the presence of sequences adjacent to, sometimes lying between, genes that they control. An operon may be defined as that genetic region carrying one or more genes and the associated sequence responsible for its/their expression. Regulation that acts at the genetic level, that is on DNA or RNA, may, thus, be designated *operon control*.

8.1 Transcriptional Control

Gene expression in prokaryotes is regulated mainly at the transcriptional level. Such control is slower in response than direct action on the gene product. It has, however, the ability to conserve both the energy supply and the transcription-translation apparatus. It should be noted that though transcriptional control functions on *de novo* RNA synthesis, pre-formed mRNA species are subject to rapid degradation (Section 2.2.1 (b)). This efficient scavenging system, thus, prevents unnecessary translation after imposition of a block on RNA chain initiation.

The movement of RNA polymerase molecules along a DNA strand can be likened to the flow of water down a pipe: both represent steady-states. The amount of water coming out at any one time can be no greater nor less than that flowing in. That is, the rate of RNA chain initiation must be equivalent to the rate of RNA chain termination once the steady state is reached (Figure 8.1). Overall, the number of molecules initiating (or terminating) at any one time is dictated by the flow rate, the RNA chain elongation rate, since RNA polymerase binding is dependent upon translocation of the previously bound molecule (there is no apparent 'storage' site for multiple polymerase copies to await initiation). The elongation rate must, therefore, be always greater than the on/off rate to accommodate initiating molecules. The RNA chain elongation rate is, however, largely invariant.

The maximal initiation rate may, therefore, be determined from this constant and the space on a DNA duplex over which one molecule of RNA polymerase extends. Since the distance travelled by the enzyme in one second (about 55-60 bp) is approximately equivalent to the space it occupies, the maximal initiation rate physically possible is about once per second. Only in the transcription of rRNA genes under optimal growth conditions (i.e. nutrient broth) is this rate achieved. Transcription of structural genes is generally initiated at significantly lower frequencies (30- to 60-fold less).

Synthesis of RNA is controlled both at the initiation and termination level. In the first, the on-rate is directly controlled. The second, termination control, does not block release of a completed transcript at the end of an operon. Rather, it acts at sequences preceding genes or at intercistronic regions to vary transcriptional readthrough, i.e. elongation of initiated RNA chains. In this manner, the number of productive initiation events is modulated. In summary, the number of RNA polymerase molecules engaged in transcription at any one time and, thus, the total RNA synthetic capacity is dictated by the on-rate *(promoter control)* and the off-rate *(attenuator control)*.

The present discussion is concerned with a general description of control circuits. It should be stressed that prokaryotic gene expression frequently involves numerous levels of regulation. Control of bacterial gene expression is considered in Chapter 9, and Chapter 10 deals with regulation of genetic elements such as plasmid and phage replicons.

8.1.1 Control at RNA Chain Initiation

By far the most efficient means of gene control is that involving the transcriptional initiation stage. Such *promoter control* makes use of *negative* and/or *positive* regulatory elements, factors that either inhibit RNA chain initiation or promote it.

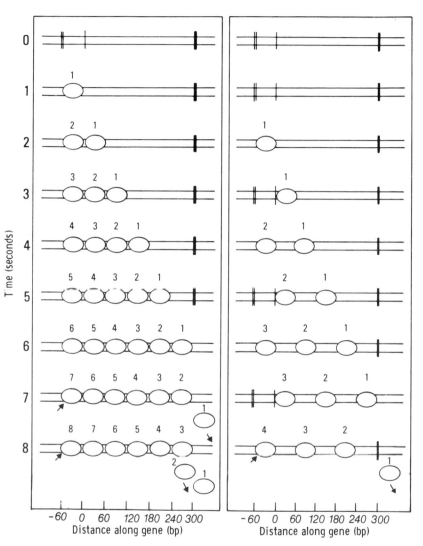

Figure 8.1: Steady-State Polymerase Kinetics: The Equality between the On/Off Rates

The number of RNA polymerase molecules initiating at any one time is equal to the number terminating under steady-state conditions. In the present examples, on/off rates are 1 (left) and 0.5 (right) per second (steady-state having been achieved after 7 and 8 seconds, respectively)

(a) Promoter Control

(i) *The Elements of Promoter Control.* On the operon model of Jacob and Monod, first introduced in 1961 and now widely substantiated, gene regulation is genetically built into an organism. That is, expression of a gene (and other linked genes) is dictated by specific sequences adjacent to the coding region. These *regulatory regions* are responsible for both transcription of the genes they control and for the response to environmental changes. Control of expression is mediated through diffusible elements. The activity of these *regulatory proteins* is often modified by *effectors*, small organic molecules such as sugars or ribonucleotides. The proteins are allosteric in that they are multimeric complexes with distinct sites for the recognition of DNA and effector. They are generally regarded to be synthesised at a constant rate, expression being independent of the gene(s) they control. (Some regulatory elements are themselves controlled (cf. *recA-lexA*, Section 6.2.1 (a i); the F *tra* operon, Section 10.1; the *lambda* regulatory genes, Section 10.2).) Regulatory proteins may either prevent transcription *(negative control)*, or promote it *(positive control)*. The active species are termed *repressors* and *activators*, respectively. A negative-control protein that requires an effector, a *co-repressor*, for function — one that is unable to inhibit transcription *per se* — is referred to as an *apo-repressor*. In keeping with this nomenclature, *apo-activators* only promote RNA production in the presence of their respective *co-activator* (see Table 8.1).

The site of action of these controlling elements, the regulatory region, appears to consist of up to three main elements: a *promoter (P)* at which transcription initiates; a repressor-binding region, the *operator (O)*; and, a *promoter-potentiation* site at which the activator interacts. Regulatory regions need not contain all three of these sites although a promoter is clearly necessary for specific transcriptional initiation. Moreover, these sites are not necessarily physically distinct. It will be seen later, for example, that in certain operons dependent upon an activator for their expression, the promoter-potentiation site is strictly speaking part-and-parcel of the promoter (see *lac*, Section 9.1.2 (b)). *Operator-promoter interpenetration* has also been known for some time though this need not apply to all repressor-controlled operons. The order of these three sites when present in a single operon is likely to be, promoter-potentiation site — P — O — gene. The operator generally either overlaps the promoter or lies between the promoter and the gene it controls since this orientation allows a simple blockade mechanism for repressor-mediated transcriptional inhibition.

Promoter control — regulation at the transcriptional initiation stage — may function in two main ways: on different genes within the same operon or on genes of unlinked operons. In the first, genes involved in the same metabolic pathway are closely clustered such that they are transcribed into a single message. This *polycistronic* transcript carries, therefore, the genetic information for a number of different genes. Since each gene in a polycistronic operon is physically linked, an increase or decrease in the rate of synthesis of the polycistronic message results in a comparable increase or decrease in the expression of all the gene products encoded (but see Polarity, Section 8.1.2 (a) and Attenuator Control, Section 8.1.2 (b)). *Coordinate regulation* of a polycistronic operon is, thus, effected through a dependence upon a common regulatory region that supplies the necessary structural elements for transcriptional initiation and its control. Promoter control may also act on dispersed genes thanks to the diffusible nature of the regulatory elements concerned. Clearly, for a single regulatory protein to interact at a number of regions, the control sequences must be similar, if not identical. In conclusion, control of RNA chain initiation through specific regulatory regions and diffusible regulatory elements permits coordinate regulation of both linked and unlinked genes.

(ii) *Promoter Control Circuits.* Expression of promoter-controlled operons is mediated by both regulatory proteins and effectors. It is the effector that is sensitive to physiological changes (positive- and negative-controlled systems are unchanging in the absence of this second controlling element). Effectors are, in general, small organic molecules closely related to the metabolites of the biochemical pathway controlled. Effectors mediate in gene expression also in a negative or positive manner by either inhibiting or activating a specific regulatory protein.

Effector-dependent 'switching' on and off of gene expression is referred to as *induction* and *repression*, respectively. Clearly, both negatively- and positively-controlled operons may be inducible or repressible (Table 8.1). There are, therefore, two main types of negatively-controlled systems: in the inducible one, the *repressor* protein is inactivated by an *inducer*; and in the repressible one, the repressor is a complex of an *apo-repressor* protein and *co-repressor* (Table 8.1 and Figure 8.2). Similarly, in positively controlled systems, the functional activator species either consists of an *apo-activator* gene product and *co-activator* (inducible) or it is a single entity, inactivated by a specific inhibitor (repressible) (Table 8.1 and Figure 8.3).

A theoretical analysis of control systems suggests that negative control is employed where there is a low demand for expression and positive control where a high level is required. Furthermore, as a broad generalisation, inducible systems are catabolic, allowing the use of energy sources other than glucose through *de novo* synthesis of specific degradative enzymes. Repressible systems, on the other hand, are generally associated with biosynthetic processes and prevent unnecessary flux through a pathway 'under conditions of plenty'. Lactose breakdown and tryptophan synthesis, examples of these two processes, are discussed in Chapter 9.

Finally, it should be noted that a separate regulatory gene is not always necessary. In *autogenous control*, the expression of a gene is regulated by the gene product, itself (in either the negative or positive mode, see *lambda* cI and *cro*, Section 10.2.1 (a)). Since synthesis of the regulatory protein/gene product is not invariant in this case, no effector molecules are required to alter its

Table 8.1 Characteristics of Negative- and Positive-Control Systems

Mode	Regulatory Elements[a]		Operon Function		Operon Expression in the Absence of Regulatory Protein
	Regulatory Protein	Effector	Operational	Inoperative	
negative, inducible	repressor	inducer	repressor + inducer	repressor	constitutive
negative, repressible	apo-repressor	co-repressor	apo-repressor	apo-repressor + co-repressor	constitutive
positive, inducible	apo-activator	co-activator	apo-activator + co-activator	apo-activator	absent
positive, repressible	activator	inhibitor	activator	activator + inhibitor	absent

a   There is a varied terminology for these regulatory elements (the co-activator has been termed for example , 'inducer'). The system presented, though not necessarily general, distinguishes between the different types of regulatory proteins and effectors. The prefix 'apo' or 'co' indicates that the particular protein or effector is not active *per se*. Thus, repression is achieved by the presence of both apo-repressor and its specific co-repressor in the case of a negatively-controlled repressible operon.

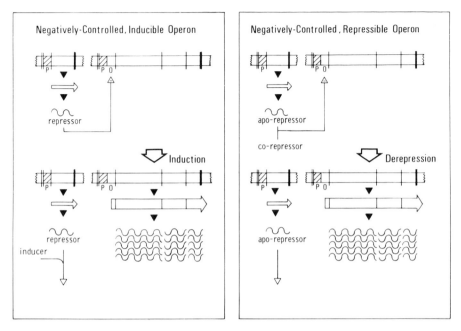

Figure 8.2: Negative Control of Gene Expression; The Involvement of a Repressor

Regulation of a negatively controlled, inducible operon is mediated by a single entity, a repressor protein (left). Induction involves inactivation of this control element by an inducer molecule. In a negatively controlled, repressible operon, the active regulatory element is a complex of apo-repressor and co-repressor, the absence of the latter leading to derepression (right). In each case, the regulatory complex interacts at the operator (note that the regulatory gene, for repressor or apo-repressor, may be unlinked from the locus it controls). See also Table 8.1/Figure 8.3

function. Such parsimony is of particular interest.

(b) Cis- and Trans-Acting Regulatory Mutations. Regulation at the level of transcriptional initiation is mediated by proteins acting on specific DNA sequences and, hence, two types of regulatory mutation are possible: in structural genes encoding the regulatory proteins themselves, and in the regulatory region. Regulatory mutations have, characteristically, pleiotropic effects. However, whereas lesions in a regulatory gene result in the altered expression of all genes under its control (whether linked or dispersed), changes in the region at which the diffusible elements interact is responsible for perturbation in the synthesis of only those structural genes directly adjacent to, and co-transcribed from, that region (Figure 8.4).

A regulatory protein can be supplied both in *cis* and *trans* (cf. the *cis-trans* test, Section 7.1.1 (a)). In fact, complementation in *trans* of a pleiotropic-acting lesion has been taken as evidence for a structural locus that encodes a protein regulatory element (note that certain regulatory mutations exhibit

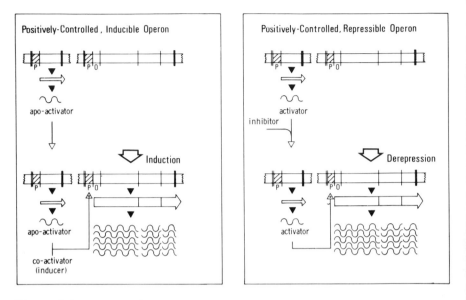

Figure 8.3: Positive Control of Gene Expression; The Involvement of an Activator

An inducible, positively controlled operon requires both the protein apo-activator and co-activator (inducer) for induction (left). In a positively controlled, repressible operon, derepression is mediated by inactivation/loss of the inhibitor (right). In each case, the regulatory complex interacts at a site within the promoter (note that the regulatory gene, for apo-activator or activator, may be unlinked from the locus it controls). See also Table 8.1/Figure 8.2

negative complementation in that the mutant, rather than the wild-type, allele is dominant in *trans*, Section 7.1.1 (a)). On the other hand, a physiological defect that stems from lesions in a control region cannot be complemented in *trans*. Moreover, such regulatory site lesions are always *cis*-specific. Thus, while regulatory mutations *per se* exhibit pleiotropic effects, only lesions in the regulatory sequences themselves are *cis*-specific.

That the four general control circuits outlined above (Figures 8.2 and 8.3) are quite distinct, is aptly demonstrated by lesions that affect the function of the regulatory protein involved (that is, mutations in the regulatory gene or at the site at which it interacts). For, in operons under negative control, inactivation of the repressor (or apo-repressor) structural gene imparts uncontrolled, constitutive expression (Figure 8.5). In a positively-controlled system, conversely, a non-functional activator results in a permanent loss of expression (unless a wild-type copy of the activator gene is supplied in *trans*). Moreover, if a functional regulatory gene is supplied in *trans* to a constitutive strain carrying a mutant allele, operons subject to negative control should regain their inducible property, while the phenotype of positively-regulated systems should remain unchanged (although this is by no means a hard-and-fast rule, due, for example, to negative

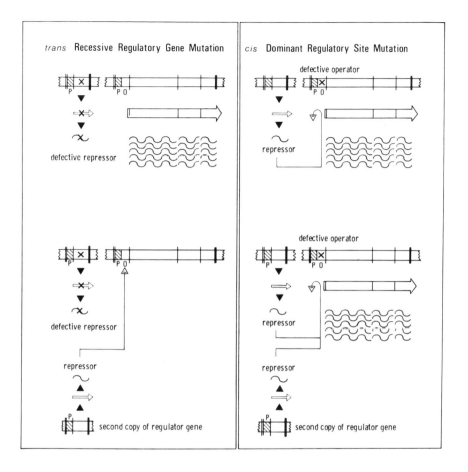

Figure 8.4: *Trans-* and *Cis-* **Acting Regulatory Elements**
A scheme showing the effect of the presence of a second copy of a regulatory gene on expression of a negatively controlled, inducible operon carrying a lesion in either the regulatory gene (left) or in the operator site at which the repressor interacts (right). Whereas regulatory gene mutations are *trans*-recessive (such that the defective repressor is complemented by the second gene copy), regulatory site lesions are *cis*-specific and are not affected by a second, functional regulatory gene in *trans*. Note that both types of lesions are, however, pleiotropic in their action, altering expression of all genes subject to control by the repressor (*cis* or *trans*) or operator (*cis* only)

complementation and *cis*-specific proteins, it serves as a useful guide for merodiploid studies).

8.1.2 Control at RNA Chain Termination

Transcription termination occurs, in general, at specific DNA sequences; the accessory protein, rho (Section 2.1.1 (b)), is necessary for some sites. The location of these specific *terminators* at the end of coding sequences allows the release of a complete, functional RNA chain (Section 2.1.2 (a iii)). Termination

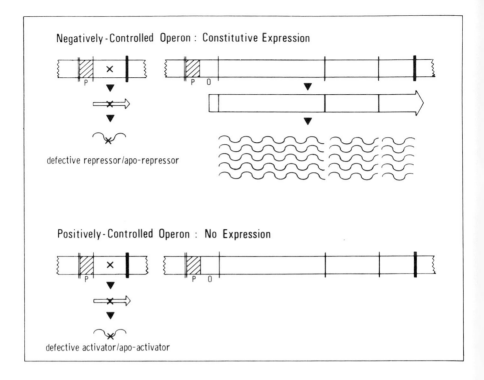

Figure 8.5: Differentiation of Regulatory Mode from the Effect of Regulatory Gene Mutations
A regulatory mutation in a structural gene (rather than in the control site) results in constitutive expression in the case of a negatively controlled operon (top) and no expression for a positively controlled operon (bottom), irrespective of whether they are inducible or repressible (see Table 8.1)

sequences may also lie internally, within or between genes, or even between a promoter and the gene(s) they control. They, therefore, prevent transcriptional readthrough when operational. It is at these internal sites that RNA chain termination is controlled.

(a) Polarity and its Suppression. Many nonsense and frameshift mutations affect expression not only of the gene in which they reside but also distal genes of the same polycistronic operon (Section 3.1.1 (c)). Not all nonsense lesions exert such a *polar* effect. The position of the mutation within a structural gene appears, in fact, to dictate the degree of expression of distal cistrons. This dependence upon genetic context for polarity generates a gradient (Figure 8.6). Each peak appears to locate the site of a translational re-start. Although there may be various polarity maxima across a single gene (represented by the 'troughs' in Figure 8.6), promoter-proximal nonsense mutations are generally most polar (see also Natural Polarity, below).

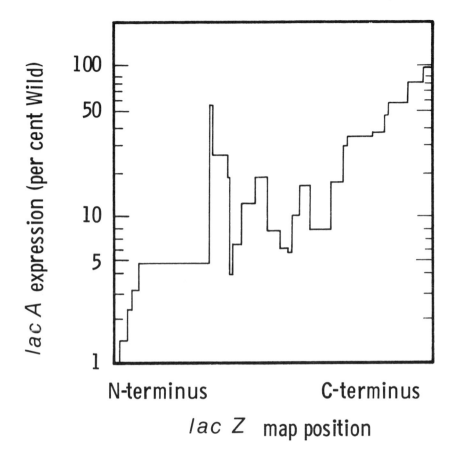

lac A expression (per cent Wild)

100
50
10
5
1

N-terminus   C-terminus

lac Z map position

Figure 8.6: Gradients of Polarity in the *lacZYA* Polycistronic Operon
The graph shows the effect of nonsense mutations in *lacZ* on expression of the distal gene *lacA*, the
lac operon being transcribed in the direction *lacZ* → *lacA* (see The *lac* Operon, Section 9.1). The
amount of β-galactoside transacetylase, the *lacA* gene product, as a percentage of the wild-type level,
is plotted on a log scale against the position of the nonsense lesions in the *lacZ* gene. The 'troughs'
are, therefore, positions of high polarity. The three clear 'peaks' are in regions with translational
reinitiation codons (internal start codons) that minimise polarity. After Zipser *et al.* (Zipser, D.,
Zabell, S., Rothman, J., Grodzicker, T. and Wenk, M. (1970) *J. Mol. Biol., 49*, 251-254)

In considering models for this phenomenon, it should be remembered that
nonsense lesions, unlike most other point mutations, interfere with the translation
process. The presence of such an internal translational stop codon (amber, ochre
or UGA, Section 2.2.1 (a)) results in premature translational termination and
the production of an incomplete polypeptide chain.

Two early models for polarity were quite disparate in nature, being based upon
mRNA degradation or transcription-translation coupling. The first proposed
that the message, bare of ribosomes from premature translation termination, is
susceptible to degradation by endogenous ribonucleases. This model is clearly in

keeping with the natural instability of bacterial mRNAs (Section 2.1.2 (b)). It successfully predicts that full-sized transcripts may be produced but are rapidly degraded. On the transcriptional-translational coupling hypothesis, translocation of RNA polymerase along the DNA template is dependent upon ribosomal movement. Since translational termination occurs at nonsense codons, it suggests that RNA polymerase is unable to transcribe far beyond a nonsense site — only a partial transcript is produced, as has been observed in some instances.

Recently, a 'unifying' model for polarity has been presented by Adhya, Gottesman and their colleagues, based upon transcriptional termination. It is now known, for example, that certain mutations that map in the *rho* gene (previously designated *suA* before the identity of the gene product was known) prevent particular polar effects. That is, failure of modified rho protein to act may give rise to *polarity suppression*.[1] On the Adhya and Gottesman model, transcription and translation are closely linked. However, it is not suggested that translation actually 'drives' transcription. Rather, premature translational termination leaves engaged RNA polymerase molecules susceptible to the transcription factor, rho.

Premature transcriptional termination and, hence, polarity, is explained by rho-dependent intracistronic (and intercistronic) terminators. It is proposed that rho is unable to interact directly with RNA polymerase to effect termination but also requires binding to 'naked' mRNA for function (Figure 8.7). The transcription factor engages with polymerase through a one-dimensional walk, making use of its inherent NTPase activity to translocate in the 5'-3' direction along the transcript. During normal events, internal transcriptional terminators are not operational since the presence of ribosomes prevents rho from binding (according to the Adhya-Gottesman hypothesis). However, rho may act when premature translational termination at a nonsense codon leaves the message distal to the nonsense site free of ribosomes (Figure 8.7).[2]

Whether premature transcriptional termination is generated by translational termination depends upon the rho-RNA polymerase complex encountering a transcriptional terminator. RNA chain termination does not take place if there is initiation of translation prior to interaction of rho and polymerase. Thus,

---

1.   Note that *polarity suppressors* are quite distinct from nonsense suppressors. Whereas the former involves derivatives of the transcription terminator factor rho, the latter is mediated by mutant tRNA species (Section 3.3.2). Thus, polarity suppression prevents premature transcriptional termination and nonsense suppression inhibits premature translational termination. Moreover, polarity suppression restores expression of distal genes of the same operon but not of the gene carrying the nonsense mutation, while nonsense suppression restores expression of the whole operon (in either case, restoration is unlikely to be complete).

2.   The Adhya-Gottesman model also allows an understanding of polarity induced by transposable genetic elements such as IS sequences and transposons (Section 6.2.2 (a iii)). The presence of translational termination codons in the three reading-frames and a rho-dependent transcriptional terminator would mimic the action of a polar nonsense mutation. The transcriptional terminator need not reside on the insertion element. Thus, whereas IS2 carries a transcriptional terminator and causes termination within the element itself, IS1-mediated polarity depends on transcriptional termination in the adjacent coding material.

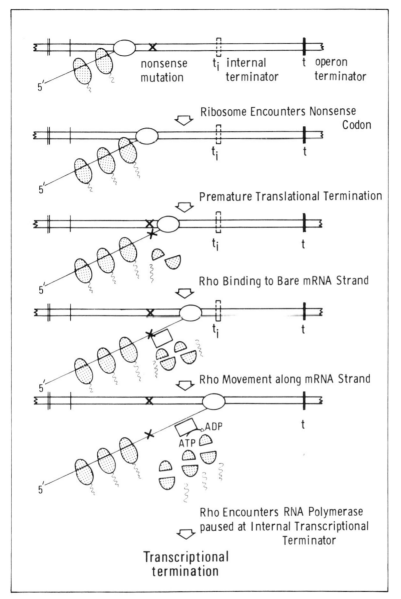

Figure 8.7: A Model for Polarity
Release of ribosomes from a message (due to the presence of a nonsense mutation) allows rho binding. If, upon NTPase-mediated translocation along the 'bared' template, the transcriptional factor encounters the RNA polymerase complex close to or at an internal transcriptional terminator, premature termination occurs and transcription of distal genes of the same operon is prevented. Such polarity is avoided when there is a translational restart site between the nonsense codon and internal transcriptional terminator. After Adhya and Gottesman (Adhya, S. and Gottesman, M. (1978) *Ann. Rev. Biochem., 47,* 967-996)

ribosome binding to the message and polypeptide chain initiation, either within the structural gene carrying the nonsense mutation or at the neighbouring gene, physically prevents rho action and allows distal gene expression. Transcription and, thus, polarity depends upon *translational re-initiation* (see below).

There is strong support for the Adhya-Gottesman model. It has been shown, for example, that rho not only binds strongly to RNA but also that this interaction is apparently necessary for function. Certainly, the integral RNA-dependent NTPase activity is required for rho-mediated termination. Moreover, the presence of intracistronic transcriptional terminators explains the phenomenon of polarity gradients (Figure 8.6). Each peak, locating the site of a translational re-start codon (and/or the absence of an internal transcriptional terminator between the nonsense mutation and next translational start point), would represent polypeptide chain initiation and, thus, transcriptional read-through. Each trough, on the other hand, would stem from translational termination and the associated transcriptional event due to the presence of an intervening rho site before the next restart point. Finally, this model elegantly explains why distal genes of some polycistronic operons are normally transcribed at a reduced rate (see, for example, the Lactose Operon, Section 9.1.2(a)). This *natural polarity* would be generated by occasional transcriptional termination events, most probably in intercistronic regions. Certainly, natural polarity in the lactose operon can be largely suppressed by mutations that alter the rho protein. Note that such natural phenomena supply a ready means for unlinking the expression of co-transcribed genes.

(b) Attenuator Control. The positioning of transcriptional termination sequences in front of a coding region offers another level of control over RNA production. Transcriptional terminators that precede or lie between genes and are actively regulated (as compared with polarity, see (a)) are called *attenuators*.[3] These sites are responsible for controlling transcriptional readthrough (Figure 8.8). At one extreme, when the attenuator is non-functional *(anti-attenuation)* the majority of initiated RNA polymerase molecules proceed from the promoter into the adjacent gene: at the other, most polymerases terminate at the attenuator *(attenuation)* and little if no RNA is produced from this adjacent region (Figure 8.8 shows only one out of ten transcripts escaping attenuation). Thus, in operons under attenuator control, the rate of production of completed RNA molecules is dictated by both the efficiency of transcriptional initiation and the proportion of engaged RNA polymerase molecules proceeding past the attenuator site. That is, the on-rate (promoter control) and the off-rate (attenuator control) regulate the number of polymerase molecules engaged in transcription. Attenuator

---

3.   The term 'attenuator' was initially coined by Kasai (Kasai, T. (1974 *Nature, 249,* 523-527) for a transcriptional barrier before the first gene of the *S. typhimurium* histidine operon, *hisG.* Chronologically speaking, this phenomenon was first discovered in *trp* by Yanofsky and his group, who showed that certain deletions increased rather than decreased distal expression (see Section 9.2.2 (b)).

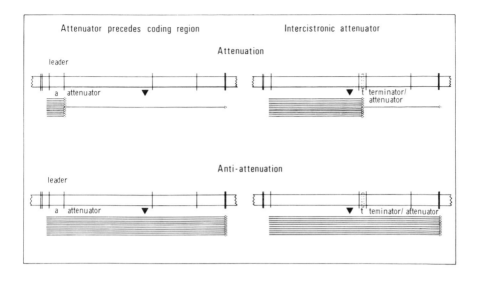

Figure 8.8: Transcriptional Regulation through Attenuation
A terminator preceding (left) or lying between genes (right) that is actively controlled is termed an attenuator. Modulation of transcriptional readthrough determines, on the one hand, the degree of expression of the entire operon and, on the other, whether promoter-proximal but not distal genes are expressed. In the present examples, attenuation allows readthrough of only one out of 10 RNA chains initiated, with 9 out of 10 chains terminating at the attenuator site. When the terminator is not functional, there may be complete transcriptional readthrough

control adds another level of finesse to gene regulation in that it implies that not all initiated RNA chains are destined for completion.

Attenuator control is well documented for biosynthetic operons (discussed in Section 9.2.2 (b)). Here, the terminator is sited in the leader sequence, proximal to the first structural gene of the operon. In the absence of anti-termination, therefore, it is mainly short incomplete transcripts that are synthesised. Attenuation control has also been observed at intercistronic regions (considered with respect to phage *lambda* expression, Section 10.2.3 (b)). In this case, termination prevents extension of a functional message rather than causing the production of short, non-functional species (Figure 8.8).

It is proposed that attenuation in biosynthetic operons is, perhaps, a rho-dependent process, modulated by ongoing translation of the leader RNA (cf. Polarity and its Suppression, above). Transcriptional termination requires a particular secondary structure in the leader transcript, formed during its use as a translational template. Anti-attenuation in such biosynthetic operons is, thus, mediated by the translational apparatus itself (see Section 9.2.2 (b) for attenuator control in the *trp* operon). Anti-attenuation in phage systems, on the other hand, results from specific proteins that prevent transcriptional termination at these intercistronic terminators through antagonism of rho action (see Section 10.2.3 (b)).

8.2 Post-Transcriptional Control

Transcriptional control is an efficient means of gene regulation since it prevents unnecessary load on the expressive machinery (Section 8.1). Such mechanisms, however, are unable to respond rapidly to physiological signals, the invariant RNA and polypeptide chain elongation rates being the limiting factor (Chapter 2).[4] Thus, although prokaryotic gene expression is controlled, in the main, at the transcriptional level, other stages in the production of RNA and protein are also targets for regulation. The interplay of these different control circuits achieves a fine level of adjustment of important cellular constituents, particularly regulatory elements themselves.

8.2.1 Control at the Level of the Transcript

Regulation at the level of the RNA product may be effected through the synthesis of a precursor molecule, with the rate of cleavage and trimming dictating the function of the mature species. *Processing* may also involve *in situ* base modification. It is sensitive, therefore, to alterations in the enzymes involved or at the sites at which they act.

Only a small number of specific transcripts are known to be susceptible to this type of post-transcriptional event. RNA processing has been observed in ribosomal and transfer RNA synthesis (discussed in Section 9.3.1 (b)), r-protein and RNA polymerase production (Section 9.3.2 (c)), and in the phage T7 early message.

8.2.2 Translational Control

There are few clear instances of translational control in *E. coli*, although recent evidence suggests that the transcription-translation apparatus is regulated at this level (discussed in Section 9.3.2 (c)). In simple genetic elements, for instance phages, translational control appears to be quite common. The RNA phage class is a case in point (Section 2.2.2 (c)). Moreover, synthesis of the *lambda* repressor in the lysogenic state is maintained at a low level through the absence of a ribosome binding sequence on the transcript (discussed in Section 10.2.2 (b)). The phage T4 gene 32 product, a DNA binding protein involved in replication, recombination and repair, regulates its synthesis in an autogenous manner. It

---

4.   Consider an average gene of 1200 nucleotides. Say, the rate of RNA chain elongation is 60 nucleotides per second. It, therefore, takes 20 seconds for one RNA polymerase molecule to traverse the gene. If the rate of initiation is changed from some value x events per second per gene to any other value, say, 10x, it takes 20 seconds for the overall rate of production to reach its new equilibrium value 10 times its previous level. The same argument applies to translational elongation.

appears to bind to its message when present in excess, to repress its own synthesis.

Finally, post-translational control, although it involves alterations in the protein product — cleavage and modification — rather than specific nucleic acid sequences, is briefly considered here for the sake of completeness (see also Section 2.2.2 (a)). Specific degradation of regulatory proteins has been observed. The *recA* gene product, for example, appears to mediate in cleavage of the bacteriophage *lambda* repressor (Sections 6.2.1 (a i) and 10.2.3 (a)). Moreover, certain positive regulatory proteins involved in *lambda* control have been shown to turnover rapidly (discussed in Section 10.2.3 (c)). With respect to protein modification, components of the transcription-translation apparatus — RNA polymerase and r-proteins (Sections 2.1.1 (a) and 2.2.1 (c)) — are subject to *in situ* alteration. In conclusion, post-translational control regulates function rather than synthesis, allowing sequential control over gene expression.

Bibliography

Adhya, S. and Gottesman, M. (1978) 'Control of transcription termination', *Ann. Rev. Biochem.*, *47*, 967-996.

Beckwith, J. and Rossow, P. (1974) 'Analysis of genetic regulatory mutations', *Ann. Rev. Genet.*, *8*, 1-13.

Franklin, N.C. (1978) 'Genetic fusions for operon analysis', *Ann. Rev. Genet.*, *12*, 193-221.

Goldberger, R.F. (ed.) (1979) *Biological regulation and development, I: Gene expression* (Plenum Press, New York).

Jacob, F. and Monod, J. (1961) 'Genetic regulatory mechanisms in the synthesis of proteins', *J. Mol. Biol.*, *3*, 318-356.

Mahon, G.A.T., McWilliam, P., Gordon, R.L. and McConnell, D.J. (1980) 'The time course of transcription', *J. Theor. Biol.*, *87*, 483-515.

Platt, T. (1981) 'Termination of transcription and its regulation in the tryptophan operon of *E. coli*', *Cell*, *24*, 10-23.

Revel, M. and Groner, Y. (1978) 'Post-transcriptional and translational control of gene expression in eukaryotes', *Ann. Rev. Biochem.*, *47*, 1079-1126.

Zipser, D. (1970) 'Polarity and translational punctuation', in J.R. Beckwith and D. Zipser (eds.), *The Lactose Operon* (Cold Spring Harbor Laboratory, New York), pp. 221-232.

9 Control of Bacterial Gene Expression

A myriad of gene products are involved in cellular metabolism, in the replicative and expressive machineries and in structural components. However, during the (normal) bacterial life cycle, certain products are required only at particular stages. Moreover, adjustment in cellular metabolism may be necessary when the bacterium is confronted with an unusual growth substance or altered external conditions. Thus, while many, if not the majority, of bacterial genes are expressed constitutively, others are actively controlled. Even when active, not all genes are expressed at the same rate, as suggested by the wide variation in the absolute levels of bacterial proteins (from 10 to 10^5 copies per cell). Control of gene expression protects against wasteful energy consumption since both RNA and protein production expend ATP at a high rate. (Indeed, control mutants that unnecessarily expend energy may be outgrown by their wild-type counterparts in liquid culture.) Gene control also, clearly, prevents monopolisation of the transcription-translation apparatus.

There are two types of genetically programmed control systems, one active and one essentially 'static': RNA production is largely their site of action. The former involves the 'switching' on and off of particular genes in response to specific physiological signals. This *adaptive response* acts at the phenotypic level (rather than causing a change in the genetic material itself). It both activates gene expression and regulates its rate. In 'static' control, on the other hand, expression is limited by predetermined rates of RNA and protein production since expression is permanently 'switched' on (regulatory proteins, for example, are generally constitutively expressed).

Bacterial gene control acts over genes involved in the same metabolic pathway and over genes of a similar function. Control may result from the linkage of genes to a common regulatory region and/or by the participation of diffusible regulatory factor(s) acting on unlinked genes. However, the fact that the target is the same, RNA synthesis, leads to an interdependence between metabolically- and functionally-related operons. The degree of expression of an individual operon *per se*, as well as its expression relative to other operons of related function, mediate in overall cellular control. Thus, bacterial gene regulation is important both for an adaptive response and for allowing integration of cellular processes (mechanisms responsible for such integration have, in many instances, the capacity to override those concerned with the expression of individual operons). It is the interplay between the various control systems that is responsible for the exquisite metabolic balance in prokaryotes.

It has been mentioned above (Section 8.1.1 (a)) that whereas catabolic operons are generally inducible, anabolic systems are repressible. Operons in each group appear also to be under a general, overriding control system. Thus, degradative operons, responsible for sugar utilisation and, hence, for energy production, appear to be subject to a common, global control element. Expression of these *catabolite-controlled* operons requires a positive-acting protein complex (promoter potentiation), the effector of which is sensitive to the intracellular level of glucose. Biosynthetic operons, on the other hand, or at least those involved in amino acid production, seem often to be both repressible (promoter control) and under attenuator control. Transcriptional readthrough is increased by polypeptide chain termination within the leader sequence, a process dependent upon a low concentration of amino acid end product. These biosynthetic operons, therefore, both supply a pool of substrate for the translational machinery and make use of this apparatus for their control.

The lactose and tryptophan systems, which exemplify catabolic and anabolic operons, are discussed in Sections 9.1 and 9.2. It should be emphasised however, that there is no such thing as a 'normal' operon: each displays its own idiosyncrasy. Expression of genes of the transcription-translation apparatus and their control are considered in Section 9.3. Much of our knowledge of the organisation of these genes stems from the isolation of specialised transducing phages (making use of secondary site insertion of phage *lambda*, Section 6.2.2 (a i)). Subsequent analysis has involved the cloning of small DNA fragments into plasmid and phage vectors (Section 7.3.3), sequencing of the bacterial region and *in vitro* transcription.

9.1 Catabolite-Controlled Operons: The Lactose System

Catabolic genes are concerned with the breakdown and use of various carbon sources for energy production (Section 1.3.1 (a)). A number of degradative operons have been studied. Although each different operon is responsible for growth on a particular sugar and, hence, each is regulated by its own specific control system, they have a common dependence upon CAP and the ubiquitous cyclic nucleotide cAMP for their expression (see also Section 2.1.1 (b)).[1] These operons are *catabolite controlled* in that the cytoplasmic cAMP concentration

---

1. Catabolite-controlled operons do not necessarily exhibit a complete dependence upon CAP-cAMP. There is some *lac* expression in the absence of cAMP (see Section 9.1.2 (b)). Moreover, of the two overlapping promoters, P_1 and P_2, of the *gal* operon, P_1-promoted transcription requires CAP-cAMP, whereas transcription from P_2 takes place in the absence of this complex and is, in fact, inhibited by it. P_2 appears to serve a biosynthetic role, to allow generation of UDP-galactose necessary for cell-wall synthesis. P_1, on the other hand, is required for the utilisation of galactose as a carbon source, when it is converted to glucose-1-phosphate and fed into the glycolytic pathway (as the 6-phosphate).

is affected by the presence of glucose. Thus, when glucose levels are high and the (normal) glycolytic pathway is able to sustain growth, these catabolite-controlled operons have a reduced potential for expression — reduced in terms of the final induced rate achieved. This rate increases greatly if a sugar other than glucose is the principal carbon source. Proteins whose synthesis is subject to catabolite control are generally inducible and non-essential: they include the enzymes for lactose, arabinose, galactose and maltose degradation as well as flagellum production and phage *lambda* lysogeny.

The *lactose system*, the case study considered, has been extensively investigated. It is of classical importance since its examination led Jacob and Monod to formulate the operon model for gene regulation. Moreover, *lac* is prototypic both as an inducible and catabolite-controlled operon (the phenomenon of glucose repression of enzyme synthesis was initially termed the 'glucose effect').

9.1.1 The lac Operon

Four genes are involved in lactose utilisation, *lacZ*, *lacY*, *lacA* and *lacI*, located at about 9 min on the bacterial map. The enzyme *β-galactosidase* cleaves the disaccharide (4-0-*β*-D-galactopyranosyl-D-glucose) between the anomeric C1 carbon and the glycosyl oxygen to generate one molecule each of glucose and galactose (Figure 9.1). It is a tetramer of four identical subunits ($4 \times 116\,400$) encoded by the *lacZ* structural gene. Entrance of the sugar into the cell is dependent upon the hydrophobic *lactose permease* protein, the *lacY* product (45 500 daltons); the active species is possibly a dimer. The *lacY* product acts also as an alternative transport protein for the related disaccharide, melibiose (6-0-*α*-D-galactopyranosyl-D-glucose). Both *lacZ⁻* and *lacY⁻* mutants are phenotypically Lac⁻ (Table 9.1). However, since the normal melibiose permease (*melB* at 93 min) is temperature-sensitive, *lacY⁻* but not *lacZ⁻* strains are unable to grow on melibiose at 42°.

As yet, no definite role has been found for the *thiogalactoside transacetylase*, a dimer of two identical subunits (30 000 daltons) encoded by *lacA*. Recent studies suggest that *lacA⁺* cells have a selective advantage by virtue of the acetylation and, hence, excretion of non-metabolisable sugar derivatives. The *lacI* gene encodes a repressor protein (38 500 daltons), active as a tetramer in inhibiting *lac* transcription (see below).

The three structural genes *lacZ, lacY* and *lacA* constitute a polycistronic operon transcribed in the direction *lacZ → lacA*, clockwise on the bacterial map (Figure 9.1). The regulatory region responsible for RNA chain initiation and its control lies between *lacI* and *lacZ*. The four *lac* genes and the control region are, therefore, tightly linked. It should be noted that the siting of the regulatory gene directly adjacent to the region with which it interacts is particularly unusual (in operon control, the regulator's site of action must, of course, neighbour the structural gene(s) it controls, Chapter 8). In many operons, regulatory genes are distant (see *trpR*, for example, Section 9.2.1).

The *lac* regulatory region represents the majority of the *lacI-lacZ* intercistronic

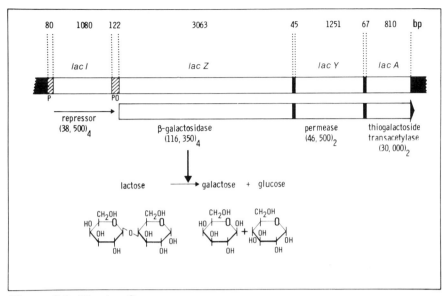

Figure 9.1: The *lac* Operon

A region of some 6.5 kb at 9 min on the bacterial map encodes the components for lactose utilisation (apart from *crp*, 73 min, and *cya*, 84 min, necessary for catabolite control, Table 9.1 and Section 9.1.2 (b)). The disaccharide, transported by the permease protein (*lacY* product), is cleaved by β-galactosidase (*lacZ* product) into glucose and galactose. Expression of the *lacZYA* operon is negatively controlled by the *lac* repressor (*lacI* product) and positively by CAP-cAMP. The thiogalactoside transacetylase (*lacA* protein) has recently been implicated in detoxification

Table 9.1 Mutations Affecting the Lac Phenotype

Locus	Map Location (min)	Genotype		Phenotype	Dominance [a]	Mechanism
lacI	8	lacI⁻	Lac⁺	constitutive *lac* expression	*trans* recessive	deficient in repressor-operator binding.
		lacI⁻ᵈ	Lac⁺	constitutive *lac* expression	*trans* dominant	negative (intracistronic) complementation.
		lacIˢ	Lac⁻	non-inducible	*trans* dominant	'super' (binding) repressor.
lacP	8	lacP⁻(classI)	Lac⁻	reduced *lac* expression	*cis*-specific	reduced affinity for CAP-cAMP.
		lacP⁻(classII)	Lac⁻	reduced *lac* expression	*cis*-specific	reduced affinity for RNA polymerase.
		lacPʳ(classIII)	Lac⁺	catabolite 'resistant'	*cis*-specific	CAP-cAMP independent.
lacO	8	lacOᶜ	Lac⁺	constitutive *lac* expression	*cis*-specific	deficient in repressor-operator binding.
lacZ	8	lacZ⁻	Lac⁻	β-galactosidase⁻	*trans* recessive	deficient in lactose cleavage.
lacY	8	lacY⁻	Lac⁻	lactose permease⁻	*trans* recessive	deficient in lactose uptake.
lacA	8	lacA⁻	Lac⁺	thiogalactoside transacetylase⁻	*trans* recessive	deficient in detoxification.
crp	73	crp⁻	Lac⁻	reduced *lac* expression	*trans* recessive	deficient in CAP.
cya	84	cya⁻	Lac⁻	reduced *lac* expression without cAMP	*trans* recessive	deficient in adenylate cyclase.

a   Refers to whether the mutation acts specifically in *cis* or *trans* and whether the mutant phenotype is expressed when a wild-type copy is also present; *cis*-specific lesions define regulatory regions (Section 8.1.1 (b)).

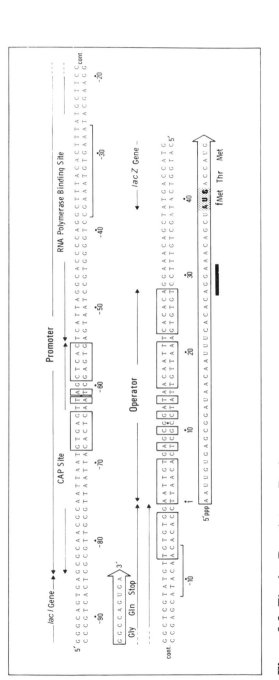

Figure 9.2: The *lac* Regulatory Region

The 122 bp intercistronic region between the *lac* repressor and β-galactosidase structural genes, *lacI* and *lacZ*, is the single control site for the entire *lacZYA* polycistronic operon. It consists of a CAP site, RNA polymerase binding site and operator (positions –84 to –55, –55 to 1 and 1 to 28, respectively). Operator-promoter interpenetration is not shown (see text). Symmetrical elements in the CAP site and operator are boxed. The sequences centred at –35 and –10 concerned with RNA polymerase function, are bracketed. The underlined region on the *lacZ* message is complementary to the 3'-end of 16S rRNA; the translational initiator triplet is stippled (note that the N-terminus of 'mature' β-galactosidase is Thr not fMET). After Reznikoff and Abelson (Reznikoff, W.S. and Abelson, J.N. (1978), in J.H. Miller and W.S. Reznikoff (eds.), *The Operon* (Cold Spring Harbor Laboratory, New York), pp. 221–243) region (112 out of 122 bp). It consists of two main sites, a promoter and operator (Figure 9.2). The operator (about 28 bp) lies adjacent to the β-galactosidase structural gene, separated from *lacZ* by about 10 bp; it has two-fold symmetry centred about position 11. There is some interpenetration of *lacO* and *lacP* in that *lac* repressor bound to *lacO* overlaps the polymerase protected region. The finding that the repressor-protected sequence falls within the polymerase-protected region, and that three of the five thymines in *lacO* lying close to the bound repressor are at positions at which RNA polymerase interacts (2, 3 and 4), initially suggested binding of repressor and polymerase is mutually exclusive. Recent studies have shown, though, that RNA polymerase is able to bind in the presence of the *lac* repressor, indicating that promoter and operator sites are two clearly defined functional domains. The placement of the operator between *lacP* and *lacZ* allows physical blockage as a means of transcriptional control. [It should be noted that RNA polymerase 'covers' a region extending out to 21. That this enzyme interacts with sequences far downstream from the transcriptional initiation point is indicated by DNase protection studies. Moreover, certain class III promoter mutations (or P^r-type, Table 9.1) that increase CAP-cAMP independent *lac* transcription are located between –16 and 10.]

On the definition that a promoter is a sequence of DNA at which RNA polymerase binds preferentially and within which RNA synthesis is initiated (Section 2.1.2 (a i), the *lac* promoter may be divided into two main regions: RNA polymerase binding site and CAP-cAMP site (Figure 9.2). The first, adjacent to (perhaps, overlapping with) *lacO* contains the information for initiation itself. The *RNA polymerase binding site* carries, therefore, the –35 region, Pribnow box and start point (Section 2.1.2 (a i)). The efficiency of RNA polymerase initiation is, however, strongly dependent upon binding of the CAP-cAMP complex at a sequence about 61 bp upstream from the initiation point. This promoter-potentiation sequence is referred to as the *CAP site* (Figure 9.2). In summary, *lac* mRNA production requires CAP-promoted RNA polymerase binding and a vacant operator site (the effect of *lac* regulatory mutations, Table 9.1, is discussed below).

9.1.2 lac Promoter Control

Synthesis of *lac* proteins other than the repressor is not constitutive (mutations in *lacI* or *lacO* may, however, impart this phenotype, Table 9.1). Rather, expression is dependent on the presence of the β-galactosidase substrate, lactose, and on CAP-cAMP. Although this control is mediated by two distinct proteins acting at different DNA sequences, RNA chain initiation is the target in each case. *lac* expression is, therefore, subject solely to promoter control.

The presence of negative and positive control elements (*lac* repressor and CAP-cAMP, respectively) serves two functions. Firstly, *lac* expression is actively controlled over a 1000-fold range such that the presence of an inducer is required (to inactivate repressor) for significant β-galactosidase synthesis. Secondly, *lac* induction is only maximal in the absence of glucose (note that lactose breakdown, itself, generates this sugar). This catabolite control, mediated by CAP-cAMP, prevents wasteful *lac* expression when a more efficient carbon source is present (expression is regulated over about a 2.5-fold range *in vivo*).

(a) Induction and Repression. Control of *lac* expression is mediated in the first instance (see also (b), below) by the operator site. Binding of the *lac* repressor prevents transcription of *lacZ* and, hence, *lacY* and *lacA* (one repressor tetramer binds to the operator, one side of the helix being favoured). As discussed above, *lac* repressor does not seem to physically exclude RNA polymerase binding. However, the complex formed in the presence of repressor is prevented from productive transcriptional initiation. The repressor protein, itself, is produced constitutively from a separate promoter; like many other regulatory proteins it is synthesised at a low rate (there are about 10 molecules per cell). It is diffusible and, hence, *trans*-acting (see Section 8.1.1 (b)).

   Binding of the *lacI* product does not totally inhibit *lacZYA* transcription. Some 10 molecules of the β-galactosidase tetramer per cell (and a comparable level of the lactose permease) are synthesised in the absence of lactose (or other suitable inducer) due to readthrough from *lacI* into *lacZ*.[2] This basal level is very important in allowing *lac* induction since it supplies the necessary transport system for the initial uptake of exogenous lactose. Moreover, the low level of β-galactosidase is responsible for the conversion of lactose to *allolactose* (6-O-β-D-galactopyranosyl-D-glucose; the galactosyl residue present on the 6 rather than 4 position of glucose), the active species in *lac* induction.

   The *lac* repressor is an allosteric complex carrying distinct sites for DNA and inducer binding (the binding site for the *lac* operator is located in the first 60 N-terminal residues). Binding of the inducer, allolactose, to the tetrameric protein takes place whether the repressor is free in solution or at the operator site. Binding of the repressor to the *lac* operator, however, occurs only in the absence of inducer (Figure 9.3). When lactose becomes the sole carbon source (but not when glucose is also present, see (b) below), rapid induction of *lac* enzyme synthesis takes place (the whole chain of events initiated by the basal *lacZ* and *lacY* expression). There is no noticeable lag in *lac* transcription (the first β-galactosidase tetramers are detected some 90 sec after induction, Figure 9.4). The final level of β-galactosidase tetramer is about 10 000 molecules per cell (approximately 5 per cent of the total bacterial protein).[3] This represents an *induction ratio* of 1000 (10 000/10).

   In conclusion, transcription of the *lac* polycistronic operon and, hence, growth

---

2.   The *gratuitous* inducer isopropyl-β-D-thiogalactoside (IPTG) is frequently employed since it is not degraded by β-galactosidase.
3.   The transacetylase is synthesised at a rate about four times less than β-galactosidase (or permease), equivalent to a sixteen-fold difference in terms of weight. Such a decrease in the expression of distal genes of a polycistronic operon relative to proximal ones has been referred to as *natural polarity* (see Section 8.1.2 (a)). It can be largely suppressed in the *lac* operon by *rho* mutations.

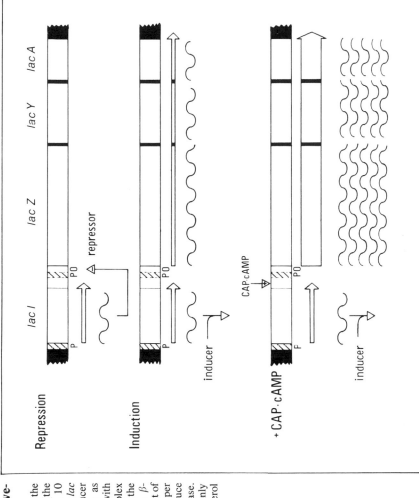

Figure 9.3: Negative- and Positive-Regulation of the *lac* Operon

Repressor (*lacI* gene product) binding at the operator prevents *lacZYA* transcription (the basal level of β-galactosidase is about 10 molecules per cell). There is low level *lac* enzyme expression in the presence of inducer (allolactose or lactose analogues such as IPTG), but maximal rate is achieved only with the positive-acting CAP·cAMP complex bound at the CAP site (strictly speaking the *lacI*-proximal region of the promoter). β-galactosidase production is only 2 per cent of the maximal rate (about 10 000 molecules per cell) in a *crp⁻ cya⁻* strain unable to produce either the CAP protein or adenylate cyclase. Growth on glucose, on the other hand, only lowers the rate to 40 per cent of the full glycerol level

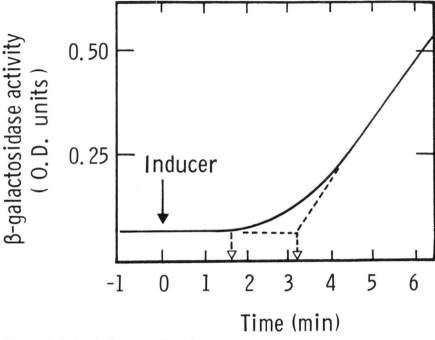

Figure 9.4: *lac* **Induction Kinetics**

β-galactosidase activity is first detected about 1.5 minutes after addition of inducer (IPTG). This represents the time taken for inducter transport/entrance, *lac* operon induction, *lacZ* transcription and translation, and monomer association. Since *lacZ* is 3063 bp long (Figure 9.1), complete transcription and translation would require a maximum of 90 seconds, indicating that there is no appreciable lag. The curvature (1.5-3.25 min) represents the time for the various stages in *lac* gene expression to reach a steady-state. Thus, transcription starts immediately in the presence of inducer with a rate that is both maximal and constant. After Kepes (Kepes, A. (1969) *Prog. Biophys. Mol. Biol., 19,* 201-236)

on lactose as the sole carbon source requires inactivation of the *lac* repressor. This negative control element is unable to bind at *lacO* in the presence of the inducer, allolactose (Figure 9.3). It is this simple on-off switch, allowing rapid induction, that imparts the adaptive facility.

It should be emphasised that the presence of allolactose *per se* is not sufficient for full *lac* enzyme expression. Rather, it requires an additional positive-controlling component, CAP-cAMP (see below).

(b) Catabolite Control. The *lac* regulatory region functions poorly as a promoter. This inefficiency would seem to be an integral component of *lac* (and other catabolite-controlled operons) since interaction of the positive regulator CAP-cAMP at the promoter permits a high level of transcription (*crp⁻ cya⁻* double mutants, lacking both CAP and adenylate cyclase, synthesise β-galactosidase at about 2 per cent of the wild-type induced rate). The sequence implicated in CAP-cAMP binding

5′ G T G A G T T A G C T C A C
C A C T C A A T C G A G T G 5′

is located between positions –68 and –55 (nucleotides from the transcriptional start point, see Figure 9.2). It consists of a tandem, staggered repeat, generating symmetrical elements centred between positions –61 and –62 (the *gal* CAP site is located between positions –25 and –50 and shows only slight homology, see below). Class I promoter mutations (see Table 9.2) that render *lac* transcription less responsive to CAP-cAMP (the presence of functional *crp cya* genes only increases expression 3-fold, rather than 50-fold), fall within this region. Moreover, CAP-cAMP binds specifically to short restriction fragments (about 200 bp long) carrying the CAP site.

The low level efficiency of the *lac* RNA polymerase binding site can be ascribed to sequences in this region. Comparison of the heptanucleotide sequences centred at –10, T A T G T T G, with the canonical sequence T A T A A T R (the Pribnow box, Section 2.1.2 (a i)) reveals a difference of two base pairs (GT rather than AA). The fact that the type III *lacP* mutation, UV5 (GT → AA), which generates the canonical sequence, renders the promoter CAP-cAMP independent (see Table 9.2), strongly supports a role for this positive regulatory complex in enhancing promoter function (RNA polymerase rapidly forms a functional complex with UV5 DNA *in vitro*). Just as such class III promoter mutations fall within (or adjacent to) one element of the RNA polymerase binding site, so class II lesions appear to define the so-called recognition site centred at –35 (Section 2.1.2 (a i)). Lesions in this region interfere with DNA-RNA polymerase interaction, either by lowering the affinity for RNA polymerase or by preventing the formation of the open promoter complex (see Section 2.1.2 (a i)).

How does the CAP-cAMP complex function? It has been suggested that CAP-cAMP promotes helix destabilisation some 30-50 bp downstream from its binding site, thereby increasing the efficiency of open promoter formation (see Section 2.1.2 (a i)). The high G + C content of the intervening sequence (and that downstream of the Pribnow box) is consonant with this contention. Such a model would, however, need to explain how the effect of CAP-cAMP binding is 'funnelled' downstream. Alternatively, the complex may stimulate *lac* promoter function by physical association with RNA polymerase holoenzyme at the promoter site. The recent evidence for polymerase protecting DNA sequences close to the CAP binding site, and the observation that CAP-cAMP and polymerase bind to the same side of the *lac* promoter upstream from the Pribnow box, support a direct interaction between the two proteins. It is, in fact, possible for CAP, RNA polymerase and σ factor to extend over the whole CAP site-start point interval if the proteins concerned are slightly elongated rather than globular (Figure 9.5). Finally, a recent suggestion, taking into consideration the CAP site at –25 to –50 in *gal*, is that the *lac* –50/–70 sequence merely represents the

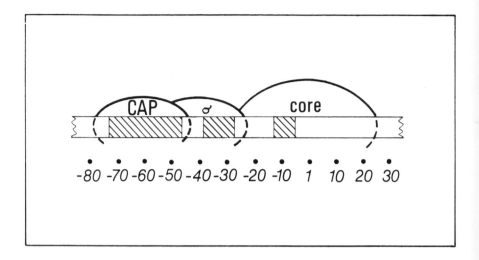

Figure 9.5: A Hypothetical Arrangement of CAP-cAMP and RNA Polymerase Holoenzyme on the *lac* Promoter
Recent evidence for RNA polymerase protecting DNA sequences close to the CAP binding site supports a direct interaction between the positive-acting factor and polymerase at this site. It is, in fact, possible for CAP and holoenzyme (Eσ) to extend over the CAP site-transcriptional start point interval if the axial ratios of the proteins are 2:1. Other models for CAP-cAMP action include CAP-mediated helix destabilisation downstream from the CAP site and secondary binding at the -50 to -30 region (see text). After Gilbert (Gilbert, W. (1976), in R. Losick and M. Chamberlin (eds.), *RNA Polymerase* (Cold Spring Harbor Laboratory, New York), pp. 193-205)

primary site of CAP binding, with CAP-cAMP promoting transcription through a secondary interaction at the -30 to -50 region. On this model, CAP-cAMP binding at the secondary CAP site requires an initial interaction at the primary one. Certainly, CAP has been shown to bind co-operatively to DNA in the absence of its effector. Moreover, other types of positive regulatory elements act at the $-30/-50$ region (see lambda P_I and P_E, Sections 10.2.1 (a) and 10.2.2 (a)).

CAP function, itself, is directly dependent upon cAMP. Moreover, cAMP levels are not invariant. Rather, they reflect the intracellular glucose concentration. Thus, when *E. coli* is grown on glucose medium (with or without lactose), internal cAMP levels are low (about 1μ M compared with 5μ M in glycerol medium). Although the structural gene for adenylate cyclase, a cytoplasmic membrane protein responsible for cAMP biosynthesis from ATP, has been identified (*cya*, mapping at 84 min), the glucose-mediated control of cAMP is not understood.[4] Certainly, regulation seems to be at the level of adenylate cyclase activity rather than cAMP degradation or export.

Since cAMP concentrations are low under high glucose conditions, and

4.   The *cya* gene product appears to be non-essential since deletion derivatives are viable. The fact that mutations at this locus but not in *crp* result in poor growth supports a role for cAMP in other cellular process.

since CAP requires this cyclic nucleotide for function, *lac* enzymes are expressed at a reduced rate in glucose medium (about 40 per cent the glycerol rate). This glucose-mediated reduction in *lac* enzyme synthesis has been referred to as *catabolite repression*. The molecular nature of this regulatory process, namely the lack of a functional activator (CAP-cAMP) rather than the presence of a repressor protein, suggests that *catabolite control* is a more correct description. (That both CAP and cAMP are important in *lac* expression is aptly demonstrated in *crp⁻ cya⁻* double mutants which synthesise β-galactosidase at only 2 per cent the wild-type rate, see Figure 9.3.) In short, the CAP-cAMP complex acts as a powerful positive transcriptional regulatory element, whose function is modulated over a 2.5-fold range *in vivo* by the presence or absence of glucose as a carbon source.

9.2 Attenuator-Controlled Operons: The Tryptophan System

Anabolic operons need only function in the absence of the products synthesised by the encoded enzymes. (Contrast this with inducible, catabolic systems whose operation is limited to the presence of substrate, Section 9.1.) Thus, a wild-type strain produces the enzymes for tryptophan synthesis when cultured on minimal medium but not when the growth medium is supplemented with the amino acid.

This economy in amino acid biosynthesis is crucial for energy conservation. It is achieved by two separate control circuits, one specific to the operon concerned and one having elements in common with certain other biosynthetic operons. The first, promoter control, tends to involve a repressor element, whereas attenuator control makes use of the translational machinery for regulating transcription.

The case study discussed in this section, the *tryptophan system*, is of classical significance in terms of the demonstration of gene-polypeptide colinearity and the identification of a repressible system. Moreover, attenuation as a means of transcriptional control in biosynthetic operons was first elucidated with *trp*.

9.2.1 The trp Operon

Five linked genes, *trpA*, *trpB*, *trpC*, *trpD* and *trpE*, mapping at 27 min, are responsible for tryptophan biosynthesis from chorismate (Figure 9.6). In addition, an apo-repressor gene, *trpR*, is sited at 100 min on the bacterial map. Tryptophanyl tRNA synthetase, tRNA[Trp] and possibly rho factor (encoded by *trpS*, *trpT* and *rho*, at 74, 84 and 84.5 min, respectively) have also been implicated in *trp* control.

Tryptophan is produced from chorismate in five main steps. The first two, the conversion of chorismate to N-5′-phosphoribosyl-anthranilate, are catalysed by the anthranilate synthetase complex, a tetramer composed of two *trpE* and two *trpD* gene products (the second component possesses both glutamine amido-transferase and phosphoribosyl-anthranilate transferase activities, the latter

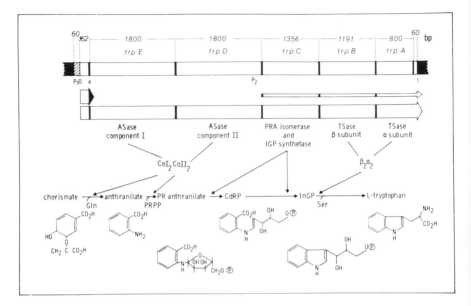

Figure 9.6: The *trp* Operon

The enzymes responsible for tryptophan production from chorismate are encoded by 5 linked genes, *trpEDCBA*, mapping at 27 min. In addition to the repressor gene (*trpR*, 100 min) responsible for negative control, tRNA[Trp] (*trpT*, 84 min) and its synthetase (*trpS*, 74 min) are involved in attenuation at *trp a* (Table 9.2). P_1 is the major promoter, with expression of *trpCBA* from P_2 representing only 3 per cent the P_1 rate under derepressed conditions. Abbreviations: ASase I and II, components I and II of anthranilate synthetase (CoII is bifunctional: the N-terminal third contributes glutamine amidotransferase activity to the ASase complex while the remainder contains phosphoribosyl anthranilate tranferase activity); PRA isomerase and IGP synthetase, phosphoribosyl anthranilate synthetase and indole glycerol phosphate synthetase (the *trpC* gene product is also a bifunctional polypeptide); TSase β and α, trptophan synthetase β and α components; PRPP, 5'-phosphoribosyl-1-pyrophosphate; PR anthranilate, N-5'-phosphoribosyl-anthranilate; CdRP, 1-(o-carboxyphenylamino)-1-deoxyribulose-5-phosphate; InGP, indole-3-glycerol phosphate. After Platt (Platt, T. (1978), in J.H. Miller and W.S. Reznikoff (eds.), *The Operon* (Cold Spring Harbor Laboratory, New York), pp. 263-302)

necessary for the second step in tryptophan synthesis, Figure 9.6). The *trpC* protein, a bifunctional polypeptide with both phosphoribosyl-anthranilate isomerase and indoleglycerol phosphate synthetase activity, is responsible for the conversion of phosphoribosyl-anthranilate to indole-3-glycerol phosphate (via 1-(*O*-carboxyphenylamino)-1-deoxyribulose-5-phosphate). Tryptophan synthetase, the final enzyme in the pathway, is a tetramer consisting of two copies of the *trpA* and *trpB* gene products, the α and β subunits. It catalyses the production of L-tryptophan from indole-3-glycerol phosphate, with indole as intermediate. The α subunit is responsible for indole production while the β component is involved in the condensation of this intermediate with serine to yield tryptophan.

Lesions in any one of the five *trp* genes prevent tryptophan biosynthesis and,

Table 9.2ᵃ Mutations that Alter the Trp Phenotype

Locus	Map Location (min)	Genotype	Phenotype	Dominance ᵇ	Mechanism
trpR	100	trpR⁻ Trp⁺	constitutive trp expression	trans recessive	deficient in repressor-operator binding.
trpP	27	trpP⁻ Trp⁻	reduced trp expression	cis-specific	reduced transcriptional initiation.
trpO	27	trpO⁻ᶜ Trp⁺	constitutive trp expression	cis-specific	deficient in repressor-operator binding.
trpL	27	trp⁻ᵃ Trp⁺	increased trp expression on high Trp	cis-specific	faulty transcriptional termination at trp a
trpL	27	trpL⁻ Trp⁻	reduced trp expression	cis-specific	enhanced transcriptional termination at trp a
trpE	27	trpE⁻ Trp⁻	grows on Trp, indole, anthranilate	trans recessive	deficient in anthranilate production.
trpD	27	trpD⁻ Trp⁻	grows on Trp, indole	trans recessive	deficient in PR anthranilate production.
trpC	27	trpC⁻ Trp⁻	grows on Trp, indole	trans recessive	deficient in InGP production.
trpB	27	trpB⁻ Trp⁻	grows on Trp	trans recessive	deficient in Trp production.
trpA	27	trpA⁻ Trp⁻	grows on Trp, indole	trans recessive	deficient in indole production.
trpS	74	trpS⁻ Trp⁺	increased trp expression on high Trp	trans recessive	faulty transcriptional termination at trp a
trpT	84	trpT⁻ᶜ Trp⁺	increased trp expression on high Trp	trans recessive	faulty transcriptional termination at trp a
rho	84	rho⁻ᵈ Trp⁺	increased trp expression on high Trp	trans recessive	faulty transcriptional termination at trp a

a   Abbreviations: Trp, tryptophan; PR anthranilate, phosphoribosyl anthranilate; InGP, indole-3-glycerol phosphate.

b   Refers to whether the mutation acts specifically in *cis* or *trans* and whether the mutant phenotype is expressed when a wild-type copy is also present; *cis*-specific lesions define regulatory regions (Section 8.1.1. (b)).

c   Inactivation of *trpT* is generally lethal since there is only one gene copy (Section 3.3.2); hence, mutant derivatives are introduced on a substituted F plasmid, F14.

d   Rho is essential for cellular growth; most *rho* derivatives studied to date are 'leaky'. There is now some doubt as to whether rho is involved in *trp* attenuation.

thus, impart a Trp⁻ phenotype (Table 9.2). The locus affected may be ascertained from the growth requirements of the mutant strain and its cross-feeding characteristics (see Pathway Analysis, Section 7.1.2 (d)). Three classes of *trp* mutants are identifiable by their growth requirements: those able to utilise only tryptophan *(trpB⁻)*, those that can survive on indole *(trpA⁻, trpC⁻, trpD⁻)*, and those that grow with either indole or anthranilate *(trpE⁻)*. These nutrients can, of course, be added to the growth medium (the phosphorylated intermediates cannot enter the cell when present exogenously). They may also be supplied by crossfeeding since inactivation of one of the *trp* enzymes results in a block in the pathway and, hence, build-up of intermediary metabolites (Figure 9.7). It is, for example, the secretion of anthranilate into the growth medium that allows cross-feeding of *trpE⁻* strains by *trpD⁻* mutants; and, crossfeeding of the indole-utilising derivatives *trpE⁻*, *trpD⁻*, *trpC⁻* and *trpA⁻* by *trpB⁻* mutants is due to accumulation of indole.

The *trp* cluster at 27 min constitutes a polycistronic operon. It is transcribed in the direction *trpE → trpA*, anti-clockwise on the bacterial map (Figure 9.6). The major *trp* promoter, *trpP₁*, is responsible for this. A second, minor promoter, *trpP₂*, situated at the distal end of the *trpD* structural gene (about 200 bp from the *trpC* translational initiation codon, Figure 9.6), is responsible for only 3 per cent of the wild-type level of *trpCBA* expression (under derepressed conditions, see Section 9.2.2 (a), below). Its physiological role is not known.

The *trp* regulatory region, responsible for transcriptional initiation and its control, consists of *trpP₁* and the operator *trpO*. It is not sited directly adjacent to the first structural gene, *trpE*, but rather separated by a *leader sequence* (162 bp). The enormous polycistronic message (about 7000 nucleotides long) encodes, therefore, the information for five structural genes and carries an additional 5′-region; it is represented *trpLEDCBA*.

The *trp* operator and major promoter have been defined by DNase protection studies with the *trp* repressor (consisting of apo-repressor and its co-repressor, see below) and RNA polymerase, as well as sequence analysis of regulatory mutants (Figure 9.8). The operator appears to extend from about position −18 to −5 (where 1 is the transcriptional start point); it contains two-fold symmetry, the centre lying between −11 and −12. The RNA polymerase protected region consists of residues −41 to 18 but a minimum of 59 base pairs preceding the start point is required for promoter function. There may be functional overlap of the *trp* operator and promoter since binding of one protein precludes productive interaction of the second (but see 9.2.2 (a), below). Certainly, operator-constitutive mutations, *trpOᶜ* (Table 9.2), fall within the sequences covered by the two proteins. The *trp* promoter P₁ is particularly interesting since the sequence centred at −10, T T A A C T A, bears little similarity to the Pribnow box, T A T A A T R (the −35 region resembles the consensus sequence, Section 2.1.2 (a i)). Evidence that the −10 sequence is nevertheless involved in transcriptional initiation comes from a promoter-down mutation — one that

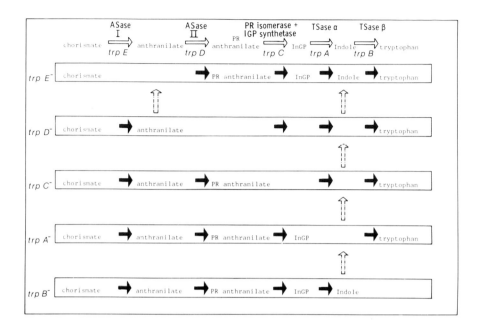

Figure 9.7: Crossfeeding of Trp Auxotrophs through the Accumulation and Secretion of Intermediary Metabolites
Anthranilate and indole, though not the phosphorylated intermediates, may enter the cell when present exogenously — through secretion by *trp*⁻ strains blocked in their breakdown (Section 7.1.2 (d)). Thus, accumulation of anthranilate in *trpD*⁻ mutants crossfeeds *trpE*⁻ strains and indole from *trpB*⁻ mutants allows the growth of *trpA*⁻, *trpC*⁻ *trpD*⁻ and *trpE*⁻ strains (abbreviations are given in Figure 9.6)

reduces *trp* expression — resulting from a T·A → G·C transversion at the upstream end of the Pribnow box (position −13).

9.2.2 trp Control

Synthesis of tryptophan is subject to feedback inhibition acting on enzyme activity. Thus, though the three aromatic amino acids tryptophan, tyrosine and phenylalanine are all derived from chorismate, the first step specific to tryptophan production, the conversion of chorismate to anthranilate, is blocked by the end product. In addition, *trp* enzyme synthesis is regulated at the transcriptional level, both through control of RNA chain initiation (promoter control) and RNA chain termination (attenuator control).

In each case, the end product tryptophan is the key element in tryptophan production. In feedback inhibition, it binds to the tetrameric enzyme complex, anthranilate synthetase, in particular the *trpE* component, to inhibit synthetase activity. With respect to transcriptional regulation, tryptophan, on the one hand, constitutes the co-repressor for inhibition of *trp* mRNA synthesis and, on the

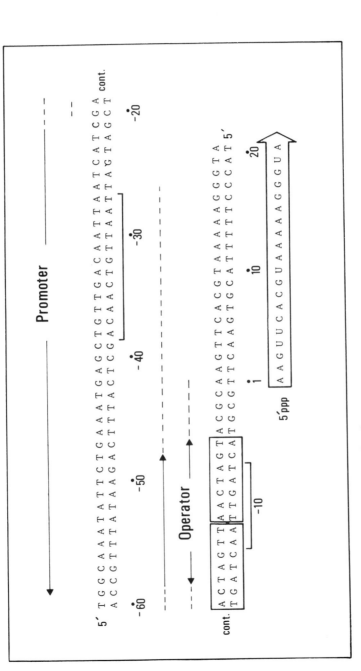

Figure 9.8: The *trpPO* Regulatory Region

The 76 bp region responsible for *trpR*-mediated negative control of the *trpLEDCBA* polycistronic operon is separated from *trpE* by the leader sequence (*trpL*, 162 bp). The regulatory region contains a promoter (−59 to −6 (or +17?)) and operator (−18 to −5 (plus about 5 bp on either side?)) that appear to interpenetrate. The sequences centred at −35 and −10 concerned with RNA polymerase binding are bracketed; symmetrical elements in the operator are boxed. After Platt (Platt, T. (1978), in J.H. Miller and W.S. Reznikoff (eds.) *The Operon* (Cold Spring Harbor Laboratory, New York), pp. 263-302) other, is required for attenuation (expression is regulated over a 70- and 10-fold range by these two separate control systems, giving an overall range of about 700-fold; see also Metabolic Control, Section 9.3.2 (a)).

(a) trp Promoter Control. trp expression is repressible (Figure 9.9). Growth in the presence of exogenous tryptophan prevents production of the *trpLEDCBA* polycistronic message by virtue of repressor binding (see also Attenuator Control, below). Transcriptional initiation from P_1 is, therefore, dependent upon a vacant operator site (promoter control). Although the regions protected by repressor and RNA polymerase overlap and despite the observation that pre-binding of the *trp* repressor at *trpO* prevents initiation of transcription, it is not known whether the repressor physically excludes RNA polymerase from binding or, as now appears to be the case with the *lacI* product (Section 9.1.2 (a)), prevents formation of the initiation complex.

The active repressor moiety in this negative control system is a complex of an apo-repressor, (possibly a homodimer, 2 x 12 500), encoded by the unlinked *trpR* locus, and a co-repressor, tryptophan. Since repressor function requires tryptophan, and since *trpR* is expressed constitutively (there are about 20-30 repressor molecules per cell), RNA chain initiation at trpP$_1$ is governed solely by the pathway's end product, tryptophan.[5] The *trans*-acting nature of diffusible elements, and the limited number of repressor molecules, explains why repressor function can be 'diluted out' by the presence of multiple copies of *trpO* in *trans* (say, through cloning the *trp* regulatory region on a multicopy plasmid).

Derepression, trp expression mediated by inactivation of the repressor-DNA complex, may result from two main causes. *Genetic derepression* refers to mutations in *trpR* or in the operator (*trpR$^-$* or *trpO^c*, Table 9.2) that leave the mutant in a permanent state of constitutive synthesis (*trp* enzyme synthesis is about 70-fold greater under these conditions than the basal repressed level). *Physiological derepression* stems from the absence of the co-repressor trypto-phan, the actual derepression ratio depending upon the cytoplasmic tryptophan concentration (see (b), below).

Although the internal promoter *trpP$_2$* is responsible for only about 3 per cent of the wild-type level of *trpCBA* expression under conditions of derepression, P_2-promoted transcription occurs independently of P_1-events and is, therefore, not subject to either promoter control or attenuator control at *trpO* or *trp a* (see below). Consequently, under repressed conditions, that is, in the presence of high tryptophan concentrations, approximately 80 per cent of *trpC, trpB* and *trpA* gene products are expressed from P_2 (Figure 9.9).

5.   The *trp* repressor also inhibits transcription of *aroH* (at 37 min). The *aroH* gene product, 3-deoxy-D-arabino-heptulosonate-7-phosphate synthase, is one of three alternative enzymes respon-sible for catalysis of the first step in the synthesis of aromatic amino acids, the condensation of erythrose-4-phosphate (itself produced in the pentose phosphate pathway) with phosphoenol-pyruvate; each enzyme is inhibited by either tryptophan, tyrosine or phenylalanine. Recent evidence, in fact, suggest that *trpR* is autoregulated. Thus, at least three unlinked operator sites are recognised by the repressor complex.

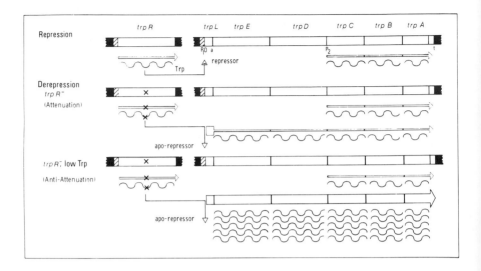

Figure 9.9: *trp* **Negative Control and its Attenuation**
Binding of the *trp* repressor at the operator site prevents transcription of the *trpLEDCBA*
polycistronic operon (that is, of both the leader sequence and structural genes) though there is low
level expression of *trpCBA* from the minor promoter, P_2 (about 3 per cent of the fully derepressed
rate). Inactivation of the repressor (through mutation in the *trpR* gene) leads to (genetic)
derepression of the entire operon, the *trp* enzymes being expressed at about 70-fold the repressed
rate. Full expression, approximately 700-fold the basal level, is achieved only under low Trp
concentrations when transcriptional readthrough past *trp a* is maximal (see A Model for *trp*
Attenuation, Figure 9.10)

(b) trp Attenuator Control. RNA polymerase molecules, after initiating at *trpP₁*
(but not the internal promoter, see above), face a further obstacle before
encountering *trpE*, the first structural gene of the polycistronic operon. The
leader sequence, lying between *trpPO* and *trpE*, carries a (rho-dependent?)
transcriptional terminator, termed *trp a*, that greatly reduces readthrough in the
presence of high tryptophan concentrations (Figure 9.9). *Attenuation* leads to
the production of a short message (140 nucleotides) that lacks any of the *trp*
structural genes. Under conditions of high tryptophan, nine out of ten RNA
chains initiated at *trpP₁* terminate at this site, some 20 bp upstream from *trpE*:
only about one out of ten messages are extended into *trpE*. *Anti-attenuation*,
readthrough past the transcription barrier, takes place when the cytoplasmic
tryptophan concentration is low.

It should be emphasised that attenuator control acts over and above *trpR*-
mediated regulation since both function at different sites (compare this with *lac*,
in which *lacI*/CAP-cAMP regulation is restricted to the PO region itself,
Section 9.1.2). Thus, inactivation of *trpR* does not prevent transcriptional
termination at the attenuator site. Similarly, repression is maintained in a *trp a⁻*
mutant strain that lacks the termination sequence, say through a deletion (though
not, of course, if the multisite lesion extends into *trpO*). Moreover, whereas the

trp repressor is a *trans*-acting complex, *trp a⁻* mutations are *cis*-specific since they affect a regulatory sequence rather than a structural gene.

Although attenuation is only a relatively recent discovery (see, for example, Bertrand *et al.* 1975), the combination of genetic and restriction technology (Chapter 7) has allowed a high level of understanding of this phenomenon. Three main observations are of importance. Firstly, synthesis of leader RNA in excess over distal *trp* mRNA and the dependence of such disproportionality upon tryptophan concentration suggests the involvement of the 'normal' transcriptional termination apparaus (rather than *trpEDCBA* mRNA degradation).[6] That termination occurs at a defined sequence in attenuation is shown by *cis*-specific mutations in the distal region of *trpL* (such deletions define the termination site *trp a*). The nucleotide sequence at the 3'-end of the *trp* leader bears marked similarity, in fact, with other transcriptional terminators (see Section 2.1.2 (a iii)). Thus, there is a run of G·C base pairs with dyad symmetry (pause sequence?), the centre of which is separated from the terminal AT-rich region by about one turn of the helix.

Secondly, sequence analysis of the leader indicates that there is an AUG codon (positions 27 to 29) with neighbouring ribosome binding sequence and an in-phase UGA translational termination codon (positions 69-71). Thus, the leader potentially encodes a small peptide of 14 residues. This *leader peptide* carries adjacent Trp residues close to the carboxy-terminus (positions 10 and 11). The importance of these tandem codons is exemplified by mutations in cistrons concerned with tRNA^{Trp} charging, either in the tRNA^{Trp} gene *(trpT)*, itself, or in the tryptophan-tRNA synthetase structural gene *(trpS)*, which reduce attenuation at *trp a*. Translation of the leader sequence has been demonstrated thanks to genetic fusions that hook up *lacI* expression to the early part of the *trp* operon and internal deletions that fuse the initial portion of the leader region to distal parts of *trpE*. In each case, novel fusion polypeptides are synthesised that carry an N-terminus apparently coded by the leader sequence.

Finally, the 3'-half of the *trp* leader RNA exhibits profound secondary structure *in vitro* (to the extent that it is resistant to single-strand specific RNases). The tandem Trp codons (region 1) can pair with a more distant portion of the *trpL* mRNA (region 2, Figure 9.10). Moreover, the attenuator site — the run of Us preceded by a GC-rich region — can form a terminator stem-and-loop (3·4). That this RNA conformation is important in attenuator control is shown by the increased readthrough past *trp a* resulting from point mutations which disrupt the secondary structure of stem-and-loop 3·4.

There is strong support for an attenuator mechanism in which translation of tandem tryptophan codons destroys secondary structure in the *trpL* transcript and activates (rho-mediated?) transcriptional termination (Figure 9.10). On this model, translation into the region encoding the distal portion of the leader

6. Certain (though not all) polarity suppressor *rho* derivatives (Section 8.1.2) relieve *trp* attenuation. Rho is not required for termination *in vitro* but it has been mooted that this protein functions in the release of the leader transcript.

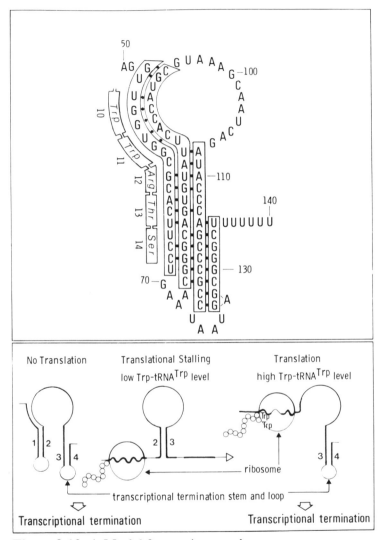

Figure 9.10: A Model for *trp* Attenuation

Transcriptional termination at the *trp* attenuator site, *trp a*, generates a 140 nucleotide transcript capable of different secondary structures (top). There are 4 stems able to base pair: the two tandem Trp codons are located in the promoter-proximal one (stem-and-loop 1·2); the region between 113 and 140 (stem-and-loop 3·4) appears to represent the terminator stem-and-loop, and carries a G·C-rich region (with dyad symmetry) preceding a polyU stretch. It is proposed that the nature of the stem-and-loop formed — dependent upon whether the leader is translated or not — dictates transcriptional termination at *trp a* (bottom). In the absence of translation, formation of the terminator stem-and-loop causes attenuation. At low Trp-tRNA[Trp] levels (either because of reduced Trp or through faulty charging), ribosomes stall at the tandem Trp codons, preventing formation of stem-and-loop 1·2 and, thus, allowing the generation of stem-and-loop 2·3. Transcriptional readthrough takes place. When Trp-tRNA[Trp] levels are high, translation into stem-and-loop 1·2 prevents 2·3 pairing and so gives rise to the terminator 3·4 again. After Oxender *et al.* (Oxender, D.L., Zurawski, G. and Yanofsky, C. (1979) *Proc. Nat. Acad. Sci. USA,* **76,** 5524-5528) peptide inhibits pairing of regions 2 and 3, and so allows formation of the stem-and-loop (3·4) responsible for transcriptional termination. This takes place only when tryptophan is present in excess and when there is correct tRNA^{Trp} aminoacylation. Ribosome stalling at the tandem Trp codons under conditions of tryptophan starvation (or faulty tryptophanyl-tRNA charging) allows pairing of regions 2·3, and so prevents formation of the terminator stem-and-loop, thereby permitting RNA polymerase to continue into the *trp* structural genes. Note that it is the translational apparatus *per se* and not its product(s) which acts as a positive control factor; whether the leader peptide has a physiological role is not known. Similar models have been proposed for attenuator control over operons involved in the synthesis of histidine, leucine, phenylalanine and threonine.

9.3 Multivalent Control of Transcriptional-Translational Operons

In prokaryotes, transcription and translation go 'hand-in-hand' (Chapter 2). Witness, for instance, the common usage of a nascent, rather than pre-formed, RNA chain as template for protein biosynthesis. Furthermore, instances of attenuator control and polarity in polycistronic operons (Section 8.1.2 and above) clearly show the intimate connection between DNA transcription and RNA translation. That there is a mutual dependence implies a necessity for coordinate control over the components of the transcription-translation apparatus.

How is control of the transcription-translation apparatus effected? Firstly, components of the expressive machinery are subject to operon control in the classical sense, in that specific DNA sequences adjacent to coding regions are responsible for regulating the production of polycistronic transcripts. Secondly, coordination of their expression is achieved by diffusible regulatory substances that are responsive to and, hence, monitor, the intracellular state of the cell. Thirdly, the genes for the core RNA polymerase subunits and ribosomal proteins are not expressed separately but rather co-transcribed from polycistronic operons. Finally, post-transcriptional events have been implicated in r-protein control (post-transcriptional processing, namely, cleavage, trimming and base modification, is also necessary for the biogenesis of functional rRNA and tRNA species). It is the superimposition of these different mechanisms that maintains the delicate balance of synthetic and degradative rates of the transcription-translation apparatus.

9.3.1 Organisation of the Genes of the Transcription-Translation Apparatus

The transcription-translation apparatus plays a central role in the expression of genetic material. The mark of an essential gene is that its inactivation leads to cellular death in the absence of a second gene copy. This second copy may be present naturally on the bacterial chromosome. Gene duplication, in fact, is relatively common for the translation machinery. There are, for example, two unlinked genes for the translational factor EF-Tu (*tufA* and *tufB* at 73 and 89 min, respectively). Similarly, multiple copies of ribosomal RNA genes are present on the bacterial chromosome, and duplicate tRNA genes also exist (consider, for instance, the tandem tRNA^{Tyr} genes, *tyrT, V* at 27 min).

The existence of multiple gene copies may be necessary to supply a high synthetic rate for these essential gene products, of particular importance under conditions of fast growth. Moreover, it buffers against loss of function since inactivation of one gene copy is not necessarily a lethal event (certainly, viable rRNA deletion derivatives may be obtained). That these multiple copies are not truly redundant is shown by the apparent strong selective pressure for their continued presence on the bacterial chromosome (Section 6.2.3). Their stability may stem from the presence of tRNA genes in these rRNA operons, although duplicate copies also exist in most cases.

(a) RNA Polymerase Operons. In *E. coli*, the DNA-dependent RNA polymerase is responsible for most, if not all, RNA synthesis. The holoenzyme form consists of four non-identical subunits, α, β, β', and σ (Section 2.1.1 (a)). It is assembled in the order

$$\alpha + \alpha \rightarrow \alpha_2 \rightarrow \alpha_2\beta \rightarrow \alpha_2\beta\beta' \rightarrow \alpha_2\beta\beta'\sigma$$

There are several thousand core enzyme molecules per bacterium and about one-third as many σ subunits.

The genes for the four subunits of RNA polymerase holoenzyme, *rpoA*, *rpoB*, *rpoC* and *rpoD*, are dispersed on the bacterial chromosome (*rpoA*, 72 min; *rpoBC*, 89 min; *rpoD*, 67 min). The β and β' genes, *rpoBC*, constitute an operon. Genetic studies have placed the promoter to the left of the β structural gene indicating that transcription is in the direction *rpoB* $\rightarrow$ *rpoC*. It has recently been shown that *rpoBC* is co-transcribed with ribosomal protein genes (Figure 9.11). Thus, the transcript produced encodes L10, L7/L12, β and β'. The major promoter responsible for the synthesis of β and β' appears to be that used for L10 and L7/L12 expression; it is situated between the genes for L1 and L10 (termed the L10 promoter, Figure 9.11).

The α gene, *rpoA* (at 72 min), located distantly from *rpoBC*, is also present in an r-protein operon. It is co-transcribed with the r-protein genes S13, S11, S4 and L17 from the P_{S13} promoter (Figure 9.11). In this case, however, *rpoA* is not at the distal end of the polycistronic message. Rather, it lies sandwiched between the genes for S4 and L17, the latter being the 3'-terminal cistron.

The σ gene, *rpoD* (at 67 min), is separate from both *rpoBC* and *rpoA*. It is located close to the gene for the r-protein S21 (no other r-proteins, in fact, neighbour this region) but as yet there is no indication that these 2 cistrons form a polycistronic operon. It should be remembered that σ is synthesised at a lower rate than the core components (Section 2.1.1 (a)). Moreover, σ synthesis appears to be regulated independently from that of β, β' and α. It remains to be

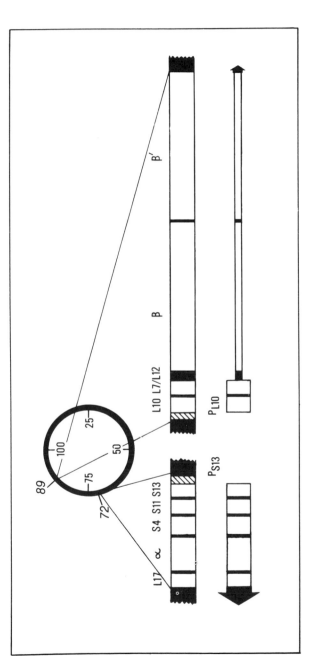

Figure 9.11: Co-Transcription of RNA Polymerase Genes with Ribosomal Genes

While the genes for β and β' *(rpoB, C)* are co-transcribed with those for L10 and L7/L12 *(rplJ, L)*, α *(rpoA)* is sandwiched between the genes for S13, S11, S4 *(rpsM,K,D)* and L17 *(rplQ)*. The promoters for these two polycistronic transcripts are designated P_L10 and P_S13. The transcript arrow is reduced in size at *rpoBC* since β. β' and α are expressed at about one-fifth the molar rate for the ribosomal proteins and there is evidence for an attenuator site between *rplL/rpoB*. After Nomura and Post (Nomura, M. and Post, L.E. (1980) in G. Chambliss, G.R. Craven, J. Davies, K. Davis, L. Kahan and M. Nomura (eds.) *Ribosomes: Structure, Function and Genetics* (University Park Press, Baltimore), pp. 671-691) seen, therefore, whether *rpoD* is subject to co-transcriptional control (see Section 9.3.2 (c)).

(b) RNA Operons. The 70S ribosome contains 3 RNA species, 5S, 16S and 23S rRNA, of which the 5S and 23S molecules reside in the large subunit (Section 2.2.1 (c)). In addition, there are between 40 and 60 different tRNAs (Section 2.2.1 (b)). These RNA species (in conjunction with r-protein messengers and mRNAs for certain outer membrane proteins) constitute the stable component of the cell: all other RNAs, and there may be several thousand different species, supply a template for translation and are turned over rapidly. Despite the fact that stable RNA represents only a very small fraction of the different RNA chains in the cell, it constitutes about one half of the RNA synthesised at any one time (in the steady-state, it represents about 97 per cent of all cellular RNA owing to the instability of mRNAs). This high concentration of rRNA reflects a high rate of synthesis due to rapid transcriptional initiation — at maximal growth rates (i.e. nutrient broth cultures), the rRNA chain initiation rate approaches the maximum of one per second — as well as multiple gene copies.

Both rRNA and tRNA species are subject to post-transcriptional events. These stable RNA species are transcribed as longer precursors and, hence, require cleavage and trimming, as well as base modification, to attain the mature form. Since both rRNAs and tRNAs appear to have defined secondary structures (witness the general cruciform shape that can be adopted by all tRNA species, Section 2.2.1 (b)) and since ribosome function requires the formation of a ribonucleoprotein complex, it seems likely that precursor formation and subsequent processing events mediate in the production of these higher order structures.

(i) Ribosomal RNA Operons. There are 7 ribosomal RNA transcription units (designated *rrnA, rrnB, rrnC, rrnD, rrnE, rrnF* and *rrnG*), each encoding the 3 rRNAs. These rRNA operons are somewhat grouped; they are confined to about one third of the bacterial chromosome (min 56-90, see Figure 2.20), distributed on either side of the origin of replication (*oriC*, 83 min). Six out of seven transcription units, in fact, lie within 10 min of *oriC* indicating that the majority of rRNA genes are duplicated early during the division cycle.

Interestingly, all four of the rRNA operons sequenced to date *(rrnA, rrnD, rrnE, rrnX)* carry tandem promoters, designated P_1 and P_2, located about 300 and 200 base pairs upstream of their respective 16S rRNA gene (see below; *rrnX* appears to be a 'hybrid' ribosomal operon generated during the isolation of the specialised transducing phage, λilv). Moreover, transcripts initiated from P_2 carry 5'-terminal pppC (the purine nucleoside triphosphates, ATP and GTP, are generally used for RNA chain initiation, Section 2.1.2 (a i)). The similarity of the sequences at the P_1 promoter in these four rRNA operons — for example the region −18 to −4

5′ T C C C T A T A A T G C G C C
A G G G A T A T T A C G C G G 5′

(the Pribnow box is underlined) is identical in each case — suggests that this regulatory site is a possible target for transcriptional control in response to certain nutritional changes (see Stringent Control, Section 9.3.2 (b)).

Ribosomal RNA is transcribed from a polycistronic operon. Each *rrn* locus contains, at the minimum, a single copy of a 16S, 23S and 5S rRNA gene (designated *rrs, rrl* and *rrf,* respectively); transcription is in the direction 16S → 23S → 5S. Ribosomal operons carry, in addition to 16S, 23S and 5S genes, certain tRNA genes (a maximum of 3 different cistrons). These are 'spacer' genes, separating the 16S and 23S coding regions: there may also be 'trailer' genes, lying at the 3′-distal end of the rRNA operon (Table 9.3). Thus, in *rrnA, alaT* and *ileT,* the genes for $tRNA_{1B}^{Ala}$ and $tRNA_1^{Ile}$, lie between *rrsA* and *rrlA* (interestingly, only three different transfer RNA species, $tRNA_2^{Glu}$, or $tRNA_{1B}^{Ala}$ and $tRNA_1^{Ile}$, have been found in spacer regions, Table 9.3). *rrnC* has an *rrs-rrl* spacer (in this case, $tRNA_2^{Glu}$) as well as $tRNA_1^{Asp}$ and $tRNA^{Trp}$ trailer genes, *aspT* and *trpT,* at the 3′-distal end. It should be noted that *trpT* represents the sole copy of the gene for tryptophanyl-tRNA and, hence, *trpT* lesions (such as the Su7 type, Section 3.3.2) are lethal to the cell when only the chromosomal copy is present.

Co-transcription of rRNA and tRNA genes results in the production of a large, polycistronic transcript (about 5000 nucleotides in length). Processing is, clearly, necessary for formation of the functional, mature species. In addition to RNA cleavage and exonuclease-mediated trimming of the smaller products, there is also extensive modification of bases during maturation of these pre-rRNA species (post-transcriptional modification of bases in tRNA is also a frequent occurrence, see Table 2.5 for a list of common modified bases).

The enzymes ribonuclease III and ribonuclease P (the *rnc* and *rnp* gene products, respectively) have been implicated in the processing of *rrn* transcripts (RNase III is also involved in the post-transcriptional processing of an r-protein-RNA polymerase transcript (Section 9.3.2 (c)) and the massive early message of phage T7. Thus, a 30S pre-rRNA accumulates in strains deficient in RNase III; it seems to contain the entire polycistronic chain including terminal 5′-triphosphate, 16S, 23S and 5S rRNA as well as spacer/trailer tRNAs. The target of RNase III appears to be specific double-stranded RNA regions generated in sequences bracketing the 16S and 23S rRNA (Figure 9.12). Complete transcription of rRNA operons is, therefore. necessary for RNase III cleavage. Precursor 16S, 23S and 5S rRNA generated by RNase III, termed p16S, p23S and p5S, lack transfer RNA. Several different activities are involved in maturation of their 5′- and 3′-termini. They appear to act subsequently to RNase III — hence, the prolonged survival of the complete 30S pre-rRNA transcript in *rnc⁻* derivatives. RNase P generates the mature 5′-ends of spacer tRNAs (Figure 9.12). This novel enzyme has an integral RNA component (about 300 nucleotides long) necessary for function.

Table 9.3[a] Components of the rRNA Polycistronic Operons[b]

r RNA Operon	Map Location (min)	tRNA (s) Encoded	
		16S – 23S Spacer	3'-Trailer
rrnA	86	$tRNA_{1B}^{Ala}$, $tRNA_1^{Ile}$	–
rrnB	89	$tRNA_2^{Glu}$	–
rrnC	84	$tRNA_2^{Glu}$	$tRNA_1^{Asp}$, $tRNA^{Trp}$
rrnD	72	$tRNA_{1B}^{Ala}$, $tRNA_1^{Ile}$	–
rrnE	90	$tRNA_2^{Glu}$	–
rrnF	74	nk	nk
rrnG	56	nk	nk

a   Abbreviations: nk, not known.
b   After Nomura and Post (1980).

(ii) *Transfer RNA Operons.* Although rRNA operons contain spacer tRNA genes (as well as 3'-distal tRNA genes in some cases), the majority of tRNA genes map outside *rrn* loci. Moreover, tRNA genes, unlike rRNA operons, are not clustered around the replication origin (see Figure 2.20). Both tandem and non-tandem duplications occur. Thus, there are three tRNA[Tyr] genes, *tyrT, tyrV* and *tyrU; tyrT,V* (at 27 min) encodes tandemly duplicated tRNAs and *tyrU*, although located distantly (89 min), codes for a tyrosyl-tRNA species that differs in only two nucleotides (Section 2.2.1 (b)).

Transfer RNAs, like ribosomal RNA, are expressed as a transcript longer than the mature form and require processing (they are short, nevertheless, in relation to the majority of cellular RNAs). This includes both ribonuclease-mediated events and base modification (Section 2.2.1 (b)). That correct processing is vital for biological function has been shown by defective tRNA derivatives carrying point mutations in the precursor region (others are mutant in the sequence encoding the mature tRNA species). Studies on these post-transcriptional events have been greatly aided by Su3 amber suppressor derivatives of *tyrT* (the gene for tRNA[Tyr]) since lesions that interfere with the production of the nonsense suppressor tRNA lead to a Su⁻ phenotype (Section 3.3.2).

Processing of the tRNA transcript appears to follow a standard pattern (see also (i), above). First, the pre-tRNA is cleaved by RNase P to generate the mature 5'-end. The common involvement of this RNA-containing enzyme in

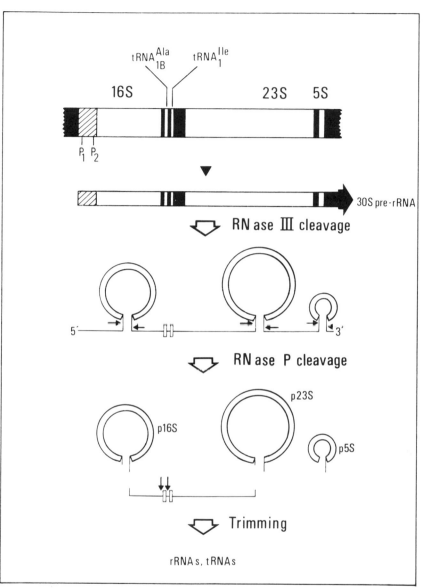

Figure 9.12: A Possible Scheme for rRNA Processing

The ribosomal operon shown has spacer genes for tRNA$_{1B}^{Ala}$ and tRNA$_1^{Ile}$ but lacks trailer genes (see Components of the rRNA Polycistronic Operons, Table 9.3). Transcription from either the P_1 or P_2 promoters would generate the large 30S pre-rRNA species (representing the entire rRNA operon) capable of forming double-stranded regions (sequences surrounding the rRNA/tRNA coding regions have been exaggerated for clarity). RNase III makes staggered breaks at these sites leaving the resultant ribosomal rRNA precursors, p16S, p23S and p5S, susceptible to trimming to give the mature form; spacer tRNAs are excised by RNase P action, functioning subsequent to RNase III, and are also trimmed for maturation. After Abelson (Abelson, J. (1979) *Ann. Rev. Biochem., 48,* 1035-1069) tRNA maturation suggests a mechanism involving (RNA·RNA-mediated?) conformational recognition rather than sequence specificity. Exonuclease activity, for instance RNase D/Q, has been implicated in the subsequent maturation of the 3'-terminus. More complex reactions are required when two or more different tRNAs are co-transcribed (for example, the *thrU-tyrU-glyT-thrT* gene cluster at 89 min which encodes $tRNA_4^{Thr}$, $tRNA_2^{Tyr}$, $tRNA_2^{Gly}$ and $tRNA_3^{Thr}$, respectively). At present, there are only a few known instances of clustering of tRNA genes (in addition to those in rRNA operons). Bacteriophage T4, on the other hand, encodes a cluster of seven different tRNAs, apparently co-transcribed from a single promoter. (The enzyme tRNA nucleotidyl transferase, the *cca* gene product, is responsible for addition of the 3'-terminal sequence, $-CCA_{OH}$, to certain of these phage tRNAs.)

(c) Ribosomal Protein Operons. The genes for the ribosomal proteins are heavily clustered on the bacterial chromosome (48 of the 52 different r-protein genes have been located, see Figure 2.20). More than half of the r-protein genes lie within 10 min of the origin of replication (the majority of *rrn* loci are similarly placed, see (b) above). There are two main gene clusters, at 72 and 89 min on the bacterial map (Figures 9.11 and 9.13). The cluster centred at 72 min encodes 27 r-proteins (and 2 translation factors, EF-G and EF-Tu). It consists of four operons, each transcribed in the same direction (anticlockwise on the circular map and in the opposite orientation to the cluster at 89 min, Figure 9.13). The four promoters responsible for these polycistronic transcripts may be designated according to the structural gene they precede, P_{S12}, P_{S10}, P_{L14} and P_{S13}.

P_{S12}-promoted expression leads to the co-transcription of the S12 and S7 genes, *rpsL* and *rpsG*, with genes for two translation factors EF-G and EF-Tu

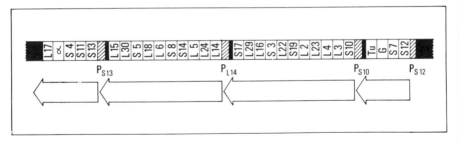

Figure 9.13: The r-Protein Gene Cluster at 72 Minutes

The gene cluster encodes 27 different r-proteins, translational elongation factors EF-Tu and EF-G, and the α subunit of RNA polymerase (see Co-Transcription of RNA Polymerase Genes with Ribosomal Genes, Figure 9.11). The promoters for the four polycistronic transcripts are designated P_{S12}, P_{S10}, P_{L14} and P_{S13}. The bottom arrows represent transcripts. After Nomura and Post (Nomura, M. and Post, L.E. (1980), in G. Chambliss, G.R. Craven, J. Davies, K. Davis, L. Kahan and M. Nomura (eds.) *Ribosomes: Structure, Function and Genetics* (University Park Press, Baltimore), pp. 671-691) (note that there is a second copy of the EF-Tu gene, *tufB*, apparently encoding a near-identical polypeptide, that maps at 89 min). Most importantly, the gene for the α subunit of RNA polymerase, *rpoA*, is co-transcribed with S13, S11, S4 and L17 (*rpsM, rpsK, rpsD* and *rplQ*, Figure 9.13). A similar instance of joint transcription of r-protein and polymerase genes, in this case *rpoBC*, has been mentioned (Section 9.3.1 (a)). The implications of co-transcription of genes for the transcription-translation apparatus is discussed in Section 9.3.2 (c), below.

9.3.2 Control of Transcriptional-Translational Operons

Many different RNA and protein species are involved in the process of gene expression, particularly at the translational level (Chapter 2); they are encoded by both dispersed and clustered genes (Section 9.3.1). Not only do these gene products function largely in multicomponent systems, but also transcription and translation are interdependent processes (Section 8.1.2). It is imperative, therefore, that both the absolute concentration and the intracellular ratio of each gene product is actively regulated to maintain the necessary balance of the transcription-translation components under a variety of physiological conditions.

(a) Metabolic Control. That the cellular environment dictates intracellular events is exemplified by *lac* induction and *trp* derepression (Sections 9.1 and 9.2). In such negative-/positive-controlled systems, gene expression is regulated at the transcriptional level through protein species and small organic effectors. The latter are commonly nutrients or their derivatives present in the growth medium. Gene expression is also dependent upon growth rate.

Generation time is dictated by the nature of the carbon source (and the presence or absence of non-essential compounds, for example, amino acids and nucleoside derivatives). It, therefore, progressively increases from nutrient broth, through glucose, succinate and acetate medium (generation time 22, 40, 67 and 120 min at 37°C). In some instances, there is a correlation between the division rate and the level of gene expression. The enzymes responsible for tryptophan synthesis, for example, appear to be regulated in this manner. Most importantly, production of the components of the transcription-translation apparatus (including aminoacyl-tRNA synthetases) increases as the doubling time decreases. The term *'metabolic' control* is used to describe these growth-related effects.

The mechanism responsible for linking growth rate with gene expression is not understood. Since metabolic regulation appears to act at the transcriptional level, and since an increase in rRNA chain initiation is associated with an increase in the number of (functional) RNA polymerase molecules, it is possible that control is mediated through the transcription complex, itself. Such a model implies, of course, that the regulatory regions of operons subject to metabolic regulation have some special properties in common. The RNA polymerase concentration may limit transcriptional initiation from these sites. Alternatively, some metabolite, as yet unknown, may mediate in their transcription in a growth-related manner.

(b) Stringent Control. When auxotrophic strains are starved of an essential amino acid, or when Ts mutants that encode a thermolabile aminoacyl-tRNA synthetase are grown at the non-permissive temperature, there is a dramatic cessation in the accumulation of stable RNA and in the synthesis of r-proteins (the reduction in bulk mRNA synthesis by about one half is largely due to alterations in the production of mRNAs for r-proteins and translation factors). This response to the build-up of uncharged tRNAs has been referred to as the *stringent response*. In effect, it buffers the cell against over-use of limiting translation potential by slowing down the production of the translation apparatus and, to a certain extent, mRNA template.

The apparent pleiotropy of the stringent response is associated with the intracellular accumulation of two unusual derivatives of GTP, ppGpp and pppGpp[7] (guanosine tetraphosphate and pentaphosphate; initially termed 'magic spot', or MS, I and II). There is an increase from 0.2 to 4 m M within seconds. ppGpp synthesis is catalysed by the *stringent protein* or (p)ppGpp synthetase I (75 000 daltons), encoded by the *relA* locus (mapping at 59 min), in response to uncharged tRNAs in the ribosome A site. The *in vitro* reaction requires both the ribosome-mRNA complex and uncharged tRNAs. That the ribosome is involved in ppGpp production *in vivo* is indicated by a *Rel⁻* mutant with altered r-protein L11. A number of other genetic loci have been implicated in the stringent response (they are prefixed '*rel*' if lesions at these sites result in a 'relaxed' response — continued stable RNA accumulation — upon starvation). In addition, the *spoT* gene (81 min) encodes a (p)ppGppase (65 000 daltons) involved in (p)ppGpp turnover; it does not require the ribosome for function *in vitro*. The ppGpp-mediated stringent response may, therefore, be reversed by stimulation of the rapid degradation of this physiological effector (Figure 9.14).

ppGpp is the common effector in the regulation of stable RNA production, expression of both rRNA and tRNA genes, and of r-protein biosynthesis. Genes contained within these operons, for instance, those encoding EF-G and EF-Tu, are similarly controlled (though not RNA polymerase synthesis, see (c) below). How does this unusual nucleotide prevent expression of these operons? Although it is known that this *stringent control* acts at the transcriptional level — for both rRNA/tRNA and r-protein operons — to prevent further *de novo* RNA synthesis under stringent conditions (rather than through promoting degradation

---

7.   A variety of metabolic processes, in addition to RNA production, are subject to the stringent response. These include synthesis of carbohydrates, fatty acids, lipids, peptidoglycans and polyamines. Moreover, ppGpp has been implicated in the expression of attenuator-controlled operons such as *trp* or *his*. The fact that the mistranslation in *rel⁻* strains is associated with the absence of ppGpp (see also Phenotypic Suppression, Section 3.3.3) and that only starvation for tryptophan (and other amino acids encoded by codons close to the tandem Trp triplets) stimulates *trp* expression (witness the translation model for *trp* attenuation, Section 9.2.2 (b)) has led to the suggestion that ppGpp acts at the translational level, the site of action being the two GTP-dependent steps in polypeptide chain elongation (see Section 2.2.2 (a ii)).

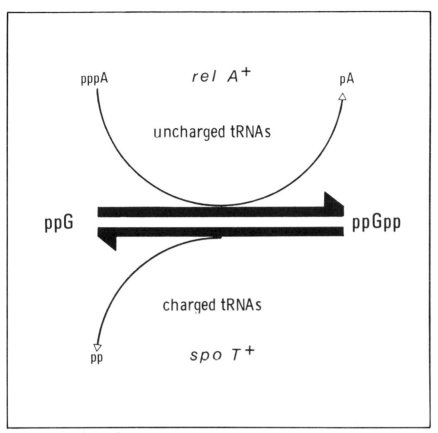

Figure 9.14: Synthesis and Degradation of ppGpp
The presence of uncharged tRNAs in the ribosome A site stimulates production of ppGpp (magic spot, or MS, I) in a *relA*+-mediated reaction. The *spoT*+-directed breakdown of MSI is, on the other hand, ribosome independent. After Richter (Richter, D. (1980), in G. Chambliss, G.R. Craven, J. Davies, K. Davis, L. Kahan and M. Nomura (eds.), *Ribosomes: Structure, Function and Genetics*, (University Park Press, Baltimore), pp. 743-765)

of existing RNA species), the site of action of ppGpp is not clear. An attractive hypothesis, supported by certain *in vitro* studies using defined DNA templates, is that ppGpp specifically inhibits RNA polymerase transcription at rRNA, tRNA and r-protein operons.

Whether this nucleotide acts through direct association with RNA polymerase itself (or by initially binding to the promoter site) is not known. It would be expected that if stringent control functions at promoters, all genes co-transcribed from such sites would be similarly affected. Certainly, this contention is supported by ppGpp-related reduction in EF-G and EF-Tu synthesis. Models that propose a promoter-specific effect readily explain the insensitivity of RNA polymerase synthesis to stringent conditions (see below). However, although the P_1 promoters of rRNA operons have sequences in common (a putative ppGpp-dependent control site), the r-protein promoter sites show similarities with non-stringently controlled operons.

In vivo studies on stringent control are complicated by the multiple control circuits involved in the regulation of the transcriptional-translational apparatus (see (c), below). Clarification of the ppGpp mechanism awaits purified transcription systems that correctly mimic the *in vivo* response.

(c) Co-Transcriptional and Post-Transcriptional Control. Since the expression of structural genes requires both transcription and translation, the placement of genes of the expressive machinery into polycistronic operons would supply a ready means for their coordinate control (as well as imposing a strong pressure for continued preservation of the multiple rRNA loci). Despite the apparent co-transcription of RNA polymerase and r-protein genes (Section 9.3.1 (a)), and their common response to increased growth rate (see (a), above), the core components of RNA polymerase are synthesised at one-fifth the molar rate of r-protein production. Moreover, RNA polymerase biogenesis is largely unaffected by the stringent response and reacts specifically to perturbations that inhibit its function (addition of the antibiotic rifampicin, for example, to a wild-type strain results in a transient increase in core enzyme synthesis, but r-protein production is unaffected).

To explain the reduced expression of RNA polymerase genes in relation to r-protein cistrons, it has been suggested that *rpoBC* and *rpoA* are under attenuator control (Section 8.1.2 (b)). On this model, partial transcriptional termination reduces production of the distal polymerase messages without affecting r-protein synthesis. The attenuator site would need to be placed between the most distal r-protein gene and the *rpo* locus. That is, in the case of *rpoBC* attenuator control, the termination site would lie between the genes for L7/L12 and β (see Figure 9.11). Evidence supporting this contention includes a rho-dependent five-fold reduction in *rpoBC* transcription relative to *rplJL* transcription, and a terminator-like sequence in the intercistronic region (321 bp) between L7/L12 and β genes.[8] Moreover, reduced expression from *rplJL* can apparently be compensated by anti-attenuation prior to *rpoBC*. The α gene, however, is flanked by two r-protein cistrons, those for S4 and L17 (see Figure 9.14). Whether attenuator control applies to *rpoA* requires resolution. Certainly, it is possible to reconcile a single mechanism of attenuation with the fact that β and β' synthesis but not that of EF-G and EF-Tu is apparently independently regulated (witness their insensitivity to the stringent response, Section 9.3.2 (b)).

An additional control mechanism can act at the post-transcriptional level. It is proposed that unassembled ribosomal proteins promote the functional inactivation of their own transcripts. This feedback mechanism could cause mRNA

8.   There appear to be weak promoters preceding *rpoB* (both between *rplJ* and *rplL* and between *rplL* and *rpoB*). Although the physiological significance, if any, of these sites is not known, they may be important in dissociating control of RNA polymerase subunits and r-proteins. RNase III may also play a role in this, albeit at the post-transcriptional level.

degradation directly or through translational inhibition (see Polarity, Section 8.1.2 (a)). The model readily explains why a doubling in gene dosage does not cause a comparable doubling in r-protein synthesis and why the message, synthesised at a two-fold higher rate, is rapidly degraded. Certainly, there is a precedent for feedback translational control (Section 8.2.2). Strong support for feedback regulation as a means of controlling r-protein synthesis comes from recent *in vitro* and *in vivo* studies showing that S4 and S8 specifically regulate operons that encode these r-proteins (the P_{S13} and P_{L14} operons, Figure 9.13).

It can be concluded that production of the transcription-translation apparatus is subject to multiple control circuits acting both at the transcriptional and post-transcriptional level (see also (a) and (b), above). Superimposition of these different regulatory mechanisms may be necessary to maintain accurately the relative concentrations of the different components. It may, for instance, compensate for gene dosage effects during the cell cycle owing to the dispersed nature of several of the transcriptional-translational genes. Moreover, the presence of different control circuits would provide for both coordinate and independent expression of these vital cistrons.

Bibliography

9.1 Catabolite-Controlled Operons: The Lactose System

Beckwith, J.R. and Zipser, D. (eds.) (1970) *The Lactose Operon* (Cold Spring Harbor Laboratory, New York).

Kepes, A. (1969) 'Transcription and translation in the lactose operon of *Escherichia coli* studied by *in vivo* kinetics', *Prog. Biophys. Mol. Biol., 19*, 201-236.

Miller, J.H. and Reznikoff, W.S. (eds.) (1980) *The Operon*, 2nd edn (Cold Spring Harbor Laboratory, New York).

Pastan, I. and Adhya, S. (1976) 'Cyclic adenosine 5'-monophosphate in *Escherichia coli', Bacteriol. Rev., 40*, 527-551.

Siebenlist, U., Simpson, R.B. and Gilbert, W. (1980) '*E. coli* RNA polymerase interacts homologously with two different promoters', *Cell, 20*, 269-281.

9.2 Attenuator-Controlled Operons: The Tryptophan System

Bertrand, K., Korn, L., Lee, F., Platt, T., Squires, C.L., Squires, C. and Yanofsky, C. (1975) 'New features of the regulation of the tryptophan operon', *Science, 189*, 22-26.

Borer, P.N., Dengler, B., Tinoco, I. and Uhlenbeck, O.C. (1974) 'Stability of ribonucleic acid double-stranded helices', *J. Mol. Biol., 86*, 843-853.

Crawford, I.P. (1975) 'Gene rearrangements in the evolution of the tryptophan synthetic pathway', *Bacteriol. Rev., 39*, 87-120.

Crawford, I.P. and Stauffer, G.V. (1980) 'Regulation of tryptophan biosynthesis', *Ann. Rev. Biochem., 49*, 163-195.

Platt, T. (1978) 'Regulation of gene expression in the tryptophan operon of *Escherichia coli*', in J.H. Miller and W.S. Reznikoff (eds.), *The Operon* (Cold Spring Harbor Laboratory, New York), pp. 263-302.

Platt, T. (1981) 'Termination of transcription and its regulation in the tryptophan operon of *E. coli', Cell, 24*, 10-23.

Umbarger, H.E. (1978) 'Amino acid biosynthesis and its regulation', *Ann. Rev. Biochem., 47*, 533-606.

Yanofsky, C. (1981) 'Attenuation in the control of expression of bacterial operons', *Nature, 289*, 751-758.

9.3 Multivalent Control of Transcriptional-Translational Operons

9.3.1 Organisation of the Genes of the Transcription-Translation Apparatus

Abelson, J. (1979) 'RNA processing and the intervening sequence problem', *Ann. Rev. Biochem.,* *48*, 1035-1069.

Altman, S. (1978) 'Biosynthesis of tRNA', in S. Altman (ed.), *Transfer RNA* (The M.I.T. Press, Cambridge), pp. 48-77.

Isono, K. (1980) 'Genetics of ribosomal proteins and their modifying and processing enzymes in *Escherichia coli*', in G. Chambliss, G.R. Craven, J. Davies, K. Davis, L. Kahan and M. Nomura (eds.), *Ribosomes: Structure, Function and Genetics* (University Park Press, Baltimore), pp. 641-669.

Nomura, M., Morgan, E.A. and Jaskunas, S.R. (1977) 'Genetics of bacterial ribosomes', *Ann. Rev. Genet., 11,* 297-347.

Yura, T. and Ishihama, A. (1979) 'Genetics of bacterial RNA polymerases', *Ann. Rev. Genet., 13,* 59-97.

See also references in Section 9.3.2 (c)

9.3.2 Control of Transcriptional-Translational Operons

Gausing, K. (1980) 'Regulation of ribosome biosynthesis in *E. coli*', in G. Chambliss, G.R. Craven, J. Davies, K. Davis, L. Kahan and M. Nomura (eds.), *Ribosomes: Structure, Function and Genetics* (University Park Press, Baltimore), pp. 693-718.

Nierlich, D.P. (1978) 'Regulation of bacterial growth, RNA, and protein synthesis', *Ann. Rev. Microbiol., 32,* 393-432.

Travers, A. (1976) 'RNA polymerase specificity and the control of growth', *Nature, 263,* 641-646.

(a) Metabolic Control

See references above.

(b) Stringent Control

Cashel, M. (1975) 'Regulation of bacterial ppGpp and pppGpp', *Ann. Rev. Microbiol., 29,* 301-318.

Gallant, J.A. (1979) 'Stringent control in *E. coli*', *Ann. Rev. Genet., 13,* 393-445.

Kjeldaard, N.O. (1979) 'Control mechanisms of the formation of ribosomal RNA and transfer RNA and the synthesis of guanosine tetraphosphate', in J.E. Celis and J.D. Smith (eds.), *Nonsense Mutations and tRNA Suppressors* (Academic Press, London), pp. 191-205.

Richter, D. (1980) '*In vitro* synthesis and decay of guanosine 3', 5'-bis (diphosphate) (ppGpp)', in G. Chambliss, G.R. Craven, J. Davies, K. Davis, L. Kahan and M. Nomura (eds.), *Ribosomes: Structure, Function and Genetics* (University Park Press, Baltimore), pp. 743-765.

(c) Co-Transcriptional and Post-Transcriptional Control

Friesen, J.D., Fiil, N.P., Dennis, P.P., Downing, W.L., An, G. and Holowachuk, E. (1980) 'Biosynthetic regulation of *rplJ, rplL, rpoB* and *rpoC* in *Escherichia coli*', in G. Chambliss, G.R. Craven, J. Davies, K. Davis, L. Kahan and M. Nomura (eds.), *Ribosomes: Structure, Function and Genetics* (University Park Press, Baltimore), pp. 719-742.

Nomura, M. and Post, L.E. (1980) 'Organisation of ribosomal genes and regulation of their expression in *Escherichia coli*', in G. Chambliss, G.R. Craven, J. Davies, K. Davis, L. Kahan and M. Nomura (eds.), *Ribosomes: Structure, Function and Genetics* (University Park Press, Baltimore), pp. 671-691.

10 Control of Extrachromosomal Genetic Elements

A bacterial cell survives by virtue of expression of its genetic material, making use of the encoded transcription-translation apparatus. What of an extrachromosomal genetic element that invades the cell? Clearly, the mere fact that autonomous elements are inherited upon cell division indicates their replication. Although multiplication is usually catalysed by mainly host-coded functions, genetic elements that replicate in $E.$ $coli$ generally contain a definite origin site as well as supplying specific replication proteins.

Since propagation demands multiplication and transmission (not merely inheritance), the fate of these foreign DNAs is not just one of (passive) replication in the host cell. Conjugative plasmids, for example, can mediate both in their own transfer and that of the host chromosome (as well as mobilise nonconjugative plasmids). Whereas plasmid transfer relies on cell-to-cell contact, a process that requires the formation of a conjugation bridge and entrance of a single strand of DNA, viral nucleic acid is transmitted as a double- or single-stranded molecule of nucleic acid packaged in a protein coat. Nevertheless, whether plasmid or phage, transfer is mediated by numerous gene products, the information for which is specific to, and contained in, the nucleic acid of the respective genetic element. In addition to replication and transfer loci, genes that impart a selective advantage to the host — such as drug resistance — may also be present. The immunity to superinfection inherited by a cell upon lysogenisation is, on the other hand, part and parcel of the phage regulatory system that permits this stable relationship.

When considering expression of extrachromosomal material, three main points should be remembered. Firstly, these elements may be present in many copies. Consider, for example, multicopy plasmids or the hundreds (and in some instances thousands) of phages produced in a single infection (Chapters 4 and 5). This high gene dosage is in marked contrast to genetic loci confined to the bacterial chromosome where only the rRNA operons (and certain tRNA genes) are found in multiple copy number (Section 9.3.1 (b)). Secondly, in view of the number of genes involved (Chapter 2), it is not surprising that these extrachromosomal elements do not encode their own transcription-translation machinery. Rather, they make use of, sometimes even annex, the host's system.

The lack of symbiotic relationship between invader and cell is particularly true of virulent bacterial viruses. They commonly encode factors which, after expression by the host system, prevent synthesis of bacterial products to selfishly maximise their own multiplication. Thirdly, the limited coding capacity of small genetic elements means that economy is necessary to encode the essential information for survival. The presence of overlapping genes — where one stretch of DNA encodes up to three different protein products through use of the three translational reading-frames — is sometimes used to increase coding potential. However, to divorce expression of the extrachromosomal element from that of the natural division cycle of its host, more sophisticated replicons such as phages are controlled by a series of elaborate interlocking circuits. Economy, in this instance, is sacrificed for versatility.

The present chapter considers the control of only two extrachromosomal genetic elements, the F plasmid (Section 10.1) and bacteriophage *lambda* (Section 10.2). Although F and *lambda* may be considered prototypes, they are unique in many respects. F, for example, is a stringently controlled plasmid that is replicated bidirectionally from a specific origin whereas the replication fork in the relaxed plasmid ColE1 moves in only one direction (Section 6.1.3 (a)). Moreover, a large part of the coding potential of F is taken up with transfer functions. There is some homology between F and conjugative F-like R plasmids but this lies mainly in the transfer region. With respect to bacteriophage *lambda*, the majority of the known bacterial viruses (apart from members of the lambdoid phage group) are quite distinct both in terms of their genetic make-up and their propagation mechanisms. The use of the host expressive machinery is particularly pertinent in this regard. *Lambda*, for instance, encodes a specific transcriptional anti-terminator that interacts with the host RNA polymerase whereas other phages may modify the host enzyme (phage T4) or even encode new, albeit simpler versions (phage T7). Moreover, virulent phages, unlike their temperate counterparts, do not generally require accessory host functions for replication.

Why then the F plasmid and *lambda*? These two genetic elements, partially for historical reasons, have been extensively studied. As a conjugative plasmid, F is a model system for transmission of genetic material via cell contact. It allows, furthermore, study of stringent control and the property of incompatibility. The ability of F to insert into other DNAs may be considered as an extra perk! Certainly, substituted plasmids have served as a useful source of DNA for investigation of this phenomenon. Coliphage *lambda*, on the other hand, is a beautiful example of organised complexity. Genes of related function are clustered together, thereby supplying a ready means for their joint control (this happens also in F). The phage serves to illustrate many forms of gene control as well as supplying a few surprises such as overlapping genes and multiple operator sites. Its ability to promote site-specific recombination is particularly interesting — does this represent a primaeval transposon? Most importantly, the phage is regulated by a developmental switch that allows either virulence or lysogeny. To understand *lambda*, therefore, is to understand the workings of a most complex extrachromosomal genetic element.

10.1 Regulation of F Plasmid Transactions

The F plasmid can exist in one of two mutually exclusive states, either as an autonomous extrachromosomal genetic element or as part of the bacterial chromosome (Chapter 4). Replication of the autonomous element is stringently controlled at no more than about two copies per cell while copy number in the integrated state is dictated by the host replicative system entirely. In addition to vegetative replication of the cytoplasmic element, multiplication may also proceed by conjugative replication. Transfer of the plasmid — and of the bacterial chromosome if the two chromosomes are covalently attached — is effected in this mode. It requires, however, direct cell-to-cell contact (unlike viral DNA, which is encapsidated in a protein coat, Section 5.2). The F plasmid is, in short, a non-lethal replicon that by itself offers no obvious benefit to its host cell other than in evolutionary terms (Section 6.2.3) and in preventing the entrance and multiplication of similar genetic elements through plasmid-encoded surface exclusion and incompatibility properties (Chapter 4). Its conjugative properties have proved invaluable for genetic analysis (Section 7.1.2 (a)).

10.1.1 Organisation of the F Plasmid

Despite the relatively large size of F (94.5 kb), the plasmid makes use of host functions for the majority of its needs. There is no evidence, for example, that it encodes a specific transcription-translation apparatus nor even that it modifies the host system for its requirements (perhaps, potentially lethal elements such as phages are unique in this regard, see Section 10.2.3 (c)). Moreover, most of the components of the bacterial replicative machinery are necessary for autonomous F replication (other than *dnaA*, see Integrative Suppression, Section 4.1.3 (a i)). Certainly, once integrated, the plasmid appears to become a passive substrate for host functions (although incompatibility is still expressed, Section 6.1.2 (c)).

Despite this reliance on its host cell, F encodes information for transfer, vegetative replication and site-specific recombination. These properties are represented by three main functional gene clusters on the circular genetic map (Figure 10.1). [T7 is a female-specific phage since growth is restricted on F+ (or Hfr) cells. The two F cistrons responsible for this restriction map at *pif* (about 33 kb on the circular element).] As yet, only the enormous *tra* operon, responsible for the transfer functions, has been extensively analysed; this reflects the difficulty of dealing with cistrons that are essential for plasmid maintenance. More recently, however, recombinant DNA techniques (Section 7.3.3) have allowed the isolation and independent maintenance of various regions of the F plasmid including the transfer origin of replication, *oriT*, and the equivalent vegetative site, *oriV*. Such methods should prove useful in determining

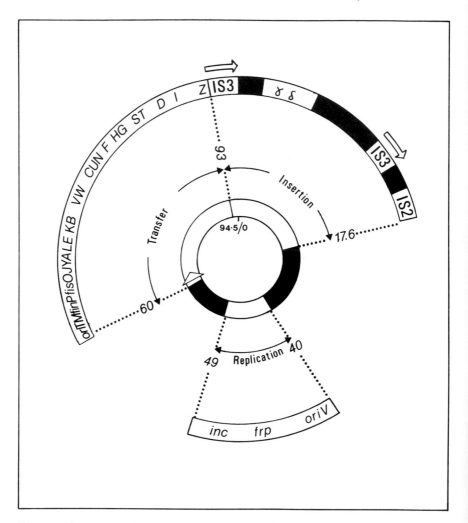

Figure 10.1: Organisation of the F Plasmid

The three main gene clusters on the 94.5 kb F plasmid are concerned with insertion, replication and transfer (93-17.6 F, 40-49 F and 60-93 F). The insertion region contains 4 insertion sequences, IS3, IS2 and γδ, the first being present in direct repeat. The replication region encodes specific replication proteins *(frp)* as well as incompatibility *(inc)* and contains the vegetative origin of replication, *oriV* (cf. *oriT*). The *tra* operon, responsible for plasmid transmission, consists of at least 22 different genes, the majority of which are concerned with F pilus production (the arrowhead representing the direction of transfer from *oriT*). Note that the coding potential of the *tra* operon and, indeed, of F, itself, is by no means saturated. After Manning and Achtman (Manning, P.A. and Achtman, M. (1979), in M. Inouye (ed.), *Bacterial Outer Membrane: Biogenesis and Functions* (John Wiley, New York), pp. 409-447 whether the so-called 'silent' regions — those to which no functions have yet been ascribed — are truly genetically silent.

(a) The Transfer Region. The *tra* operon is a 33 kb contiguous sequence of F DNA consisting of at least 22 different genes (the coding potential is by no means saturated as there are gaps in the central portion). It is sited between positions 60 and 93 on the F map (Figure 10.1). *traM, J, A* and *I* are unique to F since mutations at these loci are not complemented in *trans* by F-like plasmids.

The region from *traY* to *traZ* is transcribed as a single operon although there is a second, weak promoter between *traD* and *traI*. Since the *traJ* product is required for the transcription of all *tra* genes other than M (it may act indirectly and/or in combination with host proteins, see below), and since a single site, *fisO*, located very close to, and upstream of, *traJ*, appears to regulate *traJ* transcription directly, *fisO* is responsible for control of the entire *traJ-traI* region. It should be noted, however, that the F *tra* functions are expressed constitutively, that is, they are naturally derepressed (but see Section 10.1.2).

The majority of *tra* genes identified to date are implicated in F pilus production; these include *traJ, A, L, E, K, B, V, W, C, U, F, H* and, strangely, the promoter-proximal part of *traG*. *traA* was initially thought to encode F pilin, itself, but recent evidence suggests that *traJ* may be responsible for pilin. The *traJ* gene product may, therefore, have both a regulatory and structural role. The fact that this protein resides in the outer membrane suggests an indirect function in *tra* regulation. Mating aggregate formation, triggering of DNA transfer and transfer, itself, are mediated by *traG/traN, traI/M* and *traD* respectively; the *traYZ* endonuclease is responsible for nicking at *oriT*. The *traS* and *traT* genes, although clearly part of the *tra* operon, are not required for pilus synthesis or DNA transfer (mutations in *traS/traT* inhibit transmission only if polar on *traD* or *traI*). Rather, they are responsible for surface exclusion, one of the two barriers to conjugation whereby stable contact is inhibited between two cells harbouring plasmids of the same surface exclusion group (Section 4.1.2 (b)). Immunity to lethal zygosis (Section 4.1.3 (a ii)) also maps in the vicinity of *traT*; the locus is termed *ilzA* (a second locus, *ilzB*, appears to map outside the *tra* region).

Located just to the left of *traM*, upstream from *traY*, is the transfer origin of replication *oriT*. This locus was initially defined by deletions that give rise to a *cis*-dominant transfer defect. It has been recently narrowed down to a 250 bp stretch of DNA. Sequence analysis of this region may have identified the translational start point for the *traM* protein (through a putative ribosome binding sequence, Section 2.2.2 (a i), and neighbouring GUG codon). Most importantly, there is a decanucleotide sequence present in both the *oriT* region and adjacent to a site (termed *nic*) present on the relaxed plasmid ColE1 necessary for mobilisation (although non-conjugative plasmids are transfer-deficient *per se*, they may be mobilised by conjugative plasmids, Section 4.2).

It should be noted that since passage of F DNA proceeds in a unidirectional manner from *oriT*, the orientation of transfer (as indicated by the arrow in Figure 10.1) implies that the *tra* genes themselves are last to enter the recipient cell. Retransfer must, therefore, await the arrival of the complete molecule. In the case of an Hfr strain, when the insertion region (93 to 17.6 kb, see (c) below) is split and contains the entire bacterial chromosome, the *tra* operon and *ilzA* are similarly last to enter — they rarely do so, in fact, since random breakage prevents transfer of the entire donor chromosome (Section 4.1.3 (a ii). Thus, immunity to lethal zygosis is commonly not imparted to the recipient in an Hfr cross.

(b) The Replication Region. Three loci concerned with vegetative replication are clustered in one region of the F map (about 40-49 kb, Figure 10.1), distinct from the *tra* operon: *oriV*, the vegetative origin of replication (cf. the transfer origin, *oriT*); *frp*, the F replication gene(s); and *inc*, responsible for IncF1 incompatibility. That F-coded proteins are required for maintenance is shown by the existence of F_{TS} derivatives which are 'lost' at high temperature.

The F replication region appears to be complex since, in addition to the primary origin responsible for bidirectional replication (at 42.6 kb), there is a second site (at 44.4 kb) that will 'drive' replication of plasmid derivatives in the absence of the first. Derivatives that contain only the primary origin do not exhibit F incompatibility suggesting that it is interactions at the secondary site which are responsible for this property. The IncF1 phenotype, itself, is associated with two sites (about 45.5 and 46.6 F), the nearest of which is some 900 nucleotide pairs distant from the secondary origin.

The placement of *inc* close to *oriV* lends support to the negative control model for replication control (Sections 4.1.2 (b) and 6.1.2 (c)). Certainly, if the former locus encodes (or is responsible for the production of) the hypothetical repressor, bidirectional replication of F should allow early duplication of *inc* with the result that the product(s) encoded would be synthesised at twice the 'haploid' rate. The *inc* site proximal to the secondary replication origin may be responsible for negative control since lesions in this region both give rise to high copy number replication and destroy F incompatibility.

(c) The Insertion Region. The region implicated in F insertion (93-17.6 kb) contains four insertion sequences, three of which, IS2, IS3 and $\gamma\delta$,[1] are non-identical; IS3 is repeated directly (at 93-94.5/0 kb and 13.7-15 kb). In R plasmids, the RTF component encoding the transfer functions is separated from the resistance determinant, r, by ISs (IS1 in the case of R100). Insertion elements were initially observed through heteroduplex analysis (Section 7.3.2) of single strands of substituted F plamids: when sequences are duplicated in reverse orientation, the complementary stretches present on a single strand may anneal. The DNAs responsible for such duplex structures and for interstrand

1. $\gamma\delta$ (2.8-8.5 F), long suspected to be an insertion sequence, has recently been shown to transpose independently of *recA*, that is, to act as a transposable genetic element. It has been given the designation, Tn1000.

events were later identified by hybridisation to *lambda* derivatives carrying known IS elements.

Analysis of type II substituted plasmids (Section 4.1.3 (b)) has been particularly useful for determining insertion sites since the entire F material is retained. Significantly, IS sequences lie at the junctions between plasmid and bacterial DNA (represented by arrowed regions in Chapter 4 and thereafter). F13, for example, carries a direct repeat of IS2 (Figure 10.2) while γδ is duplicated in F14. It, therefore, seems likely that F can integrate through these ubiquitous transposable elements in a *recA*-independent event (Section 6.2.2 (a ii)). This contention is supported by direct analysis of the parental Hfr strain of F13. (The excision event responsible for the production of the substituted plasmid does not seem to have involved ISs, Figure 10.2.)

The actual number and type of IS elements on the bacterial chromosome is by no means invariant. Although there are hotspots at which Hfrs frequently form (analysed through their excision products as discussed above), integration is not necesarily restricted to a series of 'pre-ordained' sites since the IS elements are capable of translocation. In this regard, either the bacterial chromosome or F plasmid, itself, may act as a material source (Section 6.2.3).

10.1.2 Interaction Between F and F-like R Plasmids

F-like R plasmids encode both drug resistance and conjugative properties. They owe the latter to the presence of a transfer region sharing remarkable homology with the F *tra* operon (similarity between these two types of genetic elements is largely restricted to their transfer regions). In keeping with this, F-like pili show sufficient similarity to the F pilus to allow co-polymerisation of the F-like and F pilin subunits. However, whereas F is naturally derepressed for transfer functions, transmitting at high frequency between strains, F-like R plasmids are generally repressed (although genetically derepressed mutants, *drd⁻*, may be obtained, Section 4.2).

F-like R plasmids transfer at high frequency immediately upon entering a cell, but this property is rapidly lost. Does this transitory re-transfer ability and its decline represent the build-up of some repressor substance in the recipient, the delay representing the time required for the synthesis of the putative cytoplasmic inhibitor? Certainly, the fact that the presence of an F-like R plasmid in an F⁺ cell prevents high frequency F transfer — that is, that the R plasmid encodes a *trans*-acting substance — supports the involvement of a diffusible factor. Clearly, the cytoplasmic inhibitor would need to recognise a common site(s) on both plasmids since F is susceptible to this *fertility* (or *transfer*) *inhibition* (Fi⁺ phenotype, Section 4.2). F must, however, lack a gene responsible for the phenomenon (or lack the ability to express it). The story is somewhat complicated by the finding that a class of F mutants not susceptible to fertility inhibition are recessive in the presence of a wild-type F copy. This suggests that the postulated repressor consists of at least two components, one of which is supplied by F, itself.

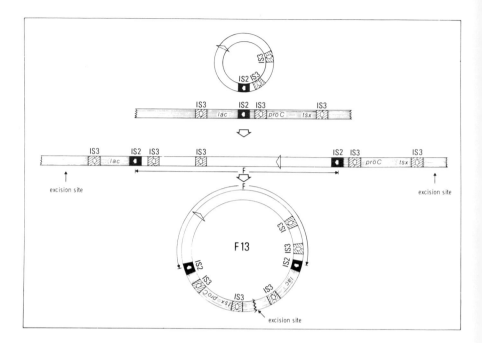

Figure 10.2: F13 Formation; F Plasmid Insertion via IS2 and Subsequent Aberrant Excision

The F plasmid may integrate into chromosomal (or extrachromosomal) material making use of one of its four insertion elements ($\gamma\delta$ is not shown, see Figure 10.1). Analysis of the type II F-prime, F13, generated after aberrant excision from the *lac-proC* interval, suggests that Hfr 13 was created through IS2 homology (note that excision appears to take place at bacterial sites lacking ISs). The *lac-proC* region seems, in fact, to be a hotspot for Hfr formation, the multiple IS elements serving as integration sites for a number of different Hfrs. After Ohtsubo and Ohtsubo (Ohtsubo, H. and Ohtsubo, E. (1977), in A.I. Bukhari, J.A. Shapiro and S.L. Adhya (eds.), *DNA Insertion Elements, Plasmids and Episomes* (Cold Spring Harbor Laboratory, New York), pp. 49-63)

An elegant model for transfer inhibition defines three loci, an operator site, *fisO*, adjacent to *traJ* and two repressor genes, *finO* and *finP*, both of which are required for inhibition (formerly termed *traO, fin* and *traP* by Finnegan and Willetts (1973)). Whereas *fisO* and *finP* are present on the two plasmids, *finO* is encoded only by F-like R plasmids that exhibit the Fi+ phenotype (Figure 10.3). The *finP* product, moreover, is plasmid specific, only acting (in conjunction with the *finO* product, see below) at its respective *fisO* site; the loci may, thus, be designated *finP*$_{R100}$ and *finP*$_F$ in the case of the plasmids R100 and F.

On this model, an F+ cell produces the F-specific *finP*$_F$ gene product but lacks the 'core' repressor component, the *finO* protein (Figure 10.3). The *fisO* operator site is, therefore, vacant allowing production of necessary transfer functions. An R+ bacterium, on the other hand, synthesises both components of the repressor within a short time after entrance of the plasmid, thereby

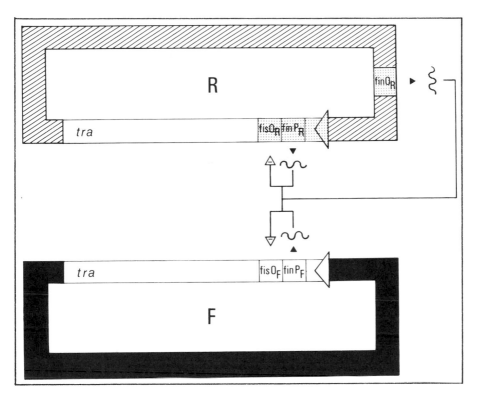

Figure 10.3: A Model for Regulation of Plasmid Transfer: The finOP-fisO System

F-like R plasmids encode the two components of the transfer repressor, *finOP*, and are transferred at low frequency (except after immediate entrance into a 'virgin' cell). The R plasmid encoded *finO_R* product can combine with both *finP_R* and *finP_F* to simultaneously repress R and F transmission (at *fisO_R* and *fisO_F*, respectively). In the absence of Fi+ R plasmids, however, F transfer is extremely efficient since only the 'core' *finP_F* component is synthesised. Mutations that knock out *finP_F* prevent fertility inhibition by Fi+ R plasmids only so long as a functional copy is not supplied in *trans* (say, in a F *finP−*/F*finP+* double male strain). *finOP* and *fisO* were formerly termed *fin*, *traP* and *traO* by Finnegan and Willetts (Finnegan, D.J. and Willetts, N.S. (1973) *Molec. Gen. Genet., 127*, 307-316)

preventing *tra* expression. When an F-like R plasmid enters the F+ cell, a proportion of the R plasmid-encoded *finO* product combines with the F plasmid-specific *finP_F* protein to inhibit expression of the F *tra* operon (the active repressor complex encoded by *finO* and *finP_R* acts on the R plasmid). Clearly, mutations in *finP_F* prevent fertility inhibition only so long as there is not a second gene copy present.

The *tra* operon, in conclusion, is subject to two regulatory circuits: positive control through *traJ* (albeit, perhaps, indirectly) and negative control of *traJ* through the *finOP-fisO* system. That a certain class of R plasmids, the F-like plasmids, has a closely related transfer system to F is demonstrated both by the homology between the respective *tra* regions and the fact that the two systems interact when the two genetic elements coexist in one cell.

10.1.3 The 'Life-Cycle' of the F Plasmid

The F plasmid, like other low copy number conjugative elements, is considerably larger than its relaxed counterparts (Section 4.2). Certainly, at least one third of F is sacrificed to transfer functions. The presence of the *tra* region creates, however, a relatively versatile genetic element — one that can coexist within the host cell or multiply through transfer to other bacteria. Its lack of protein coat is circumvented by the sophisticated conjugation bridge that permits transfer of naked(?) single-stranded DNA through the two cell envelopes (Section 4.1.3). In fact, the majority of the *tra* genes are bound over to pilus production.

Transfer, in combination with site-specific recombination, supplies a powerful evolutionary potential to plasmid and host alike (see Phylogenetic Implications of Recombination, Section 6.2.3). Such conjugative plasmids can both acquire new characteristics — arising through aberrant excision or insertion of other transposable genetic elements — and transfer these properties. In the case of F, a naturally derepressed plasmid, transfer and re-transfer occurs at high frequency.

The three gene clusters on the F plasmid allow both an autonomous and integrated existence and, unique to this element, derepressed transfer properties for either state. Despite this versatility, transfer is an elaborate process that requires intimate cell contact and is, therefore, severely limited by population density and motility of the cells concerned. Moreover, unlike the majority of temperate bacteriophages, F does not appear to possess a system for monitoring the cellular state to allow 'escape' from a doomed bacterium (if such an inducible system existed, survival of the F plasmid through transfer would be restricted by the possibility of cell-to-cell interactions).

10.2 Regulation of Phage *Lambda* Development

Coliphage *lambda* is a temperate bacterial virus and, hence, either enters the lytic cycle upon infection or lysogenises its host (Chapter 5). In the prophage state, the viral chromosome is integrated stably into the bacterial chromosome, indicating that both specific integration enzyme(s) and a system for shutting off lethal phage functions are necessary for this symbiotic relationship. The balance between the virulent and lysogenic pathway is controlled by host- and phage-coded gene products. The molecular processes responsible for lysogeny are particularly interesting since the 'choice' between the lytic and lysogenic response may be likened to a 'developmental switch'.

In considering phage *lambda* regulation, the major functional gene clusters are first described (Section 10.2.1). The events involved in prophage formation (Section 10.2.2) and induction (Section 10.2.3) are discussed next. Finally, a rationale is offered for this lysogenic-lytic decision (Section 10.2.4).

10.2.1 Organisation of the Intracellular Viral Chromosome

The *lambda* viral genome is a modestly sized double-stranded molecule of DNA (48.6 kb, Section 5.1.1 (b)) with 12 nucleotides long, single-stranded tails; the two tails are complementary to each other. The *lambda* DNA molecule can, therefore, circularise through annealing of these cohesive ends (Section 5.2.2). The viral genome can exist intracellularly as a circular extrachromosomal genetic element or as a linear prophage, integrated into the bacterial chromosome (note the circularly permuted gene order in Figures 10.4 and 10.5). There are four major functional gene clusters: those concerned with regulation of phage expression, site-specific recombination, replication and structural proteins (Table 10.1). In addition, the so-called 'silent' b region, though inessential since it can be deleted without impairing phage function, appears to code for several proteins. Finally, the Q gene is a late transcriptional regulator, and S and R are responsible for cell lysis (R encodes the 18 000 dalton phage endolysin).

There are two major 'early' promoters, P_L and P_R (the thickness of the transcript arrows in Figures 10.4 and 10.5 indicates the relative maximum levels of expression from the promoters shown). P_L is responsible for early 'leftward' transcription of regulatory genes N and $cIII$ and of various genes involved in recombination. The early 'rightward' promoter P_R controls three regulatory elements, those encoded by *cro, c*II and Q, as well as the replication proteins O and P. There are four promoters in addition to P_L and P_R, three of which, P_I, P_M and P_E, are weak and are used in lysogeny (the role of P_O is not understood). It is transcription from the 'late' promoter $P_{R'}$ that allows the massive expression of structural genes necessary for phage encapsidation.

(a) The Regulatory Region. The presence of the *lambda* repressor, the *c*I gene product, both stabilises the prophage state and stops the development of further infecting *lambda* phage. This *immunity* to superinfection occurs only if the superinfecting phage carries the same *immunity region* as the resident prophage. The boundaries of this region may be defined with respect to related, but hetero-immune, temperate phages such as 434 and 21 (lysogens of *lambda* are not immune to either of these lamboid coliphages). Thus, in genetic crosses between *lambda* mutants and, say, phage 434, all loci that fail to give wild-type recombinants possessing *lambda* immunity are said to lie in the immunity region (strictly speaking, in the 434 immunity region). The extent of this region, however, depends on whether phage 434 or 21 is employed in the cross (Figure 10.6). The 434 immunity region spans the major leftward and rightward (early) promoters, P_L and P_R, and the *cI* and *rex* genes lying between,[2] as well as the

---

2.   The role of the *rex* gene product (29 000 daltons) is unclear since λ*rex⁻* mutants are able to lysogenise or grow lytically. There is some indication that *rex* facilitates phage induction, possibly through the P_E promoter. It also has the doubtful distinction of preventing development of T4*r*II rapid lysis mutants (coliphages *lambda* and T4 are unrelated).

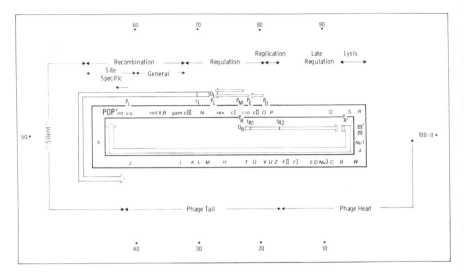

Figure 10.4: Simplified Genetic Map of Vegetative Phage *Lambda* Showing Major Transcripts

The four main gene clusters on the *lambda* genome are concerned with recombination, regulation, replication and morphogenesis (57.3-68.7, 68.7-79.9, 79.9-83.3 and 0-39.4 per cent *lambda*; 1% *lambda* = 486 bp). In addition, there is the late regulation gene, Q, and lytic functions, S and R (the latter encoding the phage endolysin). The b region is genetically 'silent' since it is non-essential. $P_L O_L$ and $P_R O_R$ designate the two major 'early' promoters-operators responsible for leftward and rightward transcription, while P_E, P_M and P_I are minor leftward promoters involved in lysogeny (P_O is apparently not necessary for phage propagation); lytic functions are expressed from the strongest *lambda* promoter, $P_{R'}$, in the rightward direction. The thickness of the transcript arrows represent the relative efficiencies of these promoters. t_L, t_{R1} and t_{R2} define rho-dependent transcriptional terminators (rho action is antagonised by lambda N protein, an anti-attenuator). Note that the origin of replication, *ori*, falls within the O gene and is not shown. See also Some *Lambda* Genes and their Function in Phage Development, Table 10.1. The *lambda* prophage map is given in Figure 10.5. The mature phage genome is a linear molecule split at *cos* (mm'), with gene order m-*Nul-J*-b-*att-int*-*c*I-*R*-m'; the left and right arms consist of genes *Nul-J* and *int-R*, respectively. After Szybalski (Szybalski, W. (1977) in J. Copeland and G.A. Marzluf (eds.), *Regulatory Biology* (State University Press, Ohio), pp. |3-45). Map positions from Echols and Murialdo (Echols, H. and Murialdo, H. (1978) *Microbiol. Rev., 42,* 577-591) and Szybalski and Szybalski (Szybalski, E.H. and Szybalski, W. (1979) *Gene, 7,* 217-270)

'second' repressor gene, *cro*. (There are, in fact, two 'back-to-back' promoters, P_M and P_R, the former being required for maintenance of lysogeny, see below.) The 21 region includes, in addition, the gene for the transcriptional anti-terminator, N, the *c*II gene, which, in conjunction with *c*III, supplies a positive regulator, and the promoter for establishment of lysogeny, P_E. (The *cro-c*II intercistronic space is called the y region, see 10.2.2 (a).) Since transcription from P_L and P_R/P_M is controlled by their respective overlapping operators, O_L and O_R (see Figure 10.7) and since the *c*I and *cro* repressors bind O_L and O_R, the immunity region (as defined by recombination with either phage 434 or 21) encodes both the two phage repressors and their sites of action.

The $5' \rightarrow 3'$ constraint on DNA transcription by RNA polymerase (Section

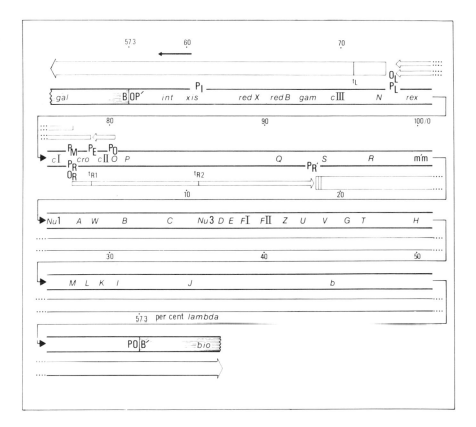

Figure 10.5: Simplified Genetic Map of Prophage *Lambda* Showing Major Transcripts

The prophage map is a circular permutation of the vegetative element since insertion and circularisation involve two different sites, att$^\lambda$ and mm' (see Figure 10.4). Note that in the absence of excision, P_L- and P_R-promoted transcription extends leftward into *gal* and rightward into *bio*. These bacterial genes may, therefore, be expressed from *lambda* promoters, a process termed escape synthesis. After Szybalski (Szybalski, W. (1977), in J.C. Copeland and G.A. Marzluf (eds.), *Regulatory Biology* (Ohio State University Press, Ohio), pp. 3-45). Map positions from Echols and Murialdo (Echoles, H. and Murialdo, H. (1978) *Microbiol. Rev., 42,* 577-591) and Szybalski and Szybalski (Szybalski, E.H. and Szybalski, W. (1979) *Gene, 7,* 217-270)

2.1.2 (a)) and the leftward and rightward orientations of transcription from the two major (early) promoters, P_L and P_R, indicate that different strands are used as template, depending upon whether transcription initiates at P_L/P_M or P_R. The sense strand utilised by P_L-promoted events is, thus, termed the *left strand, l:* similarly, that used by P_R is said to be the *right strand, r.* (Clearly, the designations 'l' and 'r' are quite arbitrary as they depend upon which phage arm of the linear, mature phage is considered 'left' or 'right', see legend to Figure

Table 10.1[a] Some Lambda Genes and their Role in Phage Development[b]

Locus[c]	Map Location[d] (per cent lambda)	Gene Product Name/Role	Gene Product Size	Function in Phage Growth
Regulatory Region				
cI	76.9/78.4	phage repressor	26,200	maintenance of lysogeny
cII, cIII	79.2/79.8,68.8/69.3	positive regulator	11,000,nk	establishment of lysogeny
cro (tof)	78.6/79.0	'second' repressor	7,400	late stage of lytic cycle
N	72.5/73.3	anti-attenuator	13,500	early development
$P_L O_L$	73.3/73.6	–	–	major leftward promoter
$P_R O_R$	78.4/78.5	–	–	major early rightward promoter
P_M (P_{RM})	78.4	–	–	promoter for maintenance of lysogeny
P_E (P_{RE})	79.2	–	–	promoter for establishment of lysogeny
t_L	71.8	–	–	terminator for P_L–N transcript
t_{R1}	79.2	–	–	terminator for P_R–cro transcript
t_{R2}	(83.3)[e]	–	–	terminator for P_R–cro–P transcript
Recombination Region				
int	57.4/59.8	'integrase'	40,300	site-specific integration and excision
xis	59.8/60.2	'excisase'	8,600	site-specific excision
P_I	60.2	–	–	promoter for int (within xis)
att	57.3	–	–	structural region for site-specific recombination
redX (exo)	64.7/66.1	'lambda' exonuclease	24,000	general recombination
redB (bet)	66.1/67.8	nd	28,000	general recombination
Replication Region				
O	79.9/81.9	nd	34,500	initiation and elongation (θ-mode)
P	81.9/83.3	nd	24,000	initiation and elongation (θ-mode)
gam	67.8/68.7	'gamma' protein	16,000	inhibits Exo V (for rolling-circle mode)
ori	80.9	–	–	origin (within O)

a   Abbreviation: nd, not designated; nk, not known.

b   After Echols and Murialdo (1978) and Szybalski and Szybalski (1979).

c   There are three main types of designations for phage *lambda* genes: lower case (regulatory loci for the lysogenic pathway), upper case (loci required for the lytic pathway) and three letter symbols (other genes). This system is not always consistent.

d   The map locations are approximate; co-ordinates are given in terms of distance from the 'left' end (see Figure 10.4), as a percentage of the total length (hence, per cent *lambda*).

e   Approximate map location.

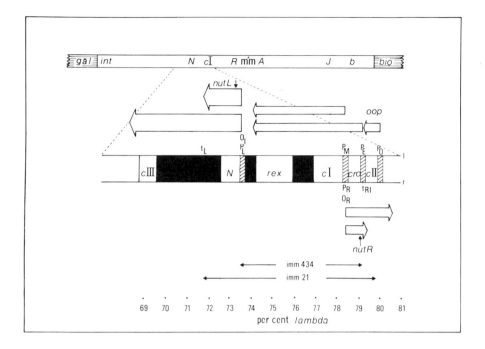

Figure 10.6: The Regulatory Region of Phage *Lambda*

About 11 per cent of the *lambda* genome (69-80 per cent *lambda*) contains the crucial regulatory elements. The region consists of the major leftward and rightward 'early' promoters, P_L and P_R (and their overlapping operator sites, O_L and O_R) as well as the two minor leftward promoters, P_E and P_M, involved in the establishment and maintenance of lysogeny (leftward and rightward transcription makes use of the *l* and *r* strands, respectively). P_E-promoted expression of the phage repressor, the *c*I gene product, requires the positive-acting *c*II-*c*III complex, transcription of which is dependent upon rho-mediated anti-termination at t_L and t_{RI} (acting through *nutL* and *nutR*). Whereas the phage repressor permits the stable lysogenic state through binding at O_L and O_R, *cro* product antagonises its action, allowing lytic growth (see Figures 10.7 and 10.8). The role of *rex* and the *oop* transcript is unclear (see text). The regulatory region is defined in crosses with hetero-immune lambdoid phages; mutations affecting *lambda* immunity that fail to recombine with, say, phage 434 fall within the 434 immunity region (and, similarly, for phage 21). Note that the phage *lambda* regulatory region encodes both the two regulatory proteins, the phage repressor and *cro* product, and their sites of action, O_L and O_R. After Szybalski (Szybalski, W. (1977), in J.C. Copeland and G.A. Marzluf, (eds.), *Regulatory Biology* (Ohio State University Press, Ohio), pp. 3-45)

10.4.) It should be emphasised that all transcription in the leftward direction makes use of the *l* strand (P_I, P_L, P_M and P_E) and all rightward transcription the *r* strand (P_R and $P_{R'}$). Thus, transcription from the two 'back-to-back' promoters, P_R and P_M, is divergent, P_M- promoted transcription on the so-called *l* strand and P_R- directed transcription on the *r* strand. Moreover, P_E is located downstream from the direction of transcription from the *cro* promoter, P_R, and hence the P_E-*c*I-*rex* transcript (but not that from P_M) contains the *cro* anti-message.

RNA chain initiation at P_L, P_R, and P_M is controlled directly by two negatively acting proteins, the *c*I repressor and *cro* inhibitor (homodimers of 26 200 and

7 400 dalton subunits). To aid distinction between these two negatively acting proteins, the *cro* gene product will, in future, be referred to as the *cro* inhibitor (it has also been termed the 'second' repressor) and the *c*I product as the (phage) repressor. Although the *lambda* repressor is an acidic protein, it has a basic amino-terminal region (residues 1 to 92), responsible for DNA binding and the effects associated with this event (negative and positive control, see below): the C-terminal domain (residues 132 to 236) dictates its quaternary structure. The *recA* protein cleaves between residues 111 and 112 upon induction (Section 10.2.3 (a)). The *cro* inhibitor, despite being a distinctive protein and negative control element in its own right, antagonises the action of the repressor through binding at the same operator sites.

Both regulatory proteins prevent P_L-promoted transcription through interaction at O_L. However, whereas binding of the phage repressor at O_R generally only inhibits rightward transcription from P_R, interaction of the *cro* inhibitor prevents, in the first instance, leftward transcription from this divergent promoter site (that is, P_M-promoted events). Transcription from P_M, the so-called *maintenance promoter*, is necessary to provide a low level of repressor in the prophage stage (Section 10.2.2 (b)). Lysogeny is ensured only when the phage repressor (rather than *cro* product) binds at O_L and O_R since the two major (early) promoters, P_L and P_R, are responsible for expression of the early functions necessary for lytic development (Section 10.2.3).

The leftward and rightward *lambda* operators, O_L and O_R, each consist of three tandem sequences at which the *lambda* repressor and *cro* product can bind; the sequences are termed O_L1, O_L2, O_L3 and O_R1, O_R2, O_R3 (with O_L1 and O_R1 overlapping P_L and P_R, respectively, Figure 10.7).[3] These sites, consisting of 17 nucleotide pairs, are separated by A+T-rich spaces 3-7 base pairs long (giving a repeat distance of 20-24 bp). They are similar but not identical and contain approximate two-fold symmetry. Binding of the two terminal sites in O_L (O_L1 and O_L2) and O_R (O_R1 and O_R2) stops expression of *N* and *cro* proteins, respectively, through inhibition of P_L- and P_R- promoted transcription. Since the RNA polymerase interaction sites and repressor sites in the rightward and leftward regulatory region overlap (Figure 10.8), the repressor may function by excluding binding of RNA polymerase (this blocking of access to polymerase is most effective when both O_R1/O_R2 or O_L1/O_L2 are bound).

Both phage repressor and *cro* inhibitor interact with the major groove along one side of the helix. Whereas the promoter-proximal sites, O_R1 and O_L1, have the greatest affinities for the repressor, the order is reversed with the inhibitor such that O_L3 and O_R3 are first bound. This differential binding plays a significant role in regulation of the divergent promoter site P_R/P_M.

The affinity order of the rightward operator for the *lambda* repressor is $O_R1 \simeq O_R2 > O_R3$ (the order for *cro* is $O_R3 > O_R1 \simeq O_R2$). The repressor binds co-

3.   Virulent mutants of *lambda*, λ*vir*, that grow on *lambda* lysogens carry, in general, multiple lesions in the leftward or rightward operator sites. Clearly, it is necessary for both O_R1 and O_R2 (or O_L1 and O_L2) to be inactive before repression is completely eliminated.

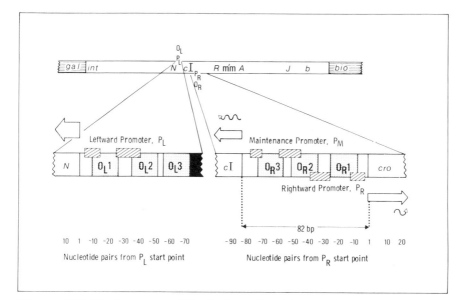

Figure 10.7: The Structure of the Leftward and Rightward Regulatory Sites, P_L/O_L and $P_R/O_R/P_M$.

The two main operators in the regulatory region (Figure 10.6) each consist of three 17 bp sites (O_L1, O_L2, O_L3 and O_R1, O_R2, O_R3, stippled) separated by AT-rich spacers about 3-7 bp long. These sites are similar in structure and have approximate two-fold symmetry. Leftward and rightward transcription from P_L and P_R (that is, N and cro expression) is controlled through the terminal sites O_L1/O_L2 and O_R1/O_R2 overlapping these promoters (the −10 and −35 sequences are in hatched boxes). Binding at O_R3 blocks the maintenance promoter, P_M (see also Figure 10.8). The transcriptional start points for these promoters and the translational initiation sites for the cI and cro genes are shown. Note that the 5′-terminus of the P_M-cI-rex message encodes the repressor's terminal residue. After Ptashne *et al.* (Ptashne, M., Jeffrey, A., Johnson, A.D., Maurer, R., Meyer, B.J., Pabo, C.O., Roberts, T.M. and Saurer, R.T. (1980) *Cell, 19,* 1-11)

operatively to O_R1 and O_R2 to inhibit P_R-promoted transcription. It is co-operative in that a filled O_R1 site promotes binding of a second repressor dimer at O_R2. In binding to O_R2, moreover, repressor stimulates leftward transcription from P_M. Thus, the *lambda* repressor is autogenously controlled on two counts: positively by binding of repressor to O_R2 (low repressor concentration) and negatively by binding of repressor to O_R3 (high repressor concentration). The former represents the situation in the normal lysogenic state (about 200 monomers per cell) as the concentration required for O_R3 binding is not generally attained under these conditions.

The stimulation of P_M-promoted transcription by binding of repressor at O_R2 is not understood. It is intriguing since O_R2 is only one nucleotide pair nearer to the P_R transcriptional start point than that of P_M — and yet the presence of repressor at O_R2 stimulates P_M transcription while inhibiting initiation at P_R. Since binding at O_R1 is generally a prerequisite for O_R2-mediated stimulation, the different effects associated with this event may reflect the free and bound

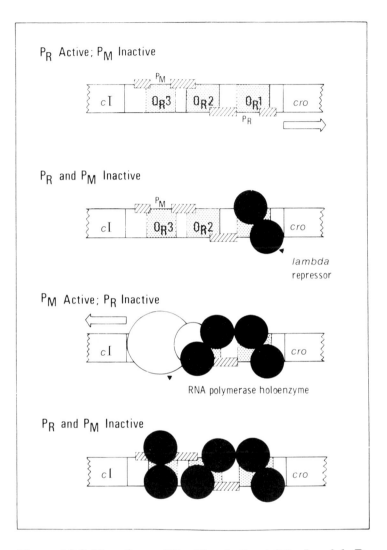

Figure 10.8: Negative and Positive Action of the *Lambda* Repressor at the Rightward Regulatory Site, $P_R/O_R/P_M$.

In the absence of the repressor, transcription may proceed from P_R for *cro* protein production (the RNA polymerase molecule has been omitted for clarity). Binding of a repressor dimer at O_R1 prevents this event. The presence of a second dimer at O_R2 through co-operative interaction promotes, in an as yet unspecified manner, transcription from P_M for *cI* production (see below). Under high repressor concentrations, conditions not normally present in a lysogenic bacterium, binding of a further dimer at O_R3 inhibits P_M function. Thus, the *cI* repressor can regulate its own synthesis both positively and negatively. The affinity order of these three sites for repressor and *cro*, respectively, is $O_R1 \simeq O_R2 > O_R3$ and $O_R3 > O_R1 \simeq O_R2$, indicating that while repressor generally inhibits P_R events, and stimulates P_M-promoted transcription, *cro* antagonises repressor action to block P_M. After Ptashne *et al.* (Ptashne, M., Jeffrey, A., Johnson, A.D., Maurer, R., Meyer, B.J., Pabo, C.O., Roberts, T.M. and Saurer, R.T. (1980), *Cell, 19,* 1-11) states of the operator sites flanking O_R2. It has been suggested that the repressor boosts P_M transcription in a co-operative manner through direct interaction with RNA polymerase on the DNA template (cf. Promoter Potentiation by CAP, Section 9.1.2 (b)). On this model, specificity in O_R2-mediated P_M stimulation by repressor may stem from the orientation of the repressor at O_R2, the position of the domain necessary for interaction favouring stimulation from P_M. Alternatively, the repressor may enhance formation of an open promoter complex (Section 2.1.2 (a i)), perhaps through helix destabilisation. The presence of a repressor dimer at O_R1 could, therefore, prevent DNA melting at P_R.

The *cro* inhibitor is an antagonist of the *lambda* repressor. It binds at the same sites as the repressor to inhibit transcription from P_L, P_R, P_M (and P_E). However, whereas repressor binds co-operatively to O_R1 and O_R2 and stimulates P_M-promoted production of the *cI-rex* transcript (at the same time as preventing initiation from P_R),[4] the *cro* inhibitor binds initially to O_R3, thereby blocking P_M events, without altering rightward transcription from P_R. Subsequent to this, but at higher inhibitor concentrations, it turns off its own gene, *cro*, by binding at the repressor-free O_R1 and O_R2 sites. The *cro* inhibitor is, thus, subject to negative autogenous control. The ability of two clearly distinct protein dimers to recognise the same DNA sequences — but elicit different responses — is particularly interesting. Synthesis of these two regulatory proteins is considered with respect to lysogeny and phage induction (Sections 10.2.2 to 10.2.4).

(b) The Site-Specific Recombination Region. The phage attachment site *att*λ*P*, consisting of the common core O and flanking sequences PP′, supplies the necessary structural elements for site-specific recombination (Section 5.2.2). Integration at primary or secondary sites (Sections 5.2.2 and 6.2.2 (a i)) requires integrase, the *int* gene product (40 300 daltons), as well as certain, as yet unspecified, host-coded proteins. The reverse reaction, phage excision, requires both the *xis* protein (8000 daltons) and integrase, though much lower levels of the latter are necessary at this stage.

The functions involved in site-specific recombination, *att*λ, *int* and *xis*, are clustered in a 1.5 kb segment of the phage chromosome (Figure 10.9). Transcription of *int* may initiate from either P_L or P_I (integrase, like the phage repressor, is, therefore, encoded in two different transcripts, but see below). The P_L-promoted transcript is a massive molecule carrying the information for the genes *N, cIII, redX, B*, as well as *int* and *xis*. Transcription from P_I, on the other hand, appears to initiate within the N-terminal region of the *xis* gene such that the P_I-*int* transcript lacks the first five codons for the *xis* gene product. Thus, whereas the P_L transcript allows both *int* and *xis* production (albeit with

---

4.   It has been suggested that since the concentration required for repressor to bind at O_R3 is not generally achieved, the physiological role of this O_R3 site is one of *cro*-, rather than *cI*-, mediated control. However, the fact that significant gene dosage effects are not observed for repressor synthesis in multiple *lambda* lysogens — isogenic bacterial strains carrying, say, one, two or three phage copies — suggests an active, negative control over *cI* production even in the absence of *cro* protein. Whether this reflects employment of O_R3 or some other regulatory mechanism is not known.

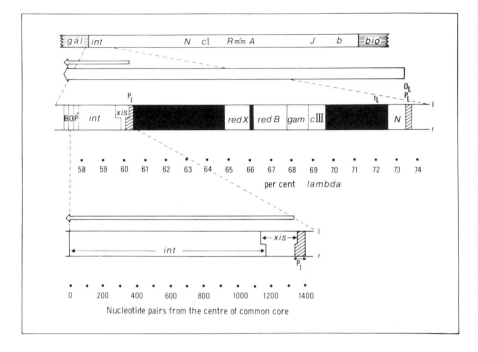

Figure 10.9: The Site-Specific Recombination Region of Phage *Lambda*
Packed into a 1.5 kb stretch of DNA is the phage attachment site, POP′, and the *int* and *xis* genes (whereas *int* is sufficient for site-specific integration, both *int* and *xis* are necessary for the excisive reaction, see Figure 5.13). The positioning of the *c*II-*c*III dependent promoter, P_I, within the N-terminal region of *xis* results in the P_I-*int* message lacking the first 5 *xis* codons. Thus, only P_L-promoted events allow expression of *xis* — and readthrough of t_L requires the anti-terminator *N* protein (*c*II-*c*III production is also dependent upon readthrough past t_{RI} and t_L). *int* and *xis* genes overlap, suggesting that translation of *xis* from the P_L-*int*-*xis* message may inhibit translational initiation at the *int* start triplet. Note that the *l* strand supplies the template for transcription for both P_L and P_I events

integrase synthesised at a significantly lower rate), P_I-promoted transcription directs only integrase synthesis.

Transcription from P_L is inhibited by binding of the phage repressor to O_L (see (a), above). Moreover, in the absence of the *N* protein, the terminator t_L prevents transcriptional readthrough into *c*III and, hence, into the other remaining genes of this long polycistronic operon. The P_L-*int* transcript is, therefore, dependent upon both a vacant operator site and a transcriptional anti-terminator.

Initiation from P_I is under positive control by *c*II-*c*III (expression of which requires free O_L and O_R sites and the *N* gene product). Investigation of the DNA sequence neighbouring the N-terminal region of the *xis* gene has led to the identification of a putative −35 sequence (Section 2.1.2 (a i)) that shows remarkable homology with the proposed *c*II-*c*III interaction site for the establishment promoter P_E (see Section 10.2.2 (b)). Deletions that extend into this sequence prevent *int* transcription from P_I but not from P_L. A potential Pribnow box (the –10 sequence, Section 2.1.2 (a i)) is situated downstream from this site and helps to define a possible transcriptional start point. Certainly, mutations which render *int* expression constitutive, independent of *c*II-*c*III, fall within the box.

Finally, it should be noted that the *int* coding region overlaps the *xis* gene by about 20 base pairs. Translation of the C-terminus of the *xis* protein from the P_L-*xis*-*int* transcript may prevent efficient ribosome loading for *int* translation. The disproportionate amounts of *xis* and *int* gene products observed might, therefore, stem from translational control.

(c) Replication Elements. The two major phage genes involved in *lambda* replication, *O* and *P*, and the origin, *ori*, are tightly linked in a 2.5 kb region on the phage map (map position 79.9 to 83.3%, see Figure 10.4). The origin lies within the coding sequence for the *O* protein (this feature is common for the lambdoid phages but has not been observed in other origin regions). Initiation of replication seems to involve two components, the origin (carrying a series of tandem inverted repeats), and an 81-base-pair sequence (21 bp of which show homology with other lambdoid phages), situated several hundred nucleotide pairs distant, in the centre of the *c*II gene. The latter sequence, termed the *inceptor*, is sufficient for driving replication of circular segments of DNA (that lack a functional origin of replication), provided a promoter is present in the correct orientation. The inceptor is thought to be necessary for terminating the primer to allow initiation of DNA synthesis.

The unlinked *gam* gene (map position 67.8 to 68.7% *lambda*) is necessary for transition from the θ to the rolling-circle mode of *lambda* replication (see Section 6.1.3). The phage general recombination genes *redX* and *redB*, sited adjacent to *gam*, are also required for this transition (note that they are responsible for non-reciprocal recombination and do not participate in site-specific events). A number of host functions have also been implicated in *lambda* replication; these include DNA polymerase III, *dnaB* and *dnaG* gene products and RNA polymerase but not *dnaA* and *dnaC* (Section 6.1.1).

It was initially proposed that the *oop* transcript, a 4S (81 nucleotides) RNA species initiated from a promoter called P_o (between *c*II and *O*/*ori*), supplied the RNA primer for replicative initiation. That there is no direct involvement of *oop* RNA in replication has recently been shown by deletions that remove the entire *oop* region without preventing initiation of replication at *ori*. The similarity of the *lambda* origin region with that for phage G4 (SS → RF mode) suggests a role for DNA primase in *lambda* initiation (Section 6.1.2 (a i)). Since initiation of *lambda* replication requires rightward *c*I/*cro*-controlled transcription from P_R into the *O*/*ori* region, *transcriptional activation* may be necessary to expose a single-stranded stretch for primase action (this enzyme does not appear to act on double-stranded DNA *in vitro*, Section 6.1.1 (b)). Transcription through *ori* is necessary for *O* protein production *per se* because *ori* lies within the *O* gene.

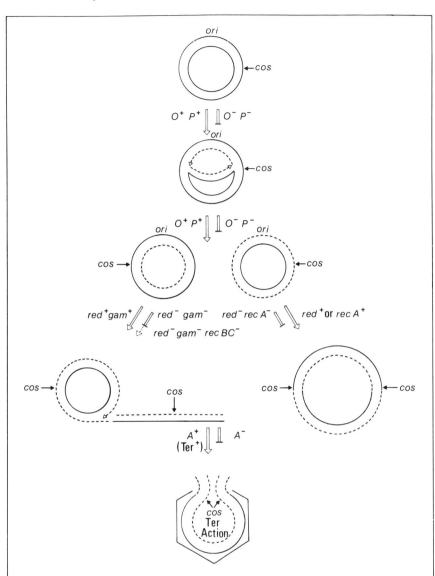

Figure 10.10: Alternative Modes of Phage *Lambda* Replication

The phage *O* and *P* gene products are necessary for initiation of replication at *ori* (sited within the *O* gene) and for chain elongation. This θ mode of replication (see Figure 6.7) produces a small number of circular molecules. The phage *gam* gene product is necessary to permit initiation of rolling-circle replication (see figure 6.13) by binding to the *recBC* nuclease, Exo V, an inhibitor of this transition. Since the *lambda* packaging enzyme, Ter (gene *A* protein), cleaves only at *cos* in concatemeric DNA, mutations that affect the transition to rolling-circle replication — such as *red⁻ gam⁻* — prevent encapsidation. This block on phage particle production may be circumvented in two ways; by host *recBC⁻* lesions that compensate for the Gam⁻ phenotype; and by the presence of a functional recombination system — the host Rec system, the phage Red system or to a lesser extent *int* — that allows the production of dimeric circles susceptible to Ter action

However, mutations that create a new rightward promoter to the right of the origin sequence circumvent the P_R requirement. Clearly, it is transcription of the adjacent *ori* region and not *O* expression which is a prerequisite for initiation of *lambda* replication.

The *P* and *O* proteins (34 500 and 24 000 daltons, respectively) are required for the early, θ mode of *lambda* replication (Figure 10.10). The unstable *O* protein interacts specifically with the *lambda* origin. The *P* gene product appears to associate with the host *dnaB* protein at the initiation stage as well as with the *O* protein. In addition to their role in the initiation of θ-type replication, the two phage-coded proteins are also required for bidirectional elongation from the *lambda* origin. There is not a defined terminator: termination seems to occur through convergence of the two replication forks.

The vegetative *lambda* chromosome replicates once or twice in the θ mode. The few circular daughter chromosomes produced in this way are not encapsidated (Figure 10.10). Rather, the switch from the θ to rolling-circle mode of replication (about 10 minutes after its start) supplies concatemeric *lambda* DNA — linear molecules up to eight times the length of the viral genome — that serves as substrate for the Ter enzyme (gene *A* product). A staggered, double-stranded cut at the *cos* (mm') site, the 12 bp sequence responsible for circularisation of the viral genome, produces monomers (the length dependence of this process has been discussed in Section 5.2.2). This ATP-dependent reaction appears to take place within the partially formed capsid (see also (d), below).

The transition from θ to rolling-circle mode requires the Red system and *gam* gene product; the latter specifically inhibits the action of the host *recBC* nuclease, ExoV, through direct interaction (Figure 10.10). Although the circular monomers that accumulate with *red⁻ gam⁻* phages are not susceptible to the encapsidation process (they contain only a single *cos* site), dimeric circles generated through homologous recombination are a suitable substrate. The host Rec system, the phage Red system and, to a much lesser extent, the phage integrase, catalyse dimerisation. Infection of a *recA⁻* host by λ*red⁻ gam⁻* phages is, thus, largely abortive due to lack of encapsidation (Figure 10.10). Introduction of a second chromosomal mutation, *recBC⁻*, allows some maturation of phage particles through removal of the ExoV block on the transition to rolling-circle replication.

(d) The Morphogenetic Region. There are two main groups of morphogenetic genes, clustered in the left arm of the linear chromosome. These are the phage head and tail genes (map positions 0 to 16.3 and 16.3 to 39.4 per cent *lambda*, Figure 10.4). Host functions are also involved. The *mop* locus (formerly designated *groE/tabB*), for example, is required for head assembly (for both *lambda* and T4). The *lambda* head and tail are made in separate pathways and, hence, mutations perturbing one process have no effect on the other (see Figure 10.11).

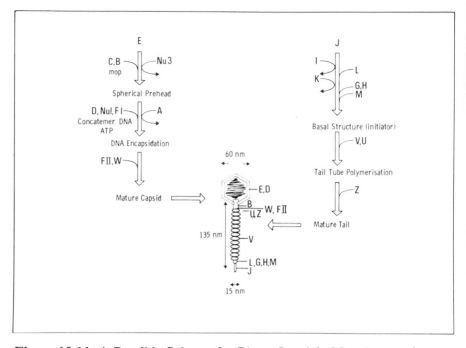

Figure 10.11: A Possible Scheme for Phage *Lambda* Morphogenesis
The *lambda* head and tail are made in separate pathways. After DNA encapsidation, the tail is connected through the action of the *F*II, *W* and *Z* gene products. In addition to the 10 head and 11 tail gene functions encoded by the phage (the tail cistron *T* is not shown), the host supplies the *mop* protein (and, as yet, unspecified elements) for morphogenesis. After Eiserling (Eiserling, F.A. (1979), in H.F. Fraenkel-Conrat and R.R. Wagner (eds.), *Comprehensive Virology, 13; Structure and Assembly* (Plenum Press, New York), pp. 543-580)

Of the 10 *head genes*, four, *B, C, Nu3* and *E*, are implicated in the production of the capsid precursor, the *prehead* (Figure 10.11). The *Nu3* protein (19 000 daltons) directs the assembly of the major capsid protein, the gene *E* product (38 000 daltons), by supplying a 'scaffolding' function.[5] It has, however, a transitory role in prehead formation since it is subsequently lost through cleavage. The *C* protein (58 000 daltons) is apparently required at an early stage in this process. It is both covalently linked to a small proportion of *E* protein as well as subjected to several proteolytic reactions. The gene *B* product and a cleaved derivative (60 000 and 54 000 daltons, respectively) are minor components of the mature prehead. In addition to these prehead genes, a number of phage functions are involved in DNA packaging (see also (c), above). These include the Ter enzyme, the gene *A* product (79 000 daltons), which acts at the

5.   Interestingly, the *Nu3* gene lies within *C* such that the C-terminal region of the *C* gene supplies the information for the entire *Nu3* protein (the scaffolding protein is one third the size of the gene *C* product). Interaction between the *Nu3* protein and the C-terminal region of the *C* protein through their identical amino acid sequences may be an important factor in prehead formation.

cos site on concatemeric *lambda* DNA, the *Nu*1 and *F*I proteins, and the *D* major capsid protein. The *F*II gene product (11 500 daltons) is present in head-filled structures and supplies a 'connector' point for joining the phage tail to one vertex of the mature isometric capsid; the *W* protein (about 7000 daltons) is required for *F*II action.

The tail is made in a separate pathway, a process requiring 11 gene functions, and then attached (via the *F*II/*W* proteins). The so-called *basal structure* consists of the *J, I, L, K, G, H* and *M* gene products although the *I* and *K* proteins are not apparently present in mature phage particles (Figure 10.11). It serves as an initiator for tail tube formation. Polymerisation of the major tube protein, the *V* product (about 28 000 daltons) and, thus, the final length of the tail is determined by the *U* protein (15 000 daltons). Finally, positioning of the complete tail structure on the mature capsid is mediated by the *Z* protein (20 000 daltons). Note that the *J* product is responsible for the single *tail fibre* and, hence, dictates host specificity (a hybrid $\lambda/\phi80$ phage in which the *J* gene is replaced by the $\phi80$ equivalent, designated $\lambda h80$, makes use of the $\phi80$ receptor rather than *lamB*, see Section 5.12).

10.2.2 Lysogen Formation

The linear viral genome rapidly circularises in the host's cytoplasm upon infection through base pairing of its cohesive ends and ligation (making use of the bacterial DNA ligase). Integration of this circular, double-stranded molecule at the chromosomal attachment site *att\B* requires the action of the phage-coded protein integrase and host functions. Expression of the *int* gene is positively-regulated by the phage *c*II and *c*III gene products acting at the promoter for *int*, P_I. Production of *c*II-*c*III is, itself, dependent upon *N*-mediated transcriptional anti-termination at t_L and t_{R1}.

Immediately after infection, no phage proteins are present in the host's cytosol. The fate of the infected bacterium is dictated, in fact, by which phage products attain supremacy — in terms of both their relative levels and time of production (see The Lysogenic-Lytic 'Decision', Section 10.2.4). The stability of the union between these two diverse, covalently-closed circular molecules of double-stranded DNA — the viral and bacterial chromosomes — is maintained by a single gene product, the phage repressor. Expression of all other genes, with the single exception of *rex* for which a role has not as yet been ascribed, is inhibited in the prophage state (Figure 10.12). The *c*I product is, therefore, responsible for blocking lytic functions, production of which would lead to cellular death.

(a) Establishment of Lysogeny: P_E-promoted Transcription. The phage repressor can be efficiently expressed soon after infection. Two promoters, P_E and the more distal P_O site, have been implicated in this event. P_O, the *oop* promoter, is located a short distance upstream from P_E. It is responsible for the production of the 4S, 81 nucleotides long *oop* transcript and may, therefore, act

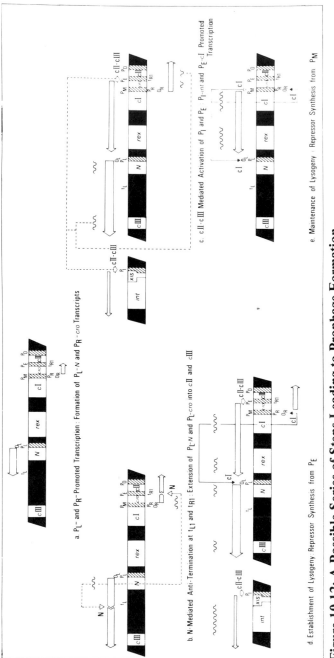

Figure 10.12: A Possible Series of Steps Leading to Prophage Formation

Production of the functionally unstable *N* protein from P_L (a) allows extension of the P_L-*N* and P_R-*cro* transcripts past t_L and t_R1 due to *N*-mediated anti-attenuation (b). *c*III and *c*II expressed from these messages promote transcriptional initiation at P_I and P_E (c), thereby supplying the means for establishing expression of the phage repressor (d). Finally, repressor binds to O_L and O_R to prevent further leftward and rightward transcription from the two major 'early' promoters (and, thus, inhibits production of *N*, *cro*, *c*II-*c*III and *int* proteins) at the same time as promoting *c*I transcription from P_M (e). These events are antagonised by the *cro* product, which inhibits, in the first instance, P_L and P_M. See The Lysogenic-Lytic 'Decision', Section 10.2.4

as a leader for *cI-rex* transcription through anti-attenuation in the y region (see Attenuator Control, Section 8.1.2 (b)). Recent evidence, however, indicates that the promoter responsible for this high level *cI* expression is, in fact, the previously termed *establishment promoter* P_E, situated in the y region, some 200 nucleotide pairs to the right of O_R (Figure 10.12).

The unstable *cII* and *cIII* gene products are required for the *establishment mode* of transcription, both for P_I-promoted integrase production and for repressor synthesis. They may, therefore, act as anti-attenuation factors (akin to N and Q protein function, Section 10.2.3) or as positively acting regulators, depending upon whether P_O or P_E is responsible for establishment of *cI* expression. Support for the latter contention comes from deletion studies which indicate that the y region is required for *cII-cIII* dependent repressor synthesis. It is hard to reconcile the P_O model with these results since removal of the hypothetical attenuator should lead to *cII-cIII* independent *cI* transcription, rather than prevent *cI* transcription. (Note that *oop* RNA is terminated in the absence of rho.)

Further credence for a positive *cII-cIII* role and, hence, for the P_E model, stems from a class of rare mutations affecting *cI* expression, termed *cy*, which map in the y region several hundred nucleotide pairs upstream from P_M (these derivatives are unable to lysogenise and, thus, give *c*lear rather than turbid plaques). The mutations lie within the presumptive −10 and −35 sequence for the P_E promoter. The Pribnow box shows poor homology with the consensus sequence (Section 2.1.2 (a i)). The type of base changes observed in the potential −35 region suggests, moreover, that these lesions affect interaction of a positively-acting element (i.e. *cII-cIII*) rather than RNA polymerase. Certainly, the −35 region has already been implicated in other operons as a potential site for positive-acting regulatory elements (Section 9.1.2 (b)). Moreover, the second *cII-cIII* dependent promoter P_I contains a very similar sequence centred 35 nucleotide pairs upstream from the transcript start point. The P_E start site, so defined, lies 360 and 10 nucleotide pairs to the right of P_M and t_{R1}, respectively.[6]

Initiation at P_E during establishment results in the production of a *cI-rex* transcript that carries a leader several hundred nucleotides long (the anti-message, in fact, of the *cro* gene, Figure 10.12). The putative ribosome binding site, beginning 12 base pairs upstream from the *cI* translational start point, appears to allow efficient translation of the repressor gene. This initial high rate of repressor synthesis provides for effective shut-down of leftward and rightward transcription from P_L and P_R, thereby inhibiting expression of phage lytic functions. *cII* is unstable and, hence, P_E- and P_I-promoted events cease after repressor blocks O_R.

In summary, establishment of lysogeny is controlled by three separate regulatory elements, *cII-cIII*, *cro* and N products (in addition to *cI*, Section

---

6. If, as seems likely, P_O is neither implicated in *lambda* replication (Section 10.2.1 (c)) nor establishment expression of *cI*, it remains to be seen what roles P_O and its 4S *oop* transcript have in phage development. Perhaps they supply a back-up mechanism.

10.2.1 (a)). *c*II-*c*III act positively, apparently by promoting transcription from P_E; they also allow P_I-dependent integrase synthesis (Figure 10.12). The *cro* inhibitor, on the other hand, negatively controls repressor production both indirectly by inhibiting *c*III and *c*II transcription (by binding to O_L and O_R) and directly by preventing P_E-promoted *c*I-*rex* transcription. Finally, the *N* protein is a transcriptional anti-terminator that through interaction at the *nut* sites permits extension of the leftward and rightward transcripts into *c*III and *c*II (see Section 10.2.3 (c)).

(b) Maintenance of Lysogeny: P_M-promoted Transcription. Once established, the prophage stage is stably maintained by *c*I transcription from P_M (only about 4 per cent of *lambda* DNA, map positions 74.2 to 78.4, is transcribed in the prophage state due to termination upstream from P_L, Figure 10.12). The maintenance promoter is located only about 30 nucleotide pairs upstream from the *c*I translational initiation codon with the result that this P_M-*c*I-*rex* transcript lacks entirely a leader sequence (see Figure 10.7). The 5'-terminal pppAUG encodes, in fact, the N-terminal amino acid of the *lambda* repressor. Translation of *c*I from the P_M-*c*I-*rex* message is inefficient (about one tenth of that from the P_E-*c*I-*rex* message), presumably because of the absence of a ribosome binding site.

P_M-promoted transcription requires no accessory factors (although it is inhibited by the *cro* product, Section 10.2.3 (c)). It is, however, autogenously regulated by the *c*I gene product since at low concentrations of the *lambda* repressor, binding to O_R1 and O_R2 stimulates P_M-*c*I-*rex* transcription (at high concentrations, conditions not generally attainable in the lysogenic state, binding at O_R3 blocks transcription from P_M). The maintenance mode of repressor synthesis is only attainable subsequent to establishment expression (see (a), above) since P_E but not P_M functions independently of the *c*I product (Figure 10.12).

10.2.3 Prophage Induction

Upon prophage *lambda* induction (or during the lytic pathway after phage infection), there is finely controlled expression of phage functions, resulting in the sequential synthesis of three main groups of genes. These are termed the immediate-early, delayed-early and late gene functions according to the time, after induction, at which they are expressed (Figure 10.13).

Transcription of the immediate-early genes, *N* and *cro*, occurs approximately one minute after heat induction (making use of a thermolabile phage repressor, see (a) below). The *N* protein allows expression of the delayed-early genes through anti-termination at t_L, t_{R1} and t_{R2} (about 3-5 minutes after induction). There is only a burst of expression of these early genes since the immediate-early protein, the *cro* inhibitor, blocks further transcription from P_L, P_R, P_M and P_E. The *N* protein, moreover, is functionally unstable and any residual RNA polymerase molecules proceeding from P_L and P_R stop at the early terminators.

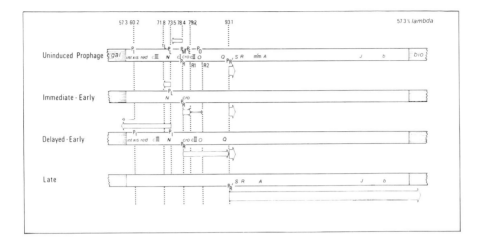

Figure 10.13: Temporal Sequence of Transcription Events upon Prophage
Lambda **Induction**

In the prophage state, only the maintenance promoter, P_M, is functional due to repressor binding at O_L and O_R (transcripts initiated from $P_{R'}$ terminate shortly afterwards). The presence of vacant operator sites upon repressor inactivation gives rise, in the Immediate-early stage, to production of the short P_L-N and P_R-*cro* transcripts because of termination at t_L and t_{R1} (there is some readthrough past t_{R1} to generate the larger P_R-*cro*-*c*II-*O*-*P* message). In the delayed-early period, these two transcripts are extended by N-mediated anti-termination, to produce functions necessary for lysis (or lysogeny, see Figure 10.12). While the Q gene product prevents premature termination past $P_{R'}$, *cro* protein inhibits P_L (and P_M) at low concentrations, and later P_R (and, thus, its own expression). Massive, late transcription from $P_{R'}$ is responsible for phage morphogenesis. After Szybalski (Szybalski, W. (1977), in J. C. Copeland and G.A. Marzluf (eds.), *Regulatory Biology* (Ohio State University Press, Ohio), pp. 3-45)

At the end of the delayed-early period, all early genes are 'silent' due to binding of *cro* inhibitor at the major early leftward and rightward promoters (as well as the establishment promoter); the absence of *c*II-*c*III also indirectly prevents initiation from either P_I or P_E. The only functional promoter remaining is $P_{R'}$. Synthesis of the Q protein (23 000 daltons) during the latter portion of the delayed-early period is responsible for anti-termination of $P_{R'}$-promoted transcripts. Massive transcription from this late promoter (occurring some 10-20 minutes after induction) supplies the phage lytic functions.

(a) Repressor Inactivation. Interference with host DNA synthesis or the presence of a damaged replicon following mutagenesis results in increased levels of the *recA* protein (see SOS Response, Sections 3.2.1 (b i), 6.2.1 (a)) and, hence, cleavage of the phage repressor. This specific proteolytic activity of the *recA* gene product, dependent upon polynucleotide and ATP, gives rise to two inactive repressor fragments. Since the target of this attack is between residues 111 and 112, the N-terminal DNA binding domain (Section 10.2.1 (a) is unable to dimerise and presumably binds poorly. The observation that the *lambda*

repressor is a less efficient substrate at high concentrations and the fact that there is a concentration-dependent monomer-dimer transition suggests that it is the monomer which is susceptible to attack in *indirect induction.*

λcI(Ts) lysogens, in which the *c*I gene encodes a thermolabile repressor, can be induced by heat treatment. This *direct induction* is *recA*-independent. It possibly functions by thermal dissociation of the repressor dimer into its individual subunits. Heat induction is very convenient for producing *lambda* lysates, particularly in conjunction with mutations that prevent cellular lysis (such as the S7 amber derivative) since there is a build-up of mature phage particles within the cell.

Finally, the introduction by conjugation (or transduction) of prophage *lambda* into a non-lysogenic host — mediated, for example, through an Hfr cross (Section 4.1.3 (a)) in which the donor is lysogenic for *lambda* — kills the majority of recipient cells. The rationale for this *zygotic induction* is that upon entrance into the repressor-free recipient cell, the maintenance promoter, the only site functional in the prophage state (Section 10.2.2 (b)), is unable to synthesise the repressor at sufficient rate to establish lysogeny. Zygotic induction is a useful property for screening for cointegrates between *lambda* and conjugative plasmids since strains harbouring such plasmid derivatives give only few, if any, transconjugants in crosses with a λ^- recipient.

(b) Immediate-Early Transcription: P_L-and P_R-promoted Events. Repressor inactivation releases O_L and O_R, and allows leftward and rightward transcription from the two major early promoters, P_L and P_R (Figure 10.13). Rho-mediated transcriptional termination, occurring at the two early terminators in these leftward and rightward operons, t_L and t_{R1}, generates the 12S P_L-N and 9S P_R-*cro* transcripts (the former is some 850 nucleotides in length; the *cro* message is about 310 nucleotides long consisting of an 18 nucleotide leader, 198 residue coding region and about 94 nucleotides that are not translated). There is some readthrough past t_{R1} but apparently efficient, rho-dependent termination at t_{R2}, resulting in a small amount of the large P_R-*cro*-*c*II-*O*-*P* message (further terminators may precede t_{R2} since the N protein is required for significant P expression even in the absence of t_{R1}). Transcription of the late genes is prevented since there is a strong N protein-resistant (rho-independent) terminator(s) downstream from $P_{R'}$ (termination occurs at both the 15th and 198th nucleotide).

(c) Delayed-Early Transcription: P_L- and P_R-promoted Readthrough Past the Early Terminators t_L, t_{R1} and t_{R2}. The functionally unstable N protein acts as an anti-terminator for P_L- and P_R-promoted transcription (Figure 10.13). In its presence, the host polymerase reads through the early terminators, t_{L1}, t_{R1} and t_{R2}, to produce the two large transcripts P_L-N-*c*III-*gam*-*redX,B*-*xis*-*int* and P_R-*cro*-*c*II-*O*-*P*-*Q* (the major leftward transcript terminates, in fact, about 7500 nucleotides downstream, Figure 10.4). Rightward transcription into *O/ori*

allows the initiation of the θ mode of *lambda* replication (Section 10.2.1 (c)).

For *N* protein to prevent transcriptional termination at these early terminators, the transcripts must have originated from the major leftward or rightward early promoters. Two *cis*-dominant sites (17 bp long) in these operons, designated *nutL* and *nutR*, have been implicated in *N* action; they are identical in all but one base and contain the same region of dyad symmetry. Whereas *nutL* is promoter-proximal (about 50 nucleotides downstream from the P_L start point), *nutR* lies close to the first rightward terminator (about 50 nucleotides upstream from t_{R1}). It is not known how recognition of these sequences by *N* gene product prevents transcriptional termination. That the protein functions on the transcription complex, itself, or at least the β subunit of RNA polymerase, is demonstrated by mutations in *rpoB* which prevent growth of the wild-type phage, and their suppression by lesions in *lambda N* gene; an involvement of the translational machinery has also been suggested.

Extension of the leftward transcript through *N* action allows *int* and *xis* expression. The disproportionate production of the two proteins (Section 10.2.1 (b)) catalyses phage excision through reciprocal recombination between *attL* and *attR* (Section 5.2.2). Conversely, production of the two components of the positive *c*II-*c*III regulator, after readthrough past t_{L1} and t_{R1}, activates P_I- and P_E-promoted events to create the conditions necessary for lysogeny. The latter is antagonised and over-ridden (if lysis is to occur) by the synthesis of the *cro* inhibitor, which shuts off immediate-early and delayed-early expression from P_L, P_R, P_M and P_E.

(d) Late Transcription: $P_{R'}$-promoted Readthrough. The delayed-early *Q* gene product stops premature termination of the P_R-transcript (Figure 10.13). It appears to encode an anti-terminator (although the site(s) at which it functions is rho-independent). $P_{R'}$ is the most powerful *lambda* promoter. Massive transcription of the circular vegetative chromosome from this site seems to be responsible for all the lytic functions, head and tail formation, DNA encapsidation and lysis of the host cell from within (multiplication is through the rolling-circle mode at this stage, Section 10.2.1 (c)). Some 100-200 progeny phages are released.

10.2.4 The Lysogenic-Lytic 'Decision'

What are the conditions that channel a wild-type *lambda* phage into the lysogenic or lytic pathway after infection? That the physiological state of the host cell is important in resolving this dichotomy is indicated by induction of the prophage through alterations in the host's DNA metabolism and the fact that conditions of starvation or high cAMP concentrations favour the lysogenic state. Although it has been shown that UV-mediated induction of prophage *lambda* is not linked to a specific event in the cell cycle (making use of synchronised cultures), the cycle may be an important factor after infection. That is, after injection of phage DNA, there may be commitment to either the lytic or lysogenic pathway depending on the particular stage in the cell cycle of the infected bacterium.

Expression of the genes required for formation of the lysogenic state (N,[7] cII, $cIII$, *int* and cI) may permit transcription of genes (*cro*, O, P, *redX,B, gam, Q,* etc.) necessary for productive growth. There are, however, two main, mutually interdependent controlling factors.[8]

(i) *cI/cro Balance.* The phage repressor and *cro* inhibitor act at the same sites, O_L and O_R, affecting P_L, P_R and P_M. Whereas *cro* prevents all early transcription, including that of its own gene, the cI product activates maintenance expression of the repressor. Factors that affect the balance of these two antagonistic regulatory elements will 'switch' the newly infected cell to either the lysogenic state (*cI/cro* ratio high) or the lytic state (*cI/cro* ratio low). The rate of synthesis of these two proteins upon infection is, therefore, crucial in deciding the fate of the cell. Moreover, the temporal sequence of events is also important since cI expression must occur sufficiently early to block lytic development (through prevention of *cro* transcription).

(ii) *cII-cIII/cro Balance.* The high level expression of the *lambda* repressor necessary for establishment can be supplied only from P_E (witness the occurrence of zygotic induction, Section 10.2.3 (a)). P_E-promoted transcription of cI is, itself, positively controlled by cII-$cIII$. Since *cro* prevents cII-$cIII$ transcription, the balance of these negatively- and positively-acting regulators controls the possible outcome. If the cII-$cIII$/*cro* ratio is low at the time of *cro* autorepression, P_E-transcription of cI (and production of the P_I-*int* message) cannot be efficient and the lytic cycle is favoured. However, a high cII-$cIII$/*cro* ratio permits efficient transcription from P_E (and P_I), thus escaping a possible block on lysogeny imposed by the *cro* inhibitor.

7.　The importance of the N product in the lytic response is exemplified by λN^- derivatives that carry a functional O and P gene. Such mutants express their essential replication functions due to transcriptional readthrough past t_{R1} but lethal loci are shut off. Thus, upon infection, circularisation and replication may occur without lysis, the phage existing in this lysogenic state as an autonomous plasmid. (λdvs also replicate as plasmids. These phage are only about 10-20 per cent of the wild-type's size. The presence of the P_R-P region is sufficient to allow multiplication of the circular genetic element. The minimal requirements of a λdv replicon appear, in fact, to be *cro*, O/*ori*, P and a suitable promoter.)

8.　The convergent nature of transcription from P_E and P_R (and the fact that t_{R1}, only about 10 bp to the left of P_E, is not a very efficient terminator) suggests that antagonism between P_E- and P_R-promoted events is of potential significance in stabilising the 'decision'.

Bibliography

10.1 Regulation of F Plasmid Transactions

Broda, P. (1979) *Plasmids* (Freeman, Oxford).

Bukhari, A.I., Shapiro, J.A. and Adhya, S.L., (eds.) (1977) *DNA Insertion Elements, Plasmids and Episomes* (Cold Spring Harbor Laboratory, New York).

Clark, A.J. and Warren, G.J. (1979) 'Conjugal Transmission of plasmids', *Ann. Rev. Genet., 13,* 99-125.

Finnegan, D.J. and Willetts, N.S. (1973) 'The site of action of the F transfer inhibitor', *Molec. Gen. Genet., 127,* 307-316.

Kahn, M.L., Figurski, D., Ito, L. and Helinski, D.R. (1978) 'Essential regions for replication of a stringent and a relaxed plasmid in *Escherichia coli'*, *Cold Spring Harb. Symp. Quant. Biol., 43,* 99-103.

Lewin, B. (1977) *Gene Expression, 3: Plasmids and Phages* (John Wiley and Sons, New York).

Manning, P.A. and Achtman, M. (1979) 'Cell-to-cell interactions in conjugating *Escherichia coli:* the involvement of the cell envelope', in M. Inouye (ed.), *Bacterial Outer Membranes: Biogenesis and Functions* (John Wiley and Sons, New York), pp. 409-447.

Willetts, N. and Skurray, R. (1980) 'The conjugation system of F-like plasmids', *Ann. Rev. Genet., 14,* 41-76.

See also references in Chapters 4 and 6.

10.2 Regulation of Phage Lambda Development

10.2.1 Organisation of the Intracellular Viral Chromosome

Echols, H. and Murialdo, H. (1978) 'Genetic map of bacteriophage *lambda', Microbiol. Rev., 42,* 577-591.

Szybalski, E. and Szybalski, W. (1979 'A comprehensive molecular map of bacteriophage lambda', *Gene, 7,* 217-270.

(a) The Regulatory Region

Pirrotta, V. (1976) 'The λ repressor and its action', *Curr. Topics Microbiol. Immunol., 74,* 21-54.

Ptashne, M. (1978) 'λ repressor function and structure', in J.H. Miller and W.S. Reznikoff (eds.), *The Operon* (Cold Spring Harbor Laboratory, New York), pp. 325-343.

Ptashne, M., Jeffrey, A., Johnson, A.D., Maurer, R., Meyer, B.J., Pabo, C.O., Roberts, T.M. and Sauer, R.T. (1980) 'How the λ repressor and Cro work', *Cell, 19,* 1-11.

Rosenberg, M., Court, D., Shimatake, H., Brady, C. and Wulff, D. (1978) 'Structure and function of an intercistronic region in bacteriophage *lambda'*, in J.H. Miller and W.S. Reznikoff (eds.), *The Operon* (Cold Spring Harbor Laboratory, New York), pp. 345-371.

(b) The Site-Specific Recombination Region

Abraham, J., Mascarentias, D., Fischer, R., Benedik, M., Campbell, A. and Echols, H. (1980) 'DNA sequence of regulatory region for integration gene of bacteriophage λ', *Proc. Natl. Acad. Sci. USA, 77,* 2477-2481.

Hoess, R.H., Foeller, C., Bidwell, K. and Landy, A. (1980) 'Site-specific recombination functions of bacteriophage λ: DNA sequence of regulatory regions and overlapping structural genes for Int and Xis', *Proc. Natl. Acad. Sci. USA, 77,* 2482-2486.

Nash, H. (1977) 'Integration and excision of bacteriophage λ', *Curr. Topics Microbiol. Immunol., 78,* 174-199.

(c) Replication Elements

Koller, R. and Helinski, D.R. (1979) 'Regulation of initiation of DNA replication', *Ann. Rev. Genet., 13,* 355-391.

Skalka, A.M. (1977) 'DNA replication — bacteriophage *lambda'*, *Curr. Topics Microbiol. Immunol., 78,* 202-237.

(d) The Morphogenetic Region

Hohn, T. and Katsura, I. (1977) 'Structure and assembly of bacteriophage *lambda'*, *Curr. Topics Microbiol. Immunol., 78,* 69-110.

Wood, W.B. and King, J. (1979) 'Genetic control of complex bacteriophage assembly', in H.

Fraenkel-Conrat and R.R. Wagner (eds.), *Comprehensive Virology, 13: Structure and Assembly* (Plenum Press, New York), pp. 581-633.

10.2.2 to 10.2.4 The Lysogenic-Lytic Pathways

Hershey, A.D. (ed.) (1971) *The Bacteriophage lambda* (Cold Spring Harbor Laboratory, New York).

Herskowitz, I. and Hagen, D. (1980) 'The lysis-lysogeny decision of phage λ: Explicit programming and responsiveness', *Ann. Rev. Genet., 14,* 399-445.

Lewin, B. (1977) *Gene Expression, 3: Plasmids and Phages* (John Wiley and Sons, New York).

Szybalski, W. (1977) 'Initiation and regulation of transcription and DNA replication in coliphage *lambda*', in J.C. Copeland and G.A. Marzluf (eds.), *Regulatory Biology* (Ohio State University Press, Ohio), pp. 3-45.

Weisberg, R.A. Gottesman, S. and Gottesman, M.E. (1977) 'Bacteriophage λ: the lysogenic pathway', in H. Fraenkel-Conrat and R.R. Wagner (eds.), *Comprehensive Virology, 8: Regulation and Genetics* (Plenum Press, New York), pp. 197-258.

Index

AB313 (Hfr) *see* genetic maps, F plasmid insertion sites

abortive transductant *see* phage P1

acetyltransferase: R plasmid-encoded enzymes 187, 189, 265; transposon-encoded enzymes 300-2; *see also* chloramphenicol

acridine orange: F plasmid curing 161; inhibition of replication 259; mutagenesis 118, 128-9

actinomycin D: inhibition of replication 259; inhibition of transcription 70; structure 72

activator *see* operon control, promoter control

adapter *see* transfer RNA

adenine *see* nucleic acid, components

adenine methylase *(dam)* 314-15; *see also* repair

adenosine *see* nucleic acid, components

adenylate cyclase *(cya) see lac* operon, control, positive

O-adenyltransferase (ATase): R plasmid-encoded enzymes 187, 189, 265; transposon-encoded enzymes 300-2; *see also* kanamycin; spectinomycin; streptomycin

adhesion point (Bayer's junction) 32, 163, 199

aerosol propellants *see* mutagens, sources

aflatoxin B *see* mutagens

agarose electrophoresis *see* electrophoresis

alaS, T, U see transfer RNA

alk 314-15; *see also* repair

alkylating agents *see* mutagens

allele 24, 110; *see also* mutation

allolactose *see lac* operon

amber (UAG) *see* nonsense triplets

amino acid: starvation 420-2; structures 17; synthesis 38; *see also trp* operon

amino acid acceptor stem *see* transfer RNA

aminoacyl-tRNA *see* Transfer RNA, amino acid acceptor stem

aminoacyl-tRNA synthetase: gene mutants *see trp* operon, *trpS, T*; genetic loci 91; genetic nomenclature 92; mechanism 84, 86-7; *see also* transfer RNA; translation

2-aminopurine (2-AP) *see* mutagens

ampC see β-lactamase, chromosomal encoded enzyme

ampicillin (penicillin, carbenicillin) *see* penicillin

amplification *see* col plasmid: R plasmid

anabolism 37

anions 29

anisometric morphology *see* phage, capsid

ansamycins *see* rifampicin; streptovaricin; streptolydigin

antibiotic: aminoglycoside *see* kanamycin; spectinomycin; streptomycin; bacterial production of 4; chromosomal-vs. R plasmid-encoded resistance 189; incidence of resistant *Shigella* 188; inhibitors of cell wall synthesis *see* penicillin; inhibitors of replication *see* actinomycin D; cordycepin; coumermycin; mitomycin C; nalidixic acid; rifampicin; streptolydigin; streptovaricin; inhibitors of transcription *see* actinomycin D; cordycepin; rifampicin; streptolydigin; streptovaricin; inhibitors of translation *see* chloramphenicol; fusidic acid; kanimycin; puromycin; spectinomycin; streptomycin; tetracycline; resistance of Gram-positive vs. Gram-negative 29; R plasmid-encoded resistance to 187, 189, 265; transposon-encoded resistance to 300-2

anticodon (loop) *see* transfer RNA

antiserum 354

anthranilate *see trp* operon, enzymes

anthranilate synthetase *(trpE, D) see trp* operon, enzymes

apo-activator *see* operon control, promoter control

apo-repressor *see* operon control, promoter control

argF, G, I see gene, duplication

asbestos *see* mutagens

asn see oriC, locus

ATase *see O*-adenyltransferase

ATP 37-8; *see also* DNA-dependent ATPase; transcription

attachment site *(att^P/POP') see* phage *lambda*, lysogeny/site-specific recombination; phage *lambda*

transcription, *int-xis*; phage P2, lysogeny
attenuator: phage *lambda* attenuator sites,
    see phage *lambda* genes and mutants,
    $P_{R1}/t_L/t_{R1}/t_{R2}$; *trp* attenuator sequence
    (trpa) 408-10; *see also* operon control;
    terminator
attenuator control *see* operon control
att^λ*B* (BOB') *see* phage *lambda*, site-
    specific recombination; primary bacterial
    attachment site; phage *lambda*,
    transduction
attL, attR (BOP', POB') *see* phage *lambda*,
    site-specific recombination, primary
    attachment site/secondary attachment
    site; phage *lambda*, transduction
att^λ*P* (POP') *see* phage *lambda*, lysogeny/
    site-specific recombination; phage
    lambda transcription, *int-xis*
att^{P2} *see* phage P2, lysogeny
autogenous control *see* operon control,
    promoter control, circuits
autoradiography *see* radiolabelling
autotroph 36
auxotroph 41; *see also* mutant, prototroph

B7 (Hfr) *see* genetic maps, F plasmid
    insertion sites
B_{12} uptake *see* Col plasmid, colicin,
    receptor; phage BF23, receptor
Bacillus subtilis: chloramphenicol-resistant
    189; chromosome size 11; restriction
    endonuclease 367; spore formation 4
bactericidal 110, 134
bacteriocin *see* Col plasmid, colicin
bacteriophage *see* phage
bacteriostatic 110
base analogues *see* mutagens,
    2-aminopurine/5-bromouracil
base pairing *see* DNA, double helix
base substitution *see* mutation
bet see phage *lambda*, genes and mutants,
    redB
BF23 *see* phage BF23
bifunctional polypeptide *see trp* operon,
    enzymes
bio see phage *lambda*, site-specific
    recombination, primary bacterial
    attachment site; phage *lambda*,
    transduction, $λ$*pbio.*
BOB' *(att*^{λ}*B) see* phage *lamba*, site-specific
    recombination, primary bacterial
    attachment site; phage *lambda*,
    transduction
BOP', POB' *(attL, attR) see* phage *lambda*,
    site-specific recombination, primary
    attachment site/secondary attachment site
5-bromouracil (5-BU) *see* mutagens
Bsu R1 *see* DNA restriction, restriction
    endonucleases

btuB see Col plasmid, colicins, receptor;
    phage BF23, receptor
burst size *see* phage

^{14}C *see* radiolabelling
cI see phage *lambda*, regulatory proteins
cII-cIII see phage *lambda*, regulatory
    proteins
Cairns-type replication *see* replication,
    mode
Campbell event *see* F plasmid, site-specific
    recombination; phage *lambda*, site-
    specific recombination
CAP (CRP, CGA) *see* transcription factors
capR (lon, deg) see lon
capsid *see* phage
carbenicillin (penicillin, ampicillin) *see*
    penicillin
1-(*O*-carboxyphenylamino)-1-
    deoxyribulose-5-phosphate (CdRP) *see*
    trp operon, enzyme
carcinogenic sources 122; *see also*
    mutagens
catabolism 37
catabolite activator protein (CAP) *see*
    transcription factors
catabolite control (repression) *see lac*
    operon, control, positive
cations *see* metal ions
cca see transfer RNA, nucleotidyl
    Transferase
cell composition 28-9, 55; *see also* cell
    envelope
cell culture: cell counting 43; colony
    formation 42-3; culture inoculation 42,
    44-5; mutant isolation *see* mutant;
    requirements 39-42; viability estimation
    see cell culture, cell counting; *see also*
    cell growth
cell division *see* cell growth; cell culture;
    cell envelope, synthesis; replication,
    control; SOS functions
cell envelope: cell wall (peptidoglycan,
    murein) 29-33, 35, 358; *see also* cell
    envelope, synthesis; function 29; Gram-
    negative vs Gram-positive 29, 31; inner
    (cytoplasmic) membrane 30-3, 35, *see
    also* replication, control; inhibitors of
    synthesis *see* cell envelope, synthesis;
    layers *see* cell envelope, cell wall/inner
    membrane/outer membrane/periplasmic
    space; lipoprotein 32, 34; outer
    membrane 29-35, *see also* receptor;
    periplasmic space 29-32, 35;
    permeability 29, 70, 244-6; synthesis 39,
    134, 358; *see also* cell lysis
cell filamentation *see* SOS functions,
    regulation of cell division
cell growth: characteristics 40, 42, 44-5;

generation time 39-40, 44-5, 439; mating aggregation *see* F plasmid transfer; mutant proliferation *see* mutant; septum formation 39-40, *see also* cell envelope, synthesis; replication, control, and cell division

cell lysis: from without 198-9; on cellophane discs 246; with chloroform 359; with EDTA/lysozyme 354, 358; with sodium dodecyl sulphate 356, 359; with sonication 357; *see also* cell envelope

cellophane disc *see* replication, study of

central dogma *see* gene expression

centrifugation: DNA 358-9; ribosomes 87-8

chemotaxis 35-6

chi *see* general recombination

chiasma *see* general recombination

chimaera *see* DNA restriction, cloning

chloramphenicol: Col plasmid amplification 186, 265, 359; incidence of resistant *Shigella* 188; inhibition of translation 105; phage P1-encoded resistance (P*lcml*) 215; R plasmid-encoded resistance 187, 189, 265; structure 106; transposon-encoded resistance 300-2; *see also* peptidyl transferase

chorismate *see trp* operon, enzymes

chromosome: composition *see* DNA; nucleic acids; multimerisation 307-8, *see also* gene, duplication; rearrangement *see* mutation, multisite; transposable genetic element; segregation 159, 162, 212, 255, *see also* replication; size 11, 14-15, 185, 194-5; supercoiling 14-15, *see also* DNA gyrase; replication

circular permutation (of chromosome(s)) *see* phage(s) *lambda*, T4; F plasmid

cis-specific (dominant) *see* operon control, promoter control, mutants

cis-trans test *see* genetic mapping, by complementation *in vivo*

clone *see* cell culture, colony formation; DNA restriction

cloning *see* DNA restriction

cloverleaf *see* transfer RNA

codon *see* genetic code

cohesive termini (*cos*/m,m′) *see* phage, chromosome; phage *lambda*, chromosome circularisation/replication

cointegrate 167, 196, 217, 223, 262, 310, 347, 454; *see also* (general recombination) F-prime plasmid, transfer, integration vs reciprocal recombination; phage *lambda*, dilysogen formation/lysogen formation (site-specific recombination) F plasmid; phage *lambda*/Mu transposable genetic element

Col *see* Col plasmid

cold-sensitive mutants *see* mutant, conditional

colicin (A1, E1, E2, E3, K) *see* Col plasmid

coliphage *see* phage

Col plasmid: amplification 186, 265, 359; characteristics 185; colicins: bacterial immunity 190-1, bactericidal action 34, 190, colicin A 191, colicins E1/E2/E3 34, 185, 190-1, 198, colicin K 191, 198, colicin I 185, induction 273, *see also* SOS functions, phage cross-resistance 190, 198, receptors 34, 190, 198, synthesis 190, 273, *see also* SOS functions; copy number 185-6, *see also* Col plasmid, vegetative replication, control; DNA cloning 263; *see also* Col plasmid, DNA vector; DNA heteroduplex analysis 359-61; DNA isolation 358-9; DNA restriction analysis 360-4, 367; DNA vector 364-9; incompatibility 185-7; *nic* site *see* col plasmid, transfer replication; pilus type 184-5; size 11, 185; transfer: chromosome mobilisation 184, 186, non-self transmission vs. self-transmission 159, 185-6; transfer replication 186, 263; vegetative replication: and protein synthesis *see* Col plasmid, amplification; Col plasmid, vegetative replication, control; control 186, 263-5; fork movement 262; origin 263; requirements 262-3; *see also* F-duction; F plasmid; F plasmid transfer; genetic mapping; R plasmid; transposable genetic element

complementation analysis *see* genetic mapping

concatemer *see* DNA

concatenane *see* DNA

conditional streptomycin dependence *see* suppression, phenotypic

confluent lysis *see* phage

conjugation *see* plasmid, transfer; F-duction; Col/F/R plasmids, transfer

conjugation bridge (tube) *see* F plasmid, transfer

consensus sequence *see* DNA gyrase; promoter; terminator; translation, initiation sequence

constitutive expression *see* operon control, promoter control, mutants

conversion *see* phage

Coomassie blue 356

coordinate regulation (of transcription) *see* operon control

copy number *see* plasmid; Col/F/R plasmid

cordycepin: inhibition of replication 259; inhibition of transcription 71; structure 72

cordyceps militaris see cordycepin

co-repressor *see* operon control, promoter control

cos (m,m') *see* phage *lambda*, chromosome circularisation; phage *lambda*, replication

co-transcription *see* RNA polymerase, operons

cotransduction *see* genetic mapping; phage P1

cotransduction frequency 342-4; *see also* phage P1, transduction

cou (gyrB) see DNA gyrase

coumermycin 259; *see also* DNA gyrase

counter-selection *see* F-prime plasmid, transfer

cro (tof) see phage *lambda* regulatory proteins

crossfeeding *see* pathway analysis

crossover event *see* general recombination

crp see lac operon, CAP protein; transcription factors

cryptic prophage (Rac) *see* phage *lambda*

curing *see* F plasmid

cyasin *see* mutagens

cycad nuts *see* mutagens, sources

cyclic AMP *see* lac operon, control, positive; phage *lambda* transcription, lysogenic vs. lytic decision

cytidine *see* nucleic acid, components

cytoplasmic membrane *see* cell envelope

cytosine *see* nucleic acid, components

cytosine methylase *(dcm/mec):* function 131-2, 313, *see also* DNA restriction; genetic locus 314-15; role in hotspot formation 131-2, *see also* uracil-DNA glycosylase; *see also* repair

dam see adenine methylase

daughter strands *see* DNA

deceleration sequence *see* terminator, general sequence

deg (lon, capR) see lon

deletion *see* mutation, frameshift/multisite

deletion analysis *see* genetic mapping, by recombination analysis/F-prime plasmid transfer

deletion interval *see* genetic mapping, by recombination analysis/F-prime plasmid transfer

deletion (insertion) loop *see* DNA analysis, heteroduplexing

dem (mec) see cytosine methylase

denV (v) see repair, enzymes, *v* protein

deoxyribonuclease (DNase) *see* DNA restriction; endonuclease(s); exonuclease(s)

deoxyribonucleic acid *see* DNA

deoxyribonucleoside *see* nucleic acid, components

deoxyribonucleotide see nucleic acid, components

deoxyribose see nucleic acid, components

derepressed (for transfer replication) *see* R plasmid, transfer

derepression *see trp* operon; operon control, promoter

2', 3'-dideoxynucleoside triphosphate *see* replication, inhibitors

diffusion pore *see* receptor

D(ihydrouridine) arm *see* transfer RNA, secondary structure

dihydrouridine *see* Transfer RNA, modified nucleosides of

disinfectants *see* mutagens, sources

displacement loop *see* general recombination

DNA: A, B, Z forms 13; antiparallel nature 8, 11, 12, *see also* general recombination; repair; replication, transcription; base pairing 8, 10-11; cell content 29, 55; composition *see* nucleic acids; concatenane vs. concatenane 21, *see also* phage, chromosome, packaging; phage *lambda*, morphogenesis, DNA encapsidation; phage ϕX174, replication; daughter vs. non-sister strands 283, *see also* general recombination; double helix 8, 11-14, *see also* general recombination; operator; promoter; repair; replication; transcription; extrachromosomal *see* Col/F/R plasmid(s); phage(s) *lambda*/Mu; transposable genetic element; gap vs. nick 231, *see also* DNA ligase; repair; insertion sequence *see* transposable genetic element; Okazaki fragment 19-20, *see also* replication, elongation; proof-reading *see* DNA polymerase(s); -RNA hybrid 65, *see also* DNA analysis; sense vs. anti-sense strand 51, *see also* transcription; sequence homology *see* DNA analysis, heteroduplexing; general recombination; single-stranded *see* phage; size range 8, 11, 185, 194, 300; supercoiling 14-15, *see also* DNA gyrase; Eco DNA topoisomerase I; termini *see* phage, chromosome; uracil-containing *see* dUTPase; *see also* general recombination; mutation; repair; replication; transcription

DNA analysis: fractionation: by centrifugation 360, by electrophoresis 359, 362-3, 365; heteroduplexing 359-61; hybridisation 357, 366, 368; isolation 358-9; visualisation 359; *see also* DNA restriction

DNA-dependent ATPase 278; *see also dnaB*; DNA gyrase; Eco topoisomerase I; exonuclease V; phage ϕX174, replication (*rep* involvement); protein X

DNA-dependent DNA polymerase *see* DNA polymerase

DNA-dependent RNA polymerase *see* RNA polymrase

DNA glycosylase *see* hypoxanthine-DNA glycosylase; 3-methyladenine glycosylase; uracil-DNA glycosylase; repair, pathways, excision repair

DNA gyrase (*gyrA,B*): antibiotic inhibitors *see* replication, inhibitors (coumermycin/ nalidixic acid); consensus sequence 243; function 240-3, 245; genetic locus 244, 280; role in recombination 287, 297; role in replication 250-1; role in transcription 61-2, 243, *see also* promoter

DNA ligase (*lig*): cellular concentration 240; function 240-1, 245; gene mutants 240; genetic locus 244-5; role in recombination 289; role in repair 319, 320, 321; role in replication 251-2; role in phage *lambda* circularisation 215-16, 449, *see also* phage *lambda*; sealing rate 240-1; use in cloning 365

DNA polymerase I: elongation rate 237; exonuclease V association *see* DNA polymerase I, gene mutants, *polA recBC⁻*; fidelity *see* DNA polymerase III; functional domains 237; functions 234-7, 245, 3′-5′ exonuclease activity 234-6, 245, *see also* DNA polymerase I, proof-reading, 5′-3′ exonuclease (*polAex*) activity 234-7, 251, 319, 320, 321 *see also* DNA polymerase I, gene mutants/ nick translation, polymerase activity 234-7, 245,*see also* DNA polymerase I, genes mutants/nick translation; general recombination, stages, strand invasion; replication, elongation, primer excision; gene mutants: *polAex⁻* 237, 251, 262, *polA⁻ recBC⁻* (*recA⁻*) 237, 286, *polA⁻ polB⁻* 237, *see also* DNA polymerase I, mutator alleles; genetic locus 235, 244-5; mutator/anti-mutator alleles 118-19; nick translation 237-8; 251; proof-reading 247-8, *see also* DNA polymerase I, mutator alleles; requirements 234-5; role in recombination 286; role in repair 319, 320; role in replication 251, 262; vs. RNA polymerase 239

DNA polymerase II: elongation rate 237; function 234-7, 245; gene mutants 237; genetic locus 235, 244-5; requirements 234-7; role 237; vs. RNA polymerase 239

DNA polymerase III: elongation rate 233, 246, 250; fidelity 246-8, *see also* DNA polymerase III, mutator alleles; functions 234-7, 245; gene mutants *see* DNA polymerase III, mutator alleles; genetic loci 235, 244-5; holoenzyme 235, 237-9, 245; mutator/anti-mutator alleles 118-19; proof-reading 247-8, *see also* DNA polymerase III, mutator alleles; requirements 234-7; role in recombination *see* general recombination, stages, breakage and reunion/chiasma formation/strand invasion; general recombination, mechanistic models; role in repair *see* repair, pathways, error-prone induced repair; role in replication *see* replication, elongation/initiation/ termination; subunits 235, 237-9, 245; vs. RNA polymerase 239

DNA primase (*dnaG*): function 56, 239-40, 245; genetic locus 244-5; role in repair 322; role in replication 249, 251, 266

DNA repair *see* repair

DNA restriction: applications 4, 59, 117, 227, 248, 330, 334, 350, 358, 391, 445; cleavage maps 362-5; cloning 364-9; host-controlled restriction-modification 360-3; restriction endonucleases 131, 360-2, 364, 367

DNase (deoxyribonuclease) *see* DNA restriction; endonuclease(s); exonuclease(s)

DNase-protection studies *see* operator; promoter

dnaA 244-5; *see also* F plasmid, integrative suppression; replication, initiation

dnaB,C 244-5; *see also* replication, elongation/initiation

dnaE (*polC*) *see* DNA polymerase III

dnaG see DNA primase

dnaI,P 244-5; *see also* replication, initiation

dnaX,Z see NDA polymerase III, subunits

donor (male) cell *see* plasmid, transfer; F plasmid

double helix *see* DNA

downstream *see* transcription

drd see R plasmid, transfer

Drosophila see DNA restriction, cloning

duplication *see* gene; mutation, multisite

dut see dUTPase

dUTPase (*dut*): function 314, *see also* replication, leading strand discontinuities; gene mutants 317; genetic locus 314-15; *see also* repair

eclipse period *see* phage

Eco DNA topoisomerase I (ω protein) 243; *see also* DNA gyrase

Eco DNA topoisomerase II *see* DNA gyrase

EcoRI, Eco RII see DNA restriction,

restriction endonucleases
Eex (Sex) *see* F plasmid transfer, entry exclusion
efficiency of plating (EOP) *see* phage, plaque assay
EF-G *see* translation, elongation factors; translation inhibitors, inhibitors (fusidic acid)
EF-Tu *see* translation, elongation factors
electron microscopy *see* DNA analysis, heteroduplexing
electrophoresis: DNA 359, 362-3, 365, in agarose 357, 359, 362-3, 365, in polyacrylamide 153, 353-7; protein 153, 353-7; protein 153, 353-6; RNA 357
elongation factor(s) *see* translation; replication
EMS (ethylmethane sulphonate) *see* mutagens, alkylating agents
Endo *see* endonuclease III; endonuclease IV; endonuclease V; exonuclease(s)
endolysin *see* phage; phage *lambda*
endonuclease II/VI *(xthA)*: function 314, 316, 317; genetic locus 314-15
endonuclease III, IV, V 316, 317
Enterobacteriaceae 4
Entry exclusion (Eex) *see* F plasmid, transfer
EOP (efficiency of plating) *see* phage, plaque assay
episome *see* plasmid
equilibrium sedimentation centrifugation *see* DNA analysis, isolation
escape synthesis 437
Escherichia coli: base composition 11; chromosome size 28, 55; composition 28-9; cytoplasmic milieu 55; electron micrograph 30-1, 33; gene number 16; polypeptide species 28-9; rifampicin sensitivity 29; size 29-30; strains B, B/r, C, W 4, 194, 311; supercoil number 15; vs. *Salmonella typhimurium* 3-4, 311-12
ethanol derivatives *see* replication, role of cell envelope
ethidium bromide: DNA isolation 15, 359; inhibition of replication 259
ethylmethane sulphonate (EMS) *see* mutagens, alkylating agents
eukaryote 1
'excisase' *(xis) see* phage *lambda*, site-specific recombination; phage *lambda* transcription
exclusion *see* phage, mutual/superinfection exclusion
exo see phage *lambda* genes and mutants, *redX*
Exo *see* DNA polymerase(s) I, II, III; exonuclease(s) I, II, III, V, VII, VIII; endonuclease(s)

exonuclease I *(sbcB)*: function 277, 280, 316; gene mutants 277-81; genetic locus 279-80; role in recombination 277-81
exonuclease II *see* DNA polymerase I, functions, 3'-5' exonuclease activity
exonuclease III *(xthA) see* endonuclease II
exonuclease V *(recBC)*: endonuclease activity 280, 317-19, 322; function 277, 280, 316; gene mutants 277-82, 287, 318, *recBC⁻* 277-82, 287, 446-7, *recBC⁻ polA⁻* 237, 286, 318, *see also* general recombination pathways; genetic locus 279-80; role in recombination 277-82, 286-7, 319; role in repair 319, 322; role in replication 446-7
exonuclease VI *see* DNA polymerase I, functions, 5'-3' exonuclease activity
exonuclease VII *(xseA)*: function 314, 316, 317-18; genetic locus 314-15; role in repair *see* repair, pathways, excision repair
exonuclease VIII *(recE)*: function 277-8, 280, 316; genetic locus 279-80; role in recombination 281
exponential phase *see* cell growth, characteristics
extrachromosomal genetic element *see* plasmid(s); phage(s) *lambda*, Mu, P1, P2; transposable genetic element, insertion sequence/transposon

FI, FII *see* plasmid, incompatibility groups
f1, f2 *see* phage(s) f1, f2
F13, F14 431-2; *see also* F-prime plasmid, formation
F104, F111, F112, F116, F126, F128, F129, F140, F142, F143, F152, F254, F500, F506 *see* genetic maps, F-prime plasmids
facultative anaerobe 4
fd *see* phage fd
F-duction 166-7; Hfr recombination 171-2, *see also* general recombination; Hfr transfer: frequency 171, orientation 169, 171, 174, 176, origin *(oriT)* 169-70, 175, *see also* F plasmid transfer, replication, schematic model 170; in genetic mapping 171, 173-5, 337-40, *see also* genetic mapping; lethal zygosis 169, 428-30; *see also* Col plasmid; F plasmid; F plasmid transfer; genetic mapping; R plasmid
feedback inhibition *see* trp operon
fermentation 37
fertility inhibition (Fi) *see* plasmid; F plasmid transfer
Fi *see* plasmid, fertility inhibition
figure-eight structure *see* general recombination
filamentation *see* SOS functions

fimbriae *see* pili, common

fin (finO) see F plasmid genes and mutants, *finO*

finP see F plasmid genes and mutants

fine structure analysis *see* genetic mapping, by recombination; DNA restriction, cleavage maps

fisO see F plasmid genes and mutants

flaF see flagellin

flagella 30, 34-5

flagellin 35

flame retardants *see* mutagens, sources

flavin adenine dinucleotide (FAD) 37

fluctuation test *see* mutagenesis, spontaneous, frequency

fluid mosaic model *see also* cell envelope

fluorography *see* radiolabelling

5-fluorouracil *see* suppression, phenotypic

'flush' ends *see* DNA restriction, cloning/restriction endonucleases

foodstuffs (amines, nitrites, pyrolysis products) *see* mutagens, sources

formylmethionine 95; *see also* translation, initiation/study of

F⁻ phenocopy *see* F plasmid transfer, entry exclusion

F pilin *see* F plasmid transfer

F plasmid: characteristics 185; circular permutation of 167-8, *see also* F-plasmid, site-specific recombination; coding potential 160, 427, *see also* F plasmid genes and mutants; copy number 160, 162, 185, 258, *see also* F plasmid, vegetative replication, incompatibility; curing 161-2, 259; F-like R plasmid 184-5, *see also* R plasmid; in genetic mapping 171, 173-5, 336-42, *see also* genetic mapping; integrated vs. autonomous 160-1; integration *see* F plasmid, site-specific recombination; integrative recombination *see* F plasmid, site-specific recombination; interaction with F-like R plasmids 431-4; 'life-cycle' 434; site-specific recombination (F insertion/Hfr formation): Campbell model 167-8, excision 176-7, 431-2, *see also* F-prime plasmid, frequency 167, integrative suppression 167-9, 186, 262, *see also* replication, initiation, primer synthesis, insertion sites 175, plasmid gene reordering 167-8, 181, role of IS sequences 167, 428, 430-2, *see also* transposable genetic element, insertion sequence, *see also* phage *lambda*/Mu, site-specific recombination; transposable genetic element; vegetative replication: abortive (unilinear) inheritance 161-2, bacterial genes implicated 262, *see also* F plasmid, site-specific recombination,

integrative suppression, DNA gyrase; DNA polymerase III; DNA primase; replication, control *see* F plasmid, vegetative replication, incompatibility, F genes implicated *(frp)* 428, 430, fork movement 262, incompatibility (Inc) 166, 185-6, 256-8, 428, 430, incompatibility genes *(inc)* 428, 430, initiation 430, *see also* F plasmid, vegetative replication, incompatibility, mode 21-2, 261, negative control 258, negative vs. positive control 161, 166, *see also* F plasmid, vegetative replication, incompatibility, origin *(oriV)* 263, 428, 430, segregation 159, *see also* F plasmid transfer, replication; *see also* Col plasmid; F-duction; F plasmid transfer; genetic mapping; R plasmid; transposable genetic element

F plasmid genes and mutants: *fin see* F plasmid genes and mutants, *finO*; *finO (fin)* 432-3; *finP, fisO (traP, O)* 428, 432-4; *frp* 428, 430; *ilz* 169, 428-30; *inc* 428, 430; insertion region 428, 430-2; *oriT* 163-4, 169, 174-5, 428-30; *oriV* 262, 428-30; *pif* 427; replication region 428, 430; *traA-traZ see* F plasmid genes and mutants, transfer region; *traO, P see* F plasmid genes and mutants, *finO, finP*; transfer region 163, 165, 428-30

F plasmid transfer: chromosome mobilisation *see* F-duction; conjugation bridge 163-4; donor vs. recipient 159-60; entry (surface) exclusion (Eex) 165-6; entry (surface) exclusion genes *(traS, T)* 429; fertility (transfer) inhibition 184-5, 431-4; *finOP-fisO* system 432-4; F-filin 163, 429; F-pilin gene 428-9; F-pilus function 161, 163-4, *see also* phage, receptors; F-pilus production genes 429; immunity to lethal zygosis *(ilz)* 169, 429; mating-aggregation (pairing) 163-4, 429; mating-aggregation (pairing) genes *(traG, D)* 429; of non-plasmid DNA *see* F-duction; F-prime, transfer; *ompA* involvement 163, 184; orientation 428-30; rate 165; recipient co-operation 165; replication: initiation 163-5, origin *(oriT)* 163-4, 170, 262, 428-30, rolling-circle mode 21-2, 261, 264, *see also* F plasmid, vegetative replication; repliconation 165, schematic model 164; *tra* operator *(fisO)* 428-9, 432-4; *tra* operon 428-34; *tra* regulator *(traJ)* 429; *traYZ* endonuclease 165, 429; *see also* Col plasmid; F-duction; F plasmid; genetic mapping, R plasmid; transposable genetic element

F-prime plasmid: composition *see* F-prime

plasmid, type I vs. type II; formation
177-8, 431-2, *see also* F plasmid, site-
specific recombination; phage *lambda*,
site specific recombination; in genetic
mapping 179, 181, 336-42; -mediated
gene fusion/transposition 179; transfer:
integration vs. reciprocal recombination
179, 182, *see also* general recombination;
site-specific recombination, merodiploid
formation 179-80, transconjugant
selection 165, 183, *see also* F plasmid
transfer; type I vs. type II 177-8, 181, *see
also* phage *lambda*, transduction,
chromosome excision; *see also* Col
plasmid; F-duction; F plasmid; F plasmid
transfer; genetic mapping; R plasmid;
transposable genetic element
frameshift *see* mutation; mutagenesis,
induced; suppression, intergenic
frp see F plasmid genes and mutants
fuc 27-8
fungi *see* mutagens, sources
fusA see translation, elongation factors
fusidic acid 106; *see also* translation,
elongation factors (EF-G)

G4 *see* phage G4
G11 (Hfr) *see* genetic maps, F plasmid
insertion sites
gal 27-8, 392; *see also* phage *lambda*, site-
specific recombination, primary
attachment site; phage *lambda*,
transduction, λ*gal*
galactose *see* lac operon, enzymes
β-galactosidase *(lacZ) see* lac operon,
enzymes
gam see phage *lambda*, replication; phage
lambda genes and mutants, Spi
'gamma' protein *(gam) see* phage *lambda*,
replication; phage *lambda* genes and
mutants, Spi
gap *see* DNA
gene: bank *see* DNA restriction, cloning;
clustering 91, 414, 418, *see also*
replication, initiation, multiple;
duplication 84-5, 144, 305, 307-9, 411-
12, 416, 418-19, *see also* chromosome,
multimerisation; essential vs.
non-essential 28, 110; number (bacterial)
51, 230; number (phage T4) 196, *see
also* F plasmid genes and mutants;
genetic maps; phage *lambda* genes and
mutants; one gene-one polypeptide 16,
see also gene, -protein colinearity;
overlapping 16, 73, 329; product 16, *see
also* (enzymes of) recombination; repair;
replication; transcription; translation;
-protein colinearity 73, 329-30, 344-6,
401; RNA 16, 18, *see also* ribosomal

RNA; transfer RNA; RNA vs. protein
see gene, structural; structural 16, 18, 53;
suppressor *see* suppression, intergenic;
transfer see F-duction; F plasmid
transfer; F-prime plasmid, transfer; phage
lambda, transduction; phage P1,
transduction; *see also* DNA; F plasmid;
genetic code; mutation; operon control;
phage *lambda*; ribosomal protein;
ribosomal RNA; RNA polymerase;
transcription; transfer RNA; translation
gene expression: central dogma 53-5;
control of *see* operon control; *lac* operon;
phage *lambda*, transcription; RNA
polymerase; *trp* operon; schematic
representation 52-3; unidirectional flow
of *see* gene expression, central dogma; *see
also* genetic code; messenger RNA;
mutation; operon control; polarity;
ribosomal protein; ribosomal RNA;
ribosome; suppression; transfer RNA;
translation
general recombination 271-92; branch
migration *see* general recombination,
stages, chiasma formation; chi 283;
chiasma *see* general recombination,
stages, chiasma formation; chiasma
movement *see* general recombination,
stages, chiasma formation; crossover 23;
displacement loop (D-loop) 273-4, 285-
6, 294, *see also* general recombination,
stages, strand invasion/strand pairing;
general recombination, mechanistic
models; enzymes 272-8, DNA gyrase
(gyrA,B) see general recombination,
stages, chiasma formation/strand
invasion; DNA gyrase, DNA ligase *(lig)
see* general recombination, stages,
breakage and reunion; DNA ligase, DNA
polymerase *see* general recombination,
stages, breakage and reunion/chiasma
formation/mismatch repair/strand
invasion; general recombination,
mechanistic models; DNA polymerase
DNA polymerase I *see* general
recombination, stages, strand invasion;
DNA polymerase I, exonuclease I *(sbcB)
see* general recombination, pathways;
exonuclease I, exonuclease V *(recBC) see*
general recombination, chi; general
recombination, stages, chiasma
formation/strand invasion; general
recombination, pathways; exonuclease V,
exonuclease VIII *(recE) see* general
recombination, pathways; exonuclease
VIII, helix destabilising protein *see*
general recombination, enzymes, single-
strand DNA binding protein, *lexA*
repressor *see* protein X, gene control;

LexA repressor; repair, pathways, daughter-strand gap repair/error-prone induced repair, protein X *(recA) see* general recombination, chi/D-loop/figure-eight structure/pathways/vs. non-homologous recombination; protein X, single-strand DNA binding protein *(ssb)*, *see* general recombination, stages, strand invasion/strand pairing; single-strand DNA binding protein; figure-eight structure 273-4; frequency (vs. non-homologous recombination) 271; genes 279-80; half-chiasma *see* general recombination, stages, strand pairing; hereoduplex intermediate 282, 291; Holliday model *see* general recombination, mechanistic models; Hotchkiss model *see* general recombination, mechanistic models; in genetic mapping *see* genetic mapping; insertion sequence promotion 302; intermediates *see* general recombination, displacement loop/figure-eight structure; interstrand vs. intrastrand 269-72; isomerisation (of half-chiasma) *see* general recombination, stages, chiasma formation; mechanistic models 292, 294, *see also* general recombination, stages; Meselson-Radding model *see* general recombination, mechanistic models; pathways 278-82, *recA*-dependent pathways 278-82, *recA*-independent pathways *see* illegitimate recombination; site-specific recombination; site-specific recombination, *recBC* pathways 279-82, *recE* pathway 281, *recF* pathway 279-82, *rpo* pathway 282; phylogenetic implications 307-12; *rec* functions *see* general recombination, pathways; *rec* genes 280; *rec* mutants *see* exonuclease I *(sbcB)*; exonuclease V *(recBC)*; LexA repressor; protein X *(recA)*; sequence homology and 269-71; sites *see* general recombination, chi; stages 282-93, breakage and reunion 289, 291, chiasma formation 287-90, mismatch repair 289-93, strand breakage 282-4, strand invasion 285-7, strand pairing 283-6; vs. non-homologous recombination 231, 271, 292-6; Whitehouse model *see* general recombination, mechanistic models; *see also* illegitimate recombination; site-specific recombination
generation time *see* cell growth
genetic code: codon-anticodon specificity *see* transfer RNA, wobble; degeneracy of 73, 77; initiation codons 73-5, 94-5; nonsense codons *see* genetic code; termination codons; sense codons 73-5;

termination codons (amber, ochre, opal) 73-5, 94, 101-2, *see also* mutant, conditional; mutation, nonsense/polar; suppression, intergenic, nonsense; universality 73, 83; *see also* messenger RNA; mutation; operon control; ribosomal protein; ribosomal RNA; ribosome; suppression; transfer RNA; translation
genetic element *see* chromosome; Col/F/R plasmid; phage(s); plasmid; transposable genetic element; insertion sequence/transposon
genetic mapping 329-30, 336-7, by complementation *in vitro* 332-4; by complementation *in vivo* (*cis-trans* test) 330-3, *see also* genetic mapping, F-prime plasmid transfer/Hfr mating/phage genes; *lac* operon, regulatory mutants; by conjugation *see* genetic mapping, F-prime plasmid transfer/Hfr mating; by deletion analysis *see* genetic mapping, by recombination analysis; genetic mapping, F-prime plasmid transfer; by function *see* genetic mapping, by complementation; by insertional inactivation 347, *see also* genetic mapping, transposable genetic element; by pathway analysis 348-50, *see also trp* operon, crossfeeding; by physical means *see* genetic mapping, heteroduplexing/restriction technology; by recombination analysis 334-8, *see also* genetic mapping, F-prime plasmid transfer/generalised cotransduction/Hfr mating/phage genes/specialised transduction; recombination; by relative distance *see* genetic mapping, by recombination analysis; F-prime plasmid transfer 179, 181, 337-42, *see also* F-prime plasmid; generalised cotransduction 212, 213, 215, cotransduction frequency and genetic distance 342-4, gene ordering vs. fine structure analysis 212, 214-15, 344-6, *see also* phage P1; heteroduplexing 359-61; Hfr mating: gradient of transmission 173, 336-40, interrupted mating 171, 173-6, 339-40, *see also* F-duction; F plasmid, site-specific recombination; phages genes 350-3, *see also* genetic mapping, by complementation in vivo/by recombination analysis/heteroduplexing; restriction enzyme technology: cleavage maps 362-5, cloning 364-9, *see also* DNA restriction, restriction endonucleases; specialised transduction 225-6, 346, *see also* phage *lambda*; transposable genetic elements: identification of polycistronic operons

347, *see also* polarity, insertion
sequences and transposons 347-8, *see
also* transposable genetic element, phages
347, *see also* phage *lambda*/Mu, site-
specific recombination; *see also* genetic
maps
genetic maps: determination of *see* genetic
mapping; *E. coli* informational
suppressors 146, 148-9; *E. coli*
recombination genes 279-80; *E. coli*
repair genes 314-15; *E. coli* replication
genes 244-5; *E. coli* transcription genes
57-8; *E. coli* translation genes 91; *e. coli*
vs. *S. typhimerium* 311; F plasmid 428;
F plasmid insertion sites 175; F-prime
plasmids 181; nomenclature 26, 27-8;
phage *lambda* 436-8; vs. physical maps
336, 352
genetic nomenclature 27-8
genetic recombination *see* general
recombination; illegitimate
recombination; site specific
recombination; F plasmid, site-specific
recombination F prime plasmid,
formation; mutagenesis, spontaneous;
mutation, multisite; phage *lambda*, site-
specific recombination/transduction;
phage Mu, site-specific recombination;
transposable genetic element
genome 5
genotype 26
ghost *see* phage
glnU,V see suppression, intergenic,
nonsense
gluconeogenesis 38
glucose 37-8; *see also* cell culture,
requirements; *lac* operon, enzymes
glucose effect *see lac* operon, control,
positive
glycerol *see lac* operon, control, positive
glycosylase *see* hypoxanthine-DNA
glycosylase; 3-methyladenine
glycosylase; uracil-DNA glycosylase;
repair, pathways, excision repair
glyT see gene, duplication; transfer RNA
Gram stain 29; *see also* cell envelope
gratuitous inducer *see lac* operon
groE (mop tabB) see phage *lambda*,
morphogenesis
growth *see* cell growth; phage, infection
growth medium *see* cell culture,
requirements
guanine *see* nucleic acid, components
guanosine *see* nucleic acid, components
guanosine tetra(penta)phosphate
(ppGpp(p)) *see* operon control, stringent
control
gyrA,B (nalA, cou) see DNA gyrase

³H *see* radiolabelling
Hae III *see* DNA restriction, restriction
endonucleases
Haemophilus see DNA restriction,
restriction endonucleases
hair dyes *see* mutagens, sources
half-chiasma *see* general recombination,
stages, strand pairing
Halobacterium halobium 32
headful model *see* phage, morphogenesis
helix destabilising (unwinding) protein *see*
phage T4, gene 32 product; single-strand
DNA binding protein
heteroduplex *see* DNA analysis
heterotroph 36
Hfr *see* F plasmid, site-specific
recombination; F-prime plasmid,
formation; F-duction
Hfr (R) *see* R plasmid
HfrC, HfrH, Hfr6 *see* genetic maps, F
plasmid insertion sites
HFT (high frequency transduction) *see*
phage *lambda*, transduction
high frequency transduction (HFT) *see*
phage *lambda*, transduction
Hind II, *Hind* III *see* DNA restriction,
restriction endonucleases
his: locus 27-8; mapping 340-1; *see also trp*
operon, control, attenuator
Holliday model *see* general recombination
Holoenzyme *see* DNA polymerase III;
RNA polymerase
homogenotisation 337; *see also* genetic
mapping
Hotchkiss model *see* general recombination
hybridisation 357, 366, 368; *see also* RNA,
-DNA hybrid
hydrogen bonding *see* DNA, double helix
hydroxylamine (HONH₂) *see* mutagens
5-hydroxymethyl cytosine *see* phage T4
hypoxanthine *see* mutagenesis,
spontaneous, deamination; hypoxanthine-
DNA glycosylase
hypoxanthine-DNA glycosylase: function
314, 316, 317; role in repair *see* repair,
pathways, excission repair; *see also* 3-
methyladenine glycosylase; uracil-DNA
glycosylase

Iα *see* plasmid, incompatibility groups
icosahedral structure *see* phage, capsid
ICR compounds *see* mutagens,
intercalators; mutagenesis, frameshift
IF-1, IF-2, IF-3 *see* translation, initiation,
factors
illegitimate recombination: abnormal phage
excision *see* phage *lambda*, transduction;
abnormal F plasmid excision *see* F-prime
plasmid, formation; deletion, duplication,

insertion, inversion *see* mutagenesis, spontaneous; mutation, multisite; general recombination vs. non-homologous recombination 231, 271, 292-6; phylogenetic implications 307-12; vs. site-specific 296

ilz see F plasmid transfer, immunity to lethal zygosis

immunity *see* phage *lambda*; phage *lambda* genes and mutants

immunity to lethal zygosis *(ilz) see* F plasmid transfer

Inc *see* plasmid, incompatibility; F plasmid, vegetative replication, incompatibility

Inc FI *see* F plasmid, vegetative replication, incompatibility

inceptor *see* phage *lambda*, replication, origin

incompatibility (Inc) *see* plasmid; F plasmid, vegetative replication

indole *see trp* operon, enzymes

indole-3-glycerol phosphate (InGP) *see trp* operon, enzymes

indoleglycerol phosphate synthetase *(trpC) see trp* operon, enzymes

inducer *see* operon control, promoter control

induction *see* Col plasmid, colicin; phage; phage *lambda*; protein X; SOS functions

induction ratio *see lac* operon

infective centre *see* phage, plaque

inhibitor *see* antibiotic; operon control, promoter control

initiation codon(s) *see* genetic code

initiation factors *see* transcription; translation

injection (of DNA) *see* phage, infection; F plasmid, transfer; DNA restriction, cloning (transfection, transformation)

inner membrane *see* cell envelope

inosine *see* transfer RNA, modified nucleosides of

insertase *see* repair, pathways, base replacement

insertion *see* mutation, frameshift/multisite

insertion sequence *see* transposable genetic element

insertional inactivation *see* polarity, and Is/Tn elements/and phages

insulation *see* mutagens, sources

int see phage *lambda*, site-specific recombination; phage *lambda* transcription

'integrase' *(int) see* phage *lambda*, site-specific recombination; phage *lambda* transcription

integration: general recombination *see* F-prime plasmid, transfer; phage *lambda*, dilysogen formation/lysogen formation;

general recombination; site-specific recombination *see* F plasmid; phage *lambda*/Mu; transposable genetic element; site-specific recombination

integrative recombination *see* site-specific recombination; F plasmid; phage(s) *lambda*, Mu; transposable genetic element

integrative suppression *see* F plasmid; replication, initiation

intercalators *see* mutagens; mutagenesis, frameshift

inversion *see* mutation, multisite

inverted repeat 24-6; *see also* transposable genetic element, insertion sequence/transposon

IS *see* transposable genetic element, insertion sequence

isomerisation *see* general recombination, stages, chiasma formation

isometric morphology *see* phage, capsid

isopycnic centrifugation *see* DNA analysis, isolation

isoschizomer *see* DNA restriction, restriction endonuclease

J4 (hfr) *see* genetic maps, F plasmid insertion sites

jumping genes *see* transposable genetic element, transposon

kanamycin: inhibition of translation 189; transposon-encoded resistance 300-2

KL14, KL16, KL25, KL96, KL98, KL99 (Hfrs) *see* genetic maps, F plasmid insertion sites

Klebsiella 4

L1-L33 *(rplA-rplY, rpmA-rpmG) see* ribosomal protein

lac operon: adenylate cyclase *(cya) see lac* operon, control, positive; allolactose 396, *see also* lac operon, induction; β-galactosidase *(lacZ) see lac* operon, enzymes; CAP protein: function 57-8, binding site 394, 398-400, synthesis 400-1, *see also* lac operon, control, positive; catabolite control (repression) *see lac* operon, negative control; class I, II, III promoter mutants 393, 398-9; control 39, 392, 395-401, negative 396-8, positive 57, 398-401, *see also* operon control; *crp (cap) see lac* operon, CAP protein; *cya see lac* operon, control, positive; cyclic AMP, and *see* lac operon, control, positive; phage *lambda* transcription, lysogenic vs. lytic decision; enzyme levels: basal vs. induced 396, 398, catabolite controlled vs. *cya⁻ crp⁻* 398-9,

401; enzymes 354, 392-3; genetic loci 28, 393; genetic mapping 173-4, 183, 334, 342; genetic nomenclature 26, 27-8, 392; glycerol and *see lac* operon, control, positive; induction 396-8; *lacI see lac* operon, *lac* repressor; *lacI-trpE* fusion 409; *lacPO see lac* operon, promoter/operator; *lacP^r see lac* operon, class I, II, III promoter mutants; *lacUV5* 399, *see also lac* operon, control, positive; *lacZYA see lac* operon, enzymes; *lac* repressor *(lacI)* 354, 392-3, 396-8; lactose permease *(lacY) see lac* operon, enzymes; mutant selection 134; natural polarity of 396, *see also* polarity; operator sequence *(lacO)* 394, *see also lac* operon, control, negative; operator-promoter interpenetration 394-5; orientation of transcription 392, 394; promoter *(lacP)* 395, 398-9, 402, *see also lac* operon, control; promoter sequence 394; regulatory mutants 342, 393, 398-9, 401; regulatory sequences 393; repressor control *see lac* operon, control, negative; structural genes *see lac* operon, enzymes/*lac* repressor; structural gene mutants 28, 334, 383, 393; thiogalactoside transacetylase *(lacA) see lac* operon, enzymes; *see also* F-prime, formation, frequency; F-prime, transfer, transconjugant selection
β-lactam ring *see* penicillin, structures
β-lactamase: chromosomal-encoded enzyme 189; R plasmid-encoded enzymes 187, 189, 265; transposon-encoded enzymes 300-2; *see also* penicillin
lactose *see lac* operon
lactose permease *(lacY) see lac* operon, enzymes
lacunae *see* Col plasmid, colicins, synthesis
Lamarkian adaptation 133
lamB see phage *lambda*, receptor
lambda (λ) see phage *lambda*
λdv see phage *lambda*, defective
λgal, λilu, λmal, λpbio, λpara, λprecA, λptrp see phage *lambda*, transduction
λvir see phage *lambda* genes and mutants
lambdoid *see* phage, classification; phage lambda (and 21, 434, φ80)
latent period *see* phage
lesion *see* mutation
lethal zygosis *see* F-duction
leu 27-8; *see also trp* operon, control, attenuator
LexA repressor: function 280; gene mutants 273-7, 322; genetic locus 279-80; role in *recA* control 273-7; role in recombination *see* protein X, gene control; role in repair 322

lexB see protein X, gene mutants
LFT (low frequency transduction) *see* phage *lambda*, transduction
lig see DNA ligase
ligation *see* DNA ligase
lipopolysaccharide *see* cell envelope, outer membrane; phage, receptors
lipoprotein *(lpp) see* cell envelope
log phase *see* cell growth, characteristics of
lon (capR, deg): gene function 314; genetic locus 314-15; suppressor of *(sulA,B)* 276, 314
low frequency transduction (LFT) *see* phage *lambda*, transduction
lpp see cell envelope, lipoprotein
lysogen *see* phage, lysogenic response; phage(s) *lambda*, Mu, P1, P2
lysozyme *see* cell lysis
lytic response *see* phage

m, m' *(cos) see* phage *lambda*, chromosome circularisation; phage *lambda*, replication
M13 *see* phage M13
magic spot *see* operon control, stringent control
major groove *see* DNA, double helix
major tRNA *see* transfer RNA
maltose uptake *see* phage *lambda* receptor
mating-pair (aggregates) *see* F plasmid, transfer
mec (dcm) see cytosine methylase
melB see melibiose permease
melibiose permease *(melB)* 392
membrane *see* cell envelope
merodiploid 28; *see also* F-prime plasmid transfer; phage *lambda*, transduction
Meselson-Radding model *see* general recombination
messenger RNA: cellular concentration 53, 55, 67; degradation 68-9, 423; leader sequence 63, 66, *see also trp* operon; polycistronic 16, 18, 51, *see also* lac operon; phage *lambda* transcription; polarity; promoter; ribosomal protein; RNA polymerase; terminator; transcription; *trp* operon; reading frame *see* translation; size 8, 51-3, 68, 415; synthesis *see* transcription
metabolic control *see* operon control
metal ions: cellular role 29; in phage infection 198; in replication and transcription 59, 234; resistance to 184
3-methyladenine glycosylase *(tag)* 314-16, 317
methylene blue 357
5-methylcytosine *see* mutagenesis, hotspots; cytosine methylase; repair; DNA restriction, restriction endonucleases
2'-O-methylguanosine *see* transfer RNA,

modified nucleosides of
7-methylguanosine *see* transfer RNA,
modified nucleosides of
methylmethane sulphonate (MMS): and
protein X induction *see* SOS functions; as
a mutagen 188, 128; sensitivity of *rec*
mutants to 273; *see also* repair
N-methyl-N'-nitro-N-nitrosoguanidine
(MNNG) *see* mutagens, alkylating agents
minor groove *see* DNA, double helix
minor tRNA *see* transfer RNA
misreading *see* translation, infidelity;
suppression
missense *see* mutation
mitomycin C: inhibition of replication 259,
315; protein X induction *see* SOS
functions; sensitivity of *rec* mutants to
273; structure 260
MNNG (N-methyl-N'-nitro-N-
nitrosguanidine) *see* mutagens, alkylating
agents
mobile replication promoter *see* phage
φX174, replication
mobilisation *see* F-duction, Hfr transfer;
Col/R plasmid(s)
modification enzyme *see* DNA restriction,
restriction endonuclease
m.o.i. *see* phage, multiplicity of infection
monomerisation *see* repair, pathways,
photoreaction
mop (groE, tabB) see phage *lambda,*
morphogenesis
mRNA *see* messenger RNA
MSI, MSII *see* operon control, stringent
control
MS2 *see* phage MS2
Mu *see* phage Mu
multimerisation *see* chromosome
multiplicity of infection (m.o.i.) *see* phage
murein (cell wall, peptiodoglycan) *see* cell
envelope, cell wall
mutagenesis: hotspots 129, 131-2, *see also*
cytosine methylase; uracil-DNA
glycosylase; induced: frameshift
(intercalators) 118, 129-30, frequency
118, 128, *see also* mutagenesis, hotspots,
localised 128, 132, mispairing (alkylating
agents, 2-AP, 5-BU, HNO$_2$, HONH$_2$)
118, 122-8, misrepair (alkylating agents,
intercalators, UV) 118-22, 128, *see also*
repair; insertional inactivation 346-8, *see
also* transposable genetic element;
spontaneous: deamination 126, *see
also* mutagens, hydroxylamine/nitrous
acid, frequency 118, 132-3, *see also*
mutagenesis, hotspots, mispairing vs.
misrepair 117-19, *mut* alleles 118-19, *see
also* mutagenesis, insertional inactivation;
study of 117; *see also* DNA; genetic

code; genetic mapping; mutant; mutation;
repair; replication; suppression;
transcription; translation; transposable
genetic element
mutagens: acridine orange *see* mutagens,
intercalators; action *see* mutagenesis;
aflatoxin B 122; alkylating agents 118,
128, 132, *see also* SOS functions; 2-
aminopurine (2-AP) 118, 125-6; asbestos
122; base analogues *see* mutagens, 2-
aminopurine/5-bromouracil;
5-bromouracil (5-BU) 118, 122, 124,
126; cyasin 122; ethylmethane
sulphonate (EMS) *see* mutagens,
alkylating agents; heat 126;
hydroxylamine (HONH$_2$) 118, 126-8;
ICR compounds *see* mutagens,
intercaltors; intercalators 118, 128-9,
153, *see also* replication, inhibitors;
methylmethane sulphate (MMS) *see*
mutagens, alkylating agents; N-methyl-
N'-nitro-N-nitrosoguanidine (MNNG)
see mutagens, alkylating agents;
N-nitroso compounds 118, 122; nitrous
acids (HNO$_2$) 118, 126-7; safrole 122;
sodium hypochlorite 122; sources 122;
use of 132, *see also* genetic mapping; UV
radiation 118-22, *see also* SOS functions;
vinyl chloride 122; visible light 117; *see
also* DNA; genetic code; genetic
mapping; mutant; mutation; repair;
replication; suppression; transcription;
translation
mutant: auxotroph 41, 110; cold sensitive
see mutant, conditional; conditional 41-2,
114-16, *see also* suppression; isolation
see mutant, selection; non-growing *see*
mutant, selection (penicillin enrichment);
penicillin enrichment *see* mutant,
selection; phage resistant 199, 203, *see
also* receptor; positive selection *see*
mutant, selection; replica-plating *see*
mutant, selection; selection 132-6;
specific *see* DNA gyrase; DNA ligase;
DNA polymerase(s); dUTPase;
exonuclease I; exonuclease V; F plasmid
genes and mutants; *lac* operon; LexA
repressor; phage *lambda* genes and
mutants; protein X; ribosomal protein,
ram mutants/spcA mutants/*strA* mutants;
RNA polymerase; R plasmid, transfer;
suppression; UVrABC system;
temperature-sensitive *see* mutant,
conditional; *see also* DNA; genetic code;
genetic mapping; mutagen; mutation;
suppression; repair; replication;
transcription; translation;
mutation: base substitution *see* mutation,
frameshfit/missense/nonsense/samesense;

cis vs. *trans* dominant (specific) *see* operon control, promoter control; class I, II, III *see lac* operon; classification 110-16; cryptic 26, 113; dominant vs. recessive 28, *see also* operon control, promoter control; forward and reverse 135, 136, 151-3, 333-5, *see also* suppression; frameshift 112-14, 153, *see also* polarity, and point mutations; suppression, intergenic; translation, reading-frame; hotspots 129-32, *see also* cytosine methylase; uracil-DNA glycosylase; identification 150-3; induced 24, 117; lethal vs. non-lethal 110-11, *see also* mutant, conditional; mechanism *see* mutagenesis; missense 113-14, *see also* suppression; multisite (deletion, duplication, insertion, inversion, substitution) 111-12; multisite vs. point 111-12, 150-3; nonsense 113-15, 152-3, *see also* polarity, and point mutations; suppression, intergenic; translation, termination; pleiotropic 19, 379, *see also* mutation, frameshift/nonsense/regulatory; point *see* mutation, frameshift/missense/samesense/nonsense; polar 114-15, *see also* polarity; regulatory 16-18, 113, *see also lac* operon; operon control; phage *lambda* genes and mutants; *trp* operon; reversion *see* mutation, forward vs. reverse; samesense 113; spontaneous 24, 117; suppressor *see* suppression; test 122-3; transition vs. transversion 112, 115, 152; *see also* DNA; genetic code; genetic mapping; mutagen; mutant; repair; replication; suppression; transcription; translation

mutator allele *(mut) see* mutagenesis, spontaneous

mutT see mutagenesis, spontaneous

mutU (uvrD, uvrE, recL) see uvrD

mutual exclusion *see* phage

N (lambda) see phage *lambda* regulatory proteins

nalA (gyrA) see DNA gyrase

nalidixic acid: inhibition of replication 258; protein X induction *see* SOS functions; structure 259; *see also* DNA gyrase

negative complementation *see* genetic mapping, by complementation *in vivo; lac* operon, regulatory mutants

nic see Col plasmid, transfer replication

nick *see* DNA

nick translation *see* DNA polymerase I

nicotinamide adenine dinucleotide (NAD) 37

nicotinamide adenine dinucleotide phosphate (NADP) 38

nif region *see Klebsiella*

N-nitroso compounds *see* mutagens

nitrous acid (HNO_2) *see* mutagens

N^6-methyladenine *see* DNA restriction, restriction endonucleases

Nocardia mediterrani see chromosome, multimerisation

non-homologous recombination *see* illegitimate recombination; site-specific recombination

nonsense triplets (amber, ochre, opal); and genetic code 73-5, 77; and translational termination 94, 101-2; as conditional mutations 116; as mutations 113-16, 153; polarity of 114, 382-4; suppression of *see* suppression, intergenic/phenotypic

non-sister strands *see* DNA

N-5′-phosphoribosyl-anthranilate (PR anthranilate) *see trp* operon, enzymes

nuclease *see* exonuclease (DNase); endonuclease (DNase); RNase; DNA restriction, restriction endonuclease

nucleic acid: components 5-6; nomenclature 5-8; polarity of 8-9, 12; size 8-11, 51, 88, 185, 194, 300; synthesis *see* replication; transcription; *see also* DNA, RNA

nucleoid 55, 244-5; *see also* replication

nucleoside *see* nucleic acid, components; phage T6, receptor

nucleoside mono-, di-, triphosphate *see* nucleic acid, components

nucleotide *see* nucleic acid, components

ochre (UAA) *see* nonsense triplets

Okazaki fragment *see* DNA; DNA ligase, role in replication; DNA replication, elongation

ompA see F plasmid, transfer; R plasmid, transfer

oop see phage *lambda* genes and mutants, Po

opal (UGA) *see* nonsense triplets

operator (transcription regulatory site): *lac* operator sequence 393-5; phage *lambda* operators *see* phage *lambda* genes and mutants, O_L/O_R; *trp* operator sequence 404, 406; *see also* operon control, promoter

operator-promoter interpretation *see* operon control, promoter control

operon 18-19; *see also lac* operon; operator; operon control; phage *lambda* transcription; polarity; promoter; ribosomal RNA; ribosomal protein; RNA polymerase; transfer RNA; terminator; transcription; *trp* operon

operon control: autogenous control *see* operon control, promoter control, circuits; phage *lambda* regulatory proteins, *c*I

repressor/*cro* repressor; ribosomal
protein, control, post-transcriptional; *trp*
operon, *trp* repressor; constitutive
expression 39, 378-81, *see also lac*
operon; regulatory mutants; *trp* operon,
regulatory mutants; coordination 373,
376-7, 390-1, 401, 411, 419, *see also*
operon control, metabolic control/
stringent control; RNA polymerase;
metabolic control 419; operator-promoter
interpenetration 376, *see also lac* operon;
phage *lambda* transcription; *trp* operon;
post-transcriptional control 405, *see also*
operon control, translational control;
post-translational control *see* operon
control, translational control; promoter
control 374, circuits 377-8, elements
376-80, *see also* operator; promoter;
transcription factor, mutants 330-1, 378-
82, *see also lac* operon; phage *lambda*
regulatory proteins, *c*I repressor/*c*II-*c*III
activator/*cro* repressor; *trp* operon;
stringent control 68, 103, 420-2, *see also*
ribosomal protein; ribosomal RNA;
transfer RNA; terminator control 374,
385-8, *see also* phage *lambda* regulatory
proteins, *N* anti terminator/*Q* anti
terminator; transcriptional control *see*
operon control, promoter control/
stringent control/terminator control;
translational control 405-7, *see also*
phage *lambda* regulatory proteins, *c*I
repressor/*c*II-*c*III activator/*N* anti-
terminator; *see also* polarity; *trp* operon,
control, feedback inhibition
ori see origin, phage *lambda*
oriC: base composition 248; function 234,
246, 248, *see also* replication, initiation;
locus 244, 248; sequence similarities: to
Pribnow box 248, *see also* promoter, to
phage G4 origin 248-9, *see also* phage
G4, replication; *see also* F plasmid,
curing; origin
oriJ see phage *lambda*, cryptic Rac
prophage origin of replication; *oriC*;
origin
oriT see F plasmid, transfer replication,
origin; *oriC*; origin
oriV see F plasmid, vegetative replication,
origin; *oriC*. origin
origin (of replication): bacterial
chromosome *see oriC*; Col plasmid:
transfer replication *(nic)* 429, vegetative
replication 263; F-plasmid: transfer
replication *(oriT)* 163-4, 169-70, 175,
263, 428-9, vegetative replication *(oriV)*
263, 428, 430; phage *lambda (ori)* 445-
6; *Rac* prophage *(oriJ)* 248
ornithine transcarbamylase *(argI/argF) see*

gene, duplication
oxidative phosphorylation *see* ATP

³²P *see* radiolabelling
P1 *see* phage P1
P1*cml see* phage P1, transduction
P1 *vir see* phage P1
P2 *see* phage P2
P4X, P801 (hfr) *see* genetic maps, F
plasmid insertion sites
PAGE *see* electrophoresis, in
polyacrylamide
palindrome 24-6; *see also* operator, *lac/trp*;
DNA restriction, restriction
endonucleases
partial diploid *see* merodiploid
pathway analysis: generalised example 348-
50; *trp* operon 402-5; *see also* genetic
mapping
pBR345 263
penicillin (ampicillin, carbenicillin):
chromosomal-encoded resistance 189;
inhibition of cell wall synthesis 29, 33; R
plasmid-encoded resistance 187-9, 265;
structures 136; transposon-encoded
resistance 300-2; use in mutant selection
134-5; *see also* cell envelope, cell wall
peptidoglycan (cell wall, murein) *see*
cell envelope, cell wall
peptidyl transferase: antibiotic inhibitor of
105; location on ribosome 90; role in
translational elongation 98-100; role in
translational termination 102
periplasmic space *see* cell envelope
pesticides *see* mutagens, sources
phage: adsorption 30, 163, 197-9, *see also*
phage, lysogenic response/lytic response;
assembly *see* phage, morphogenesis;
binding ratio *see* phage, multiplicity of
infection; burst size 200; capsid 194-7,
see also phage, morphogenesis;
capsid assembly *see* phage,
morphogenesis; capsid vs. prehead 202,
see also phage *lambda*, morphogenesis;
chromosome characteristics 11, 194-7;
chromosome packaging (DNA
encapsidation) *see* phage, morphogenesis;
classification 193-4, *see also* phage,
capsid/chromosome/virulent vs.
temperate; cointegrate 196, 347, *see also*
F plasmid, site-specific recombination;
phage(s) *lambda*/Mu, site-specific
recombination; confluent lysis 205-7;
conversion 204, *see also* phage, lysogeny;
cross *see* genetic mapping; cryptic
prophage *see* phage *lambda*; DNA
cloning *see* phage, DNA vector; phage
lambda genes and mutants, *ori*; DNA
heteroduplex analysis 359-61; DNA

isolation 358-9; DNA restriction 360-5, 367; DNA vector 364-9; eclipse period *see* phage, lytic response; efficiency of plating (EOP) *see* phage, plaque assay; encapsidation *see* phage, morphogenesis; endolysin 200, 227, *see also* phage *lambda* genes and mutants, *R*; female-specific phage *see* phage T7; filamentous phage *see* phage M13 (and fl, fd); general recombination system *see* phage *lambda*; ghost 197-9; head *see* phage, capsid; helper *see* phage *lambda*, transduction; host response *see* phage, lysogenic response/lytic response; immunity 203, *see also* phage *lambda*; infection *see* phage, adsorption; infective vs. defective 159, 185, *see also* phage(s) *lambda*, Pl; in genetic mapping *see* phage(s) lambda, Pl; isometric phage *see* phage ϕX174, G4 (and R, S13); lamboid phage *see* phage *lambda* (and ϕ80, 21, 434); latent period *see* phage, lytic response; lysate 209, 210, *see also* phage(s) *lambda*/P1, transduction; lysis from without 198-201; lysogenic response 202-6; lytic response 199-201, *see also* phage, morphogenesis; -mediated gene transfer *see* phage(s) *lambda*/P1, transduction; male specific phage *see* phage M13 (and fl, f2); phage MS2 (and f2, QB, R17); morphogenesis 202, *see also* phage *lambda*; morphology 194-7; multiplicity of infection (m.o.i) 198-9; mutual exclusion 200-1; packaging *see* morphogenesis; plaque assay 205-9; plaque forming unit (p.f.u.) 205, 207; plaque morphology 205, 207, 352-3; plaque size 205, 207; prehead assembly *see* phage, morphogenesis; prophage *see* phage, lysogenic response; (phage *lambda* and Mu, P1); receptor *see* receptor; recombination *see* genetic mapping; replication *see* phage(s) G4, *lambda*, M13, ϕX174; -resistant mutants 190-1, 198-9, *see also* receptor; RNA phage *see* phage MS2 (and f2, R17, QB); RNA polymerase *see* phage(s) T3, T7; single-stranded DNA phage *see* phage(s) ϕX174, G4 (and R, S13); phage M13 (and fl, f2); site-specific recombination *see* phage(s) *lambda*, Mu; Spi *see* phage *lambda* genes and mutants; superinfection exclusion 201, 203, *see also* phage *lambda*; tail 194-7, *see also* phage, morphogenesis; temperate phage 193, 202, *see also* phage, lysogenic response; phage(s) *lambda*, Mu, P1, P2, ϕ80, 21, 434; T-even phage *see* phage

T4 (and T6); T-odd phage *see* phage(s) T1, T3, T5, T7; titration *see* phage, plaque assay; transduction 209, *see also* phage(s) *lambda*, P1; vector *see* phage(s) *lambda*, P1; virulent phage 193, 199, 202, *see also* phage, lytic response; phage(s); *see also* phage(s); transposable genetic element

phage 21 435-6

phage 434 435-6

phage BF23 ('T5-like') 34, 190, 198; *see also* Col plasmid, colicin, phage cross-resistance; phage; phage T5

phage f1 ('filamentous') 196, 199; *see also* phage; phage M13 (and fd)

phage f2 ('RNA') 196; *see also* phage; phage MS2 (and R17)

phage fd ('filamentous') 196-9; *see also* phage; phage M13 (and f1)

phage G4 ('isometric'): characteristics 196, *see also* phage ϕX174 (and R, 513); replication 232, 248-9, 265-8, *see also* oriC; DNA primase; phage M13/ϕX174; replication, initiation, primer synthesis

phage *lambda*: as a plasmid 159, 185, 456; attachment site *see* phage *lambda*, site-specific recombination; chromosome circular permutation 215, 219, 435, 437; chromosome circularisation 215-16, 449; chromosome conformation 11, 14, 194-5, 216; chromosome size 11, 185, 194, 215; cryptic prophage 220; cryptic Rac prophage genes and locus 279-80; cryptic Rac prophage origin of replication *(oriJ)* 248; defective phage λdv 159, 203, 456; defective transducing particle *see* phage *lambda*, transduction, helper; DNA cloning 248, 445, *see also* phage *lambda*, DNA vector; DNA heteroduplex analysis 359-61; DNA isolation 358-9; DNA restriction 360-3; DNA restriction analysis 362-4, 367; DNA vector 364-9; endolysin 227, 435; general recombination system *(red)* 271, 445; genetic map 352, 436-7; genetic mapping of 350-3; immunity 203, *see also* phage *lambda* genes and mutants; inceptor *see* phage *lambda*, replication; induction 220-5, *see also* phage *lambda* transcription; SOS functions; left vs. right strand 437-9; lysis *see* phage *lambda*, endolysin; lysogeny 215-26, *see also* phage *lambda*, site-specific recombination; phage *lambda* transcription; morphogenesis 202, 365-7, 446-9; morphology 194, 197; prophage vs. vegetative phage 215, 219, 428, 432; replication 445-7, *see also* F plasmid; F plasmid transfer; phage ϕX174; replication; site-specific recombination: aberrant excision 220-2, 226, bacterial requirements 443, *see also* phage *lambda*, site-specific recombination, primary attachment site/role of DNA gyrase/secondary attachment sites, efficiency (primary vs. secondary sites) 216, 297, integration 215-17, 297-9, 347, *int-xis* involvement 216-18, 297-9, 443-5, phage attachment region (*att$^\setminus$P/POP'*) 443-5, phage gene requirements *see* phage *lambda*, site-specific recombination, phage attachment region, phage gene reordering 215, 219, 435, 437, polarity 348, *see also* polarity, precise excision 217-18, 220-1, 297-8, *see also* F-prime plasmid, formation, primary bacterial attachment site (*att$^\setminus$B/BOB'*) 194, 215-18, 296-7, 300, 443-5, role of DNA gyrase 297, *see also* DNA gyrase, secondary bacterial attachment sites (ΔOΔ') 216, 225, 297, 346-7, supercoiling and *see* phage lambda, site-specific recombination, role of DNA gyrase; superinfection 201, 218, 222; transduction (generalised) 211; transduction (specialised); and bacterial gene amplification 227, *see also* phage *lambda*, DNA vector, and gene fusion/transposition 227, chromosome excision 220-2, 226, 305, dilysogen formation 222-5, frequency 222, helper phage involvement 222-5, high frequency transducing lysate (HFT) 222-6, in genetic mapping 225-6, 346, low frequency transducing lysate (LFT) 222, 226, lysogen formation 220-3, λgal 220-6, 346, λilv 414, λmal, λpara, λptrp 225, λpbio 220-2, λprec 225, 273, vs. phage P1 transduction (generalised) 211, 225, specificity of 220, 225-7; *see also* phage(s); phage *lambda* genes and mutants; phage *lambda* regulatory proteins; phage *lambda* transcription; transposable genetic element

phage *lambda* genes and mutants: *A* (Ter) 447-9; *att* 27-8, 198, 215-25, 297-8, 300, 347, 443, 449; *attL, attR* (BOP', POB') 215, 218, 221-4, 297-9, 347, 437, 455; *B, D, E, FI, FII, Nu3 see* phage *lambda* genes and mutants, tail genes; *BOB' see* phage *lambda* genes and mutants, *att*; *BOP'/POB' see* phage *lambda* genes and mutants, *attL/attR*; *b* region 435; *b2* deletion 220; *bet see* phage *lambda* genes and mutants, *redB*; *cI* 103, 377, 388-9, 435-43, 449-56; *cII, cIII* 435-43, 449-56; *cos* (m, m') 215-

16, 352, 435, 446-9; *cy see* phage *lambda* genes and mutants, y region; endolysin gene *see* phage *lambda* genes and mutants, *R; exo see* phage *lambda* genes and mutants, *redX; G, H, I, J, K, L, M, T, V, U, Z see* phage *lambda* genes and mutants, head genes; general recombination genes 203, 271, 445-7, 454, 456; head genes 467-9; immunity region 435-43, 449, 454; *int, xis* 167, 216-18, 297-9, 435-43, 449, 453-4; integration region 435-43; m m' *see* phage *lambda* genes and mutants, *cos; N* 66, 103, 439, 450, 452-6; *nutL, nuR* 439, 452, 455-6; *O, P* 445-7, 454-6; O$_L$, O$_R$ 435-43, 449-56; *oop see* phage *lambda* genes and mutants, Po; *ori* 445-7, 454-5; POΔ', ΔOP', ΔOΔ' 297-9, 346-7; POP' *see* phage *lambda* genes and mutants, *att*; P$_E$(P$_{RE}$) 435-43, 452-6; P$_I$ 348, 435-43, 449-58; P$_L$ 435-43, 449-58; P$_M$(P$_{RM}$) 435-43, 449-58; P$_O$ 439, 445, 451; P$_R$ 435-43, 449-58; P$_R$' 435, 453, 455; P$_{RE}$ *see* phage *lambda* genes and mutants, P$_E$; P$_{RM}$ *see* phage *lambda* genes and mutants, P$_M$; *Q* 435, 453, 455-6; *R, S* 227, 435, 453, 455; recombination genes 445; *redB, X (bet, exo) see* phage *lambda* genes and mutants, general recombination genes; regulation region 435-43; replication region 445-7; *rex* 203, 353, 435, 449, 451-2; Ter *see* phage *lambda* genes and mutants, *A*; t$_L$, t$_{R1}$, t$_{R2}$ 435-43, 449-58; tail genes 202, 447-9; tail fibre gene 449; *tof see* phage *lambda* genes and mutants, *cro; vir* 202, 440; y region 436, 439, 451

phage *lambda* regulatory proteins: cI repressor: binding sites 440-2, *cro* antagonism 443, 451-2, 456, gene mutants 454, immunity region 203, 435-6, 439, inactivation 103, 276, 389, 440, 453-4, *see also* protein X, control; SOS functions, synthesis 96, 388-9, 435-43, 449-52, 455-6, *see also* operon control, promoter control/translational control; cII-cIII activator: activator vs. anti-terminator 443, 451-2, mutants 451, stability 451, synthesis 439, 450, 454, 456, *see also* operon control, promoter control/translational control; *cro* repressor: binding sites 440-2, cI antagonism 443, 452, 456, synthesis 435-43, 454-6, *see also* operon control, promoter control; *N* anti-terminator: function 66, 450, 453-5, gene mutants 455-6, sites of action *(nutL, R)* 439, 452, 455-6, stability 103, 353, 454-5, synthesis 439, 450, 454-5, *see also*

operon control, translational control/
terminator control; *Q* anti-terminator 66,
453, 455, *see also* operon control,
terminator control; size and loci 428,
432, 438; *see also* attenuator; operator;
operon control; promoter; transcription
factors; terminator
phage *lambda* transcription: delayed-early
genes 453-5, *see also* phage *lambda*
regulatory proteins, *N* anti-terminator/*c*II-
*c*III activator/*cro* repressor; immediate
early genes 453-4; *int-xis* 443-5, 450,
454, *see also* phage *lambda*, site-specific
recombination; phage *lambda* regulatory
proteins, *c*II-*c*III activator; late genes
453, 455, *see also* phage *lambda*,
morphogenesis/replication; left vs. right
strand 360, 437-9; lysogenic
establishment 449-52, *see also* phage
lambda regulatory proteins, *N* anti-
terminator/*c*I repressor/*c*II-*c*III activator/
cro repressor; phage *lambda*
transcription, lysogenic vs. lytic pathway;
lysogenic maintenance 450, 452, *see also*
phage *lambda* regulatory proteins, *c*I
repressor; lysogenic vs. lytic pathway
450, 452-5, *see also* phage *lambda*
regulatory proteins, *N* anti-terminator/*c*I
repressor/*c*II-*c*III activator/*cro* repressor;
oop 439, 445, 451; operator-promoter
interpenetration 436, 441, *see also*
operon control; sequence after induction
452-3, *see also* phage *lambda*, lysogeny;
phage *lambda* regulatory proteins; *see
also* operon control; phage, infection;
phage *lambda*, lysogeny/site-specific
recombination; phage *lambda* regulatory
proteins
phage M13 ('filamentous'): characteristics
194, 196; commensal nature 199;
morphology 30, 196; receptor 30, 163,
198; repair *see* repair, pathways, excision
repair; replication 232, 265-8, *see also*
phage G4; replication, initiation, primer
synthesis/transcriptional activation; RNA
polymerase; *see also* phage; phage(s) f1,
fd
phage MS2 ('RNA'): burst size 200;
chromosome 11, 194; genetic mapping
349; morphology 30, 194, 197; receptor
30, 163, 198; replication 230; *see also*
phage; phage(s) f2, R17
phage Mu: chromosome 194, 300, 304;
lysogeny 203; morphology 194; site-
specific recombination 300, 303-4; *see
also* phage; phage *lambda*, site-specific
recombination; transposable genetic
element
phage P1: as a plasmid 159, 162;

characteristics 185, 194; defective 212;
lysogeny 203-4; P1*vir* 202-3, 211, 222;
prophage 203; receptor 198; transduction
(generalised): abortive transductant 162,
212, cotransduction 210-15, *see also* phage
P1, transduction, in genetic mapping,
frequency 211-12, 342-4, in genetic
mapping 210-15, 342-6, mechanism 211-
12, P1*cml* 215, phenotypic lag 212,
phenotypic mixing 211, stages 213,
transductional shortening 211-12, vs.
phage *lambda* transduction (specialised)
209, 225; *see also* phage; transposable
genetic element
phage P1*cml see* phage P1, transduction
phage P1*vir see* phage P1
phage P2: characteristics 194; DNA
encapsidation 202; induction 203;
lysogeny 203; *see also* phage; transposable
genetic element
phage ϕ80 ('lambdoid') 199; *see also* phage;
phage *lambda* (and 21, 434)
phage ϕX174 ('isometric'): chromosome 11,
14, 194, 196; host specificity 194;
morphology 194, 196-7; receptor 194;
repair *see* repair, pathways, excision repair;
replication 232, 240, 265-8, *see also* DNA
primase; phage G4; replication, initiation
primer synthesis; *see also* phage; phage(s)
G4, R, S13
phage Qβ('RNA') *see* phage MS2 (and f1, fd)
phage R ('isometric') 196; *see also* phage;
phage ϕX174 (and G4, S13)
phage R17 ('RNA') 104, 196; *see also* phage;
phage MS2 (and f2)
phage S13 ('isometric') 196; *see also* phage;
phage ϕX174 (and G4, R)
phage T1 ('T-odd'): characteristics 196;
latent phase 201; receptor 199; survival to
dehydration 211; transduction 196; *see
also* phage; phage(s) T3, T5, T7
phage T3 ('T-odd'): characteristics 196;
RNA polymerase 239; *see also* phage;
phage(s) T1, T5, T7
phage T4('T-even'): base composition 7, 129,
194-5; chromosome 194-5; DNA
encapsidation 202; DNA unwinding
protein *see* phage T4, gene 32 product;
eclipse and latent periods 201; gene 32
product 118, 240, 389, *see also* single-
strand DNA binding protein; gene cluster
(tRNA) 418; gene *rII* mutants 353; gene *v*
(denV) product 318; genetic mapping of
352; 5-hydroxymethyl cytosine and *see*
phage T4, base composition; DNA
restriction, host-controlled restriction-
modification; ligase 240, 365; morphology
194, 196-7; transduction 211; *see also*
phage; phage T6

phage T5 ('T-odd'): chromosome 195-6; latent period 201; morphology 196-7; receptor 199; *see also* phage; phage BF23; phage(s) T1, T3, T7

phage T6 ('T-even'): DNA encapsidation 202; lysis from without 198-9; receptor 198; *see also* Col plasmid, colicins, phage cross-resistance; phage; phage T4

phage T7 ('T-odd'): chromosome 11, 194-6; DNA encapsidation 202; messenger RNA processing 417; receptor 198; RNA polymerase 55, 239; *see also* phage; phage(s) T1, T3, T5

phenocopy mating *see* F plasmid, transfer, entry exclusion

phenol extraction 356, 359

phenotype 26

phenotypic lag 134, 212

phenotypic mixing *see* phage P1, transduction

phospholipids *see* cell envelope, inner membrane/outer membrane

phosphoribosyl-anthranilate isomerase *(trpC) see trp* operon, enzymes

phosphoribosyl anthranilate transferase *(trpD) see trp* operon, enzymes

O-phosphotransferase (PTase): R plasmid-encoded enzymes 187-9, 265; transposon-encoded enzymes 300-2; *see also* kanamycin; spectinomycin; streptomycin

photolysis *see* repair, pathways

photoreactivating enzyme *(phr)*: function 313-15; genetic locus 314, 315; role in repair 317-21

photoreactivation *see* repair, pathways

phr see photoreactivating enzyme

pif see F plasmid genes and mutants

pili 30, 36; *see also* F-duction, Hfr transfer; F plasmid transfer

PK3, PK19 (Hfr) *see* genetic maps, F plasmid insertion sites

plaque *see* phage

plaque forming unit (p.f.u.) *see* phage

plasmid: amplification 186, 265, 350, *see also* plasmid, vegetative replication, control; as an extrachromosomal genetic element 159; cellular location 14, 159; classes 184-5; cointegrates 167, *see also* phage *lambda*/Mu, site-specific recombination; transposable genetic element; commercial uses 4, 184; conjugation bridge *see* F plasmid transfer; conjugative vs. non-conjugative *see* plasmid, transfer, non-self transmission vs. self-transmission; copy number 185-6, *see also* plasmid, vegetative replication, control; curing *see* F plasmid; determinants 184-5, 187; DNA cloning *see* plasmid, DNA vectors; DNA heteroduplex analysis 359-61; DNA isolation 358-9; DNA restriction 362-5; DNA vector 364-

9; episome 159, *see also* F plasmid, R plasmid; fertility inhibition (Fi) 184-5, 431-4; incompatibility groups 166, 185-6; in genetic mapping *see* F plasmid: integrative recombination *see* F plasmid, site-specific recombination; phage(s) *lambda*, Mu; transposable genetic element; integrative suppression 167-9, 186, 262, *see also* replication, initiation; -mediated antibiotic resistance *see* plasmid, determinants; chloramphenicol; streptomycin; tetracycline; -mediated chromosome transfer *see* plasmid, transfer, chromosome mobilisation; F-duction; mobilisation *see* plasmid, transfer; pili *see* plasmid, determinants; F plasmid transfer; representative elements 185; self-transmissible vs. non-transmissible *see* plasmid, transfer; site-specific recombination *see* F plasmid; phage(s) *lambda*, Mu; transposable genetic element; size 11, 14, 159, 185, 194; transfer: chromosome mobilisation *see* F-duction; Col plasmid; R plasmid, conjugative vs. non-conjugative 159, 184, non-self transmission vs. self-transmission 159, 184-5, *see also* Col plasmid; F plasmid; R plasmid; transfer replication *see* F plasmid transfer; vegetative replication: and protein synthesis *see* plasmid, amplification; plasmid, vegetative replication, control, control (relaxed vs. stringent) 186, 263-5, Cairns replicative model 21-2, 261, *see also* F plasmid, vegetative replication; replication, Cairns vs. rolling-circle replication modes 21-2, 261, 263-4, *see also* F plasmid, transfer replication/ vegetative replication; phage(s) G4, M13, φX174, cellular requirements 262, *see also* plasmid, integrative suppression, fork movement 262, origin 263, primer 263, rolling-circle mode 21-2, 261, 263-4, *see also* F plasmid, transfer replication; phage(s) G4, M13, φX174 replication, *see also* Col/F/R plasmid, replication; phage(s) G4, M13, φX174; replication; viral *see* phage *lambda*; phage P1; vs. phage 159; *see also* Col plasmid; F-duction; F plasmid; F plasmid transfer; genetic mapping; R plasmid; transposable genetic element

plasmolysis *see* cell lysis, EDTA/lysozyme

pleiotropic *see* mutation

polarity: and phage(s) 347, *see also* phage *lambda*, site-specific recombination; and point mutations 114, 347, *see also* mutations; polarity, mechanistic models for; and IS/Tn elements 300, 303, 347-8,

384, *see also* transposable genetic
element; gradients of 114, 383, 386;
mechanistic models for 383-6; natural
386, 396, 422-3; nonsense vs. polarity
suppressors 384; *see also* operon control,
attenuation; phage *lambda* regulatory
proteins, *N* anti-terminator/*Q* anti-
terminator

polA see DNA polymerase I

polAex see DNA polymerase I, functions,
5'-3' exonuclease activity

polB see DNA polymerase II

polC (dnaE) see DNA polymerase III

polyacrylamide gel electrophoresis *see*
electrophoresis

polycistronic (operon/transcript) 16, 18, 51-
3; *see also* messenger RNA; operon
control; polarity; RNA polymerase;
transcription; translation

polylysogen *see* phage *lambda*, lysogeny

polymerase *see* DNA polymerase(s); RNA
polymerase

POP' *(att*$^\lambda$*P) see* phage *lambda*, lysogeny/
site-specific recombination; phage
lambda transcription, *int-xis*

POΔ'/ΔOP'/ΔOΔ' *see* phage *lambda*, site-
specific recombination, secondary
bacterial attachment sites

post-transcriptional control *see* operon
control

post-translational control *see* operon control

ppGpp(p) (MSI, MSII) *see* operon control,
stringent control

ppGppase *(spoT) see* operon control,
stringent control

prehead (prohead) *see* phage

Pribnow box (–10 sequence) *see* promoter,
consensus sequences

primary bacterial attachment site *(att*$^\lambda$*B/*
BOB') *see* phage *lambda*, site-specific
recombination; phage *lambda*,
transduction

prmA, B 87

proC 177-9; *see also* F-duction, in genetic
mapping

processing *see* operon control, post-
transcriptional control; ribosomal RNA;
transfer RNA

processive enzyme *see* DNA polymerase(s);
RNA polymerase

proflavin *see* mutagens, intercalators

prokaryote 3

promoter (transcription initiation site) 51;
consensus sequence 59, 63; *gal*
promoters (P$_1$, P$_2$) 391; *lac* promoter
(lacP) sequence 394, 399; open vs.
closed 62; phage *lambda* promoters *see*
phage *lambda* genes and mutants,
P$_E$/P$_I$/P$_L$/P$_M$/P$_O$/P$_R$/P$_R$'; Pribnow

box (–10 sequence) *see* promoter,
consensus sequence; recognition site (–35
sequence) *see* promoter, consensus
sequence; ribosomal protein promoters
412, 417-19; ribosomal RNA promoters
P$_1$ sequence 415; ribosomal RNA
promoters (P$_1$, P$_2$) 413-15, 417;
RNA polymerase promoters 412, 413,
418-19; *trp* major and minor promoters
(trpP$_1$*, trpP*$_2$*)* 402, 404; *trp* major
promoter sequence 404, 406; *see also*
operon control

promoter control *see* operon control

proof-reading *see* DNA polymerase(s)

prophage *see* phage; phage, lysogenic
response; (temperate) phage(s) *lambda,*
Mu, P1, P2

protein: composition 17; degradation 103,
389; modification 87, 101-3, 389;
synthesis *see* translation; structure 17

protein analysis: fractionation: antiserum
354, centrifugation 87, electrophoresis
116, 153, 353-6; tryptic mapping 363-4;
visualisation 356

protein X *(recA/recH):* and SOS functions
see protein X, gene control; SOS
functions; function 273-7, 280; gene
control 273-6; gene mutants: *lexB, tif,*
zab 276-7, *recA* 120-1, 273-8, 282, 296,
305, 322, 330, 344, 446-7, *recA polA*
237, *see also* general recombination,
pathways/phylogenetic implications;
genetic mapping; genetic locus 279-80;
Gudas-Pardee model *see* protein X, gene
control; induction *see* protein X, gene
control; SOS functions; isolation 273,
see also phage *lambda* transduction; role
in phage *lambda* replication 446-7; role
in recombination *see* general
recombination, chi/D-loop/figure-eight
structure/pathways/vs. non-homologous
recombination; role in repair 322-5

Proteus mirabilis see R plasmid,
amplification

prototroph 41; *see also* auxotroph; mutant

Pseudomonas 4, 11

pseudouridine (ψ) *see* transfer RNA,
modified nucleosides of

PTase *see* O-phosphotransferase

pulse-chase 354-6

purine *see* nucleic acid, components

puromycin *see* translation, inhibitors

pyrimidine *see* nucleic acid, components

pyrimidine dimer *see* mutagens, UV
radiation; repair

pyronine 357

ϕ80 *see* phage ϕ80

ϕX174 *see* phage ϕX174

Qβ *see* phage MS2 (and f1, fd)
queuosine (Q) 148-9; *see also* transfer RNA, modified nucleosides

R *see* phage R; R plasmid
R1 *see* plasmid, representative elements; R plasmid
R17 *see* phage R17
R64 *see* plasmid, representative elements; R plasmid
r component *see* R plasmid
Ra-2 (Hfr) *see* genetic maps, F plasmid insertion sites
Rac prophage *see* phage *lambda*, cryptic Rac prophage
radiolabelling: autoradiography 356-7; fluorography 356; protein (^{14}C, ^{3}H, ^{35}S) 354-6; RNA (^{14}C, ^{3}H, ^{32}P) 356-7
ram (rpsD) *see* ribosomal protein (S4)
ras 314-15; *see also* repair
reading context *see* suppression, intergenic, efficiency
recA (recH) see protein X
recBC *see* exonuclease V
recE see exonuclease VIII; general recombination, pathways; phage *lambda*, cryptic Rac prophage
recF (uvrF) see general recombination, genes/pathways; repair, pathways daughter-strand gap repair
recG 279-80; *see also* general recombination
recH (recA) see protein X
recK see general recombination, pathways, *recA* dependent
recL (uvrD, uvrE, mutD): role in general recombination 279-81; role in repair 313-14
rec mutants *see* exonuclease I *(sbcB)*; exonuclease V *(recBC)*; LexA repressor; protein X *(recA)*
receptor: colicin 34, 190, 198; maltose 198; mating aggregation 163, 184-5; nucleosides 198; phage 30, 34, 190, 198-9; vitamin 34, 190, 198; vs. diffusion pore 198
recipient (female) cell *see* plasmid, transfer; F plasmid
reckless *see* protein X, gene mutants
recognition site (−35 sequence) *see* promoter, consensus sequence
recombinant DNA *see* DNA restriction, cloning
redB, X see phage *lambda* genes and mutants
redundant termini *see* phage, chromosome
redX, B (exo, bet) see phage *lambda*, general recombination system/replication
regulatory effector *see* operon control, promoter control
regulatory protein: *finOP-fisO* system 432-4; *lac* repressor *(lacI)* 392-3, 396-8; phage *lambda see* phage *lambda* regulatory proteins; *trp* repressor *(trpR)* 401, 403, 407-8; *see also* operon control
regulatory sequence(s): in replication *see* origin; in transcription *see* attenuator; operator; promoter; terminator; in translation *see* translation, initiation
relA see operon control, stringent control
relaxed control (of plasmid replication) *see* plasmid, vegetative replication; Col/F/R plasmid
relaxed response (to amino acid starvation) *see* operon control, stringent control
release factors *see* translation, termination
rep see phage φX174, replication
repair: enzymes: DNA ligase *(lig) see* repair, pathways, excision repair; DNA ligase, DNA polymerase I *(polA) see* repair, pathways, excision repair; DNA polymerase I, *polAex* activity, DNA polymerase III *see* repair, pathways, error-prone induced repair; DNA polymerase III, DNA primase *(dnaG) see* repair, pathways, daughter-strand gap repair; DNA primase, dUTPase *see* replication, leading-strand discontinuities; dUTPase, endonuclease activities *see* repair, pathways, daughter-strand gap repair/excision repair; endonuclease II; endonuclease III; endonuclease IV; endonuclease V, exonuclease activities *see* repair, pathways, excision repair; exonuclease V; exonuclease VIII; DNA polymerase I, *polAex* activity, glycosylase activities *see* repair, pathways, excision repair; hypoxanthine-DNA glycosylase; 3-methyladenine glycosylase; uracil-DNA glycosylase, insertase *see* repair, pathways, base replacement, LexA repressor, *see* repair, pathways, error-prone induced repair; LexA repressor, υ protein 313, 317, 318, photoreactivation enzyme *(phr) see* repair, pathways photoreactivation; photoreactivation enzyme, protein X *see* repair, pathways, daughter-strand gap repair/error-prone induced repair; protein X, RecF product *see* general recombination, genes/pathways; repair, pathways, daughter-strand gap repair, UvrABC system *see* repair, pathways, excision repair; UvrABC system; genes 314-15; monomerisation *see* repair, pathways, photoreactivation; of pyrimidine dimers *see* repair, enzymes, υ protein; repair, pathways,

photoreactivation; pathways 320, 318-25, base replacement (pre-replication) 320, 321, daughter-strand gap repair (post-replication) 322-3, error-prone induced repair (post-replication) 322-5, error-prone long patch repair (pre-replication) 120-1, 324-5, excision repair (pre-replication) 320, 319-21, long patch repair, *see* repair, pathways, error-prone long patch repair, photolysis 321, photoreactivation (pre-replication) 320, 321, short patch repair *see* repair, pathways, excision repair, pre-replication vs. post-replication 318, study of 318; transdimer synthesis *see* repair, pathways, error-prone induced repair; Weigle reactivation *see* repair, pathways, error-prone induced repair

repair synthesis *see* repair, pathways, error-prone induced repair

replica-plating *see* mutant, selection

replicase 104, 230

replication: antibiotic inhibitors 259-60, *see also* acridine orange; actinomycin D; cordycepin; coumermycin; ethidium bromide; mitomycin C; nalidixic acid; rifampicin; streptolydigin; streptovaricin; bidirectional movement 247-50; chromosome segregation *see* replication, role of cell envelope; control: and cell division 255-7, negative (repressor) control 256-9, positive control (membrane attachment) 256, *see also* Col/F plasmid, replication, control; elongation 250-1, lagging and leading strand discontinuities 250, *see also* dUTPase, Okazaki fragment ligation 251-2, Okazaki fragment synthesis 250-2, primer excision 251, rate 246-7, 255, requirements 251, replication fork 233, 250, supercoiling 251; enzymes: DNA gyrase *(gyrA,B) see* replication, elongation, supercoiling; DNA gyrase, DNA ligase (lig) *see* replication, elongation, Okazaki fragement ligation; DNA ligase, DNA polymerase I *(polA) see* replication, elongation, primer excision; DNA polymerase I, DNA polymerase II *(polB) see* DNA polymerase II, DNA polymerase III *(polC/dnaE, dnaX, dnaZ) see* replication, elongation/initiation/termination; DNA polymerase III, DNA primase *(dnaG) see* replication, initiation, primer synthesis; replication, elongation, supercoiling; DNA primase; phage(s) G4, M13, φX174, DnaA protein *see* replication, initiation, primer synthesis; F plasmid, integrative suppression, DnaB,

DnaC proteins *see* replication, initiation, primer synthesis; replication, elongation, requirements, DnaI, DnaP proteins *see* replication, initiation, primer synthesis, helix destabilising proteins *see* replication, enzymes single-strand DNA binding protein, PolAex activity *see* replication, elongation, primer excision; plasmid, replication, RNase H *see* replication, elongation, primer excision, RNA polymerase *see* replication, initiation, primer synthesis/transcriptional activation; RNA polymerase, single-strand DNA binding protein *(ssb) see* replication, elongation, requirements; single-strand DNA binding protein; fidelity 246-8, *see also* DNA polymerase III, proof-reading; fork *see* replication, elongation; genes 244-5; gene mutants *see* DNA ligase; DNA polymerase I, III; RNA polymerase; initiation: multiple 251-4, *see also* replication, control, and cell division, origin of 248-9, *see also* oriC, primer synthesis 249, *see also* DNA primase; RNA polymerase, transcriptional activation 249, *see also* phage *lambda*, replication; leading strand vs. lagging strand 232-3, *see also* replication, elongation; modes: Cairns (θ) mode 21-2, 248, 261, *see also* Col/F plasmid, vegetative replication, Cairns mode vs. rolling-circle mode 21-2, 261, rolling-circle (σ) mode *see* F plasmid, replication, transfer; phage(s) G4, M13, φX174, replication; negative supercoiling *see* DNA gyrase; replication, elongation; phages *see* phage(s) G4, *lambda*, M13, φX174; plasmid *see* Col/F plasmid; polarity 232, 246; positive supercoiling *see* DNA gyrase; replication, elongation; role in recombination *see* general recombination, stages, breakage and reunion/chiasma formation/mismatch repair/strand invasion; general recombination, mechanistic models; transposable genetic element, transposition mechanism; role in repair *see* repair, pathways, excision repair/error-prone induced repair; role of cell envelope 254-5, *see also* replication, control; semi-conservative nature 246; semi-conservative vs. conservative 19-20; study of 231-2, 243-6, 260-1, 265-6, 332-4; supercoiling *see* DNA gyrase; replication, elongation; termination 251-4; unwinding of DNA 240-3, *see also* DNA gyrase; replication, elongation, supercoiling; single-strand DNA binding protein replicon 21, 261-2; *see also* phage(s), plasmid(s)

repliconation 165

repressor *see* operon control, promoter control

restriction endonuclease *see* DNA restriction

restriction-modification (R-M) *see* DNA restriction

reversion analysis *see* mutation, multisite vs. point

revertants *see* mutation, forward and reverse

rex see phage *lambda* genes and mutants

RF (replicative form) *see* phage ϕX174 (G4, M13), replication

RF-1, RF-2, RF-3 *see* translation, termination, release factors

rho factor *see* transcription factors

ribonuclease *see* RNase

ribonuclease H *see* RNase

ribonucleic acid *see* RNA

ribonucleoside *see* nucleic acid, components

ribonucleotide *see* nucleic acid, components

ribose *see* nucleic acid, components

ribosomal ambiguity mutant *(ram/rpsD) see* ribosomal protein

ribosomal protein: antibiotic action and *see* translation, inhibitors; antibiotic resistance and *see* phenotypic lag; control: metabolic 419, post-transcriptional 422-3, stringent 68, 420-2; copies per ribosome 87-8; function in binding messenger RNA 96-8; function in translational fidelity 93, 146, 150-1; function in translational termination 101; genetic loci 91, 412, 413, 418-19; genetic nomenclature 92; methylation 87; nomenclature 87, 103; *ram (rpsD)* mutants *see* ribosomal protein, function in translational fidelity; size 87; *spcA (rpsE)* mutants 189, *see also* ribosomal protein, function in translational fidelity; *strA (rpsI)* mutants 134, 183, 188-9, 212, *see also* ribosomal protein, function in translational fidelity; *see also* genetic code; messenger RNA; mutation; operon control; ribosomal RNA; ribosome; suppression; transfer RNA; translation

ribosomal RNA: and transfer RNA binding 93; base pairing with messenger RNA (Shine-Dalgarno sequence) 96; cellular concentration 55, 414; control: metabolic 419, stringent 68, 420-2, gene clusters 305, 307, 414; genetic loci 91, 414, 416; genetic nomenclature 92; nonsense codon interaction 148; orientation of transcription 415, 417; processing 415-17; promoter initiation rate 374, 414; promoter sequence 415; size 87-8; spacer-trailer tRNA genes 415-17; *see also* genetic code; messenger RNA; mutation; operon control; ribosomal protein; ribosome; suppression; transfer RNA; translation

ribosome: A and P sites 98-101, 105; antibiotic action and *see* translation, inhibitors; cellular concentration 55; composition 87-90; cycle 93-5; function 53-5, 87; maturation 92; peptidyl transferase centre 88, 90; polyribosome 52-5; role in attenuator control *see trp* operon; size 87-8; stalling *see trp* operon, control, attenuator; three-dimensional model 87-90; transfer RNA binding sites 98; *see also* genetic code; messenger RNA; mutation; operon control; ribosomal protein; ribosomal RNA; suppression; transfer RNA; translation

ribothymidine *see* transfer RNA, modified nucleosides

rif (rpoB) see RNA polymerase (β subunit); rifampicin

rifampicin: F plasmid curing 259; inhibition of replication: chromosomal 249, F plasmid 259, phage M13 266; inhibition of transcription 56, 70-1; resistance of Gram-positive vs. Gram-negative 29; structure 72; *see also* RNA polymerase

rim see ribosome, maturation

$r_K m_K$ *see* DNA restriction

RNA: cell content 29, 55; components of *see* nucleic acids; -DNA hybrid 65, *see also* DNA analysis; in plasmid 265, *see also* plasmid, vegetative replication; primer RNA *see* DNA primase; RNA polymerase; size 8, 51, 68, 78, 88, 415; stable vs. unstable 67-8; structure *see* nucleic acids; synthesis *see* transcription; vs; DNA 8, 51; *see also* messenger RNA; ribosomal RNA; RNA analysis; transfer RNA

RNA analysis 356-7

RNA phage *see* phage, classification; phage MS2 (and f2, Qβ, R17)

RNA polymerase: antibiotic inhibitors 70-2, *see also* rifampicin; streptolydigin; streptovaricin; assay 59, 354; cellular concentration 56-7, 412; control: attenuator 422, metabolic 419, post transcriptional 422-3, stringent 420-2; function 56-7; gene mutants *see* RNA polymerase, operons; genetic loci 57-8; holoenzyme 56; operons: cotranscription 412, 413, 418-19, *see also* RNA polymerase, control, gene mutants 28, 70, 455; phage T3/T7 55, 239; processiveness 59; role in gene control *see* operon control; attenuator; operator;

promoter; ribosomal protein; ribosomal
RNA; terminator; translation; transfer
RNA; *trp* operon; role in recombination
see general recombination, pathways, *rpo*
pathways; role in replication *see*
replication, initiation, primer synthesis/
transcriptional activation; phage G4,
replication; replication, elongation,
Okazaki fragment synthesis; nuclease;
role in transcription *see* transcription;
sigma (σ) subunit: cycling 64-5, function
56-7, role in transcriptional initiation 61-
2, *see also lac* operon, control, positive;
subunits *see* RNA polymerase, genetic
loci/holoenzyme; transcriptional
specificity *see* RNA polymerase, sigma
subunit; vs. DNA polymerase 239
rna-rnp see RNase
RNase (ribonuclease): genetic loci 57, 91-2;
RNase III *(rnc)*: function 57, 68, role in
phage T7 messenger RNA processing
415, role in stable RNA processing 415-
17, 422; RNase D/Q 418; RNase H
246, 251; RNase P*(rnp)* 416-18
RNA synthesis *see* transcription
R plasmid: amplification 265, 307;
antibiotic resistance determinants 185,
187, 189, 265, 430; characteristics 185;
components 265, *see also* R plasmid,
antibiotic resistance determinants;
transposable genetic element, insertion
sequence/transposon; copy number 185-
6, *see also* R plasmid, vegetative
replication, control; fertility inhibition
(Fi) 184-5, 431-4; F-like 184-5; Hfr(R)
186; incidence of antibiotic resistance
188; pilus type 184-5; RTF and r
components *see* R plasmid, components;
size 185; transfer: chromosome
mobilisation 186, *drd* mutants 184, 431,
ompA involvement 184, *see also* F
plasmid transfer, self-transmission vs.
non-self-transmission 184-6; vegetative
replication: control 263-5, fork movement
262, requirements 262-3; *see also* Col
plasmid; F-duction; F plasmid; F plasmid
transfer; genetic mapping; transposable
genetic element
rpoA,B,C,D see RNA polymerase
rpo pathway *see* general recombination
r-protein *see* ribosomal protein
rp1A-rp1V (L1-L25) *see* ribosomal protein,
genetic nomenclature
rpmA-rpmG (L27-L33) *see* ribosomal
protein, genetic nomenclature
rpsA-rpsU (S1-S21) *see* ribosomal protein,
genetic nomenclature
rpsD,E,L (ram, spcA, strA) see ribosomal
protein (S4, S5, S12), *ram* mutants/*spcA*

mutants/*strA* mutants
rrfB,C (5S rRNA) *see* ribosomal RNA,
genetic nomenclature
rrl B,C,D,G (23 rRNA) *see* ribosomal
RNA, genetic nomenclature
rRNA *see* ribosomal RNA
rrnA-rrnG (rRNA operon) *see* ribosomal
RNA, genetic nomenclature
rrsA,E,G (16S rRNA) *see* ribosomal RNA,
genetic nomenclature
RTF component *see* R plasmid
rule of the ring 196; *see also* chromosome

S *see* sedimentation coefficient
[35]S *see* radiolabelling
S1-S21 *(rpsA-rpsU) see* ribosomal protein
S4, S5, S12 *(rpsD,E,F) see* ribosomal
protein, *ram* mutants/*spcA* mutants/*strA*
mutants
S13 *see* phage S13
S-adenosylmethionine *see* DNA restriction,
restriction endonucleases
safrole *see* mutagens
Salmonella typhimurium: chromosome
evolution 4, 309-12; electron micrograph
of 31; frameshift suppressors 138-40,
143, *see also* suppression; lactose
utilisation 3, *see also lac* operon;
mutation test 122-3, *see also*
mutagenesis; vs. *E. coli* 3-4, 309-12
samesense *see* mutation
sassafras oil *see* mutagens, sources
sbcA 277-80; *see also* phage *lambda*,
cryptic prophage; general recombination
sbcB see exonuclease I
SDS (sodium dodecyl sulphate) *see* cell
lysis; electrophoresis, protein
secondary bacterial attachment site (ΔOΔ′)
see phage *lambda*, site-specific
recombination; phage *lambda*,
transduction
sedimentation coefficient, 87-8; *see also*
centrifugation
segregation: of bacterial chromosome 255;
of F plasmid 161-2; of temperate phage
203; *see also* Col/F/R plasmid,
vegetative replication; phage(s) *lambda*,
Mu, P1, replication; replication
self-transmission *see* plasmid, transfer;
Col/F/R plasmid(s)
sense codon(s) *see* genetic code
sensitiser *see* repair, pathways, photolysis
septum 39-40; *see also* cell envelope,
synthesis; replication, control, and cell
division
serine *see trp* operon, enzymes; translation
sex plasmid *see* F plasmid
sexual conjugation *see* plasmid, transfer; F-
duction; Col/F/R plasmid(s), transfer

sfiA,B (sulA,B) see repair, genes; *lon*,
   suppressor of
Shigella 188
Shine-Dalgarno sequence *see* translation,
   initiation
'shot-gun' approach *see* DNA restriction,
   cloning
sigma factor *(rpoD) see* RNA polymerase;
   transcription, initiation; transcription
   factors; promoter
sigma (σ) structure *see* replication, modes,
   rolling-circle
single-strand DNA binding protein *(ssb)*:
   function 240-5; genetic locus 244-5; role
   in recombination 278, 283-7; role in
   replication 250-1, 266-8; *see also* phage
   T4, gene 32 product
site-specific recombination: general
   recombination vs. non-homologous
   recombination 231, 271, 296; insertion
   sequence *see* transposable genetic
   element; phage *see* phage(s) *lambda*/Mu;
   phylogenetic implications 307-12;
   plasmid *see* F plasmid; transposable
   genetic element *see* transposable genetic
   element; transposon *see* transposable
   genetic element; vs. illegitimate
   recombination 296
sizing factor *see* RNase, RNase III
sodium dodecyl sulphate (SDS) *see* cell
   lysis; electrophoresis, protein
SOS functions: cell filamentation *see* SOS
   functions, regulation of cell division;
   colicin induction 273; DNA degradation
   273; error-prone repair 120-1, 322-5;
   phage induction 220-5; protein X
   induction 273-6; regulation of cell
   division 120-2, 273, 276, 314; *see also*
   methylmethane sulphonate; mitomycin C;
   nalixidic acid
spacer gene *see* ribosomal rRNA, operons
spcA (rpsE) see ribosomal protein (S5)
spectinomycin: inhibition of translation 189;
   R plasmid-encoded resistance 187, 189,
   265
spheroplast *see* cell lysis, EDTA/lysozyme
Spi *see* phage *lambda* genes and mutants
spore formation *see Bacillus subtilis*
spoT see operon control, stringent control
spr see LexA protein, gene mutants; protein
   X, gene control
srl see protein X, isolation
SS (single-strand) form *see* phage ϕX174
   (G4, M13), replication
ssb see single-strand DNA binding protein
Staphylococcus aureus 29
stationary phase *see* cell growth,
   characteristics of
'sticky' (cohesive) ends *see* DNA

restriction, cloning/restriction
   endonucleases; phage *lambda*,
   chromosome circularisation
strA (rpsL) see ribosomal protein (S12)
Streisinger model *see* mutagenesis, induced,
   frameshift; mutagenesis, hotspots
streptolydigin: inhibition of replication 259;
   inhibition of transcription 70; structure
   72; *see also* RNA polymerase
Streptomyces 4
Streptomyces coelicolour see chromosome,
   multimerisation
streptomycin: incidence of resistant
   Shigella 188; inhibition of translation
   105-6; phenotypic suppression 146, 150-
   1; R plasmid-encoded resistance 187-9,
   265; structure 106; transposon-encoded
   resistance 300-2; use in counter-selection
   179, 183; *see also* phenotypic lag
streptovaricin: inhibition of replication 259;
   inhibition of transcription 70; mode of
   action 70; structure 72; *see also* RNA
   polymerase
stringent control (of gene expression) *see*
   operon control
stringent control (of plasmid replication) *see*
   plasmid, vegetative replication; Col/F/R
   plasmid
stringent protein *(relA) see* operon control,
   stringent control
substituted plasmid *see* F-prime plasmid
substitution *see* mutation, multisite/point
substitution loop *see* DNA analysis,
   heteroduplexing
substrate level phosphorylation *see* ATP
substrate transition 126; *see also*
   mutagenesis, induced, mispairing
suf see suppression, intergenic, frameshift
suicide techniques *see* mutant, selection
Su (1, 2, 3, 6, 7, B, C, G, 9) *see*
   suppression, intergenic, nonsense
sulA, B (sfiA, B) see repair, genes; *lon*,
   suppressor of
sup (B, C, D, E, F, G, P, U, 9) see
   suppression, intergenic, nonsense
supercoiling *see* chromosome; DNA gyrase;
   DNA isolation, replication, elongation
superinfection exclusion *see* phage
suppression: informational *see* suppression,
   intergenic; integrative *see* F plasmid, site-
   specific recombination; intergenic: amber/
   ochre/opal *see* suppression, intergenic,
   nonsense; efficiency 145, 148-9,
   frameshift 138-40, 143, 153, missense
   138-40, 145, 153, 277, nonsense 140,
   142-9, 151, 153, 384, 416, nonsense vs.
   polar 347, 384, reading context 145,
   wobble effect 144-5; intergenic vs.
   intragenic 137-8; intragenic: frameshift

138-9, 153, missense 138, 153;
phenotypic 146-51, 420; *see also* genetic
code; messenger RNA; mutation;
ribosomal protein; ribosome; transfer
RNA
suppressor loci 146, 148-9; *see also*
suppression, intergenic
SupT see suppression, intergenic, missense
surface exclusion (Sex) *see* F plasmid,
transfer, entry exclusion (Eex)
swivelase *see* DNA gyrase; Eco
topoisomerase I

T1, T4, T5, T6, T7 *see* phage(s) T1, T4,
T5, T6, T7
T-even vs. T-odd *see* phage, classification
tabB (groE, mop) see phage *lambda*,
morphogenesis
tag see 3-methyladenine glycosylase
TψC arm *see* transfer RNA
template transition 126
Ter enzyme *(A) see* phage *lambda*,
morphogenesis
terminal redundant *see* phage T7,
chromosome
termination codon(s) *see* genetic code
termination factor(s) *see* transcription,
termination; transcription factors, rho;
terminator; translation; termination,
release factors
terminator: general sequence 66; phage
lambda attenuators *see* phage *lambda*
genes and mutants, t$_L$/t$_{R1}$/t$_{R2}$; *trp* operon
attenuator sequence 409-10; *trp* operon
terminator *(trp t)* 402; *see also* operon
control
terminator (of replication) *see* replication,
termination
terminator control *see* operon control
tet proteins *see* tetracycline, R plasmid-
encoded resistance
tetracycline: incidence of resistant *Shigella*
188; inhibition of translation 189; R
plasmid-encoded resistance 187, 189,
265; transposon-encoded resistance 300-
2
theta (θ) structure *see* replication, modes,
Cairn's
thiogalactoside transacetylase *(lacA) see lac*
operons, enzymes
2-thiouridine 148-9; *see also* transfer RNA,
modified nucleosides
4-thiouridine *see* transfer RNA, modified
nucleosides
thr 27-8; *see also trp* operon, control,
attenuator
thymidine *see* nucleic acid, components
thymine *see* nucleic acid, components;
mutagens, UV radiation; mutagenesis,

hotspots; repair, of pyrimidine dimers;
SOS functions; transcription
thymine dimer *see* mutagenesis, induced,
misrepair; mutagens, UV radiation;
repair, of pyrimidine dimers
thymine starvation *see* SOS functions
tif see protein X, gene mutants
Tn *see* transposable genetic element,
transposon; site-specific recombination
tobacco tar *see* mutagens, sources
tof (cro) see phage *lambda* regulatory
proteins
tolA see Col plasmid, colicins, bacterial
immunity
toluene *see* replication, study of
tonA see phage(s) ϕ80, T1, T5, receptor
tonB see Col plasmid, colicins, receptors;
phage(s) ϕ80, T1, receptor
topoisomerase *see* DNA gyrase; Eco DNA
topoisomerase I (ω protein)
tra gene(s) *see* F plasmid genes and
mutants; F plasmid transfer
trailer gene *see* ribosomal rRNA, operons
transconjugant *see* F-prime plasmid,
transfer
transcript 16, 18; *see also* messenger RNA;
operon control; polarity; RNA
polymerase; transcription; translation
transcription: antibiotic inhibition 70-2, *see
also* actinomycin D; cordycepin;
rifampicin; streptolydigin; streptovaricin;
control *see* operon control; RNA
polymerase; convergent 51, 456; cycle
59-60; divergent 51, 436-40; elongation
60, 62-6; elongation rate 52-3, 64-5;
fidelity 59; initiation 60-4, 243, *see also*
operon control, promoter; initiation rate
70, 374-5, 414; mapping 357; polarity of
movement 52-3, 58-9, 64, 139; rate-
limiting step 70; RNA-DNA hybrid in
65; selectivity of 56-7, *see also* promoter;
sense vs. antisense strand 51, 63; study of
59, 353, 356-7; termination 66-7, *see
also* operon control; polarity; terminator;
upstream vs. downstream 58-9; UV-
induced termination 354; *see also* operon
control; polarity; promoter; RNA
polymerase; terminator
transcription factor: catabolite activator
protein (CAP): function 57-8, role in *lac*
operon control 396-8, *see also* promoter;
rho: function 58, 60, 66, role in phage
lambda transcription 453-4, role in
polarity 303, 384-6, role in RNA
polymerase control 422, role in *trp*
operon control 403, 408-9, *see also*
terminator; size and loci 57-8; *see also
lac* operon, *lac* repressor; phage *lambda*
regulatory proteins; RNA polymerase;

RNase; *trp* operon, *trp* repressor
transcriptional activation 249, 445; *see also*
  replication, initiation
transcriptional control *see* operon control
transcriptional mapping 357
transdimer synthesis *see* repair, pathways,
  error-prone induced repair
transducing fragment *see* phage P1
transduction: generalised *see* phage P1;
  generalised vs. specialised (restricted)
  209; specialised *see* phage *lambda*
transductional shortening *see* phage P1
transfection *see* DNA restriction, cloning
transfer genes *see* F plasmid
transfer RNA: amino acid acceptor stem
  78, 79-82, 85-6; and ribosomal RNA 93;
  anticodon loop 78, 79-82, 85-6; cellular
  concentration 55; control: metabolic 419,
  stringent 68, 420-2; dihydouridine arm
  78, 79-82, 85-6, 93, 140, 144; function
  53, 78; gene duplication 84-5, 144, 412,
  416; gene mutants *see* suppression,
  intergenic; *trp* operon, control, attenuator;
  genetic loci 91, 416; genetic
  nomenclature 85, 92; initiator 84-5, 93-8;
  isoaccepting species 83; major vs. minor
  84; methyltransferase 92; modified
  nucleosides 78-9, 85, 148-9;
  nomenclature 78; nucleotidyl transferase
  92; number of 55, 84; processing 78,
  413-18, role in attenuator control *see trp*
  operon; secondary structure 78,
  aminoacyl-transfer RNA 86, *E. coli*
  transfer RNAs 85, generalised cloverleaf
  79; size 78, 88; spacer-trailer genes 415-
  17; suppressor *see* suppression,
  intergenic; tertiary structure: yeast
  transfer RNA 78-82, structure-function
  relationships 78, 93; Tψ/C arm 78, 79-82,
  85-6, 93; variable arm 78, 79-82, 85-6;
  wobble pairing and 83-4; *see also* genetic
  code; messenger RNA; mutation; operon
  control; ribosomal protein; ribosomal
  RNA; ribosome; suppression; translation
transformation *see* DNA restriction, cloning
transition *see* mutation, point
translation: antibiotic inhibitors 105-6, 190,
  see also Col plasmid, colicins;
  chloramphenicol; spectinomycin;
  streptomycin; tetracycline; control *see*
  operon control; elongation: factors 92,
  98-9, 106, peptidyl transferase centre and
  98-9, ribosomal translocation 98-100,
  speed of 52-3, 106; fidelity 93; infidelity
  (misreading) 93, 146, 150-1; initiation:
  complexes 93-8, factors 93-8, reinitiation
  see polarity, gradients of/mechanistic
  models for, ribosome subunits and 97,
  sequence (Shine-Dalgarno) 95-6, *see also*

suppression, intergenic, efficiency;
  polarity of movement 52-3, 90; post-
  translational modification 101-3; rate-
  limiting step 103; reading-frame 73, 76;
  release factors *see* translation,
  termination; study of 90-3, 353-6;
  termination 101-2; Tψ/C arm and 93; *see
  also* genetic code; messenger RNA;
  mutation; operon control; polarity;
  ribosomal protein; ribosomal RNA;
  ribosome; suppression; transfer RNA
translational control *see* operon control
transposable genetic element: classification
  296, 298-302; copies *see* transposable
  genetic element, insertion sequence/
  transposon; deletion promotion 303; in
  genetic mapping 346-8, *see also* genetic
  mapping; insertion frequency *see*
  transposable genetic element,
  transposition frequency; insertion
  sequence: as regulatory switches 302,
  copies per chromosome 302, copies per F
  plasmid 302, 428, 430, 432, copies per
  R100 plasmid 430, F-plasmid integration
  see F plasmid, site-specific
  recombination, heteroduplex analysis
  360, IS1 298-302, 304, IS2 300, 302-4,
  IS3 300, 302, γδ 428, 430, promoter
  300, 303, size 300, structure 300-1,
  terminator (transcriptional) 300, 303,
  384; orientation of insertion 303; phage
  see phage(s) *lambda*, Mu, P1, P2;
  phylogenetic implications 309-12;
  plasmid *see* F plasmid; polarity of 300,
  303, 346-7, 384, *see also* polarity;
  promoter *see* transposable genetic
  element, insertion sequence; phage
  lambda, site-specific recombination,
  polarity; recombination promotion 310,
  see also transposable genetic element,
  insertion sequence; reversion (precise
  excision) 303; sequence duplication *see*
  transposable genetic element, terminal
  duplication; sites of insertion 302; site-
  specific chromosomal deletion 303; size
  300; structural relationship 298-302;
  structure 300; terminal repetition 300-1,
  304; terminator (transcriptional) *see*
  transposable genetic element, insertion
  sequence; translocation *see* transposable
  genetic element, transposition;
  transposition frequency 302;
  transposition mechanism 304; transposon
  300-2, 304, 309, *see also* R plasmid,
  antibiotic resistant determinants; *see also*
  phage(s) *lambda*, Mu; plasmid(s); site-
  specific recombination
transposase *see* transposable genetic
  element, transposon

transposition (translocation) *see*
transposable genetic element
transposon *see* transposable genetic element
transversion *see* mutation
tricarboxylic acid (TCA) cycle 37
triplet-binding assay *see* translation, study
of
tritium (^{3}H) *see* radiolabelling
tritium suicide *see* mutant, selection
trmA-trmD see transfer RNA,
methyltransferase
tRNA *see* transfer RNA
trp operon: amplification of expression 227,
407; and *aroH* 407; and *rho* 403, 408-9;
anthranilate synthetase *(trpED) see trp*
operon, enzymes; apo-repressor *see trp*
operon, *trp* repressor; attenuator *see trp*
operon, control; bifunctional proteins *see*
trp operon, enzymes; control 38-9, 391,
401, 405-11, attenuator 387, 405-11,
feedback inhibition 38, 401, metabolic
419, negative 39, 405-8, stringent 420,
see also operon control; co-repressor *see*
trp operon, *trp* repressor; crossfeeding
404-5, *see also* genetic mapping,
pathways analysis; derepression 407;
enzyme levels: major vs. minor
promoter 404, 407, negative vs.
attenuator control 405-7; enzymes 38,
401-3, 405; genetic loci 27, 403; genetic
mapping 334, 344-7; genetic
nomenclature 25-8, 401, 403;
indoleglycerol phosphate synthetase
(trpC) see trp operon, enzymes; leader
404, 410, *see also trp* operon, control,
attenuator; major promoter *(trpP₁)* 402,
404, 406-7; major promoter sequence
404, 406; minor promoter sequence
(trpP₂) 402, 404, 407; mutant selection
134; operator sequence *(trpO)* 404-6, *see*
also trp operon, control, negative;
operator-promoter interpenetration 404-6;
orientation of transcription 402, 404;
phosphoribosyl-anthranilate isomerase
(trpC) see trp operon, enzymes;
phosphoribosyl-anthranilate transferase
(trpD) see trp operon, enzymes;
regulatory mutants 403, 407; regulatory
sequences 406, 416; repressor control *see*
trp operon, control, negative; structural
gene mutants 28, 334, 344-7, 403;
structural gene products *see trp* operon,
enzymes; *trp-POL see trp* operon, major
promoter/operator/leader; *trpR see trp*
operon, *trp* repressor; *trp* repressor *(trpR)*
401, 403, 407-8; *trpS,T* 403, 407, *see*
also suppression, intergenic, nonsense;
tryptophan synthetase *(trpAB) see trp*
operon, enzymes

trpT see suppression, intergenic, nonsense;
trp operon; transfer RNA, genetic
nomenclature
tryptophan *see trp* operon; translation
tryptophan synthetase *(trpAB) see trp*
operon, enzymes
tsl see LexA protein, gene mutants; protein
X, gene control
tsx see Col plasmid, colicins, phage cross-
resistance; phage T6, receptors
tufA,B see translation, elongation factors;
gene duplication
tyrR,S see aminoacyl-tRNA synthetase
tyrT see suppression, intergenic, nonsense;
transfer RNA

ultraviolet (UV) radiation *see* mutagenesis,
induced, misrepair; mutagens; repair,
enzymes, υ protein/UvrABC system;
repair, genes; repair, pathways; SOS
functions; transcription, UV-induced
termination
ultraviolet visualisation 357, 359
uncB see oriC, locus
ung see uracil-DNA glycosylase
unilinear inheritance *see* F plasmid,
replication; phage P1, transduction,
abortive
untwistase *see* DNA gyrase; Eco
topoisomerase I
upstream *see* transcription
uracil *see* nucleic acid, components;
replication, elongation, leading strand
discontinuities
uracil-DNA glycosylase *(ung)* 132, 314-17
uridine *see* nucleic acid, components
UvrABC system: function 313-14; gene
mutants 120-1; genetic loci 314-15; role
in repair 319, 320
uvrD (mutU, recL, uvrE): role in general
recombination 279-81; role in repair 313-
14
uvrE (mutU, recL, uvrD) see uvrD
uvrF (recF) see general recombination,
genes/pathways; repair, pathways,
daughter-strand gap repair

valS,T see transfer RNA
variable arm *see* transfer RNA
vector *see* DNA restriction, cloning; genetic
mapping, generalised transduction/F-
prime transfer/specialised transduction;
phage *lambda*/P1, transduction
Vibrionaceae 4
vinyl chloride *see* mutagens
virus (bacterial) *see* phage(s)

Weigle reactivation *see* repair, pathways,
error-prone induced repair

Whitehouse model *see* general
   recombination
wobble pairing *see* transfer RNA;
   suppression, intergenic
ω protein *see* Eco DNA topoisomerase I

Xanthine *see* mutagens, nitrous acid
xis see phage *lambda*, site-specific
   recombination; phage *lambda*
   transcription
xseA see exonuclease VII
xthA see endonuclease II

y *see* plasmid, incompatibility groups

zab see protein X, gene mutants